Surface and Interfacial Aspects of Cell Adhesion

Surface and Interfacial Aspects of Cell Adhesion

Edited by

A. Carré and K. L. Mittal

CRC Press
Taylor & Francis Group
Boca Raton London New York

CRC Press is an imprint of the
Taylor & Francis Group, an **informa** business

CRC Press
Taylor & Francis Group
6000 Broken Sound Parkway NW, Suite 300
Boca Raton, FL 33487-2742

First issued in paperback 2017

© 2011 by Taylor & Francis Group, LLC
CRC Press is an imprint of Taylor & Francis Group, an Informa business

No claim to original U.S. Government works

ISBN 13: 978-1-138-11621-4 (pbk)
ISBN 13: 978-90-04-19078-8 (hbk)

This book contains information obtained from authentic and highly regarded sources. Reasonable efforts have been made to publish reliable data and information, but the author and publisher cannot assume responsibility for the validity of all materials or the consequences of their use. The authors and publishers have attempted to trace the copyright holders of all material reproduced in this publication and apologize to copyright holders if permission to publish in this form has not been obtained. If any copyright material has not been acknowledged please write and let us know so we may rectify in any future reprint.

Except as permitted under U.S. Copyright Law, no part of this book may be reprinted, reproduced, transmitted, or utilized in any form by any electronic, mechanical, or other means, now known or hereafter invented, including photocopying, microfilming, and recording, or in any information storage or retrieval system, without written permission from the publishers.

For permission to photocopy or use material electronically from this work, please access www.copyright. com (http://www.copyright.com/) or contact the Copyright Clearance Center, Inc. (CCC), 222 Rosewood Drive, Danvers, MA 01923, 978-750-8400. CCC is a not-for-profit organization that provides licenses and registration for a variety of users. For organizations that have been granted a photocopy license by the CCC, a separate system of payment has been arranged.

Trademark Notice: Product or corporate names may be trademarks or registered trademarks, and are used only for identification and explanation without intent to infringe.

**Visit the Taylor & Francis Web site at
http://www.taylorandfrancis.com**

**and the CRC Press Web site at
http://www.crcpress.com**

Contents

Preface

Cell adhesion comes into play in almost all domains of life. The range of situations in which it occurs, involving organisms, living tissues, microorganisms or single cells, is endless. Cell adhesion is involved in the binding of a cell to a surface, extracellular matrix, or another cell using cell adhesion molecules. It is crucial in the formation and maintenance of coherent multicellular structures. Cell surface adhesion molecules (integrins, for example) which transmit information from the extracellular matrix to the cell play vital roles in numerous cellular processes. Some of these include: cell growth, differentiation, embryogenesis, immune cell transmigration and response, and cancer metastasis. Also cell adhesion is involved in most of pathological situations.

Cell adhesion is an outstanding example of multidisciplinary science — at the frontiers of physics for understanding surfaces and interactions, chemistry to assemble molecules of interest on a surface, and biology to understand cascade of chemical signals transducing in cells after adhesion.

Thus, cell adhesion has been attracting the attention of physicists and materials scientists for the last twenty years. However, complexity of biological interfaces has meant that progress in the theory of cell adhesion has been slow. For biologists, delineating and understanding the wide variety of mechanisms involved in cell adhesion is crucial. Many important processes in cells and their cycles are indeed triggered by adhesion. For instance, cells divide or differentiate only when they adhere. A healthy cell that cannot adhere commits suicide (apoptosis) within a short time. This self-regulatory mechanism prevents sick or imperfect cells from invading and disrupting the entire organism. Adhesion is thus a key player in the control of cell proliferation and differentiation.

Adhesion process requires close proximity. To adhere to substrate, cells must be located within a few nanometers of the host surface. This involves transport mechanisms which concern not only cells, but also molecules and macromolecules present in biological media. Being smaller than cells, these entities can reach the substrate material faster and form a surface film or primary film. The importance of protein adsorption lies in the widely-held view that protein adsorption is among the first steps in biological response to artificial materials, especially in

Surface and Interfacial Aspects of Cell Adhesion
© Koninklijke Brill NV, Leiden, 2010

cell adhesion according to the biological view. Thus, the early steps of cell adhesion are mediated by non-receptor interactions. Recent studies converge to describe the receptor-mediated adhesion phase required for regulation of the cell function and fate, distinct from the initial cell–substrate interaction phase. Indeed, integrin receptors that mediate attachment between a cell and the tissues surrounding it change in structure and molecular properties from dot-like focal complexes to stress fiber-associated focal contacts. The development of novel 2D or 3D surfaces, with hydrophilic or superhydrophobic properties or with particular patterning, has allowed new theoretical advances, while at the same time offering numerous and varied technological applications.

The complexity, diversity and continuous progress in the cell adhesion field challenge any attempt to cover the topic of cell adhesion in its entirety. This book is based on the two-part Special Issue of the *Journal of Adhesion Science and Technology (JAST)*: Vol. 24, No. 5, pp. 811–1030 (2010) and Vol. 24, Nos 13–14, pp. 2027–2334 (2010) dedicated to this topic. Based on the widespread interest and tremendous importance of cell adhesion, we decided to make this book available as a single and easily accessible source of information. The papers as published in the above-mentioned Issue have been re-arranged in a more logical fashion in this book.

With this book our intent is to offer the reader a survey on selected aspects of cell adhesion. It contains a total of 26 papers covering many different aspects of cell adhesion and is divided into four parts as follows: Part 1: Fundamentals of Cell Adhesion; Part 2: Methods to Study Cell Adhesion; Part 3: Surface Treatments to Control Cell Adhesion and Behavior; Part 4: Cell Adhesion in Medicine and Therapy.

The panel of authors who have contributed to this book represent a wide variety of disciplines, ranging from materials science to surface science, and from physics to biology. We certainly hope that our initiative to bring out this book will provide new perspectives to anyone interested in surface and interfacial aspects of cell adhesion.

We further hope that this book containing bountiful up-to-date information will be of great interest and value to anyone interested in this wonderful field of cell adhesion. This book should serve as a gateway for the neophyte and a commentary on current research for the veteran researcher. All papers included in this book *Surface and Interfacial Aspects of Cell Adhesion* were invited. We believe that this book will be a significant and very timely addition to the literature on this complex subject.

Acknowledgements

Now comes the pleasant task of thanking those who helped in materializing this book. First and foremost our thanks go to the authors for their interest, enthusiasm, cooperation and contribution without which this book would not have seen the light of day. We profusely thank the reviewers for their time and efforts in providing many valuable comments as comments from peers are *sine qua non* to maintain the highest standard of a publication. Finally, our appreciation goes to the appropriate individuals at Brill (publisher) for giving this book a body form.

ALAIN CARRÉ
Research Fellow
Retired from Corning, Inc.
Le Chatelet en Brie, France

K. L. MITTAL
P.O. Box 1280
Hopewell Jct., NY 12533, USA

Part 1
Fundamentals of Cell Adhesion

How Substrate Properties Control Cell Adhesion.
A Physical–Chemical Approach

Alain Carré * and Valérie Lacarrière

Corning SAS, Corning European Technology Center, 7 bis, Avenue de Valvins, 77210 Avon, France

Abstract

Most living cells derived from solid tissues require an adhering surface to live *in vitro* conditions. A good understanding of the relationships between the behavior of cells and the physicochemical properties of substrates such as the surface free energy, the surface polarity, the presence of functional groups and surface charges is of prime importance for the optimization of adhesion, spreading and proliferation of cells.

Polystyrene and treated polystyrene surfaces were characterized by determining their surface free energies using wettability measurements. The knowledge of the surface properties of the culture substrates provides a good view of the influence of the substrate properties on cell adhesion. However, this study shows that it is not directly the surface free energy of materials that controls cell adhesion but rather the interfacial free energy between the culture medium and the substrate. The interfacial free energy between the culture medium and the solid surface controls the adsorption of serum components that may inhibit or promote cell adhesion. One of the components inhibiting cell adhesion is serum albumin. The results indicate that the adsorption of serum albumin is related to the interfacial free energy between the culture medium and the substrate. Hydrophilic substrates, such as plasma treated polystyrene substrates, have a lower interfacial free energy with water than hydrophobic polystyrene leading to a lower adsorption of proteins inhibiting cell adhesion. In addition, it is observed that there is a competition between proteins inhibiting (serum albumin) and proteins promoting cell adhesion (such as fibronectin).

Keywords

Cell adhesion, cell culture, surface free energy, surface polarity, interfacial free energy, protein adsorption, polystyrene, serum albumin, fibronectin

1. Introduction

Cell culture is one of the major tools used in cell and molecular biology. Cells derived from solid tissues require an adhering surface to live and to proliferate *in vitro* conditions. The first step after exposure of any biomaterial to a biological environment results in the rapid adsorption of proteins to its surface [1]. The composition, type, amount, and conformation of adsorbed proteins regulate the secondary

---

* To whom correspondence should be addressed. Tel.: 33 16 469 73 71; Fax: 33 164 69 74 55; e-mail: carrea@corning.com

Surface and Interfacial Aspects of Cell Adhesion
© Koninklijke Brill NV, Leiden, 2010

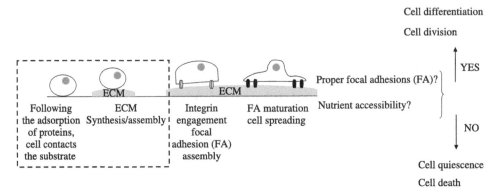

Figure 1. Scheme showing how cell adhesion (in the dotted rectangle) precedes other events such as cell spreading, cell proliferation, cell differentiation. The ECM is the Extra Cellular Matrix (Scheme by courtesy of Dr. M. Hervy, Corning Inc.).

phenomena such as cellular adhesion and protein exchange and also the following cellular reactions such as migration, proliferation and differentiation. As shown schematically in Fig. 1, cell adhesion precedes other events such as cell spreading, cell proliferation and eventually cell differentiation. A good understanding of the relationships between the behavior of cells and the physicochemical properties of the substrates such as the surface free energy, the presence of functional groups and surface charges is of prime importance for the optimization of cell cultures.

Under physiological conditions, the adhesion of mammalian cells to a solid surface is mediated mainly by the presence of a protein adlayer [2] whose properties depend on the substrate and on the composition of the liquid phase. Anchorage-dependent cells are cultivated in nutritive media supplemented with serum, necessary to support cell growth by provision of hormonal factors, attachment and spreading factors, and carrier proteins. Most cell culture media are indeed multi-component systems containing a wide range of proteins which may be involved in competitive or sequential adsorption.

In this work, we describe the effect of protein adsorption on cell adhesion to polystyrene substrates of different surface properties. We will show that proteins may inhibit or promote adhesion depending on their nature and that one major factor controlling the adsorption of serum components, and subsequently controlling cell adhesion, is the interfacial free energy between the culture medium and the solid substrate. This argument is significantly different from other studies [3–5] aiming to relate cell adhesion to the surface free energy or surface free energy components of solid substrates because the interfacial free energy between the culture medium and the solid substrate is not solely related to the solid surface free energy. The surface properties of the liquid medium and the nature of interfacial interactions are also involved.

2. Experimental Section

2.1. Polymer Substrates

Sterilized substrates of polystyrene products considered were: untreated polystyrene (35 mm culture dish, Nalge Nunc International (PS)), Costar® 6-Well Clear Not Treated Microplates (gamma irradiated (g-PS), Corning Inc., USA), tissue culture treated polystyrene (TCT PS) (Costar® 6-Well Clear TC-Treated Microplates) and Corning® CellBIND® 6-Well Clear Microplate (Corning® CellBIND® Surface, Corning Inc.).

Natural, unmodified polystyrene (PS) surfaces are hydrophobic and bind cells and biomolecules only through hydrophobic interactions. However, if these un-treated vessels are sterilized by gamma irradiation (g-PS), it slightly increases the wettability of the surface and, thus, its binding characteristics [6, 7]. Untreated polystyrene vessels may work for some cell lines. Many transformed cell lines (Chinese hamster ovary cells, for example) and macrophages will readily attach and grow on untreated, radiation sterilized polystyrene.

Corning® polystyrene cell culture treated vessels (TCT PS) are surface modified using either corona discharge (flasks, dishes and microplates) or gas–plasma (roller bottles and culture tubes). These processes generate highly energetic oxygen ions and free radicals which graft onto the surface polystyrene chains so that the surface becomes hydrophilic and negatively charged [8–10]. The more the oxygen is incorporated onto the surface the more hydrophilic it becomes and the better it is for cell attachment and spreading.

The Corning® CellBIND® Surface enhances attachment making it easier to grow cells under difficult conditions, such as reduced-serum or serum-free medium, resulting in higher cell yields. This patented technology [11] uses microwave plasma for treating the culture surface. The process improves cell attachment by incorporating significantly more oxygen into the cell culture surface, rendering it more hydrophilic. X-ray photoelectron spectroscopy (XPS) shows that the amount of oxygen incorporated into the Corning® CellBIND® Surface is more than 60% higher than with the traditional tissue culture surface treatment (29 *versus* 17.2% of oxygen atoms [11]).

2.2. Biological Materials and Cell Culture Conditions

Three cell lines were chosen: Chinese hamster ovary (CHO) cells, human fetal lung fibroblasts (MRC5) and human umbilical vein endothelial (HUVE) primary cells. The cells were purchased from ATCC (Manassas, VA, USA).

CHO cells were seeded at a concentration of 200 000 cells per well in 2 ml HamF-12 culture medium (Gibco® Cell Culture Products, Invitrogen Corporation, Carlsbad, CA, USA) containing 20 µl antibiotic (penicillin–streptomycin (10 000 units of penicillin (base) and 10 000 µg of streptomycin (base)/ml), Gibco® Cell Culture Products, Invitrogen Corporation), with 10% or without FBS (Fetal Bovine Serum, Gibco® Cell Culture Products, Invitrogen Corporation) in the medium.

MRC5 cells were seeded at a concentration of 50 000 cells per well in 2 ml of IMDM medium (Iscove's Modified Dubelcco's Medium, Gibco® Cell Culture Products, Invitrogen Corporation) with 20 μl antibiotic, with 10% or without FBS in the medium.

HUVE cells were seeded at a concentration of 90 000 cells per well in 2 ml of HamF-12 medium containing 20 μl antibiotic, 0.1 mg/ml heparin (H3393, Sigma Aldrich, Inc., St. Louis, MO, USA), 0.05 mg/ml endothelial cell growth factor (ECGF, E2759, Sigma Aldrich) and 10% FBS in the medium. Adhesion experiments were conducted in presence of FBS, heparin and ECGF under normal culture conditions or in absence of these components.

After seeding, the cells were grown at 37°C in an incubator with 5% CO_2. The physiological pH of cell medium was 7.4.

The number of adhering cells was measured at the end of the adhesion step after 1.5 h of incubation with CHO cells and 3 h for the other cells. Three methods of counting were used: counting with the Malassez cell, colorimetry (staining of cells with crystal violet) or flow cytometry (Z1 Coulter Counter, Beckman, Fullerton, CA, USA). First, counting of adhering cells involves the removal of non-adhering cells and the rinsing of adhering cells with fresh culture medium. The principle of the colorimetric method consists in fixing adhering cells with glutaraldehyde (1% in water). Then the cells are colored with 0.1% crystal violet in water. Finally, the colorant is released by lysing the tinted cells in sodium dodecylsulfate (SDS, 1% in water). With calibration curves, the optical density measured at 595 nm gives the number of adhering cells. For counting with Malassez cell or Coulter Counter, the adhering cells are detached with trypsine–EDTA ($1 \times$, 0.25% trypsine, 5.3 mM EDTA, Invitrogen Corporation). Then the number of cells per unit volume is determined by manual counting with a Malassez cell or by flow cytometry (Coulter Counter).

Bovine serum albumin (BSA) was obtained from Sigma Aldrich (99% BSA, Sigma Aldrich). Fibronectin was kindly supplied by the ERRMECe Laboratory (Equipe de Recherche sur les Relations Matrice ExtraCellulaire Cellules, Université de Cergy-Pontoise, Cergy-Pontoise, France).

2.3. Adsorption of Proteins on Polystyrene Substrates

Experimentally, the adsorption of proteins on polystyrene substrates was revealed by staining of proteins with a colloidal gold dispersion (Colloidal Gold Total Protein Stain, Bio-Rad Laboratories, Hercules, CA, USA). The samples exposed to proteins were dipped in the gold staining dispersion for 18 h, after which they were rinsed with pure water and blow-dried under nitrogen. The optical density of the gold coating was measured at 550 nm. As an alternative, a method based on fluorescence measurement was developed that consists in adsorbing biotinylated BSA (29130, Pierce, Thermo Fisher Scientific, Rockford, IL, USA) stained with Cy5-streptavidin (PA45001, Amersham, Little Chalfont, UK).

3. Theoretical Section

3.1. Adsorption at Solid–Liquid Interfaces

In a recent study [12], we have shown that the presence of FBS in a culture medium has a strong negative influence on the adhesion of cells, especially on hydrophobic PS. In absence of FBS during the adhesion step of cells, the number of adhering cells on PS is dramatically increased to reach 90% or more (however, serum is necessary for the proliferation of cells). The main argument of the theory supporting our experimental results is summarized below.

Proteins are natural surface active substances [13]. The thermodynamics of adsorption of a surface active compound at a bulk concentration c at the solid/water interface can be described by the Gibbs equation for adsorption [13, 14] as

$$d\gamma_{SW} = -RT\Gamma d \ln c, \tag{1}$$

where γ_{SW} is the solid/water medium interfacial free energy (varying with c), R is 8.32 J/mol · K, T is temperature (in K) and Γ is the excess concentration of the surface active molecules at the interface. For a dilute solution of protein at a concentration c, taking pure water as a reference ($c = 0$, $\Gamma = 0$), we can consider that $d \ln c = \frac{\Delta c}{c} = \frac{c-0}{c} = 1$ [15]. At $c = 0$, γ_{SW} has its maximum value ($\gamma_{SW}(0)$). At a concentration c in protein (surfactant), $\gamma_{SW}(c) < \gamma_{SW}(0)$. The surface excess concentration is given by:

$$\Gamma(c) = \frac{\gamma_{SW}(0) - \gamma_{SW}(c)}{RT}. \tag{2}$$

Thus, the maximum amount of molecules that can be adsorbed at the water/solid interface cannot be greater than Γ_{max}, given by:

$$\Gamma_{max} \approx \frac{\gamma_{SW}(0)}{RT}. \tag{3}$$

The maximum amount of molecules of a surface active compound that can be adsorbed at the substrate/water interface is proportional to the interfacial free energy, γ_{SW}. A direct proportionality between the interfacial free energy and the capacity for adsorption at an interface was already proposed by Lucassen–Reynders with usual surfactants [16].

Thus, the interfacial free energy between the culture medium and the solid surface, γ_{SW}, controls the adsorption of serum components that may inhibit or promote cell adhesion. However, γ_{SW} cannot be directly determined. Nevertheless, this parameter can be related to the surface free energies of the 2 phases in contact by using the Young's equation

$$\gamma_{SW} = \gamma_{SV} - \gamma_W \cos\theta, \tag{4}$$

where θ is the contact angle of water (surface tension γ_W) on the solid of surface free energy γ_{SV} in presence of liquid vapor. The difference $\gamma_S - \gamma_{SV}$, or spreading pressure, is the reduction of γ_S due to vapor adsorption. The spreading pressure is negligible for a solid that possesses a low energy surface such as polymers, 100 mJ/m^2 being generally considered as the cutoff value between high and low energy surfaces.

To obtain the solid–water interfacial free energy, γ_{SW}, it is necessary to know the surface free energy of the substrate. The determination of this parameter for the three polystyrene substrates is discussed in the next section.

3.2. Determination of the Surface Free Energies of Polymer Substrates

There are various approaches to evaluate the surface free energy of solid materials from wetting experiments [17]. Contact angle measurements with different probe liquids form the basis for the calculation of the surface free energy of solids. However, the solid surface free energy obtained can be different depending on the model chosen for the calculation. The problem is that the Young equation, describing the equilibrium of a liquid drop, L, having a contact angle θ on a solid surface, S, involves two unknowns: γ_{SV} and γ_{SL}. As one equation having two unknowns cannot be solved, several theories taking the Young equation as the starting point have been proposed [18–24]. However, none of these theories is fully satisfactory.

Interestingly, it was mentioned in 1972 by Fowkes [25], on the basis of data published by Dann [26], that the polar energy of interaction between polar liquids and polymers, including polystyrene, was proportional to the liquid polarity quantified by the polar component of the liquid surface free energy. We also made similar observations, and an alternative to describe polar interactions between polar liquids and polar polymers has been recently proposed [27]. Our hypothesis is that the polar liquid/polymer interactions result from polar forces induced by the liquid at the surface of the polarizable solid.

As usual, the experimental method to determine the solid surface free energy consists of measuring the contact angles of a series of probe liquids of known surface free energy and known dispersion (non-polar) and polar components: water, glycerol, formamide, ethylene glycol, diiodomethane and tricresylphosphate (TCP). The liquid properties and their contact angle values on polystyrene substrates are given in Table 1.

Considering two types of intermolecular interactions between liquids and polymer surfaces, the work of solid–liquid adhesion, W_{SL}, is written as

$$W_{SL} = \gamma_S + \gamma_L - \gamma_{SL} = \gamma_L(1 + \cos\theta) = I_{SL}^D + I_{SL}^P, \tag{5}$$

where I_{SL}^D and I_{SL}^P represent the dispersion and polar (solid–liquid) interactions energies, respectively. The dispersion energy satisfies the classical relationship involving the geometric mean of the dispersion components of the surface free energies [18, 19]:

$$I_{SL}^D = 2\sqrt{\gamma_S^D \gamma_L^D}. \tag{6}$$

However in this new model [27], I_{SL}^P is given by:

$$I_{SL}^P = k_S \gamma_L^P, \tag{7}$$

Table 1.
Liquid surface tension and its components (mJ/m²) and equilibrium contact angles (°) on polystyrene substrates used in this study

Liquid						
	Water	Glycerol	Formamide	Ethylene glycol	Diiodomethane	TCP
γ_L	72.8	63.4	58.2	48.0	50.8	40.9
γ_L^D	21.8	37.0	39.5	29.0	48.5	39.2
γ_L^P	51	26.4	18.7	19.0	2.3	1.7
Contact angles on polystyrene substrates						
PS	85.4 ± 0.7	73.4 ± 1.3	65.3 ± 1.6	59.5 ± 0.7	41.4 ± 1.5	19.3 ± 1.2
g-PS	75.9 ± 0.6	70.5 ± 1.7	59.4 ± 1.0	50.2 ± 0.6	32.4 ± 0.5	13.3 ± 1.7
TCT PS	54.3 ± 0.9	48.9 ± 0.8	15.6 ± 1.5	21.3 ± 1.1	32.4 ± 1.2	9.8 ± 0.7
Corning® CellBIND® Surface	37.4 ± 1.0	35.4 ± 0.8	5.0 ± 1.0	11.6 ± 1.1	37.3 ± 0.7	17.2 ± 0.8

the polar contribution to the surface free energy of the solid being a function of its surface polarizability k_S, and of the polar component of the liquid surface free energy, γ_L^P. When a polar liquid is in contact with a polarizable solid surface, the induced solid/liquid polar interaction can also be expressed with the geometric mean as [27]:

$$I_{SL}^P = 2\sqrt{\gamma_S^{Pi}\gamma_L^P}.$$
(8)

But the polar component of the solid surface free energy, γ_S^{Pi}, is induced by the contacting liquid. Equations (7) and (8) show that the induced polar contribution to the surface free energy of the solid is related to the solid surface polarizability, k_S, and to the liquid polarity by:

$$\gamma_S^{Pi} = \frac{k_S^2}{4}\gamma_L^P.$$
(9)

When in contact with a liquid, the 'apparent' solid surface free energy becomes liquid dependent as:

$$\gamma_S = \gamma_S^D + \gamma_S^{Pi}.$$
(10)

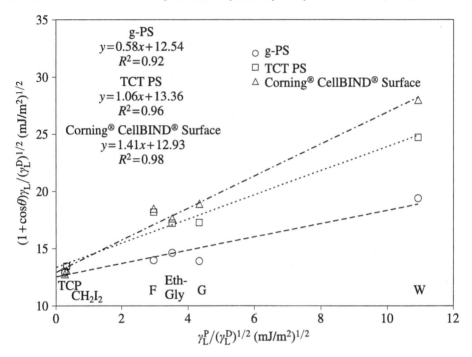

Figure 2. Characterization of surface properties of the g-PS, TCT PS and Corning® CellBIND® Surface by contact angle measurements (probe liquids: W = water; G = Glycerol; Eth–Gly = Ethylene glycol; F = Formamide; CH_2I_2 = Diiodomethane; TCP = Tricresylphosphate). The dispersion component γ_S^D is deduced from the ordinate at the origin $(2\sqrt{\gamma_S^D})$ of the linear regression and the polarizability (k_S) is deduced from the slope.

With the hypothesis of the induced polarity, the surface properties of polymer surfaces can be determined by using contact angle measurements and the following linear form of equation (5) written as:

$$\frac{\gamma_L}{\sqrt{\gamma_L^D}}(1 + \cos\theta) = 2\sqrt{\gamma_S^D} + k_S\frac{\gamma_L^P}{\sqrt{\gamma_L^D}}. \tag{11}$$

An example of application of equation (11) is presented in Fig. 2 for the g-PS, TCT PS and the Corning CellBIND® Surface. The dispersion component γ_S^D is deduced from the ordinate at the origin $(2\sqrt{\gamma_S^D})$ of the linear regression of equation (11), and k_S is deduced from the slope. The R^2 parameter in Fig. 2 is the coefficient of linear correlation, equal to 1 when the correlation is perfect. The surface properties of the polymer substrates (γ_S^D, k_S) are summarized in Table 2. The interfacial free energy between the substrates and water, γ_{SW}, are also reported in Table 2. It is obtained by expressing equation (5) with the following form:

$$\gamma_{SW} = \gamma_S + \gamma_W - 2\sqrt{\gamma_S^D\gamma_W^D} - k_S\gamma_W^P. \tag{12}$$

Table 2.
Interfacial free energy, γ_{SW} between water and the different polystyrene substrates obtained from the surface polarizability (k_S) approach of polar interactions. The dispersion component (γ_S^D) of polymer substrates is also given

Polymer substrate	γ_S^D (mJ/m^2)	k_S	γ_{SW} (mJ/m^2)
PS	37.6 ± 0.5	0.39 ± 0.14	35.2 ± 5.7
g-PS	39.3 ± 0.2	0.58 ± 0.09	28.3 ± 3.2
TCT PS	44.6 ± 0.3	1.06 ± 0.10	15.3 ± 2.6
Corning® CellBIND® Surface	41.8 ± 0.3	1.41 ± 0.06	7.8 ± 0.9

It has to be noted that it is the surface polarizability of the substrate which varies considerably for PS (0.39), g-PS (0.58), TCT PS (1.06) and Corning® CellBIND® Surface (1.41) and has a strong impact on the interfacial free energy with water. The dispersion component of the surface free energy of polystyrene substrates stays around 40 mJ/m^2, irrespective of the surface preparation.

4. Results: Cell Adhesion and Protein Adsorption

Following the protocol developed by Vogler and coworkers [28, 29], the steps of cell adhesion and proliferation were clearly identified for each cell line. The adhesion step is typically of a few hours and precedes the exponential proliferation step. Cell proliferation stops at confluence, i.e., when the entire available surface of the substrate is covered by the cells. Figure 3 shows the percentage of adhering MRC-5 cells to polystyrene substrates in the 6-well plate formats. It is very clear that the number of adhering cells is substrate dependent. The graph indicates also that the adhesion step is over after 2 h with MRC5 cells.

Figures 4–6 show the percentages of adhering cells to PS, TCT PS and to the Corning® CellBIND® Surface at the end of the adhesion step, in presence or absence of fetal bovine serum (FBS). Cell adhesion appears to be substrate dependent in presence of FBS in the culture medium. In presence of FBS, the number of adhering cells increases with the polarizability of the substrate (here quantified by k_S) for all three cell lines. For each cell line, Fig. 7 shows that the percentage of adhering cells increases with the surface polarizability (k_S) of the polymer substrate in presence of serum. However, for a given substrate, the percentage of adhering cells is also cell line dependent. In absence of FBS, the percentage of adhering cells is notably increased, especially on the PS substrate and with all three cell lines (Figs 4–6).

It was already reported that the presence of serum in culture medium might inhibit or delay cell adhesion compared to serum-free medium [2, 30–34]. In many cases, it seems likely that cell adhesion in the presence of serum is governed by

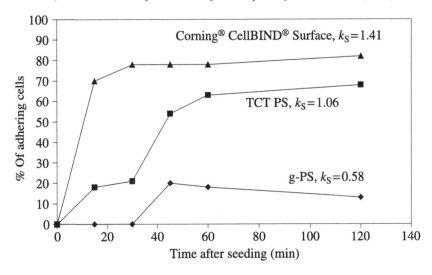

Figure 3. Adhesion of MRC-5 cells to polystyrene substrates (6-well plates). The coefficient k_S is the polarizability of the substrate.

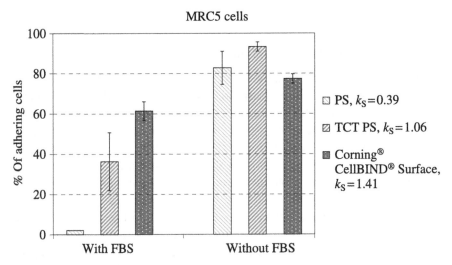

Figure 4. Percentage of MRC5 adhering cells to PS (35 mm dish), TCT PS and the Corning® CellBIND® Surface (6-well plates) with and without FBS in the culture medium, after 3 h of incubation. The coefficient k_S is the polarizability of the substrate.

competition between adsorption of Extra Cellular Matrix (ECM) proteins secreted by the cells or already present in the serum, mainly fibronectin, vitronectin, and collagen, and adsorption of the two most abundant proteins in serum: albumin and globulin.

The adsorption of proteins contained in FBS, and in particular of bovine serum albumin (BSA), on polystyrene substrates was revealed by staining of proteins with the colloidal gold dispersion. The samples exposed to BSA were dipped in the

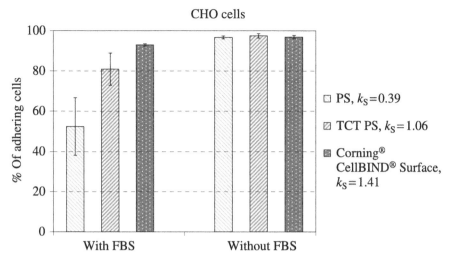

Figure 5. Percentage of CHO adhering cells to PS (35 mm dish), TCT PS and the Corning® CellBIND® Surface (6-well plates) with and without FBS in the culture medium, after 1.5 h of incubation. The coefficient k_S is the polarizability of the substrate.

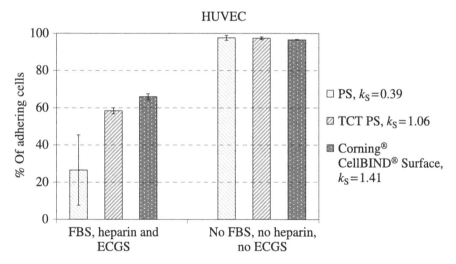

Figure 6. Percentage of HUVE adhering cells to PS (35 mm dish), TCT PS and the Corning® CellBIND® Surface (6-well plates) with and without FBS, heparin and ECGS in the culture medium, after 3 h of incubation. The k_S is the polarizability of the substrate.

gold staining dispersion for 18 h, after which they were rinsed with pure water and blow-dried under nitrogen. The optical density of the gold coating was measured at 550 nm. The results are presented in Table 3 for a BSA concentration of 1 g/l. In [12], we have shown that the optical density of the gold stain on polystyrene substrates increases with the BSA concentration. The optical density increases with the amount of proteins retained on the solid substrates. When exposed to a BSA so-

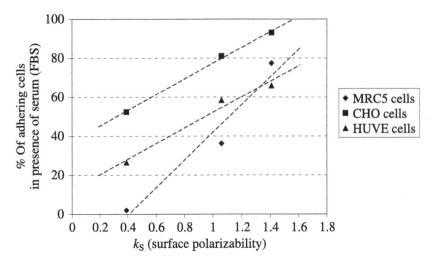

Figure 7. In presence of serum (FBS) the percentage of adhering cells increases with the surface polarizability of the polymer substrate.

Table 3.

Adsorption of BSA on polystyrene substrates from HamF12. The protein was stained with colloidal gold dispersion and the optical density (OD, no unit) was measured at 550 nm. The results are confirmed by fluorescence using biotinylated BSA (0.1 g/l) stained with Cy5-streptavidin

BSA concentration and measurement	PS	TCT PS	Corning® CellBIND® Surface
1 g/l. OD at 550 nm after gold staining	0.197	0.125	0.091
0.1 g/l. Cy5 signal in RFU*	15 000	4000	400

RFU = Relative Fluorescence Unit.

lution at a concentration of 1 g/l in the HamF-12 medium, the optical density after staining is about twice on PS than on the Corning® CellBIND® Surface. With TCT PS, the optical density is between the value for PS and for the Corning® CellBIND® Surface. Thus, it is clear that there is a higher adsorption of BSA on PS than on the other substrates. This result was confirmed by fluorescence using biotinylated BSA (0.1 g/l) stained with Cy5-streptavidin (Cy5-SA). The data are gathered in Table 3. The adsorption of BSA on the culture substrates was also studied by measuring the contact angle of an octane drop in presence of the culture medium containing different concentrations of BSA [12]. The results of the three methods (gold staining, fluorescent staining with Cy5-SA, and contact angle measurements [12]) indicate at least semi-quantitatively that BSA is significantly more adsorbed on PS than on TCT PS and than on the Corning® CellBIND® Surface.

The adhesion of CHO cells to PS was also characterized by counting the number of adhering cells on PS as a function of the concentration of BSA added to the FBS-

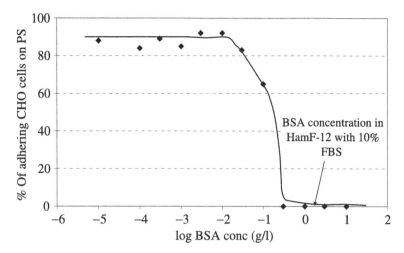

Figure 8. Percentage of adhering CHO cells to PS (35 mm dish) as a function of the BSA concentration in the culture medium.

free medium after 1.5 h of incubation. The quantitative results are reported in Fig. 8. At a concentration of 1.5 g/l of BSA corresponding to the actual concentration of BSA in the culture medium supplemented with 10% FBS (FBS contains 15 g/l of BSA), no more cells adhere to PS. The pictures in Fig. 9 show the morphology of adhering CHO cells on PS which changes considerably as a function of the BSA concentration. Increasing the BSA concentration reduces cell spreading. Cells become rounded and finally only a few cells adhere to PS from a BSA concentration of 0.3 g/l.

However, with a medium supplemented with 10% FBS, the percentage of adhering CHO cells is about 50% in 35 mm PS dish (Fig. 5). Thus, in the current culture medium supplemented with 10% FBS, the negative impact of BSA seems somehow to be balanced by other components of FBS, promoting cell adhesion. One of these could be fibronectin which is an important protein for cell adhesion to substrates in spite of its quite low concentration in the serum (0.2 g/l [32]). This possibility is compatible with the data presented in Fig. 10 where the percentage of adhering CHO cells deposited on PS is plotted as a function of the amount of fibronectin added to the HamF12 medium containing 1 g/l BSA. We see that the addition of fibronectin, even at a very low concentration (from 0.001 g/l) compensates the negative effect of BSA on CHO cell adhesion. Thus, we clearly observe that BSA inhibits cell adhesion on hydrophobic PS whereas fibronectin enhances cell adhesion and, thus, a competitive adsorption process may regulate cell adhesion (it must be noted that BSA and fibronectin are not the only proteins regulating cell adhesion).

Surface charges are often invoked as playing also a key role in protein adsorption. BSA (inhibiting cell adhesion) and fibronectin (promoting cell adhesion) have isoelectric points of about 5 (i.e., they are negatively charged at pH $= 7.4$), and as

BSA: 0 g/l   0.0001 g/l   0.0003 g/l   0.003 g/l   0.01 g/l

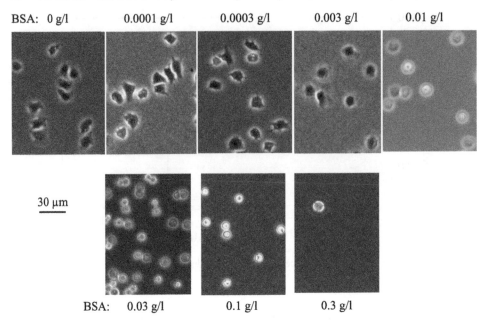

30 µm

BSA: 0.03 g/l   0.1 g/l   0.3 g/l

Figure 9. Morphology of CHO cells on PS in FBS-free medium as a function of the added BSA concentration (incubation period: 1.5 h).

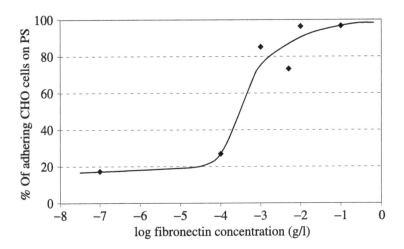

Figure 10. Percentage of adhering CHO cells to PS as a function of fibronectin addition in HamF12 medium containing 1 g/l of BSA.

PS surfaces are also usually negatively charged [2], an electrostatic repulsion between the proteins and the substrates can be expected. However, the relatively high concentration of salts in a culture medium (about 0.12 mol/l) can inhibit electrostatic interactions as the screening or Debye length is less than 1 nm at this salt concentration [35].

5. Conclusion

Cell adhesion is a very complicated process. Adhesion of anchorage-dependent mammalian cells is traditionally viewed through at least four major steps that precede proliferation: protein adsorption, cell-substrate contact, cell-substrate attachment and cell spreading. Protein adsorption itself is a complex process involving molecular-scale interactions with the substrate in presence of the aqueous culture medium.

For a long time, it has been shown that a determining parameter controlling the adhesion of cells to polystyrene substrates is their hydrophilicity [36]. In the present study, we observed that the presence of FBS in a culture medium has a strong negative influence on the adhesion of cells to hydrophobic PS and g-PS. In absence of FBS during the adhesion step, the number of adhering cells to PS, TCT PS and to the Corning® CellBIND® Surface is dramatically increased to reach 90% or more of adhering cells. In presence of FBS, the number of adhering cells is always higher on the more hydrophilic substrate (Figs 4–6) having the highest polarizability.

Our interpretation of these observations is that the interfacial free energy between the culture medium and the substrate controls the adsorption of serum components, especially proteins that may inhibit or promote cell adhesion. Thus, it is not directly the surface free energy of materials that determines cell adhesion but the interfacial free energy between the culture medium and the substrate. As observed in literature, we found that one component inhibiting cell adhesion is BSA. This protein is more adsorbed on hydrophobic substrates.

The second important result is the evidence of the positive role of fibronectin in cell adhesion. The competition between proteins that promote or inhibit cell adhesion seems also to be controlled by the interfacial free energy between the solid surface and the culture medium.

Of course, more work is needed to draw a better 'picture' of the phenomenon of cell adhesion. If the physical–chemical properties of substrates are able to influence the amount of adsorbed proteins as shown in this study, they are also able to influence their conformation as emphasized in [32] for fibronectin.

References

1. M. Jäger, C. Zilkens, K. Zanger and R. Krauspe, *J. Biomed. Biotechnol.*, Article ID 69036 (2007).
2. J.-L. Dewez, V. Berger, Y.-J. Schneider and P. G. Rouxhet, *J. Colloid Interface Sci.* **191**, 1 (1997).
3. M. Morra and C. Cassinelli, *J. Biomater. Sci. Polymer Edn.* **9**, 55 (1998).
4. J. Vitte, A. M. Benoliel, A. Pierres and P. Bongrand, *Eur. Cells Mater.* **7**, 52 (2004).
5. P. Prokopovich and S. Perni, *J. Mater. Sci. Mater. Med.* **20**, 195 (2009).
6. E. C. Onyiriuka, L. S. Hersh and W. Hertl, *Appl. Spectros.* **44**, 808 (1990).
7. E. C. Onyiriuka, L. S. Hersch and W. Hertl, *J. Colloid Interface Sci.* **144**, 98 (1991).
8. M. Hudis, in: *Techniques and Applications of Plasma Chemistry*, J. R. Hollahan and A. T. Bell (Eds), pp. 113–147. Wiley, New York, NY (1974).
9. C. F. Amstein and P. A. Hartman, *J. Clin. Microbiol.* **2**, 46 (1975).
10. W. S. Ramsey, W. Hertl, E. D. Nowlan and N. J. Binkowski, *In Vitro* **20**, 802 (1984).

11. M. D. Bryhan, P. E. Gagnon, O. V. LaChance, Z. H. Shen and H. Wang, US Patent 6617152 B2 (2003).
12. A. Carré and V. Lacarrière, in: *Contact Angle, Wettability and Adhesion*, K. L. Mittal (Ed.), Vol. 5, pp. 253–267. VSP/Brill, Leiden (2008).
13. V. N. Izmailova and G. P. Yampolskaya, in: *Proteins at Liquid Interfaces*, D. Möbius and R. Miller (Eds), pp. 103–147. Elsevier Science, B. V., Amsterdam (1998).
14. M. J. Rosen, *Surfactants and Interfacial Phenomena*, 2nd edn. Wiley, New York, NY (1989).
15. V. Thoreau, L. Boulangé and J. C. Joud, *Colloids Surf. A* **261**, 141 (2005).
16. E. H. Lucassen-Reynders, *J. Phys. Chem.* **67**, 969 (1963).
17. F. M. Etzler, in: *Contact Angle Wettability and Adhesion*, K. L. Mittal (Ed.), Vol. 3, pp. 219–264. VSP, Utrecht (2003).
18. F. M. Fowkes, *J. Phys. Chem.* **66**, 382 (1962).
19. F. M. Fowkes, *Ind. Eng. Chem.* **56 (12)**, 40 (1964).
20. F. M. Fowkes, D. C. McCarthy and M. A. Mostafa, *J. Colloid Interface Sci.* **78**, 200 (1980).
21. A. W. Neumann, R. J. Good, C. J. Hope and M. Sejpal, *J. Colloid Interface Sci.* **49**, 291 (1974).
22. A. W. Neumann, O. S. Hum, D. W. Francis, W. Zingg and C. J. van Oss, *J. Biomed. Mater. Res.* **14**, 499 (1980).
23. C. J. van Oss, R. J. Good and M. K. Chaudhury, *J. Colloid Interface Sci.* **111**, 378 (1986).
24. C. J. van Oss, M. K. Chaudhury and R. J. Good, *Adv. Colloid Interface Sci.* **28**, 35 (1987).
25. F. M. Fowkes, *J. Adhesion* **4**, 155 (1972).
26. J. R. Dann, *J. Colloid Interface Sci.* **32**, 321 (1970).
27. A. Carré, *J. Adhesion Sci. Technol.* **21**, 961 (2007).
28. E. A. Vogler and R. W. Bussian, *J. Biomed. Mater. Res.* **21**, 1197 (1987).
29. J. Y. Lim, X. Liu, E. A. Vogler and H. J. Donahue, *J. Biomed. Mater. Res. A* **68**, 504 (2004).
30. Y. Tamada and Y. Ikada, *J. Colloid Interface Sci.* **155**, 334 (1993).
31. J. A. Witkowki and W. D. Brighton, *Expl. Cell Res.* **70**, 41 (1972).
32. F. Grinnell and M. K. Feld, *J. Biol. Chem.* **257**, 4888 (1982).
33. A. S. G. Curtis and J. V. Forrester, *J. Cell Sci.* **71**, 17 (1984).
34. P. Knox, *J. Cell Sci.* **71**, 51 (1984).
35. J. Israelachvili, *Intermolecular and Surface Forces*, 2nd edn, p. 238, Academic Press (1992).
36. A. S. G. Curtis, J. V. Forrester, C. McInnes and F. Lawrie, *J. Cell Biology* **97**, 1500 (1983).

Is the Mechanics of Cell–Matrix Adhesion Amenable to Physical Modeling?

Alice Nicolas [a,*], **Achim Besser** [b] **and S. A. Safran** [c]

[a] Laboratory of Technologies of Microelectronics, Université Joseph Fourier-CNRS, France
[b] Harvard Medical School, Department of Cell Biology, 240 Longwood Avenue, Boston, MA 02115, USA
[c] Department of Materials and Interfaces, Weizmann Institute of Science, 76100 Rehovot, Israel

Abstract
We review the characteristics and differences of the adhesion of living cells to surfaces compared to the adhesion of inert matter. In particular, we show that while mimicking cell adhesion by vesicles sheds light on the role of membrane fluctuations and ligand phase separation, it neglects the role of active processes that in live cells control the number of adhesion proteins and their spatial organization. Recent models that incorporate active cell contractility are shown to predict some of the features. In this review, we mainly focus on thermodynamic modeling approaches and outline challenges for future theoretical work.

Keywords
Cell adhesion, mechanosensor, extracellular matrix, elasticity, dynamics, physical modeling

1. Introduction

Cell adhesion has been attracting the attention of physicists and materials scientists for the last fifteen years. The adhesion of cells to fibrous extracellular matrices, such as those found in the connective tissue in the dermis, is characterized by discrete, micrometer-sized domains of adhesion proteins (see Fig. 1). For biologists, delineating and understanding the rich set of mechanisms involved in cell adhesion is crucial. Many important processes in cells and their cycles are indeed triggered by adhesion. For instance, cells replicate their DNA, divide or differentiate only when they adhere [1]. A healthy (not cancerous) cell that cannot adhere commits suicide within a short time. This self-regulatory mechanism prevents sick or imperfect cells from invading and disrupting the entire organism. Adhesion is thus a key player in

---

* To whom correspondence should be addressed. Laboratory of Technologies of Microelectronics, Université Joseph Fourier-CNRS UMR5129, 17 av des Martyrs, 38054 Grenoble cedex, France. Tel.: 33 4 38 78 94 04; Fax: 33 4 38 78 58 92; e-mail: alice.nicolas@cea.fr

Surface and Interfacial Aspects of Cell Adhesion
© Koninklijke Brill NV, Leiden, 2010

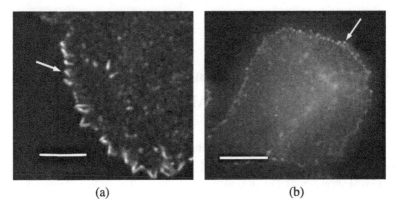

(a)   (b)

Figure 1. Adhesion of normal rat kidney epithelial cell to a stiff (40 kPa (a)) or a soft (3.5 kPa (b)) substrate coated with type I collagen. Adhesion domains are visualized by phosphotyrosine immunostaining. The shape of the adhesion domains is elliptical on rigid substrate and circular on soft substrate. Biochemical composition also differs. Bar: 10 µm. Adapted from Pelham *et al.*, *Proc. Natl Acad. Sci. USA* **94**, 13661 (1997).

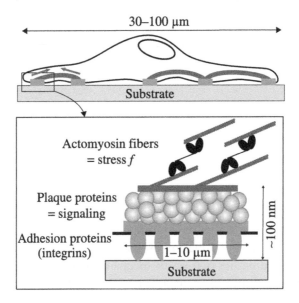

Figure 2. Schematic view of cell–matrix adhesions named focal adhesions. Plaques of adhesion proteins are constantly subjected to actomyosin stress f.

the control of cell proliferation and differentiation, and in the response of the cell to soluble ligands such as hormones [2].

From the physical point of view, adhesion domains show, at first glance, some similarities with phase separation on a surface in multicomponent systems. Adhesion domains are indeed characterized by the condensation of transmembrane proteins (e.g., integrins), that support a complex, intracellular gel-like complex of signaling proteins [3] (see Fig. 2). A well-formed adhesion site nucleates from a seed of transmembrane proteins [4, 5] that coexists with a solution of non-adhering proteins that are found in dilute solution in the cytoplasm or in the plasma membrane. The condensation of the transmembrane proteins in an adhesion domain has some similarity with the adsorption of a dilute solution of molecules on an interface (with the difference that the cytoplasm or the plasma membrane are crowded environments so that the diluted proteins do not behave as an ideal gas) and one might expect that this behavior might be explained by a generalization of the Langmuir isotherm. The relative stability of the domains in time (tens of minutes) suggests that these phases might be at equilibrium. However, such micrometer-sized domains are unexpected from a thermodynamic point of view [6], since the growth of these domains is energetically favorable due to the reduction in the line energy. Therefore, in thermal equilibrium a single condensed domain and complete phase separation is expected. Furthermore, it has been shown that in addition to diffusion, energy consuming, active processes can transport free proteins from the dilute phase in the cytoplasm or membrane to the adhesion sites. These active processes involve molecular motor proteins and are highly regulated by the cell [7]. All this suggests that some of the mechanisms behind the formation of the adhesion domain might differ from thermodynamic condensation processes. As described below, individual cellular adhesion domains have specific shapes (disk-like, elliptic, or fiber-like) that relate to their composition, and, more uniquely, to their connection to the contractile machinery of the cell. These features cannot be understood without considering, to some degree, the complexity of the intracellular architecture of the adhesions. Nevertheless, the combination of the great interest of biologists in cell adhesion and the puzzling similarity with phase separation processes has gained the attention of physicists in the attempt to identify some physical, generic mechanisms despite the complexity of the molecular events. Consequently, several physics-inspired scenarios have emerged that may indeed account for features of adhesion domains, such as their shape or growth velocity [8–10]. We review here recent steps in these directions. We focus only on cell–matrix adhesions. Cell–cell adhesions are also of great biological interest but their description and understanding is far from being as advanced.

In Section 2, we introduce the reader to the mechanics of cell adhesion. Adhesion domains connect the cell both to the extracellular environment as well as the internal cellular cytoskeleton (that contains the elastic and contractile parts of the cell). Adhesions are thus the regions in which the mechanical properties of that environment are probed by the cellular contractile machinery to allow the cell to adapt to the rigidity and chemical environment of its surroundings. In other words, cellular adhesions are anchors to the extracellular matrix as well as sensors of the physical properties of the cellular environment. We summarize the essential biology for the reader and highlight distinction between physical adhesion and cellular adhesion. The two processes are similar in their focus on molecules that adsorb on a substrate; for cells, these result in the formation of "lock and key" bonds between transmembrane proteins and extracellular proteins [1]. The formation of these bonds is followed by the assembly of a complex, intracellular architecture of proteins [5, 11]. The chemical interactions between these many proteins was at first identified as the origin of the formation of the adhesion domains. Along these lines, decorated vesicles were used as model systems to mimic the formation of cell–matrix adhesions and their response to physical cues. As detailed in Section 3, these studies demonstrated that despite some similarities, cells and vesicles show different adhesion behavior; the vesicle studies were indeed useful since they highlighted that the morphology and the mechanosensitivity of cell adhesion domains do not resemble vesicle adhesion domains and, therefore, are most probably not simply due to the phase separation of the many adhesion proteins, as occurs in vesicles [12]. Subsequent studies thus focused on the response of cellular adhesion domains to active, mechanical cues (not present in vesicles), as reviewed in Section 4. These approaches emphasize the role of active processes in the regulation of cell adhesion, and show that while very complicated on the molecular scale, cell adhesion may be understood in a generic manner, with some promising results.

2. Focal Adhesions are Stressed Adhesion Domains

As mentioned in the Introduction, adhesion domains consist of transmembrane proteins that connect a complex, intracellular, assembly of proteins [3, 13] to the external environment (see Fig. 2). In addition, these domains are connected in the interior of the cell to the cytoskeleton [1]. The cytoskeleton acts as both an elastic skeleton and an active, contractile muscle for the cell. Semi-flexible polymers such as actin filaments or microtubules form a dynamic scaffold that can either stabilize the cell shape or allow part of the cell to move by a polymerisation/depolymerisation process that takes place near the cell membrane [14]. The connection to the cytoskeleton is a major ingredient that distinguishes cellular adhesion domains from domains in inert, adhesive systems. Actin filaments self-assemble along with molecular motors (myosin II for instance) into stress fibers; the motors (which are proteins that convert chemical energy into motion) contract the stress fibers and cause them to exert shear forces on the adhesion domains. Compared with non-biological materials, large, stable cell adhesion domains form only when acted upon by the cytoskeleton-induced tension. This tension has a directionality (generally towards the nucleus) which results in an anisotropy in the shape of the adhesion domains [15] (see Fig. 1(a)). Two mechanisms contribute to these nN scale forces [16]: (i) The polymerization of individual cytoskeletal fibers, *via* the addition of a monomer at one end which leads to a force of order of 10 pN. (ii) The effective sliding of two parallel actin polymers in opposite directions caused by molecular motors; this contractility effect leads to forces of order of 1.5 pN per motor. Since hundreds of actin filaments can terminate at a single adhesion the total force sums up to the order of nN which yields a shear stress of the order of kPa [17, 18].

The connection of the adhesions to the intracellular, contractile apparatus allows cells to probe their physical environment. Adhesion domains exhibit different molecular compositions depending on the mechanical properties of the extracellular matrix (elasticity [19–21] and plasticity [22]) or geometrical features (2D or 3D [23, 24], the roughness of the substrate [25, 26], or the disorder of the nanoscale topography of the substrate [27, 28]). The domains nevertheless share some basic, common compositional features including the transmembrane proteins that join the cell to the extracellular matrix and the connection to the contractile cytoskeleton. In addition to geometrical constraints on the grafting of the adhesion proteins that largely influence the detailed composition and structure of the intracellular part of adhesion domains [29], the shape of adhesion domains can change in response to the physical characteristics of the extracellular environment. For instance, focal adhesions are adhesion domains that form on fairly rigid and firmly cross-linked networks of molecules from the matrix (see arrow in Fig. 1(a)). Focal adhesions are characterized by elongated, anisotropic domains that are localized mainly at the periphery of the cell. Although mobile, their lifetime is long, of the order of an hour [30, 31], and their size increases with the rigidity of the matrix [19]. On the other hand, fibrillar adhesions are observed when the molecules of the matrix can more easily disengage their crosslinks and the matrix can be remodeled by the cell itself. The adhesion domains then organize as lines that extend throughout the cell–matrix interface and not just at the periphery [22]. Finally, focal complexes are smaller, circular, transient adhesion domains that form on soft matrices (see arrow in Fig. 1(b)).

Cells therefore adapt their adhesions (and thereby regulate many biochemical signaling pathways that are initiated at these adhesions, as discussed in the Introduction) to the rigidity of the extracellular matrix in which or on which they are placed. In a sense, cells are rigidity sensors, and can resolve rigidity differences to the order of kPa [32]. The precise mechanisms that are at the origin of cell adhesion mechanosensitivity are still under debate. Experiments show that in addition to the sensitivity to the mechanical properties of the extracellular matrix, cell adhesions respond to externally applied, intra or extracellular stresses. Adhesion domains grow in the direction of forces, such as pipette induced shear [33], hydrodynamic flow [34], and stretching forces on the substrate [35]. Recently, local perturbations of stresses within the cytoskeleton by mechanical micro-devices, laser nanosurgery and other measurement techniques [33, 36–41] have supplemented biochemical tools to elucidate the basis of cell mechanosensitivity. We review below several physical and materials science approaches to understand cellular adhesion with the help of engineered vesicles; we then present theoretical models that aim at elucidating the force and rigidity sensing of cellular adhesion domains.

3. Can Adhesive Vesicles Serve as a Model System for Cell Adhesion?

If the formation of adhesion domains were due only to phase separation of the adhesion ligands into condensed domains in a manner that is independent of cell activity, one could model cell adhesion with a biomimetic, inert system. The challenge is then to determine the essential ingredients that this model system must contain. The fact that vesicles were well studied naturally led to the development of biomimetic vesicles for studying cell adhesion and de-adhesion (see [42] for a review). Such model system would indeed mimic cell adhesion by the following aspects: (1) Adhesive linkers that are mobile in the fluid bilayer behave like a 2D liquid [43], (2) Minimization of the curvature energy of the membrane induces the aggregation of the bound linkers [44], (3) The presence of thermal undulations of the bilayer adjacent to the adhering surface prevents the formation of large domains and thus stabilizes metastable, micrometer-sized, adhesion domains [45, 46].

The presence of membrane undulations or fluctuations can cause the distance between embedded adhesion proteins and the attractive surface to be large in some cases (see Fig. 3); this can prevent or at least slow down the formation of large domains [47]. The lifetime of micrometer-sized metastable domains can be increased by decorating the vesicle with an outer, polymeric coat whose steric interaction prevents the membrane from reaching the surface and mimics the cellular glycocalix coating [12, 48] (see Fig. 3). The polymeric repeller molecules prevent the membrane from approaching the substrate and thus inhibit adhesion of these ligands. These repellers can result in a long-range repulsion between two adhesion domains, as long as the osmotic pressure of the repellers is slightly larger than the van der Waals attraction between the adhesion proteins. The growth of adhesion domains is then controlled by the diffusion of the repellers outside the contact area [48]. The time required to form an adhesion domain extends from few seconds in absence of a polymeric coat [47] to few hundreds seconds in the presence of polymers [49]. Domains that coexist with a polymeric coat remain stable over several hours without fusing, in contrast to a lifetime of only a few minutes in the absence of repellers. This is to be compared to the characteristic time for growth of fo-

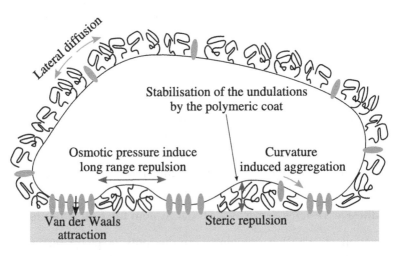

Figure 3. Engineered vesicle with adhesion ligands and artificial glycocalix that limits the formation of adhesion domains.

cal adhesions which is of the order of 30 min on glass substrate [31], and to their overall lifetime which is on the order of an hour, although this is highly dependent on the contractile activity of the cytoskeleton [50]. From these considerations we see that a model for adhering vesicles must minimally include a short-range, adhesion-induced attraction among the adhering ligands as well as a long-range repulsive contribution [49]. Even if the equilibrium state is still characterized by a macroscopic phase separation, these ingredients ensure the existence of phases of long-lived, metastable, micrometer-sized domains [51]. Such decorated vesicles adhere to substrates coated with various adhesion proteins and poly(ethylene glycol) repellers (reviewed in [52]). Submicrometer-sized domains are obtained by using the appropriate concentration of adhesion proteins [53, 54].

However, it is not at all clear if these models of phase separation of adhesion ligands in vesicles shed light on the mechanosensitivity of cellular adhesion. As mentioned earlier, cellular adhesion domains are observed to grow in the direction of external shear forces. On the other hand, they seem to decrease in size when the normal component of the stress is increased: their average size is smaller in a 3D environment, where internal stresses have a larger normal component than in 2D geometries where the stress is mainly in the plane [23, 24]. Force-induced domain growth is also observed in vesicle systems, but only in response to a stretch of the vesicle in the normal direction, where the vesicles are pulled off the substrate [54]. In this case, growth of the domains requires that the ligands and the receptors are mobile, so that the ligand–receptor link can migrate toward a smaller contact area, instead of being stretched by the normal force. The growth of the domains therefore occurs simultaneously with a decrease in the contact area. The osmotic pressure of the larger domains then contributes together with the curvature energy of the deformed vesicle to balance the work of the normal force. Whether such forces exist in cells is an open question but it is unlikely that they can be produced by the actomyosin fibers that exert lateral forces on adhesion domains with probably only a small normal component; the directionality of these forces is inferred from the arrangement of stress fibers that are usually observed to be parallel to the adhesive surface.

In addition, the size of cellular adhesion domains increases with the rigidity of the extracellular matrix [19]. To our knowledge, similar experiments have not been conducted with adhering vesicles. Nevertheless, one expects that adhesion domains in biomimetic vesicles should be larger on softer surfaces, since the elastic interaction of inclusions in a semi-infinite, deformable substrate is attractive [55, 56]. Soft substrates are thus expected to favor large adhesion domains, compared with more rigid substrates. Cellular adhesions, however, show the opposite trend.

To summarize, adhesion domains in vesicles do not respond to mechanical cues as do cellular adhesions. However, domain formation in vesicle resembles adhesion domains in altered cells, in which the connection between adhesion proteins and the actin cytoskeleton has been eliminated [57]. Vesicle systems then might offer some clearer understanding of the very initial formation of the cellular adhesions, for timescales at which the forces exerted by the cytoskeleton do not contribute.

4. Scenarios for Mechanosensitivity of Adhesion Domains

Because of the aforementioned limitations of adhering vesicles as a model system for cell adhesion, several groups have focused on mechanisms that could account for the response of adhesion domains to forces as well as their sensitivity to the mechanical properties of the extracellular matrix. Several approaches have emerged. A first group of models [58–60] assumes that adhesion molecules are strained by the cytoskeleton (which constantly pulls on the adhesion site) with a *constant strain rate*. For a fixed number of bound adhesive molecules, the stress in the links increases with time which, in turn, increases the rate of detachment according to a well-known experimental observation that is summarized in Bell's law[1] [62]. In these models, sensitivity of the adhesion domains to the rigidity of the extracellular matrix originates from the coupling between the stretching energy imposed by the cytoskeleton and the force-dependent molecular detachment of the adhesion proteins. The rigidity of the substrate determines whether the strain-induced stretching is localized in the adhesive bonds (subjected to Bell's law) or in the substrate [60]. The rigidity also impacts the time it takes the system to build up a certain force level; the dynamics is faster on stiff substrates. On soft substrates, these approaches predict a cascade of rupture events of the bonds of the adhesion proteins [58, 60], or a limitation on the magnitude of the cytoskeleton force applied to the adhesive bonds [63].

A second group of models assumes that the cytoskeleton pulls on the adhesions with a *constant stress* (in contrast with the first set of models that assume constant strain rate) [8–10]. The assumption of constant stress is consistent with experiments [17, 18] that show that the size of the adhesion domains increases linearly with the total force to which they are submitted. This is true even on the timescale of seconds [17] which implies that the stress may be constant at all timescales. In these models, heterogeneity in the stress field is assumed (consistent with the observed heterogeneity of the anchoring of the cytoskeleton along the adhesion), which accounts for the directional growth of the adhesion domain in response to a force. Ref. [9] assumes that new proteins join the stretched domain in order to restore the local concentration (which is reduced by the stretching of the protein "gel") and release the elastic energy of adhesion proteins. This mechanism resembles the formation of domains in the adsorption of molecules in solution on a surface; the condensation process is driven by the minimization of the energy of the adsorbing proteins. At the present stage of development of this model, the contribution of the rigidity of the

---

[1] This phenomenological relationship between the rate of detachment of proteins and the force they are submitted to suffers however some exceptions, namely the catch bonds, for which the bond is reinforced under tension [61].

substrate to the dynamics of growth of the adhesions has not been predicted. A very naive extension of this model would lead to predict that adhesions are insensitive to the flexibility of the substrate, since stress is imposed and deforms the elastic elements in series independently of each other. More work is therefore needed to allow this model to predict the effects of substrate flexibility. An alternative model that focuses on both the proteins and the substrate is presented in Ref. [8]. This model assumes that the response of the adhesions results from a tendency of the system to minimize the free energy of the linker proteins including the work done by the cytoskeleton to maintain the constant stress. Equivalently, the dynamics of the adhesions is given by the minimization of the energy the cell provides for maintaining the adhesions under constant stress. This differs from inert systems such as adhering vesicles, where the self-assembly of adhesive molecules results from the minimization of their free energy alone. Considering this new thermodynamic approach leads to the prediction of a linear relationship between the size of the adhesion domain and the rigidity of the substrate [64], consistent with experiments in [65] as long as constant stress is assumed [17]. In addition, the model predicts that the characteristic timescale to reach a stationary state is proportional to the rigidity of the substrate [66].

Some testable differences between the models in Refs [9] and [8] result from the underlying mechanical description of focal adhesions and the different deformation modes to which adhesion assembly or disassembly are coupled. In [8] and also in the subsequent work in Refs [10, 64, 66] the mechanical adhesion model predicts deformations (dilatation and compression modes) localized at the front and rear of the adhesion (front and rear being defined with respect to the direction of the stress that pulls on the adhesion), whereas, according to [9], the adhesions are deformed much more homogeneously along their length. Similarly, the inward or outward directed flux of proteins are predicted to be either localized to the front or rear of the adhesion [8, 10, 64, 66] or to be more homogeneously distributed along the entire adhesion [9]. Measuring the local (as opposed to integrated over the entire adhesion) protein flux during focal adhesion growth can thus differentiate between these models. As suggested in [10], such measurements are feasible by fluorescent recovery after photobleaching (FRAP) if the recovery of the adhesion protein under study is slow compared to the growth of the adhesion. Fluorescent speckle microscopy is yet another powerful technique that can be used to investigate adhesion dynamics. It has already been applied to measure the motion of proteins within adhesions [67] and it can be used to measure assembly and disassembly of polymeric structures with high spatiotemporal resolution [68]. Its spatial resolution is likely sufficient to map the protein fluxes over several micrometers large adhesions.

The models discussed so far focused on the response of focal adhesions to mechanical cues such as substrate stiffness or changes in the actomyosin stresses. Thus, these models treat active processes in an effective manner, in the sense that they do not explicitly account for biochemical signaling at focal adhesions that may be induced by the mechanical cues. For instance, it is known that the Rho-GTPases,

Rac1 and RhoA which mainly regulate the morphology of the actin network as well as the actomyosin stress, are themselves regulated by proteins that reside in the adhesion plaque [69] (see Fig. 2). It has been hypothesized that the activation of these regulatory pathways may also be modulated by the applied actomyosin stress. Thus, a complex mechanical and biochemical feedback loop is formed that relates adhesion and stress fiber assembly. Modeling of the complete regulatory system necessitates a higher degree of simplification, thus, adhesions are usually treated as a simple component in a reaction diffusion system. Such models have been used to describe the spatial distribution of adhesions within cells [70], the response of cells to applied shear stress [71] or the inhomogeneous contraction of stress fibers upon perturbation with a contractile drug [72]. However, rigidity sensing has not been addressed yet by these models and remains a challenge for the future.

5. Conclusion

With the currently available experimental data, it is not easy to show that one of these models is more appropriate than the others. Each model has different regimes in which it compares well with the observations and we cannot distinguish between the various theories at this point. For instance, there is no experimental proof of whether cells on substrates sense strain or stress (see the discussion in [64]). Only indirect results, such as the kinetics of growth of the domains [31] or its dependence on the rigidity of the matrix, can prove the assumptions made. In any case, the important feature of these models is that they augment the thermodynamic analysis focused on the linker proteins to include active processes. Some models do it by modeling the stress–strain relationship in the cytoskeleton by the analysis of the molecular events [60] with a phenomenological relationship (Hill's law [73]). Other models [8] suggest that active processes can be accounted for in a more macroscopic and traditional thermodynamic analysis by reconsidering the thermodynamic ensemble to be chosen.

Acknowledgements

S.A.S. is grateful for the support of the Israel Science Foundation, the Clore Center for Biological Physics, the Kimmelman Center for Biomolecular Structure and Assembly, and the Schmidt Minerva Center.

References

1. B. Alberts, D. Bray, J. Lewis, M. Raff, K. Roberts and J. D. Watson, *Molecular Biology of the Cell*, 3rd ed. Garland Publishing, New York (1994).
2. F. G. Giancotti and E. Ruoslahti, *Science* **285**, 1028 (1999).
3. E. Zamir and B. Geiger, *J. Cell Sci.* **114**, 3577 (2001).
4. S. Miyamoto, H. Teramoto, O. A. Coso, J. S. Gutkind, P. D. Purbelo, S. K. Akiyama and K. M. Yamada, *J. Cell Biol.* **131**, 791 (1995).

5. C. Cluzel, F. Saltel, J. Lussi, F. Paulhe, B. A. Imhof and B. Wehrle-Haller, *J. Cell Biol.* **171**, 383 (2005).

6. P.-F. Lenne and A. Nicolas, *Soft Matter* **5**, 2841 (2009).

7. K. Kawakami, H. Tatsumi and M. Sokabe, *J. Cell Sci.* **114**, 3125 (2001).

8. A. Nicolas, B. Geiger and S. A. Safran, *Proc. Natl Acad. Sci. USA* **101**, 12520 (2004).

9. T. Shemesh, B. Geiger, A. D. Bershadsky and M. Kozlov, *Proc. Natl Acad. Sci. USA* **102**, 12383 (2005).

10. A. Besser and S. A. Safran, *Biophys. J.* **90**, 3469 (2006).

11. R. Zaidel-Bar, C. Ballestrem, Z. Kam and B. Geiger, *J. Cell Sci.* **116**, 4605 (2003).

12. R. Bruinsma and E. Sackmann, *C. R. Acad. Sci. Paris, t. 2, Serie IV*, 803–815 (2001).

13. E. Zamir and B. Geiger, *J. Cell Sci.* **114**, 3583 (2001).

14. J. A. Cooper, *Ann. Rev. Physiol.* **53**, 585 (1991).

15. W. M. Petroll, L. Ma and J. V. Jester, *J. Cell Sci.* **116**, 1481 (2003).

16. J. Howard, *Mechanics of Motor Proteins and the Cytoskeleton.* Sinauer Press, Sunderland, Massachusetts (2001).

17. N. Q. Balaban, U. S. Schwarz, D. Riveline, P. Goichberg, G. Tzur, I. Sabanay, D. Mahalu, S. A. Safran, A. Bershadsky, L. Addadi and B. Geiger, *Nature Cell Biol.* **3**, 466 (2001).

18. J. L. Tan, J. Tien, D. M. Pirone, D. S. Gray, K. Bhadriraju and C. S. Chen, *Proc. Natl Acad. Sci. USA* **100**, 1484 (2003).

19. J. M. Goffin, P. Pittet, G. Csucs, J. W. Lussi, J.-J. Meister and B. Hinz, *J. Cell Biol.* **172**, 259 (2006).

20. T. Yeung, P. C. Georges, L. A. Flanagan, B. Marg, M. Ortiz, M. Funaki, N. Zahir, W. Ming, V. Weaver and P. A. Janmey, *Cell Motil. Cytoskeleton* **60**, 24 (2005).

21. G. Jiang, A. H. Huang, Y. Cai, M. Tanase and M. P. Sheetz, *Biophys. J.* **90**, 1804 (2006).

22. B. Z. Katz, E. Zamik, A. Bershadsky, Z. Kasu, K. M. Yamada and B. Geiger, *Mol. Cell. Biol.* **11**, 1047 (2000).

23. K. A. Beningo, M. Dembo and Y.-L. Wang, *Proc. Natl Acad. Sci. USA* **101**, 18024 (2004).

24. E. Cukierman, R. Pankov, D. R. Stevens and K. M. Yamada, *Science* **294**, 1708 (2001).

25. J. Y. Wong, J. B. Leach and X. Q. Brown, *Surface Sci.* **570**, 119 (2004).

26. M. J. Dalby, *Med. Eng. Phys.* **27**, 730 (2005).

27. M. J. Dalby, N. Gadegaard, R. Tare, A. Andar, M. O. Riehle, P. Herzyk, C. D. W. Wilkinson and R. O. C. Oreffo, *Nature Mater.* **6**, 997 (2007).

28. J. Huang, S. V. Grater, F. Corbellini, S. Rinck, E. Bock, R. Kemkemer, H. Kessler, J. Ding and J. P. Spatz, *Nano Lett.* **9**, 1111 (2009).

29. M. Arnold, E. A. Cavalcanti-Adam, R. Glass, J. Blummel, W. Eck, M. Kantlehner, H. Kessler and J. Spatz, *ChemPhysChem* **5**, 383 (2004).

30. L. B. Smilenov, A. Mikhailov, R. J. Pelham Jr, E. E. Marcantonio and G. G. Gundersen, *Science* **286**, 1172 (1999).

31. D. Raz-Ben Aroush, R. Zaidel-Bar, A. D. Bershadsky and H. D. Wagner, *Soft Matter* **4**, 2410 (2008).

32. U. Schwarz, *Soft Matter* **3**, 263 (2007).

33. D. Riveline, E. Zamir, N. Q. Balaban, U. S. Schwarz, T. Ishizaki, S. Narumiya, Z. Kam, B. Geiger and A. D. Bershadsky, *J. Cell Biol.* **153**, 1175 (2001).

34. R. Zaidel-Bar, Z. Kam and B. Geiger, *J. Cell Sci.* **118**, 3997 (2005).

35. R. Kaunas, P. Nguyen, S. Usami and S. Chien, *Proc. Natl Acad. Sci. USA* **102**, 15895 (2005).

36. A. Saez, M. Ghibaudo, A. Buguin, P. Silberzan and B. Ladoux, *Proc. Natl Acad. Sci. USA* **104**, 8281 (2007).

37. M. Allioux-Guerin, D. Icard-Arcizet, C. Durieux, S. Henon, F. Gallet, J.-C. Mevel, M.-J. Masse, M. Tramier and M. Coppey-Moisan, *Biophys. J.* **96**, 238 (2008).
38. N. J. Sniadecki, A. Anguelouch, M. T. Yang, C. M. Lamb, Z. Liu, S. B. Kirschner, Y. Liu, D. H. Reich and C. S. Chen, *Proc. Natl Acad. Sci. USA* **104**, 14553 (2007).
39. D. Choquet, D. P. Felsenfeld and M. P. Sheetz, *Cell* **88**, 39 (1997).
40. C. G. Galbraith and M. P. Sheetz, *Proc. Natl Acad. Sci. USA* **94**, 9114 (1997).
41. J. Colombelli, A. Besser, H. Kress, E. G. Reynaud, P. Girard, E. Caussinus, U. Haselmann, J. V. Small, U. S. Schwarz and E. H. K. Stelzer, *J. Cell Sci.* **122**, 1665 (2009).
42. A.-S. Smith and U. Seifert, *Soft Matter* **3**, 275 (2007).
43. P.-G. deGennes, P.-H. Puech and F. Brochard-Wyart, *Langmuir* **19**, 7112 (2003).
44. S. Komura and D. Andelman, *Eur. Phys. J. E* **3**, 259 (2000).
45. W. Helfrich, *Z. Naturforsch* **33A**, 305 (1978).
46. R. Lipowsky and S. Leibler, *Phys. Rev. Lett.* **56**, 2541 (1986).
47. P. H. Puech, V. Askovic, P.-G. de Gennes and F. Brochard-Wyart, *Biophys. Rev. Lett.* **1**, 85 (2006).
48. R. Bruinsma, A. Behrisch and E. Sackmann, *Phys. Rev. E* **61**, 4253 (2000).
49. A. Albersdörfer, T. Feder and E. Sackmann, *Biophys. J.* **73**, 245 (1997).
50. X. D. Ren, W. B. Kiosses, D. J. Sieg, C. A. Otey, D. D. Schlapfer and M. A. Schwartz, *J. Cell Sci.* **113**, 3673 (2000).
51. N. Destainville, *Phys. Rev. E* **77**, 011905 (2008).
52. E. Sackmann and R. Bruinsma, *ChemPhysChem* **3**, 262 (2002).
53. V. Marchi-Artzner, B. Lorz, C. Gosse, L. Jullien, R. Merkel, H. Kesslev and E. Sackmann, *Langmuir* **19**, 835 (2003).
54. A.-S. Smith, K. Sengupta, S. Goennenwein, U. Seifert and E. Sackmann, *Proc. Natl Acad. Sci. USA* **105**, 6906 (2008).
55. H. Wagner and H. Horner, *Adv. Phys.* **23**, 587 (1974).
56. I. B. Bischofs, S. A. Safran and U. S. Schwarz, *Phys. Rev. E* **69**, 021911 (2004).
57. H. Delano-Ayari, R. A. Kurdi, M. Vallade, D. Gulino-Debrac and D. Riveline, *Proc. Natl Acad. Sci. USA* **101**, 2229 (2004).
58. U. S. Schwarz, T. Erdmann and I. B. Bischofs, *Biosystems* **83**, 225 (2006).
59. V. S. Deshpande, M. Mrksich, R. M. McMeeking and A. G. Evans, *J. Mech. Phys. Solids* **56**, 1484 (2007).
60. C. E. Chan and D. J. Odde, *Science* **322**, 1687 (2008).
61. B. T. Marshall, M. Long, J. W. Piper, T. Yago, R. P. McEver and C. Zhu, *Nature* **423**, 190 (2003).
62. G. I. Bell, *Science* **200**, 618 (1978).
63. V. S. Deshpande, R. M. McMeeking and A. G. Evans, *Proc. Natl Acad. Sci. USA* **103**, 14015 (2006).
64. A. Nicolas and S. A. Safran, *Biophys. J.* **91**, 61 (2006).
65. A. Saez, A. Buguin, P. Silberzan and B. Ladoux, *Biophys. J.* **89**, L52 (2005).
66. A. Nicolas, A. Besser and S. A. Safran, *Biophys. J.* **95**, 527 (2008).
67. K. Hu, L. Ji, K. T. Applegate, G. Danuser and C. M. Waterman-Storer, *Science* **315**, 111 (2007).
68. G. Danuser and C. M. Waterman-Storer, *Ann. Rev. Biophys. Biomol. Struct.* **35**, 361 (2006).
69. R. Zaidel-Bar, S. Itzkovitz, A. Ma'ayan, R. Iyengar and B. Geiger, *Nat. Cell Biol.* **9**, 858 (2007).
70. I. L. Novak, B. M. Slepchenko, A. Mogilner and L. M. Loew, *Phys. Rev. Lett.* **93**, 268109 (2004).
71. G. Civelekoglu-Scholeya, A. Wayne Orr, I. Novak, J.-J. Meister, M. A. Schwartz and A. Mogilner, *J. Theor. Biol.* **232**, 569 (2005).
72. A. Besser and U. S. Schwarz, *New J. Phys.* **9**, 425 (2007).
73. A. V. Hill, *Proc. R. Soc. Lon. Ser-B* **126**, 136 (1938).

Modulation of Cell Structure and Function in Response to Substrate Stiffness and External Forces

Martial Hervy [*]

Corning S. A., Corning European Technology Center, 7 bis avenue de Valvins, 77210 Avon, France

Abstract

The mechanical balance maintained between the cell and its environment forms the basis of tissue formation, cohesion and homeostasis. Cells are able to sense the mechanical characteristics of the extracellular medium and to modulate their function accordingly. Furthermore, cells are able to modulate the mechanical properties of their environment in response to intracellular signals.

In this article we review recent papers describing the regulators and signaling pathways involved in these processes. We describe how the adhesion sites of the cell to the Extra Cellular Matrix are of primary importance in sensing force and how the actin network *via* actomyosin contractility and the associated Rho-family dependent regulatory paths also play a key role in mechanotransduction. We then focus on the different types of cellular responses to forces: modulation of adhesion, shape, migration properties, proliferation rate or regulation of differentiation. Cellular response can also be communicated to neighboring cells and be integrated to modulate processes such as development and can lead to diseases if misregulated.

Keywords

Substrate stiffness, mechanosensing, mechanotransduction, Rho GTPases, focal adhesions

1. Introduction

Cells in living organisms are subjected to various types of forces generally associated with tissue functions. Vascular cells, for example, are subjected to shear stress from blood flow and to tension generated by blood pressure; pulmonary cells are cyclically stretched as a consequence of respiration; and striated and smooth muscle cells are able to generate and support strong contractile forces.

Cells adhere specifically to a substrate *via* the interaction of transmembranes receptors: the integrins, with proteins of the extra-cellular matrix (ECM), present on the substrate or produced by the cells. On the intra-cellular side, integrins interact with multiprotein complexes in structures called focal adhesions. Focal adhesions connect the ECM to the actin cytoskeleton and in particular to contractile actomyosin fibers called stress fibers. Adherent cells exert tension forces *via* stress fibers

---

[*] Tel.: 011 33 1 64 69 70 73; e-mail: hervym@corning.com

Surface and Interfacial Aspects of Cell Adhesion
© Koninklijke Brill NV, Leiden, 2010

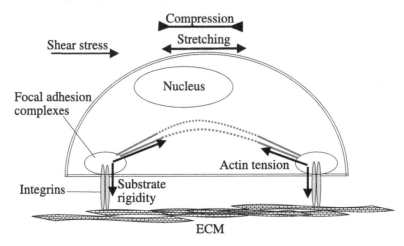

Figure 1. Schematic representation of the different types of forces impacting cell function. Cells can be subjected to and sense various types of forces including shear stress, compression, or stretching. The generation of forces *via* actomyosin contractility on an elastic ECM is another way for the cell to be subjected to forces.

on the focal adhesions. The forces they are subjected to are thus self-generated and the intensity of these forces is dependent on the rigidity and elastic nature of the tissue or surface they adhere to (Fig. 1). In an organism, cells may encounter a wide variety of tissue stiffnesses ranging from liquid to solid such as blood and bone, respectively. The elastic modulus of different tissues spans a wide range of values: 0.1–1 kPa for brain, 8–17 kPa for muscle and 25–40 kPa or higher for crosslinked collagen matrices or bones [1].

The intensity of forces generated by cells has been evaluated in several studies [2, 3]. Forces with a magnitude of 1 to 75 nN per focal adhesion were, for example, measured in smooth muscle cells, using microfabricated arrays of elastomeric needles [4]. Force generation can occur *via* actomyosin contractility; muscle cells rely on specialized structures and specific myosin isoforms whereas non-muscle cells generate forces through non-muscle types of myosins (e.g., myosin II) associated with actin in stress fibers [5].

The variation of substrate rigidity and the ability of cells to adapt their structure accordingly is important for the proper regulation of cell function. It was shown that cells are able to sense and respond to forces as low as 5 pN [6]. Interestingly, many processes are modulated in response to direct forces or substrate stiffness. The regulation of cell function including motility, proliferation, and differentiation is a crucial component of organism development and homeostasis.

In this article we review recent papers deciphering the diverse aspects of force sensing and mechanotransduction. We have mainly focused on two points: how is a cell able to translate forces into biochemical events leading to modulation and adaptation of cell function and how extensive is the mechano-dependent regulation of organism function?

2. Force Sensing and Molecular Responses to Substrate Stiffness

2.1. Focal Adhesions and Actin Cytoskeleton Reorganization

Interaction sites of the cell with the substrate are structures of choice to exert a mechanosensitive function. It is intuitive that the regions of the cell linking the external environment to the cell cytoskeleton, such as cell/ECM adhesion sites (focal adhesions), act as mechanosensors or host mechanosensor molecules.

The molecular changes occuring in focal adhesions in response to substrate rigidity or external forces, as well as the involvement of these protein complexes in mechanosensitivity and mechanotransduction have been extensively studied and reviewed [7–9]. In the chronology of the formation and maturation of the multi-proteins complexes associated with the intracellular domain of integrins, proposed by Zaidel-Bar *et al.* [10], the authors describe the maturation of focal complexes (FX) in focal adhesions (FA), characterized by the recruitment of proteins such as zyxin or tensin and then the maturation in fibrillar adhesions (FB) which are more complex and morphologically distinct from FA. This process requires continuous application of forces to adhesion points and the maturation of the different structures (FX to FA, FA to FB) depends on actomyosin contractility and ECM pliability [10]. Accordingly, several papers describe the strengthening of focal adhesions and stress fibers on stiff substrates or in response to external forces. It is also apparent that the modulation of focal adhesion complexes composition in response to the substrate modulus is a process regulated by internal cell signaling, including phosphorylation events. The focal adhesions of cells grown on flexible substrates are indeed irregularly shaped, highly dynamic and they contain decreased amounts of phosphotyrosine [11]. Treatment of cells on flexible substrates with tyrosine phosphatase inhibitors induces the formation of normal stable focal adhesions. Conversely, the treatment of cells on firm surfaces with myosin inhibitors causes the reduction of vinculin and phosphotyrosine at adhesion sites [11]. Studies of this type suggest that the translation of forces into biochemical signals does not necessarily depend on a distinct specific pathway but rather involves the modulation of existing processes such as cell adhesion, migration and shape maintenance.

2.2. Force Induced Signaling

2.2.1. The Role of Integrins
Complex cascades of events involved in cell migration and the associated signaling are modulated in response to substrate flexibility or external forces. The retrograde flux of actin and focal adhesion proteins, inversely proportional to cell migration, which takes place at focal adhesions in the lamellipodium region appears to be regulated by mechanical signals. This kind of effect could be a way to translate mechanical information into a biochemical regulation of cell function [12]. With this view of mechanosensing and transduction it is likely that the mechanosensor function depends on the integration of information from several key points. By

directly linking ECM molecules to the actomyosin network, the transmembrane proteins integrins are potentially implicated in such a process.

Integrin alpha5/beta3 in association with RPTP-alpha (Receptor-like Protein Tyrosine Phosphatase) was recently demonstrated to be critical to sense fibronectin matrix rigidity [13]. This cellular response to fibronectin rigidity which increases cell spreading and growth involves phosphorylation of a major signaling protein at focal adhesions: p130Cas by c-src kinase protein family, which is a type of regulation also described in regular cell adhesion and motility processes [14]. On soft surfaces, p130Cas would be displaced from these kinases preventing further phosphorylation [15]. One could then have a regulation of rigidity-dependent signaling *via* a spatial regulation of proteins. It was also observed that the dynamic formation of new connections between integrins and their specific ligands was critical in relaying the signals induced by shear stress to intracellular pathways [16]. This last study illustrates how a slight modulation of normal integrin function can be sufficient to initiate or propagate force-dependent regulation.

It is known that functional activation of integrins depends on multi-step conformational changes (for review see [17]). Complete activation of integrin alpha5/beta1, for example, requires switching between relaxed and extended conformations in response to forces generated through actomyosin. This force-dependent conformational change enables the integrin to bind the synergy site of fibronectin which leads to phosphorylation of FAK and the associated signaling [18].

2.2.2. Rho GTPases Involvement

A key point in mechanosensitivity and mechanotransduction appears to be the regulation of actomyosin contractility and actin organization *via* the Rho family of GTPases. These proteins play a major role in regulating cell migration and modulating cell signaling [19, 20]. Activation of RhoA can be induced in response to several events including integrin signaling, GPCR (G-Protein Coupled Receptor) activation, growth factors or LPA (lysophosphatidic acid) stimulation [21–23]. Activated RhoA promotes the activity of its effector ROCK which, in turn, phosphorylates the MLC (Myosin Light Chain) phosphatase, releasing its inhibitory effect on myosin, thus leading to actomyosin contraction. Active RhoA also promotes the activity of mDia and LIM-Kinase proteins leading to active actin polymerization (Fig. 2) [24, 25]. As previously described, stress fiber reinforcement and contractility are required for cellular force generation which is a part of the adaptative response to substrate rigidity.

Interestingly, the generation of cellular forces is required to properly organize the extracellular matrix and is also dependent on Rho GTPases. It is now clear that fibronectin matrix assembly involves the actin cytoskeleton, myosin II and RhoA downstream signaling including ROCK I and II [26]. This force-dependent unfolding of fibronectin and the signaling induced by the accessibility and engagement of new binding sites was shown to participate in sensing of forces on this type of ECM [18, 27].

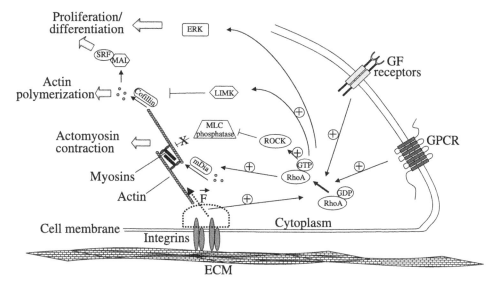

Figure 2. RhoA activation is a central event in force sensing and cellular response. RhoA activation in response to forces, focal adhesion dependent signaling, or transmembrane receptors activation modulate cell organization by the control of actin polymerization and actomyosin contractility. Cell function can be modulated accordingly *via* the activation of MAP kinases such as ERK. GF receptor: Growth factor receptor, GPCR: G-Protein coupled receptor.

2.3. Alternative Force-Dependent Signaling and Sensing

Many other processes are modulated in response to forces or substrate rigidity. It is likely that all this signals modulate the mechanosensing and fine tuning of cellular responses. Mechanosensitive ion channels, for example, were one of the first described ways for cells to translate mechanical information into biochemical signaling. They are found in many cell types and are thought to play a crucial role in vertebrate sensory receptors involved, for example, in touch and hearing (for review see [28]).

Actin polymerization state is modulated in response to substrate rigidity. On rigid substrates the ratio between free actin monomers (G-Actin) and polymerized actin (F-Actin) is balanced in the favor of F-actin. It was shown that monomeric actin and thus the G-actin/F-actin ratio directly impacted gene regulation, depending on SRF/MAL transcription factors [29, 30]. Interestingly, many of the genes regulated *via* this pathway are required for the actin cytoskeleton organization and cell motility [31, 32]. Actin thus participates in the cell response to rigidity and external forces on two distinct levels: directly *via* actomyosin contractility and *via* the modulation of gene expression, allowing structural and functional adaptations of the cells to the mechanical stimulus.

Also some proteins from the focal adhesion complexes have been shown to relocalize in response to force generation. An illustration of this process is given by fibroblasts subjected to cyclic stretching, which reorient normal to the direction of stretching and reinforce their stress fibers in response to the stimulation. Interestingly, this cellular response correlates with relocalization of the protein zyxin from focal adhesions to stress fibers [33]. Zyxin is a scaffold protein known to associate with both structural and signaling proteins at focal adhesions and could thus play an active role in cytoskeleton organization in response to force. Furthermore, many focal adhesions proteins are known to shuttle to the nucleus where they regulate gene expression [34]. It would be interesting to determine if this last type of activity could be modulated by forces exerted on cells.

It is apparent that cell mechanosensing and mechanotransduction do not rely on a single type of sensor or even on sensing mechanism but rather are the result of the integration of multiple information sources. This multi-level way to sense forces is probably the basis for the diversity of possible cellular responses through the cumulative or concerted activation of diverse signaling pathways able to impact cell fate.

3. Cells Response to Substrate Stiffness

We mainly focused in the previous section on the adaptative response of cells to substrate rigidity or external forces, which generally consists in modulating adhesive and contractile structures in order to resist the force sensed. This reorganization is associated with the initiation of signaling events and the activation of signaling pathways. In this section we describe how this information may modulate cell function and be integrated at the tissue and even at the organism scale to be a key component of developmental processes by modulating differentiation pathways, and how this cellular response can lead to diverse pathologies when altered.

3.1. Regulation of Motility: Durotaxis

One of the first short-term effects of substrate rigidity is the modulation of cell motility [11]. We saw previously that structures involved in cell migration, such as focal adhesion complexes and stress fibers, were reorganized in response to forces and many papers converge to describe durotaxis as a way for the cells to migrate from soft to stiffer regions. This phenomenon is described in various cell types including fibroblasts [35] and epithelial cells which when cultured on surfaces with an anisotropic rigidity grow and migrate along the direction of highest rigidity [36]. Mathematical modeling of this phenomenon indicates that the elasticity of the cytoskeleton is a major component of durotaxis regulation [37].

3.2. Modulation of Cell Rigidity

The reorganization of the cytoskeleton in response to forces can lead to an increase in cell rigidity. AFM measurements of the rigidity of fibroblasts grown on matrices of varying stiffness indeed demonstrate that the cell modulus is equivalent to substrate modulus, and this up to a value of 20 kPa. Fibroblasts thus tune their internal stiffness to match that of their substrate [38]. Modulation of cellular stiffness by the rigidity of the environment may be a mechanism used to direct the migration of a neighboring cell. This property may be important for development processes and for wound repair. However, cell response to substrate stiffness is dependent on cell type and environmental conditions. For example, fibroblasts and endothelial cells are reported to spread and organize actin fibers when substrate elastic modulus is higher than 3 kPa [39]. Such threshold is no longer observed when these cells organize cadherin mediated cell/cell contacts or with different cell types such as neutrophils [39].

3.3. Modulation of ECM Expression

Another effect of substrate rigidity is the modulation of the expression of genes coding for ECM proteins. It has been known for several years that the nature of the substrate influences ECM expression [40]. It is now apparent that mechanical stimulations also have an effect on ECM production. Cells adhering to a rigid substrate tend to produce increased amount of fibronectin, as observed in human mammary gland cells MCF-10A [41]. Such a modulation of ECM proteins expression is also observed in rabbit mesenchymal stem cells, which when seeded on collagen and subjected to stretching express collagen I and III [42]. In rats, acute training induces collagen expression in muscles and tendons, and this effect is dependent on the TGF-beta growth factor signaling pathway [43]. Modulation of ECM proteins expression is a good way for the cell to communicate with its close neighbors; this phenomenon associated with a control of cell motility is probably a first step allowing mechanical information perceived by a small number of cells to be transmitted to tissues and to modulate complex processes such as organogenesis or embryogenesis.

3.4. Force Generation and Sensing During Early Development

Generation of patterned forces seems to be a motor for development, connecting the execution of genetic and biochemical programs to the macroscopic tissue deformations that shape the embryo. ECM expression and orientation are thus crucial to direct stiffness-dependent cell migration and coordinate developmental processes such as gastrulation [44]. Using AFM to quantify the adhesive and mechanical properties of progenitor cells from gastrulating zebrafish embryos, it was shown that differential actomyosin dependent cell–cortex tension regulated by Nodal/TGF-beta signaling is a key factor to direct progenitor cell sorting [45]. In drosophila development it was observed that the deformation caused by germ band extension was able to upregulate the expression of the twist gene, the inactivation of which impairs differentiation of midgut cells [46].

3.5. Regulation of Proliferation and Differentiation

In the embryo, and also in the adult organism, cell proliferation is influenced by mechanical forces including substrate stiffness. Interestingly, the way this regulation occurs seems to be related to cell type and function. Both endothelial and smooth muscle cells proliferate in response to stretching, however in the case of endothelial cells this response depends on cell–cell adhesion and requires Rac1 activation, while in the case of smooth muscle cells, the effect is independent of cell–cell contacts and requires RhoA function [47]. This is another nice illustration of the central role of the Rho family of GTPases in the regulation of cell function in response to forces and modulation of cytoskeletal organization. Tissue shape and structure were also shown to regulate patterns of proliferation. It was indeed observed that regions of high traction forces were sites of increased cell proliferation. Once again, inhibiting actomyosin contractility or cell adhesion (in this case cadherin mediated cell/cell adhesion) was effective to inhibit this increased proliferation [48].

Many differentiation processes are influenced by substrate rigidity. Myoblasts in culture on collagen will fuse into myotubes independently of substrate flexibility, however the cells will only striate on substrates with an elastic modulus typical of normal muscle [49]. Interestingly, isolated embryonic cardiomyocytes cultured on flexible substrates develop optimally, with actomyosin striation and 1 Hz beating, when the elasticity of the substrate mimics the elasticity of developing myocardial environment. On more rigid surfaces that mimic post infarct fibrotic scars, the cells lack striated myofibrils and progressively stop beating. Further proteome analysis of these cells revealed differences in assembly and conformation of cytoskeletal proteins including vimentin, filamin and myosin [50].

3.6. Adult Stem Cells

Further illustration of stiffness-dependent control of cell differentiation comes from the study of adult stem cells. Naïve mesenchymal stem cells are sensitive to tissue elasticity. For this type of cells, matrices that mimic the elasticity of brain are neurogenic, matrices mimicking the elasticity of muscle are myogenic and more rigid matrices mimicking the elasticity of bone are osteogenic. This ability to differentiate as a function of substrate rigidity is dependent on myosin II activity [1]. This observation correlates nicely with previous work on human mesenchymal stem cells demonstrating that cell shape regulated lineage commitment by modulating RhoA activity and dependent signaling [51]. In this last study the authors underline the central role of RhoA in stem cell fate regulation in response to soluble factors or cell shape. At last and as previously described with other models, self-renewal and differentiation of neural stem cells can be modulated through substrate mechanical properties [52].

3.7. Associated Pathologies

The perturbation of tissue rigidity is associated with different types of pathologies. However, it is sometime difficult to determine if this variation of stiffness is the cause or the consequence of the pathology.

Tumors are generally stiffer than normal tissues. This increased rigidity could be linked to different factors including an increase in the elastic modulus of transformed cells due to perturbations of cell cytoskeleton [53] and the increased stiffness of the ECM linked to associated fibrosis [54]. The increased stiffness of tumors also seems to involve elevated Rho-dependent cytoskeletal tension promoting focal adhesions, disrupting cell–cell junctions and perturbing the regulation of tissue polarity, growth and function [55]. In contractile EGF-transformed epithelia with elevated RhoA and ERK activity, phenotypic reversion can be achieved if Rho-generated myosin II-dependent contractility or ERK activity is decreased [55]. In the case of glioblastoma, a malignant pathology of the central nervous system with a poor prognostic, the proliferation and motility of tumor cells is increased on stiff surfaces. As previously observed, this effect can be attenuated by an inhibition of myosin II-based contractility [56].

Arterial wall stiffening and cardiac hypertrophy in response to the increased mechanical load linked to hypertension are also other recurrent examples of pathologies associated with perturbation of substrate stiffness [57]. As a last example, ECM rigidity has been shown to modulate cytoskeletal organization, signal transduction and ECM deposition by human trabecular cells [58]. Variations of ECM rigidity appear to be a pathophysiologic actor in glaucoma [58].

4. Conclusion

Through these different studies it is apparent that responses and adaptation to substrate stiffness and external forces are major ways for the cells to obtain mechanical information from their environment and translate it to regulate fundamental organism level processes such as embryogenesis, organogenesis or tissue homeostasis.

Despite few specialized molecular systems for force sensing, including stretch activated channels or particular cell types such as auditive ciliated cells, the general mechanism involved in mechanosensing and force transmission appears mainly to be an adaptative way to regulate the known regulators of cell adhesion, structure and motility. The fine details of cell mechanosensing and mechanotransduction remain to be described, but it is already apparent that integrins, actomyosin network, phosphorylation pathways dependent on integrins, including FAK downstream signaling, are common themes to force sensing and transduction and the cell regulation of the balance between adhesion and migration. As for adhesion and motility, the recruitment of functional proteins is associated with the conditional assembly of signaling complexes leading to a modulation of the fundamental cell functions: proliferation, differentiation and survival. Moreover, the Rho family of GTPases seems to play a central role. These proteins coordinate the early cellular responses to substrate rigidity and external forces (actin assembly, regulation of lamellipodium and filopod activity and cell motility) as well as the longer term cellular responses. They are involved in the modulation of signaling pathways such as MAP kinases or the regulation of gene expression, leading to adaptation of cellular function to the mechanical properties of the substrate.

References

1. A. J. Engler, S. Sen, H. L. Sweeney and D. E. Discher, *Cell* **126**, 677–689 (2006).
2. N. Q. Balaban, U. S. Schwarz, D. Riveline, P. Goichberg, G. Tzur, I. Sabanay, D. Mahalu, S. Safran, A. Bershadsky, L. Addadi and B. Geiger, *Nature Cell Biol.* **3**, 466–472 (2001).
3. M. Dembo and Y. L. Wang, *Biophys. J.* **76**, 2307–2316 (1999).
4. J. L. Tan, J. Tien, D. M. Pirone, D. S. Gray, K. Bhadriraju and C. S. Chen, *Proc. Natl. Acad. Sci. USA* **100**, 1484–1489 (2003).
5. S. Pellegrin and H. Mellor, *J. Cell Sci.* **120**, 3491–3499 (2007).
6. D. Choquet, D. P. Felsenfeld and M. P. Sheetz, *Cell* **88**, 39–48 (1997).
7. A. Bershadsky, M. Kozlov and B. Geiger, *Curr. Opin. Cell Biol.* **18**, 472–481 (2006).
8. B. Geiger and A. Bershadsky, *Cell* **110**, 139–142 (2002).
9. T. Shemesh, B. Geiger, A. D. Bershadsky and M. M. Kozlov, *Proc. Natl. Acad. Sci. USA* **102**, 12383–12388 (2005).
10. R. Zaidel-Bar, M. Cohen, L. Addadi and B. Geiger, *Biochem. Soc. Trans.* **32**, 416–420 (2004).
11. R. J. Pelham and Y. Wang, *Proc. Natl. Acad. Sci. USA* **94**, 13661–13665 (1997).
12. W. H. Guo and Y. L. Wang, *Mol. Biol. Cell* **18**, 4519–4527 (2007).
13. G. Jiang, A. H. Huang, Y. Cai, M. Tanase and M. P. Sheetz, *Biophys. J.* **90**, 1804–1809 (2006).
14. D. A. Hsia, S. T. Lim, J. A. Bernard-Trifilo, S. K. Mitra, S. Tanaka, J. den Hertog, D. N. Streblow, D. Ilic, M. H. Ginsberg and D. D. Schlaepfer, *Mol. Cell. Biol.* **25**, 9700–9712 (2005).
15. A. Kostic and M. P. Sheetz, *Mol. Biol. Cell* **17**, 2684–2695 (2006).
16. S. Jalali, M. A. del Pozo, K. Chen, H. Miao, Y. Li, M. A. Schwartz, J. Y. Shyy and S. Chien, *Proc. Natl. Acad. Sci. USA* **98**, 1042–1046 (2001).
17. M. A. Arnaout, S. L. Goodman and J. P. Xiong, *Curr. Opin. Cell Biol.* **19**, 495–507 (2007).
18. J. C. Friedland, M. H. Lee and D. Boettiger, *Science* **323**, 642–644 (2009).
19. M. Fukata, M. Nakagawa and K. Kaibuchi, *Curr. Opin. Cell Biol.* **15**, 590–597 (2003).
20. K. Kaibuchi, S. Kuroda and M. Amano, *Annu. Rev. Biochem.* **68**, 459–486 (1999).
21. A. J. Ridley, *J. Cell Sci. Suppl.* **18**, 127–131 (1994).
22. A. J. Ridley and A. Hall, *Embo J.* **13**, 2600–2610 (1994).
23. C. D. Nobes, P. Hawkins, L. Stephens and A. Hall, *J. Cell Sci.* **108** (Pt 1), 225–233 (1995).
24. M. Maekawa, T. Ishizaki, S. Boku, N. Watanabe, A. Fujita, A. Iwamatsu, T. Obinata, K. Ohashi, K. Mizuno and S. Narumiya, *Science* **285**, 895–898 (1999).
25. O. Geneste, J. W. Copeland and R. Treisman, *J. Cell Biol.* **157**, 831–838 (2002).
26. A. Yoneda, D. Ushakov, H. A. Multhaupt and J. R. Couchman, *Mol. Biol. Cell* **18**, 66–75 (2007).
27. L. Li, H. H. Huang, C. L. Badilla and J. M. Fernandez, *J. Mol. Biol.* **345**, 817–826 (2005).
28. B. Martinac, *J. Cell Sci.* **117**, 2449–2460 (2004).
29. F. Miralles, G. Posern, A. I. Zaromytidou and R. Treisman, *Cell* **113**, 329–342 (2003).
30. M. K. Vartiainen, S. Guettler, B. Larijani and R. Treisman, *Science* **316**, 1749–1752 (2007).
31. D. Gineitis and R. Treisman, *J. Biol. Chem.* **276**, 24531–24539 (2001).
32. S. Medjkane, C. Perez-Sanchez, C. Gaggioli, E. Sahai and R. Treisman, *Nature Cell. Biol.* **11**, 257–268 (2009).
33. M. Yoshigi, L. M. Hoffman, C. C. Jensen, H. J. Yost and M. C. Beckerle, *J. Cell. Biol.* **171**, 209–215 (2005).
34. M. Hervy, L. Hoffman and M. C. Beckerle, *Curr. Opin. Cell Biol.* **18**, 524–532 (2006).
35. C. M. Lo, H. B. Wang, M. Dembo and Y. L. Wang, *Biophys. J.* **79**, 144–152 (2000).
36. A. Saez, M. Ghibaudo, A. Buguin, P. Silberzan and B. Ladoux, *Proc. Natl. Acad. Sci. USA* **104**, 8281–8286 (2007).

37. K. A. Lazopoulos and D. Stamenovic, *J. Biomech.* **41**, 1289–1294 (2008).
38. J. Solon, I. Levental, K. Sengupta, P. C. Georges and P. A. Janmey, *Biophys. J.* **93**, 4453–4461 (2007).
39. T. Yeung, P. C. Georges, L. A. Flanagan, B. Marg, M. Ortiz, M. Funaki, N. Zahir, W. Ming, V. Weaver and P. A. Janmey, *Cell Motility Cytoskeleton* **60**, 24–34 (2005).
40. C. H. Streuli and M. J. Bissell, *J. Cell. Biol.* **110**, 1405–1415 (1990).
41. C. M. Williams, A. J. Engler, R. D. Slone, L. L. Galante and J. E. Schwarzbauer, *Cancer Res.* **68**, 3185–3192 (2008).
42. N. Juncosa-Melvin, K. S. Matlin, R. W. Holdcraft, V. S. Nirmalanandhan and D. L. Butler, *Tissue Eng.* **13**, 1219–1226 (2007).
43. K. M. Heinemeier, J. L. Olesen, F. Haddad, H. Langberg, M. Kjaer, K. M. Baldwin and P. Schjerling, *J. Physiol.* **582**, 1303–1316 (2007).
44. R. Keller, L. A. Davidson and D. R. Shook, *Differentiation* **71**, 171–205 (2003).
45. M. Krieg, Y. Arboleda-Estudillo, P. H. Puech, J. Kafer, F. Graner, D. J. Muller and C. P. Heisenberg, *Nature Cell Biol.* **10**, 429–436 (2008).
46. N. Desprat, W. Supatto, P. A. Pouille, E. Beaurepaire and E. Farge, *Dev. Cell* **15**, 470–477 (2008).
47. W. F. Liu, C. M. Nelson, J. L. Tan and C. S. Chen, *Circulation Res.* **101**, e44–e52 (2007).
48. C. M. Nelson, R. P. Jean, J. L. Tan, W. F. Liu, N. J. Sniadecki, A. A. Spector and C. S. Chen, *Proc. Natl. Acad. Sci. USA* **102**, 11594–11599 (2005).
49. A. J. Engler, M. A. Griffin, S. Sen, C. G. Bonnemann, H. L. Sweeney and D. E. Discher, *J. Cell. Biol.* **166**, 877–887 (2004).
50. A. J. Engler, C. Carag-Krieger, C. P. Johnson, M. Raab, H. Y. Tang, D. W. Speicher, J. W. Sanger, J. M. Sanger and D. E. Discher, *J. Cell Sci.* **121**, 3794–3802 (2008).
51. R. McBeath, D. M. Pirone, C. M. Nelson, K. Bhadriraju and C. S. Chen, *Dev. Cell* **6**, 483–495 (2004).
52. K. Saha, A. J. Keung, E. F. Irwin, Y. Li, L. Little, D. V. Schaffer and K. E. Healy, *Biophys. J.* **95**, 4426–4438 (2008).
53. M. Beil, A. Micoulet, G. von Wichert, S. Paschke, P. Walther, M. B. Omary, P. P. Van Veldhoven, U. Gern, E. Wolff-Hieber, J. Eggermann, J. Waltenberger, G. Adler, J. Spatz and T. Seufferlein, *Nature Cell Biol.* **5**, 803–811 (2003).
54. M. J. Paszek and V. M. Weaver, *J. Mammary Gland Biol. Neoplasia* **9**, 325–342 (2004).
55. M. J. Paszek, N. Zahir, K. R. Johnson, J. N. Lakins, G. I. Rozenberg, A. Gefen, C. A. Reinhart-King, S. S. Margulies, M. Dembo, D. Boettiger, D. A. Hammer and V. M. Weaver, *Cancer Cell.* **8**, 241–254 (2005).
56. T. A. Ulrich, E. M. de Juan Pardo and S. Kumar, *Cancer Res.* **69**, 4167–4174 (2009).
57. A. D. Bradshaw, C. F. Baicu, T. J. Rentz, A. O. Van Laer, J. Boggs, J. M. Lacy and M. R. Zile, *Circulation* **119**, 269–280 (2009).
58. G. Schlunck, H. Han, T. Wecker, D. Kampik, T. Meyer-ter-Vehn and F. Grehn, *Invest. Ophthalmol. Visual Sci.* **49**, 262–269 (2008).

Cell/Material Interfaces: Influence of Surface Chemistry and Surface Topography on Cell Adhesion

Karine Anselme *, Lydie Ploux and Arnaud Ponche

Institut de Sciences des Matériaux de Mulhouse (IS2M), CNRS LRC7228, Université
de Haute-Alsace, Mulhouse, France

Abstract

The need to control the adhesion of cells to material surfaces plays an important role in determining the design of biomaterial substrates for biotechnology and tissue-engineering applications. As a the first step in a cascade of cellular events, adhesion affects many aspects of cell function, including spreading, migration, proliferation and differentiation. After a short description of cell adhesion and essential molecules involved in, the present knowledge on the influence of surface topography on cell behavior will be described by considering not only the amplitude of the surface topography but also its organization at all scales (micro- and nano-scale). The biological mechanisms underlying the cell response to topography will be evoked. Secondly, the influence of surface chemistry as well as surface energy on cell adhesion will be described. Thirdly, as the cells never interact with a bare material but with materials on which the proteins from biological fluids have adsorbed, some studies on the role of proteins in cell adhesion will be used to illustrate this point. Finally, the influence of substrate mechanics on cell differentiation will be described.

Keywords

Surface topography, surface chemistry, protein adsorption, cell adhesion, mechanics

1. Introduction

Cell adhesion plays an important role in cellular physiological functions (such as cell growth, differentiation and motility) and in integration of implantable biomedical devices to cells and tissues. A complete understanding of cell adhesion to materials with different textures is essential to promote integration in tissues and to minimize scar tissue formation. Moreover, the development of tissue-engineering strategies needs also to design biointerfaces between cells and biomaterials to match the anatomy and physiology of the host tissue and to meet the biochemical and biophysical requirements of specific cell types. To do so requires the control of topographical, chemical and mechanical surface properties at length scales on a par

* To whom correspondence should be addressed. Tel.: +33 389608766; Fax: +33 389608799; e-mail:
karine.anselme@uha.fr

Surface and Interfacial Aspects of Cell Adhesion
© Koninklijke Brill NV, Leiden, 2010

with the size of cells [1]. This justifies the publication of several reviews on this subject in the last ten years in the field of musculo-skeletal tissues [2–4]. In this paper, our objective is to sum up the latest developments in the field by focusing again our attention on interactions between cells and material surfaces.

2. Cell Adhesion

2.1. Proteins Involved in Cell Adhesion (Fig. 1)

2.1.1. Membrane Molecules

Integrins constitute a widely expressed family of trans-membrane receptors involved in cell–extra-cellular matrix (ECM) interactions. Integrin heterodimers, consisting of non-covalently associated α and β subunits, bind to specific amino acids sequences such as the arginine–glycine–aspartic acid (RGD) recognition motif present in many ECM proteins. The integrin family is composed of 24 heterodimers. This diversity of structures is associated with various ligand-binding possibilities [5]. The integrin spanning the cell membrane acts as an interface between the intra- and extra-cellular components. Integrins interact with the ECM through their extra-cellular domains and with components of cytoskeleton and also with signaling molecules through their intra-cellular domains. Through their interactions integrins can regulate many cellular functions, like cell adhesion, motility, shape, growth and differentiation [6, 7]. They act as transmitters of outside-in and inside-out signaling [8].

As cell–substrate adhesion is based on integrins, adherens junctions containing cadherins mediate cell–cell adhesions. Cadherins are also trans-membrane glycoproteins acting with intra-cellular partners: catenins which interact with intra-cellular proteins and are also involved in signal transduction (Fig. 1).

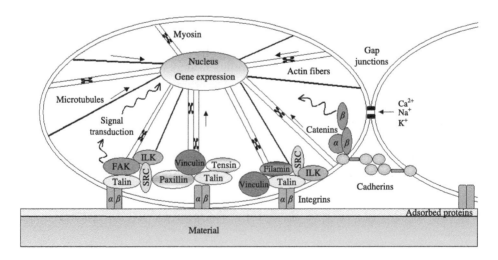

Figure 1. Schematic diagram of the proteins involved in adhesion of eucaryotic cells on materials.

Inter-cellular communication occurs through direct exchange of ions *via* gap junctions. Gap junctions are constituted of proteins called connexins. When the connexin of one cell is in register with a similar structure on a neighbouring cell, a trans-membrane channel is formed (Fig. 1).

2.1.2. Cytoskeleton Proteins

The sites of adhesion between adherent cells and material surfaces are called focal contacts, focal adhesions or adhesion plaques. Focal contacts are closed junctions where the distance between the substrates surface and the cell membrane is between 10–15 nm. Some authors make the distinction between focal complexes measuring less than 1 μm in length, the focal adhesions measuring from 1 to 5 μm in length while those measuring over 5 μm are designated as supermature adhesions [9]. A kinetic aspect can also be considered to explain these differences. During cell motion, focal complexes are continuously formed under the protruding front of the cells that is called lamellipodium. Lamellipodia are thin leaflets of cytoplasm (~200 nm thick and several micrometres in width) composed of a polarized array of thin actin filaments. Some focal complexes undergo 'maturation', transforming into focal adhesions which grow and extend centripetally, with concomitant formation of thicker actin fibers. Finally the more the force applied on focal adhesions increases, the more their size increases [10].

The external faces of focal contacts present integrins. On the internal face some proteins like talin, paxillin, vinculin, tensin, etc., are known to mediate interactions between actin filaments and integrins (Fig. 1). Many other proteins co-localize with vinculin and talin in the adhesion plaques: proteases, protein kinases, signaling molecules, etc. All these proteins are involved in signal transduction [11]. Besides actin, other cytoskeleton proteins are involved in the maintenance of cell architecture like tubulin microtubules and vimentin, lamin or keratin microfilaments. Essential proteins also involved in cell capacity to adhere, spread, divide and migrate are the myosins [12]. Myosins are actin-based motor proteins that translocate along actin, allow their contraction and, thus, the maintenance of the cellular force balance. As we will see later, all the cytoskeleton molecules and the cell adhesion receptors act together in mechano-transduction and mechano-sensitivity [13].

2.1.3. Extra-Cellular Matrix Proteins

Most of the cells in multicellular organisms are surrounded by a complex mixture of biomolecules that makes up the extra-cellular matrix (ECM). This matrix is secreted by the cells. It consists of protein fibers embedded in an amorphous mixture of huge protein–polysaccharide ('proteoglycan') molecules. The protein fibers are mainly collagen fibers but can also be noncollagen proteins. In bone for example, the ECM is composed of 90% collagen proteins (type I and type V collagens) and 10% of noncollagen proteins (osteocalcin, osteonectin, bone sialoproteins, proteoglycans, osteopontin, fibronectin, growth factors, bone morphogenic proteins, etc.). All these proteins are synthesized by osteoblasts and most are involved in adhesion. *In vitro*, other proteins such as fibronectin and vitronectin are involved in cell adhesion. They contain the RGD sequence which is specific to the fixation of cell membrane receptors like integrins. Other sequences in collagen (DGEA, GFOGER, P15), proteoglycans (KRSR) or laminin (IKVAV) have been shown to be also involved in cell adhesion [14].

2.1.4. Signal Transduction

As previously mentioned, integrin and other cell adhesion molecules regulate gene expression by a signal transduction process. After integrin clustering, cytoskeleton and signaling molecules will be recruited and activated. The signaling pathway inside the cell involves the accumulation and phosphorylation of several proteins like focal adhesion kinase (FAK), src, Rho GTPases, ERK, JNK which will be involved in the stimulation of cell proliferation, for example [7]. By means of antibodies directed towards phosphotyrosine or of molecular biology techniques like RT-PCR (reverse transcription polymerase chain reaction) and real time PCR, it has been recently possible to elucidate, for example, the signal transduction pathways involved in osteoblastic cell adhesion to materials [15–17]. However, the integrin generated signaling cascades remain incompletely understood and this is still an area of very active research [18].

2.2. Different Phases of Cell Adhesion

First of all, a very important point in the interactions of cells with a material surface is that cells never see a bare surface but a surface previously covered with water and proteins adsorbed from biological fluids (Fig. 2). It is effectively well described that implanted materials are immediately coated with proteins from blood and interstitial fluids, and that cells sense foreign surfaces through this adsorbed layer. Initially cells respond to the adsorbed proteins, rather to the surface itself [19]. It is generally admitted that cell adhesion is divided into several phases: an early phase with short-term events like physico-chemical linkage between cells and materials, involving ionic forces, van der Waals forces, etc., and a later phase involving various biomolecules, like ECM proteins, cell membrane proteins and cytoskeleton proteins, which interact together to induce signal transduction promoting the action of transcription factors and consequently regulating gene expression [2]. The different phases generally recognized by most of the authors are illustrated in Fig. 3. Firstly, when cells are inoculated on a biomaterial substrate *in vitro* or when they come in contact with an implant surface *in vivo*, the proteins respectively from the culture medium or the biological fluids adsorb on the material surface. Secondly, the cells attach on the surface, spread and, at the same time, express cytoskeleton proteins and integrins. Thirdly, there is a clustering of integrin receptors, a reorganization of cytoskeleton and an active spreading onto the substrate. Then, the cells contract their actomyosin cytoskeleton generating mechanical forces at the sites of adhesion. Indeed, the cytoskeleton of all adherent cells is always maintained in a state of mechanical tension or prestress. The cells secure their shape stability mainly through the agency of that prestress. This prestress is mainly generated by myosin motors and transmitted by the actin fibers throughout the cell body. It is balanced

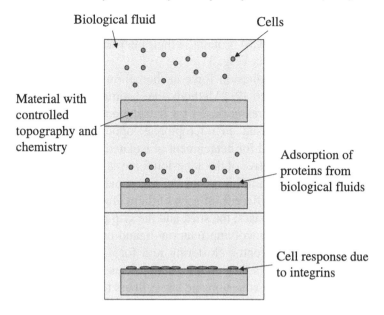

Figure 2. Schematic illustration of the different phases of interaction of biological elements with material's surface.

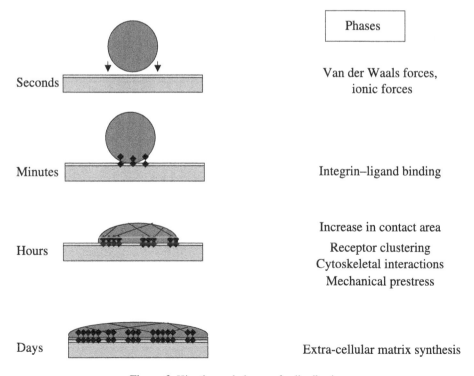

Figure 3. Kinetics and phases of cell adhesion.

by microtubules but mostly by cell adhesion to ECM or to adjacent cells [20, 21]. Finally, the cells synthesize ECM proteins at their interface with the material giving a cell/matrix/substrate interface (Fig. 3).

In order to quantify the adhesion force at the cell/material interface, different methods have been developed [2]. Methods like aspiration, centrifugation, fluid flow, and enzymatic procedures are used for detachment of cell populations. Micromanipulation methods based on micropipette, optical or magnetic tweezers, microcantilevers are also used for detachment of isolated cells [2]. The forces used for detachment are representative of the adhesion force at the cell/material interface.

Authors interested in cell adhesion consider that it can be divided into different phases but these phases are different for each of them (Table 1). Garcia and Gallant separate adhesion into two phases: the stick and the grip phases; the stick phase being the initial binding event involving integrin–ligand binding and the grip phase involving cell spreading, receptor clustering and focal adhesion assembly [22]. Murphy-Ullrich separates the adhesion into three phases: the attachment phase, the spreading phase and the focal adhesion and stress fibers formation phase [23]. Pierres et al. [24] describe a first flattening phase involving the first interaction between surface and a cell extension's tip as small as 0.01 μm^2. This flattening phase is followed by an alignment phase since the contact may, thus, extend through alignment of the cell membrane with the surface. Both flattening and alignment are necessary for membrane fitting to the surface. Finally, cell may spread with polarization

Table 1.
Definitions of adhesion phases from 4 different authors

Authors		A. J. Garcia [6, 22]	J. E. Murphy-Ullrich [23]	A. Pierres [24]	K. Anselme [25, 26]
Cell lines		Murine osteoblasts (MC3T3-E1)	Fibroblasts	Monocytes	Primary human bone-derived cells
Phases	Van der Waals forces, ionic forces	Stick	Attachment	Flattening	Short-term adhesion
	Integrin–ligand binding			Alignment	
	Increase in contact area	Grip	Spreading		
	Receptor clustering			Spreading	
	Cytoskeletal interactions, mechanical prestress		Focal adhesions + stress fibres formation		
	Extra-cellular matrix synthesis				Long-term adhesion

and increase of the contact area. These flattening, alignment and spreading phases form the fitting phase which precedes the adhesion phase. The non-covalent bonds formed during this fitting phase can be disrupted by forces as low as a few tens of piconewtons. On the contrary, the adhesion phase which last hours or days needs forces within the nanonewton range to separate bound cells from surfaces [24]. It is important to specify here that Pierres *et al.* [24] have used monocytes from blood and that these cells adhere on different and shorter timescales as compared to cells derived from other tissues like connective tissues (see later).

In our experience, we have quantified the adhesion before 24 h that we have called short-term adhesion. At the end of this first phase, all the cells were spread on the surface and their cytoskeleton was well organized but they did not already express $\beta 1$ and $\alpha 3$ integrin molecules nor synthesized matrix proteins [25]. From 24 h to 21 days we quantified what we have called long-term adhesion which represents the strength of the cell–material interface formed during 3 weeks of culture, involving at once the extra-cellular matrix proteins synthesized by the cells themselves and the cell–cell contacts [26] (Table 1).

2.2.1. Role of Cell Phenotype in Adhesion

It should be noted that all these research teams have worked with different cell types (Table 1). In our investigations, we studied the adhesion of primary human bone cells [25, 27]. These cells, as they are prepared directly from bone explants, are known to have a more differentiated phenotype than osteoblasts of cell lines obtained from human tumors. They are highly adherent, synthesize a considerable amount of ECM but their proliferation rate is low. It is evident that the mechanisms of adhesion of blood cells will be different than the adhesion of cells from connective tissues like fibroblasts or osteoblasts, or cells originated from endothelia or epithelia like endothelial vascular cells or keratinocytes. Naturally, the blood cells interact and adhere in a transitory manner on endothelial cell membranes. So, their adhesion will involve in a first step membrane adhesion proteins involved in cell/cell interactions like cadherins and selectins and then proteins involved in cell/material interactions like integrins. On the other hand, cells from connective tissues favor adhesion molecules involved in cell/ECM interactions like integrins. Epi- and endothelial cells can adhere using both adhesion molecules.

Moreover, another parameter has to be considered: the origin of cells. Because of their ease of use, many cells lines used in cell biology research are derived from tumors. The problem is that they have sometimes lost some of their phenotypic character. In particular, they are generally highly proliferative because of their tumoral character but have lost their capacity to adhere and to synthesize ECM. Moreover, cancerous cells or transformed cells are known to have a more diffuse actin cytoskeleton and, thus, to be more deformable than normal cells [28]: metastatic cancer cells are much more flexible which is in accordance with their need to move through tissues to invade other organs and tissues. We recently published an important work where osteosarcoma-derived cell lines (SaOs-2, MG-63) were cultured on poly-L-lactic acid films made by hot embossing with micropillars with a square morphology and compared to non-cancerous osteoblastic cells derived from human bone marrow [29]. The adhesion of cells was monitored using labeling of the nuclear chromatin, the nuclear membrane and the actin cytoskeleton. Labeling revealed deformation of the nuclei to an extent which had not been seen before. Very importantly, this deformation was observed using only cancerous cell lines. The shape the nuclei adopted reflected the shape of the microstructures (Fig. 4). 3-D-observation confirmed that the deformation involved the whole nucleus (chromatin + nuclear membrane) and that the nucleus was deformed and stayed within the interspaces between the pillars. Moreover, this deformation did

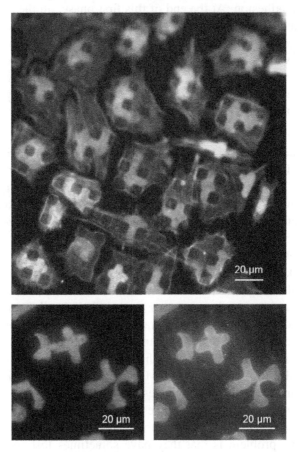

Figure 4. Top image: labeling of nuclear chromatin (light grey) and cytoskeleton (dark grey) of human osteosarcoma-derived SaOs-2 cells grown on PLLA micropillars for 7 days. Bottom images: labeling of nuclear chromatin (left) and nuclear membrane (right). Labeling revealed deformation of the nuclei to an extent which has not been seen before. The shape the nuclei (chromatin + membrane) adopted reflected the shape of the microstructures The nucleus was deformed and stayed inside the interspaces between the pillars but this deformation did not impair the future proliferation and differentiation of cells. Since this deformation is observed only using cancerous cell lines, it demonstrates that the behavior of cancerous cells *versus* topography can be abnormal and that non-cancerous cellular models should be preferred for studying the influence of surface topography.

not impair the future proliferation and differentiation of cells. This work is of pivotal importance in the understanding of cell–material interactions and illustrates clearly the importance of the cellular model used for these studies. Also, it demonstrates the need to prefer non-cancerous cellular models for studying the influence of surface topography.

3. Influence of Surface Topography on Cell Adhesion

3.1. Scale Effect

Many experiments have demonstrated that bone cells react differently with surfaces with different topographies [30–36]. However, the current difficulty in the field is the lack of comparable studies. One of the major problems is that there is no consensus for topography characterization [37–40]. In most studies, authors measure only some roughness parameters describing amplitude of topography such as R_a (arithmetic average height) or R_t (maximum height of the profile) [33, 41] but more rarely frequency parameters [40, 42]. In our investigations, we have computed more than 100 parameters describing amplitude, frequency but also fractal parameters representing the organization of surface morphology in order to find the more pertinent ones, i.e., the parameter that correlates best with cell adhesion [43]. As Meyer *et al.* [4] noted the main misunderstanding is the practice of defining a surface by its manufacturing process instead of concisely defining the topographic measurements. This practice leads to erroneous comparisons of surfaces since it is known that the same process applied in different manners or on different materials will induce different topographies and sometimes different surface chemistries. For example, use of particles of different sizes for sandblasting can induce not only different topographies but also different surface chemical modifications [27]. Moreover, it was demonstrated that the topography must be considered at the cell size scale in order to better interpret biological results [26, 44]. Indeed, it was shown that human osteoblasts adhered better on rough electro-eroded metallic surfaces than on smoother ones. However, at the scale higher than the cell size (>50 μm) these surfaces could be considered as rough whereas at the cell scale (<50 μm) they present a large number of smooth and flat areas which appear to favour cell adhesion and spreading [44]. To go deeper in the influence of cell size, the adhesion of human osteosarcoma cells (MG63) was studied on titanium surfaces displaying hemispherical cavities measuring 10, 30 and 100 μm in diameter. As the cell size was around 30 μm, a different behaviour was expected on 30 μm cavities. Effectively, cells went preferentially inside the 30 μm cavities where they divided and displayed a round morphology although their behaviour was totally different on smaller and larger cavities where they displayed rather a flat and spread morphology [45].

More recently, the adhesion of human osteoprogenitor cells derived from bone marrow was compared after 2 days of culture on a wide range of roughnesses obtained by electro-erosion processing of titanium substrates. The lower roughness amplitude (R_a) was about 1.2 μm, whereas the coarser roughness amplitude was

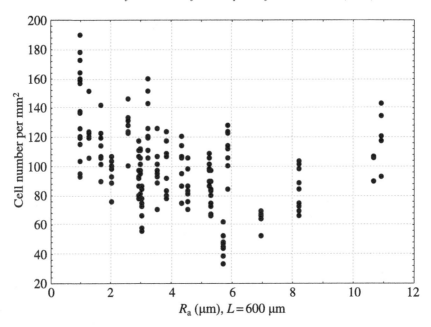

Figure 5. Variation of the number of attached human osteoprogenitor cells after 2 days on a range of roughness (R_a) measured on a length $L = 600$ µm prepared by electro-erosion on titanium substrates.

about 20 µm. The electro-erosion process produces isotropic and fractal surfaces. Thus, the increase in roughness did not modify the surface morphology. The surface with the higher R_a appeared only like a magnified picture of the surface with the lower R_a. Moreover, since these surfaces were fractal, the distance between peaks (S_m) increased also linearly with roughness amplitude in all directions. The cell number decreased when roughness amplitude increased until reaching a minimum value for $R_a \cong 6$ µm (Fig. 5). Further, the number of adhered cells increased again until reaching for the higher roughness about the same value as on the lower roughness. The conclusion of these results was that the cells rather adhere over the peaks on less rough surfaces. When the height and spacing between peaks increased, the cells were less and less able to adhere on top of peaks and the number of adhered cells decreased. For a $R_a \cong 6$ µm and an $S_m \cong 130$ µm that corresponds to the cell size, the cell number was minimum. For this roughness amplitude, the surface presented a maximum of peaks and cavities exactly at the cell size scale. Thus, it appears that when the distance between peaks or cavity size was about their own size, the cells adhered and spread with difficulties on surfaces. When the distance between peaks or cavity size increased again, the number of cells increased again (unpublished data). Kunzler *et al.* [46], with the same objectives, developed surfaces with a roughness gradient with a range of R_a from 1 to 6 µm prepared by sandblasting aluminium sheets and further smoothing using a chemical polishing process. They observed, in this range of roughness which was shorter than in the previously described study, that rat calvaria osteoblast showed an increased proliferation rate with increasing roughness, whereas the human gingival fibroblasts showed the opposite behaviour. This illustrates again the influence of the cell phenotype on the response to surfaces.

Another important parameter is the size of biological elements interacting with surfaces. Large cells like human primary bone cells (~50 μm) will react differently to the same surface topography than smaller cells like monocytes (~10 μm) or platelets (~2 μm). In the same way, we can expect a different behaviour of microorganisms like bacteria (~1 μm). Additionally, as previously described in Section 2 of this review, cell adhesion is related to adsorbed proteins. Considering the size scale of proteins, the protein adsorption will surely be influenced by the surface morphology considered at the nanometer scale.

This is likely at the origin of the cellular orientation observed on nanostructured surfaces [47–49]. Notably, the ability of cells to interact with nanofeatures with height smaller than 10 nm has been demonstrated [50]. Again in this case, not only the size and distance between nanofeatures but also the shape or organization of nanofeatures have been shown to influence the response of cells. The role of adsorbed proteins was demonstrated in the cell interactions with nanofeatures. The readers are invited for more information on nanotopography influence to read recent reviews on these aspects [51–53].

3.2. Organization of Topography

Many studies have shown that cells are able to align themselves along defined substrate surface morphologies [54]. This is known as the contact guidance phenomenon [31]. Many studies have been concerned with the effects of depth and width of grooves on cell orientation [55–58] but it appears that no consensus was found on the minimal depth and width needed for inducing cell orientation. Again this is certainly related to the cell types used by authors which were all different. Moreover, the influence of the chemistry of the substrate is certainly of high importance: the influence of chemistry on cell adhesion will be discussed in the next part of this section. It has been recently shown that cells are also able to orient themselves on surfaces presenting nano-grooves [59, 60]. However, the relation between ordered nano-topography and cell behaviour is, to a large extent, unknown so far. The possibility now to develop nano-patterned surfaces in a controlled manner will certainly boost research in this field in the next years and will allow to answer these questions. Twelve years ago, in their review on topographical control of cells, Curtis and Wilkinson [54] limited their discussion on cell reaction to random roughness as defining and quantifying random roughness is a difficult task. Since that time, we have proposed more than 100 parameters to characterize and quantify as well organized as random topographies [43]. These topographical parameters have been correlated with biological parameters describing short-term adhesion, long-term adhesion and proliferation. In several studies [43, 44], we have demonstrated that the long-term adhesion of primary human bone cells was statistically better correlated with parameters describing organization of topography than with other parameters. The best adhesion was obtained on the less organized rough surfaces. More recently, Dalby *et al.* [61] have developed surfaces presenting 120-nm-diameter nanopits with hexagonal, square, displaced square or random placements. They demonstrated in a very elegant study that the differentiation of human mesenchymal stem cells was stimulated by nano-scale surface disorder. The differentiation observed on displaced square 50 (±50 nm from true centre) was comparable to the response obtained with a stimulating medium.

All these studies have been performed *in vitro* but with the aim to better understand the future *in vivo* integration of implants. However, osseointegration is not only attained by cellular processes but also depends on continuous adaptation to functional loading. In this case, the surface roughness plays also a role in mechanical interlocking of implants in bone tissue. In a very interesting and extensive review on the role of surface topography in creating and maintaining bone at titanium endosseous implants, Cooper [62] has concluded that enhanced surface topography is associated with increased bone-to-implant contact and increased biomechanical interlocking with bone.

3.3. Mechanisms of Cell Response to Surface Topography

In 1997, Curtis and Wilkinson described that cells did react to the presence of discontinuities in the surface morphology [54], with a discontinuity having a radius of curvature less than the average length of a pseudopodium or of the distance between the sensing elements that controlled cell movements. This led them to define how cells sensed discontinuities. They considered that the reaction involved focal adhesions and cytoskeleton. The cells would be able to stretch themselves on the substratum, and this would activate stretch receptors. This would require firm adhesion at two or more points and cytoskeletal activity which would stretch the cells. This is related to the mechanism of mechano-sensitivity mediated by integrin-mediated cell–matrix adhesion [10] and the cell's contractile machinery involving actin and myosin motors [12, 63, 64]. Indeed, as described previously, after clustering of integrins with ligands and spreading across the substrate, the cells contract their actomyosin cytoskeleton generating mechanical forces at the sites of adhesion in order to maintain their cytoskeleton in a state of mechanical tension or prestress. *In vivo*, the cells use these processes as an integrated sensory system to probe the chemical and mechanical cues within the ECM. For this, they probe the mechanical compliance of ECM by locally deforming it with nanonewton-scale traction forces. This has been demonstrated using appropriately compliant substrates, containing beads or markers embedded in the matrix. The deformation applied by cells was tracked by the displacement of beads. Other approaches used poly(dimethylsiloxane) (PDMS) microstructured surfaces with microneedles-like posts. The deflection of microposts directly reported the subcellular distribution of traction forces [65].

Different hypotheses have been proposed for mechanisms underlying the identification of discontinuities by cells. One is based on thermodynamics and extracellular matrix protein adsorption. The discontinuities may differ in reactivity from nearby planar surfaces due to the existence of more unsaturated bonds. This alters the adsorption of extra-cellular matrix proteins involved in adhesion and, thus, the cell response to the surface reflects both the distribution of adhesive differences as well as the topography [66]. Another hypothesis has proposed that the discontinuities act as an energy barrier, the size of the energy barrier being dependent on both the geometry and surface chemistry [1]. Thus, the cells would try to minimize contact with the high energy discontinuities and hence modify their orientation, adhesion or spreading.

Stevenson and Donald [67] have recently published an interesting paper where they compared the attachment of cells to micro-grooved substrates with different ridge spacings, widths and heights. They observed that three distinct regimes of attachment occurred which were dependent on the ridge spacing used. At the smallest ridge spacing (~10–20 μm), the cells were able to bridge between neighbouring ridges without touching the groove floor. At moderate spacing (~30–50 μm), cells were confined to a single ridge or groove. The largest spacings ($\gtrsim$50 μm) allowed cells to connect between a ridge and a groove. Reducing the ridge height from 21 μm to 15 μm allowed ridge–groove connection at 40 μm spacing. Hence, they proposed that both a critical length and a critical angle (slope) existed for any cell protuberance connecting a ridge to a ridge or a ridge to a groove. From these results we can easily understand that depending on the morphology of cells on grooves, the cell receptors would be subjected to variable degrees of deformation or compression. Indeed, the presence under the cells of concave or convex structures would induce reorganization of attachment and cytoskeleton structures (Fig. 6). Already more than 30 years ago, Dunn and Heath [68] had shown that fibroblasts were affected by a convex ridge on a prism if the ridge angle was greater than 4°. The cells displayed discontinuities in their cytoskeletal microfilament bundle system which coincided with the discontinuity of the substratum shape. Further, the cell reaction to concave surfaces, like grooved surfaces, confirmed that cytoskeleton was also involved [69–73] but the underlying mechanisms were expected to be different since any mechanical restriction imposed onto cytoskeleton on concave or

Figure 6. Illustration of the deformation of the actin skeleton of cells adhering to concave and convex surfaces.

convex structures would be different. Similarly, Berry *et al.* [74] have noted that cells were sensitive to changes in the radius of curvature since they followed the discontinuity presented by the circumference edge of pits presenting different diameters and showed a preference for entering the larger diameter pits. Thery *et al.* [64] have observed that cells adhering on adhesive substrates presenting convex or concave micro-patterns (forming hollow triangle or letters like V, T, U or Y) always displayed a similar convex contour. This was the result of the cell contraction occurring when reaching the steady-shape state. Cells did reorganise attachment and cytoskeleton structures such that they could recover a basal stretching level. However, it should be noted that in this study the micro-patterns did not show any height.

It is widely accepted, as we have shown before, that cells react to surface topography and we will see later they are also influenced by surface chemistry. However, there is until now little information on how these factors act together in the production of biological responses. The lack of knowledge in this field is largely related to the difficulty to vary independently surface roughness and surface chemistry. Some studies have been published on the relative influence of surface chemistry and surface topography on cell behavior. Hallab *et al.* [75] demonstrated on polymeric substrates that the surface free energy was a more important surface characteristic than surface roughness for cellular adhesion strength and proliferation. We demonstrated after coating with gold–palladium of metallic substrates with different roughnesses that short-term adhesion and proliferation were more dependent on surface chemistry, whereas long-term adhesion was more dependent on surface roughness [27, 44, 76–78]. Ponsonnet *et al.* [79, 80] observed on titanium and titanium alloys that surface energy was a dominant factor for cell proliferation but that roughness could strongly disturb the relationship between surface energy and cell proliferation. In a recent review, Vitte *et al.* [81] attempted to answer the following question: Is there a predictable relationship between surface physico-chemical properties and cell behavior at the interface? There appears to be no answer to this question at the moment, considering the complexity of the processes involved. In the next part of this review, the cell response to modification of surface chemistry will be described but the reader has to keep in mind that in most cases (except self-assembled monolayers, SAMs) modifications of surface chemistry are also generally related to surface topography, at least at the nano-scale.

4. Influence of Surface Chemistry on Cell Response

Pioneer works on the influence of surface chemistry were done in 1980's. Baier *et al.* in 1984 [82], placed in the backs of New Zealand rabbits CoCrMo implants with three different surface energy states. They demonstrated an increase of cell reactivity and a stronger tissue adhesion on high-surface energy implants, although on low surface energy implants a true interfacial separation between metal and tissue did exist. In 1986, Schakenraad *et al.* [83] demonstrated in an *in vitro* study on 13 different polymers and glass substrates that the human skin fibroblasts spreading was the highest for a substrate surface energy, γ^s, of approximately 57 mJ/m^2. Moreover, in presence of serum proteins, cell spreading was similar on most materials. Yanagisawa and Wakamatsu [84] observed in 1989 that cell attachment rate and cell spreading were higher on substrates with a water contact angle below 60° and that they decreased dramatically for higher angles, whatever the time after inoculation. The zeta potential was found to influence also osteoblastic cell adhesion on a polymer/calcium phosphate composite [85]. In the same paper, the relation between attachment and proliferation of osteoblasts on biomaterials with different wettabilities was investigated. No direct relation was found between the wettability of the materials and the osteoblast reaction towards them [85].

All investigations revealed that a change in surface chemistry has significant effect with respect to cell function. This was more recently studied on the molecular scale by Zreiqat *et al.* [86]. They demonstrated using western blot analysis that the intra-cellular signaling was different in human primary bone cells cultured on titanium alloy modified by ion beam implantation of Mg or Zn ions.

Due to new techniques, it is now possible to tailor surface chemistry by synthetic tools and nano-fabrication techniques [87, 88]. In particular, self-assembled monolayers (SAMs) and polymer brushes have recently attracted considerable attention for the creation of surfaces with biospecific binding properties and minimized background interference. The SAMs have been used largely in the recent years in order to elucidate the influence of surface chemistry on cell adhesion, mainly because of their ease of processing [88]. Keselowsky *et al.* [89] demonstrated in 2005, owing to SAMs, that OH- and NH$_2$-terminated surfaces (hydrophilic surfaces) increased more osteoblasts specific gene expression, alkaline phosphatase enzymatic activity and matrix mineralization (all markers of osteoblastic differentiation) than surfaces presenting CH$_3$ groups (hydrophobic surfaces). Barbosa *et al.* [90] demonstrated *in vivo* that a significant increase in the thickness of fibrous capsules was seen around implant covered by CH$_3$-terminated SAMs in comparison with COOH- and OH-terminated SAMs.

In order to obtain more information on a complete range of wettability, techniques were developed for preparing linear gradients of surface energy. Pioneer works were done by Lee *et al.* [91] using polyethylene substrates treated by corona discharge treatment. They demonstrated that platelet adhesion was improved by a positive charge character, and that endothelial cells [92] or neural cells [93] adhesion was increased on the positions with moderate hydrophilicity on the wettability gradient surface than onto the more hydrophobic or hydrophilic positions. In 2004, Lim *et al.* [94] demonstrated that human fetal osteoblasts adhered and proliferated better on hydrophilic substrates than on hydrophobic ones. On the contrary, no effect of surface energy was demonstrated on cell differentiation. On a linear gradient in surface energy prepared by treating SAMs with UV-light, Kennedy *et al.* [95] demonstrated that surface energy did not affect murine osteoblastic cells initial adhesion nor cell spreading area at 8 h. However, the rate of proliferation was linearly dependent on surface energy and increased with increasing hydrophobicity. In this case, the preparation of surface energy gradients by UV treatment of CH_3 SAMs would have also considerably modified surface charges. This could explain these relatively contradictory observations and highlights again the need to perfectly characterize and control at each step the surface chemistry. More recently chemical gradients were produced using plasma polymers [96]. The authors developed a gradient from hydrophobic plasma polymerised hexane (ppHex) to a more hydrophilic plasma polymerised allylamine (ppAAm). They confirmed that the adhesion and proliferation of fibroblasts were higher on the hydrophilic part of the gradient. Fibroblasts adhered and proliferated preferentially on ppAAm (water contact angle: 60°) showing a gradual decreasing cell density towards the hydrophobic ppHex (water contact angle: 93°) [96]. Other gradients were made by varying the graft density of a poly(2-hydroxyethyl methacrylate) (PHEMA). On these gradients, the adsorption of bovine fibronectin and fibroblast adhesion and spreading were quantified. Both cellular response and fibronectin adsorption were found to vary sigmoidally with graft density of PHEMA, demonstrating the high degree of correlation between the two phenomena [97]. This leads us to discuss now the influence of adsorbed proteins on cell adhesion.

5. Influence of Adsorbed Proteins on Cell Adhesion

As described before, cells never see a bare surface but a surface previously covered with proteins adsorbed from biological fluids. This protein layer surely has a large influence on cell adhesion (Fig. 3). These aspects will not be developed in detail in this review since this has been well done in a very interesting paper [19]. As previously cited, Schakenraad et al. [83] have shown that the influence of substratum surface free energy on human fibroblasts adhesion was annihilated by a previous adsorption of serum proteins. Lee et al. [98] have demonstrated using SAMs previously coated with ^{125}I-labeled fibronectin that differences in adhesion strength and integrin binding of an erythroleukemia cell line on SAMs with different functional groups (OH, COOH, NH_2 and CH_3) were explained in terms of changes in the adhesion constant and binding efficiency of the adsorbed fibronectin for the integrin ($CH_3 \sim NH_2 <$ COOH $\sim$ OH). Moreover, fibronectin interacts more strongly with $\alpha 5\beta 1$ integrins when adsorbed on COOH than on OH surface suggesting that negative charges may also be a critical component of inducing efficient cellular adhesion. On ceramic substrates, it has been shown that the bioactivity of some ceramics was linked to their capacity to adsorb proteins from biological fluids, in particular fibronectin [99].

Hydrophilic substrates generally adsorb a lower quantity of proteins [100] which undergo less conformational changes than when adsorbed on hydrophobic surfaces [101, 102]. This means that adhesive proteins will rather maintain their active conformation on hydrophilic surfaces. It is generally admitted that the proteins, and particularly albumin, adsorb more on hydrophobic surfaces [103–106] and more rapidly [107]. But when albumin and fibronectin are mixed, albumin adsorbs rather on hydrophobic surfaces and fibronectin on hydrophilic surfaces [108]. This can be related to the Vroman effect, i.e., the replacement with time of initially adsorbed proteins like albumin, which presents a relatively low affinity for hydrophilic surfaces, by proteins with a higher affinity like fibronectin. Finally, a cooperation can also exist between proteins. For instance, the presence of albumin on the surface can prevent drastic conformational changes of adhesive proteins like fibronectin and render their active RGD sites more accessible to cells. A synergistic effect has also been described between fibronectin and albumin, termed albumin 'rescuing', to enhance cell adhesion and spreading on hydrophobic surfaces [105].

6. Influence of Mechanical Cues on Cell Behavior

As described in this review, cells are able to respond to different physico-chemical cues like topography and chemistry. However, another aspect will be considered more and more in the future: the mechanical cue. While surface chemistry and topography have been extensively studied, substrate mechanics has only recently been appreciated. Some years ago, it was demonstrated that cell movement was guided by the rigidity of the substrate [63, 109]. More recently, Discher *et al.* [110, 111] demonstrated in a very elegant study that the stiffness of the substrate directed stem cell differentiation in different lineages. Soft matrices that mimic brain are neurogenic, stiffer matrices that mimic muscle are myogenic, and rigid matrices that mimic collagenous bone are osteogenic. The mechanisms how cells feel their microenvironment, and particularly its elasticity, and how this affects cell structure and function is still under investigation. However, it appears that these phenomena are regulated by phosphotyrosine signaling and myosin II [109]. Focal adhesion proteins are found in small, diffuse and dynamic adhesion complexes on soft gels, although the cells display stable focal adhesion on stiff gels comparable to those seen in cells attached to glass [109]. Cytoskeletal assembly and cell tension follow the same trends as adhesion. Thus, molecular mechanisms of elasticity sensing by cells seem likely to be collective and dependent on many interacting components of the cyto-adhesion apparatus [112] but the list of molecular players will undoubtedly grow in the future.

7. Conclusions and Perspectives

Much has been learned about the response of cells to cues given by surfaces (topography, chemistry, mechanics) but several tracks must be more explored. Clearly, cells are able to recognize topography with height starting from a few nanometers to several hundred micrometers. It seems that a threshold exists in their response to topography that is close to their own dimension. This point needs further studies for elucidating the underlying mechanisms. Thus, the response of cells to topography should always be considered as functions of the cell size and of the size of features present on the surface. Cells are also able to differentiate isotropic from anisotropic surfaces with a clear preference for disordered surfaces, at least for bone-derived cells. Strong differences have been observed as a function of the cell type. However, it is still necessary to elucidate the mechanisms underlying this difference.

Cells have also shown their capacity to identify and respond selectively to surface chemical modifications. However, problems remain in the separation of the relative influences of surface topography and surface chemistry since modifications in surface topography are generally correlated with modifications in surface chemistry at least at the nano-scale. In general, the cells prefer moderate to strong hydrophilic surfaces and this has often been correlated with a higher adhesive protein adsorption like fibronectin from biological medium.

Finally, all the works reviewed in this paper have been concerned with 2-D material surface effects on cell adhesion. However, as the cells are influenced *in vivo* by interactions with 3-D microenvironment, it appears more and more important to develop *in vitro* models dealing with the 3-D aspect of surface effects, since it has been shown that the cell adhesion was different in three-dimensional systems compared with two-dimensional systems [113–117]. Moreover, one major application of the research on cell–material interactions is the tissue-engineering field: in the most cases, the scaffolds used for these applications will display a 3-D architecture for allowing growth factor and/or cell entrapping. Future studies in this field will need to consider the third dimension.

Acknowledgements

The authors acknowledge the financial support given by CNRS and ANR (ANR2007 EUKA 006 01, ANR06-NANO-022) and Ms P. Davidson for contribution to some figures.

References

1. A. W. Feinberg, W. R. Wilkerson, C. A. Seegert, A. L. Gibson, L. Hoipkemeier-Wilson and A. B. Brennan, *J. Biomed. Mater. Res.* **86A**, 522 (2008).
2. K. Anselme, *Biomaterials* **21**, 667 (2000).
3. B. D. Boyan, T. W. Hummert, D. D. Dean and Z. Schwartz, *Biomaterials* **17**, 137 (1996).
4. U. Meyer, A. Büchter, H. P. Wiesmann, U. Joos and D. B. Jones, *Eur. Cell. Mater.* **9**, 39 (2005).
5. R. O. Hynes, *Cell* **110**, 673 (2002).
6. A. J. Garcia, *Biomaterials* **26**, 7525 (2005).
7. M. C. Siebers, P. J. Ter Brugge, X. F. Walboomers and J. A. Jansen, *Biomaterials* **26**, 137 (2005).
8. M. G. Coppolino and S. Dedhar, *Int. J. Biochem. Cell Biol.* **32**, 171 (2000).
9. M. J. Biggs, R. G. Richards, S. McFarlane, C. D. Wilkinson, R. O. Oreffo and M. J. Dalby, *J. Roy. Soc. Interface* **5**, 1231 (2008).
10. A. Bershadsky, M. Kozlov and B. Geiger, *Curr. Opin. Cell Biol.* **18**, 472 (2006).
11. M. A. Wozniak, K. Modzelewska, L. Kwong and P. J. Keely, *Biochim. Biophys. Acta* **1692**, 103 (2004).
12. J. P. Baker and M. A. Titus, *Curr. Opin. Cell Biol.* **10**, 80 (1998).

13. M. A. Schwartz and D. W. DeSimone, *Curr. Opin. Cell Biol.* **20**, 551 (2008).

14. R. G. Lebaron and K. A. Athanasiou, *Tissue Eng.* **6**, 85 (2000).

15. E. A. Cowles, L. L. Brailey and G. A. Gronowicz, *J. Biomed. Mater. Res.* **52**, 725 (2000).

16. A. Krause, E. A. Cowles and G. Gronowicz, *J. Biomed. Mater. Res.* **52**, 738 (2000).

17. M. Rouahi, E. Champion, P. Hardouin and K. Anselme, *Biomaterials* **27**, 2829 (2006).

18. R. Zaidel-bar, S. Itzkovitz, A. Ma'ayan, R. Iyengar and B. Geiger, *Nat. Cell Biol.* **9**, 858 (2007).

19. C. J. Wilson, R. E. Clegg, D. I. Leavesley and M. J. Pearcy, *Tissue Eng.* **11**, 1 (2005).

20. D. E. Ingber, *FASEB J.* **20**, 811 (2006).

21. X. Trepat, G. Lenormand and J. J. Fredberg, *Soft Matter* **4**, 1750 (2008).

22. A. J. Garcia and N. D. Gallant, *Cell Biochem. Biophys.* **39**, 61 (2003).

23. J. E. Murphy-Ullrich, *J. Clin. Invest.* **107**, 785 (2001).

24. A. Pierres, A. M. Benoliel and P. Bongrand, *Eur. Cell. Mater.* **3**, 31 (2002).

25. K. Anselme, M. Bigerelle, B. Noel, E. Dufresne, D. Judas, A. Iost and P. Hardouin, *J. Biomed. Mater. Res.* **49**, 155 (2000).

26. K. Anselme and M. Bigerelle, *Biomaterials* **27**, 1187 (2006).

27. K. Anselme, P. Linez, M. Bigerelle, D. Le Maguer, A. Le Maguer, P. Hardouin, H. F. Hildebrand, A. Iost and J.-M. Leroy, *Biomaterials* **21**, 1567 (2000).

28. P. E. Fisher and C. Tickle, *Exp. Cell Res.* **131**, 407 (1981).

29. P. Davidson, H. Özçelik, V. Hasirci, G. Reiter and K. Anselme, *Adv. Mater.* **21**, 3586 (2009).

30. B. D. Boyan, R. Batzer, K. Kieswetter, Y. Liu, D. L. Cochran, S. Szmuckler-Moncler, D. D. Dean and Z. Schwartz, *J. Biomed. Mater. Res.* **39**, 77 (1998).

31. D. M. Brunette and B. Chehroudi, *J. Biomech. Eng.* **121**, 49 (1999).

32. M. G. Diniz, G. A. Soares, M. J. Coelho and M. H. Fernandes, *J. Mater. Sci. Mater. Med.* **13**, 421 (2002).

33. J. Links, B. D. Boyan, C. R. Blanchard, C. H. Lohmann, Y. Liu, D. L. Cochran, D. D. Dean and Z. Schwartz, *Biomaterials* **19**, 2219 (1998).

34. C. H. Lohmann, R. Sagun, V. L. Sylvia, D. L. Cochran, D. D. Dean, B. D. Boyan and Z. Schwartz, *J. Biomed. Mater. Res.* **47**, 139 (1999).

35. B. Nebe, F. Lüthen, A. Baumann, U. Beck, A. Diener, H.-G. Neumann and J. Rychly, *Mater. Sci. Forum* **426–432**, 3023 (2003).

36. X. Zhu, J. Chen, L. Scheideler, R. Reichl and J. Geis-Gerstorfer, *Biomaterials* **25**, 4087 (2004).

37. M. Bigerelle and A. Iost, *C. R. Acad. Sci. Paris* **323**, 669 (1996).

38. M. Bigerelle, D. Najjar and A. Iost, *J. Mater. Sci.* **38**, 2525 (2003).

39. P. F. Chauvy, C. Madore and D. Landolt, *Surf. Coat. Technol.* **110**, 48 (1998).

40. W. MacDonald, P. Campbell, J. Fisher and A. Wennerberg, *J. Biomed. Mater. Res.* **70B**, 262 (2004).

41. J. Y. Martin, Z. Schwartz, T. W. Hummert, D. M. Schraub, J. Simpson, J. Lankford, D. D. Dean, D. L. Cochran and B. D. Boyan, *J. Biomed. Mater. Res.* **29**, 389 (1995).

42. A. Wennerberg, C. Hallgren, C. Johansson, T. Sawase and J. Lausmaa, *J. Mater. Sci. Mater. Med.* **8**, 757 (1997).

43. K. Anselme and M. Bigerelle, *Acta Biomater.* **1**, 211 (2005).

44. M. Bigerelle, K. Anselme, B. Noël, I. Ruderman, P. Hardouin and A. Iost, *Biomaterials* **23**, 1563 (2002).

45. O. Zinger, K. Anselme, A. Denzer, P. Habersetzer, M. Wieland, J. Jeanfils, P. Hardouin and D. Landolt, *Biomaterials* **25**, 2695 (2004).

46. T. P. Kunzler, T. Drobek, M. Schuler and N. D. Spencer, *Biomaterials* **28**, 2175 (2007).

47. M. J. Dalby, S. J. Yarwood, M. O. Riehle, H. J. H. Johnstone, S. Affrossman and A. S. G. Curtis, *Exp. Cell Res.* **276**, 1 (2002).
48. M. J. Dalby, M. O. Riehle, D. S. Sutherland, H. Agheli and A. S. G. Curtis, *Biomaterials* **25**, 5415 (2004).
49. M. J. Dalby, M. O. Riehle, D. S. Sutherland, H. Agheli and A. S. G. Curtis, *Eur. Cell. Mater.* **9**, 1 (2005).
50. M. J. Dalby, M. O. Riehle, H. Johnstone, S. Affrossman and A. S. Curtis, *Cell Biol. Int.* **28**, 229 (2004).
51. R. Kriparamanan, P. Aswath, A. Zhou, L. Tang and K. T. Nguyen, *J. Nanosci. Nanotechnol.* **6**, 1905 (2006).
52. K. T. Nguyen, K. P. Shukla, M. Moctezuma and L. Tang, *J. Nanosci. Nanotechnol.* **7**, 2823 (2007).
53. E. K. F. Yim and K. W. Leong, *Nanomedicine* **1**, 10 (2005).
54. A. S. G. Curtis and C. Wilkinson, *Biomaterials* **18**, 1573 (1997).
55. W. A. Loesberg, X. F. Walboomers, J. J. W. A. Van Loon and J. A. Jansen, *J. Biomed. Mater. Res.* **75A**, 723 (2005).
56. X. F. Walboomers, H. J. E. Croes, L. A. Ginsel and J. A. Jansen, *Biomaterials* **19**, 1861 (1998).
57. X. F. Walboomers, L. A. Ginsel and J. A. Jansen, *J. Biomed. Mater. Res.* **51**, 529 (2000).
58. X. F. Walboomers, W. Monaghan, A. S. G. Curtis and J. A. Jansen, *J. Biomed. Mater. Res.* **46**, 220 (1999).
59. S. Lenhert, M. B. Meier, U. Meyer, L. Chi and H. P. Wiesmann, *Biomaterials* **26**, 563 (2004).
60. C. A. Mills, M. Navarro, E. Engel, E. Martinez, M. P. Ginebra, J. A. Planell, A. Errachid and J. Samitier, *J. Biomed. Mater. Res.* **76A**, 781 (2006).
61. M. J. Dalby, N. Gadegaard, R. S. Tare, A. Andar, M. O. Riehle, P. Herzyk, C. D. W. Wilkinson and R. O. C. Oreffo, *Nat. Mater.* **6**, 997 (2007).
62. L. F. Cooper, *J. Prosthet. Dent.* **84**, 522 (2000).
63. C. M. Lo, H. B. Wang, M. Dembo and Y. L. Wang, *Biophys. J.* **79**, 144 (2000).
64. M. Thery, A. Pepin, E. Dressaire, Y. Chen and M. Bornens, *Cell Motil. Cytoskeleton* **63**, 341 (2006).
65. J. L. Tan, J. Tien, D. M. Pirone, D. S. Gray, K. Bhadriraju and C. S. Chen, *Proc. Natl. Acad. Sci. USA* **100**, 1484 (2003).
66. J. Chen, S. Mwenifumbo, C. Langhammer, J. McGovern, M. Li, A. Beye and W. O. Soboyejo, *J. Biomed. Mater. Res.* **82B**, 360 (2007).
67. P. M. Stevenson and A. M. Donald, *Langmuir* **25**, 367 (2009).
68. G. A. Dunn and J. P. Heath, *Exp. Cell Res.* **101**, 1 (1976).
69. J. M. Rice, J. A. Hunt, J. A. Gallagher, P. Hanarp, D. S. Sutherland and J. Gold, *Biomaterials* **24**, 4799 (2003).
70. C. Oakley and D. M. Brunette, *J. Cell Sci.* **106**, 343 (1993).
71. J. C. Grew, J. L. Ricci and H. Alexander, *J. Biomed. Mater. Res.* **85A**, 326 (2008).
72. A. Sorensen, T. Alekseeva, K. Katechia, M. Robertson, M. O. Riehle and S. C. Barnett, *Biomaterials* **28**, 5498 (2007).
73. W. T. Su, I. M. Chu, J. Y. Yang and C. D. Lin, *Micron* **37**, 699 (2006).
74. C. C. Berry, G. Campbell, A. Spadiccino, M. Robertson and A. S. G. Curtis, *Biomaterials* **25**, 5781 (2004).
75. N. J. Hallab, K. J. Bundy, K. O'Connor, R. L. Moses and J. J. Jacobs, *Tissue Eng.* **7**, 55 (2001).
76. K. Anselme and M. Bigerelle, *J. Mater. Sci. Mater. Med.* **17**, 471 (2006).
77. M. Bigerelle and K. Anselme, *J. Biomed. Mater. Res.* **72A**, 36 (2005).

78. K. Anselme and M. Bigerelle, *Surf. Coat. Technol.* **200**, 6325 (2006).
79. L. Ponsonnet, V. Comte, A. Othmane, C. Lagneau, M. Charbonnier, M. Lissac and N. Jaffrezic, *Mater. Sci. Eng. C* **21**, 157 (2002).
80. L. Ponsonnet, K. Reybier, N. Jaffrezic, V. Comte, C. Lagneau, M. Lissac and C. Martelet, *Mater. Sci. Eng. C* **23**, 551 (2003).
81. J. Vitte, A. M. Benoliel, A. Pierres and P. Bongrand, *Eur. Cell. Mater.* **7**, 52 (2004).
82. R. E. Baier, A. E. Meyer, J. R. Natiella, R. R. Natiella and J. M. Carter, *J. Biomed. Mater. Res.* **18**, 337 (1984).
83. J. M. Schakenraad, H. J. Busscher, C. R. H. Wildevuur and J. Arends, *J. Biomed. Mater. Res.* **20**, 773 (1986).
84. I. Yanagisawa and Y. Wakamatsu, *J. Oral Implantol.* **15**, 168 (1989).
85. K. Möller, U. Meyer, D. H. Szulczewski, H. Heide, B. Priessnitz and D. B. Jones, *Cells Mater.* **4**, 263 (1994).
86. H. Zreiqat, S. M. Valenzuela, B. B. Nissan, R. Roest, C. Knabe, R. J. Radlanski, H. Renz and P. J. Evans, *Biomaterials* **26**, 7579 (2005).
87. D. Falconnet, G. Csucs, H. M. Grandin and M. Textor, *Biomaterials* **27**, 3044 (2006).
88. W. Senaratne, L. Andruzzi and C. K. Ober, *Biomacromolecules* **6**, 2427 (2005).
89. B. G. Keselowsky, D. M. Collard and A. J. Garcia, *Proc. Natl. Acad. Sci. USA* **102**, 5953 (2005).
90. J. N. Barbosa, P. Madureira, M. A. Barbosa and A. P. Aguas, *J. Biomed. Mater. Res.* **76A**, 737 (2006).
91. J. H. Lee, G. Khang, J. W. Lee and H. B. Lee, *J. Biomed. Mater. Res.* **40**, 180 (1998).
92. J. H. Lee, S. J. Lee, G. Khang and H. B. Lee, *J. Colloid Interface Sci.* **230**, 84 (2000).
93. S. J. Lee, G. Khang, Y. M. Lee and H. B. Lee, *J. Colloid Interface Sci.* **259**, 228 (2003).
94. J. Y. Lim, X. Liu, E. A. Vogler and H. J. Donahue, *J. Biomed. Mater. Res.* **68A**, 504 (2004).
95. S. B. Kennedy, N. R. Washburn, C. G. Simon and E. J. Amis, *Biomaterials* **27**, 3817 (2006).
96. M. Zelzer, R. Majani, J. W. Bradley, F. R. Rose, M. C. Davies and M. R. Alexander, *Biomaterials* **29**, 172 (2008).
97. Y. Mei, J. T. Elliott, J. R. Smith, K. J. Langenbach, T. Wu, C. Xu, K. L. Beers, E. J. Amis and L. Henderson, *J. Biomed. Mater. Res.* **79A**, 974 (2006).
98. M. H. Lee, P. Ducheyne, L. Lynch, D. Boettiger and R. Composto, *Biomaterials* **27**, 1907 (2006).
99. P. Ducheyne and Q. Qiu, *Biomaterials* **20**, 2287 (1999).
100. M. Lestelius, B. Liedberg and P. Tengvall, *Langmuir* **13**, 5900 (1997).
101. L. Baujard-Lamotte, S. Noinville, F. Goubard, P. Marque and E. Pauthe, *Colloids Surfaces B* **63**, 129 (2008).
102. D. J. Iuliano, S. S. Saavedra and G. A. Truskey, *J. Biomed. Mater. Res.* **27**, 1103 (1993).
103. N. Faucheux, R. Schweiss, K. Lützow, C. Werner and T. Groth, *Biomaterials* **25**, 2721 (2004).
104. S. N. Rodrigues, I. C. Gonçalves, M. C. L. Martins, M. A. Barbosa and B. D. Ratner, *Biomaterials* **27**, 5357 (2006).
105. C. A. Scotchford, C. P. Gilmore, E. Cooper, G. J. Leggett and S. Downes, *J. Biomed. Mater. Res.* **59**, 84 (2002).
106. H. M. W. Uyen, J. M. Schakenraad, J. Sjollema, J. Noordmans, W. L. Jongebloed, I. Stokroos and H. J. Busscher, *J. Biomed. Mater. Res.* **24**, 1599 (1990).
107. P. Roach, D. Farrar and C. C. Perry, *J. Am. Chem. Soc.* **127**, 8168 (2005).
108. J. Wei, M. Yoshinari, S. Takemoto, M. Hattori, E. Kawada, B. Liu and Y. Oda, *J. Biomed. Mater. Res.* **81B**, 66 (2007).
109. R. J. Pelham and Y. L. Wang, *Proc. Natl. Acad. Sci. USA* **94**, 13661 (1997).
110. D. E. Discher, P. Janmey and Y. L. Wang, *Science* **310**, 1139 (2005).

111. A. J. Engler, S. Sen, H. L. Sweeney and D. E. Discher, *Cell* **126**, 677 (2006).
112. A. L. Zajac and D. E. Discher, *Curr. Opin. Cell Biol.* **20**, 609 (2008).
113. A. Abbott and D. Cyranoski, *Nature* **424**, 870 (2003).
114. E. Cukierman, R. Pankov and K. M. Yamada, *Curr. Opin. Biotechnol.* **14**, 633 (2002).
115. B. Geiger, *Science* **294**, 1661 (2001).
116. K. M. Yamada and K. Clark, *Nature* **419**, 790 (2002).
117. M. Ochsner, M. R. Dusseiller, H. M. Grandin, S. Luna-Morris, M. Textor, V. Vogel and M. L. Smith, *Lab on a Chip* **7**, 1074 (2007).

Bacteria/Material Interfaces: Role of the Material and Cell Wall Properties

Lydie Ploux [*], **Arnaud Ponche and Karine Anselme**

Institut de Science des Matériaux de Mulhouse (IS2M), LRC7228, Université de Haute-Alsace,
Mulhouse, France

Abstract

Most of the implant-associated infections are attributed to bacteria adhering to biomaterial surfaces as "biofilm" communities. Bacterial transport, first contact with the surface as well as some of the further developments can be considered and can be described using physical–chemical concepts. However, far from simple colloidal particles, bacteria have various macromolecular structures at their cell wall surface for interacting with their surroundings through specific and non-specific bindings. They are also able to modify composition and features of their cell wall in response to specific surrounding conditions. Therefore, bacteria/surface material interface is a complex topic, involving chemical and physical–chemical characteristics of both material surface and bacterial cell wall, as well as biological characteristics of bacteria. Furthermore, proteins and other biomolecules coming from surrounding medium influence the bacteria/material interface by adsorbing onto the material surface prior to any adhesion of bacteria. Finally, bacterial adhesion and biofilm formation phenomena occur at the same time as eukaryotic cell adhesion in an acute competition for adhering to and colonising the biomaterial surface. Therefore, developing biomaterials able to favour cell adhesion without promoting also bacterial adhesion appears still to be a challenge.

In this article, we describe briefly the common development and particularities of biofilms before focusing on what for and whether bacterial features and material surface properties are likely to be involved in bacterial adhesion, the first step in biofilm formation. The influence of adsorbed biomolecules at the bacteria/material interface is finally addressed, as well as the current knowledge about the competition between bacteria and eukaryotic cells.

Keywords

Bacteria, biofilm, bacteria/material interface

1. Introduction

Bacteria/material interfaces play an important role in many industrial and medical fields. Besides microbiologically induced corrosion, bacteria adherent to surfaces are responsible for bacterial contamination of the material environment, since their

---

[*] To whom correspondence should be addressed. Institut de Science des Matériaux de Mulhouse (IS2M), CNRS LRC7228, 15 rue Jean Starcky, BP 2488, 68057 Mulhouse Cedex, France. Tel.: +33 389608766; Fax: +33 389608799; e-mail: lydie.ploux@uha.fr

Surface and Interfacial Aspects of Cell Adhesion
© Koninklijke Brill NV, Leiden, 2010

so-called "biofilms" sessile (i.e., attached to a surface) communities constitute reserves of infectious bacteria [1]. For bacterial cells, a biofilm provides protection against many stresses coming from the surroundings, such as high or low temperatures, dehydration, detergents, or biocides [2]. Indeed, biofilm matrix is highly hydrated, stores bacteria nutrients and is able to restrict the diffusion of molecules by steric retention or by entrapping molecules through inter-molecular bindings [3]. Moreover, biofilm-associated bacteria are able to limit their sensitivity to antibacterial molecular species by appropriate metabolic adaptations [4]. Obviously, these biofilm properties are highly favourable for bacteria survival in hostile media. On the contrary, they constitute an acute disadvantage for the host surroundings and make the fight against biofilms a difficult challenge.

Bacterial adhesion and biofilm formation are critical phenomena in fields as diverse as drinking water distribution, waste water treatment, production and storage of food, and growth of plants [5–7]. Nevertheless, the biomedical field likely suffers the most dramatic consequences. Indeed, antibiotics treatment efficiency is very poor against bacteria living in biofilms [2], due to the restrictions in molecule diffusion and bacteria sensitivity evoked above. Therefore, on prostheses and other implants bacterial adhesion and colonisation lead mostly to serious infections, forcing the removal of the infected material [8]. Fighting against bacterial adhesion and biofilm formation on biomaterials is still complicated by the necessity to promote eukaryotic cell adhesion simultaneously. This competition between eukaryotic cells and bacteria for colonising the surface, known as "the race for the surface" [9], is a critical problem, largely unsolved up to now.

The development of new antibacterial and anti-biofilm strategies requires knowledge of the various factors involved in bacterial adhesion and colonisation. The influence of (i) the material properties (chemical and topographical surface properties) and (ii) the cell wall properties (physical–chemical properties, adhesion-specific structures — pili-like proteins and other adhesins, sugar receptors) is obvious but complex, depending considerably on the bacterial strain and species. In addition, chemical composition and properties of the environment are likely to impact the bacteria/material interface, by leading, for example, to the formation of molecular adsorbed layer onto the material surface (mainly composed of proteins in the biomedical field). Aiming to understand these diverse influences to be able to anticipate the bacterial adhesion onto new materials, a large number of studies have been published. They have been concerned with biofilms formed from various bacterial species on applicable surfaces or on model surfaces and they were conducted with various approaches, from colloidal description [10, 11] to microbiological analyses [12–14], in varying surrounding conditions. All the results obtained cannot be simply compared and each special case should be considered separately with regard to bacterial species and strain and to material properties. Nevertheless, common rules and contradictory trends exist which must be considered when imparting new properties to a material surface.

This review presents the most distinctive features of biofilm-associated bacterial mode of life and the cell-associated factors mostly involved in the biofilm formation on a material surface. Even if the context of this paper concerns the biomedical bio-interfaces, a good deal of knowledge and understanding of bacterial adhesion and biofilm formation coming from other fields (water research, in particular) must be taken into account for describing the role of material surface and cell membrane properties in the bacterial adhesion.

The first part of this paper describes the biofilm formation using the most commonly accepted evolution outline. Colloidal theories often used to describe and anticipate the bacterial adhesion step are briefly presented. The main features of biofilms and biofilm-associated bacteria are addressed as well. In the second and the third parts, respectively, the roles of the material surface and the cell wall properties in the bacterial adhesion and the biofilm development are reviewed. The influence of proteins adsorbed on the surface and the competition with the eukaryotic cells are finally briefly addressed.

2. The Different Phases of Biofilm Growth

Biofilm development can be described following various numbers of steps according to the accuracy of description and to the biological characteristics of the given bacteria. The presence of appendage, of specific membrane receptors, of the quantity and the nature of the exopolymeric substances synthesized by the bacteria are examples of factors highly dependent on bacterial species and able to strongly influence bacterial adhesion and its further development, the biofilm. Nevertheless, four common steps are usually distinguished. They can more or less overlap depending on bacteria species. The present description is based on this 4-step representation and is schematised in Fig. 1. It is important to note that the surface onto which bacteria are transported might have been previously chemically modified due to adsorption of molecules present in the surface environment. In biomedical surroundings, for example, adsorption of proteins and other biomolecules occurs prior to any bacteria adhesion. This point is further discussed in Section 5.

Step 1: Transport of the Bacteria onto the Surface

Bacteria are transported to the surfaces by long-range physical forces, such as gravitational and Brownian forces. Under flow conditions (for example, in water mains) hydrodynamic forces were shown to participate also in the bacteria transport to the surface [15]. In addition to this passive transport, controlled-like motion is involved in the transport of bacteria species having proteinic organelles called flagella. However, its importance for further biofilm formation is not obvious. Thus, while Pratt and Kolter [16] observed that motility promoted biofilm formation of some *Escherichia coli* (*E. coli*) strains, Prigent-Combaret *et al.* [17] showed that flagella mobility was not essential for initiating the biofilm formation of *E. coli* K12. Biofilms formed by wild-type and flagella-deficient strains after 24 h of incubation presented similar thicknesses and morphologies.

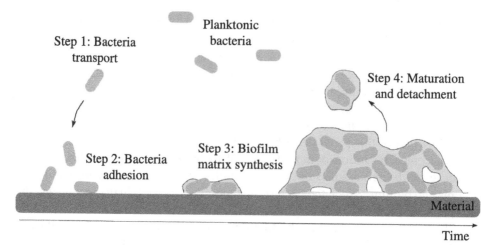

Figure 1. Scheme of the biofilm development. The description is based on a common 4-step representation including bacteria transport onto the surface, bacteria adhesion, synthesis of the biofilm matrix, and maturation of the biofilm with related detachment of biofilm parts. The frequent modifications of the substrate occurring prior to bacteria adhesion and due to adsorption of biomolecules present in the surface environment are not represented in this scheme.

Step 2: Adhesion of the Bacteria to the Surface

Usually, bacteria adhesion is described in two separate stages: the "reversible adhesion" and the "irreversible adhesion". Different bacteria/surface physical–chemical and chemical interactions are thought to be involved in these two stages (Fig. 2).

The "Reversible Adhesion"

The first stage concerns the short time period when bacterium is approaching the surface. Bacterium is subjected to physical–chemical forces occurring between cell wall and surface. These forces, resulting from typically repulsive/attractive electrostatic, van der Waals and hydrophobic/hydrophilic interactions, are usually assumed to be similar as when happening between a surface and a colloidal particle. Thus, forces are described by colloidal theories taking into account only van der Waals and electrostatic forces (Derjaguin–Landau–Verwey–Overbeek — DLVO — theory [18]), or adding hydrophobic forces (extended DLVO theory [19]), and steric repulsions (extended DLVO with steric effects [20]).

Despite the obvious relevance of DLVO theories for explaining some experimental results, it is important to note that the colloidal description although limited to the first adhesion step only provides a schematic view of bacterial adhesion. Firstly, bacteria are usually considered as rigid colloids with fixed characteristics. But, bacterium membrane has significant elasticity and porosity leading probably to variability in the colloidal characteristics. Some authors have proposed, therefore, to improve the description of bacterial adhesion by using soft-particle DLVO theory. Theoretical electrophoretic mobility values consistent with the experiments were obtained. However, this approach still fails to match the experimental interac-

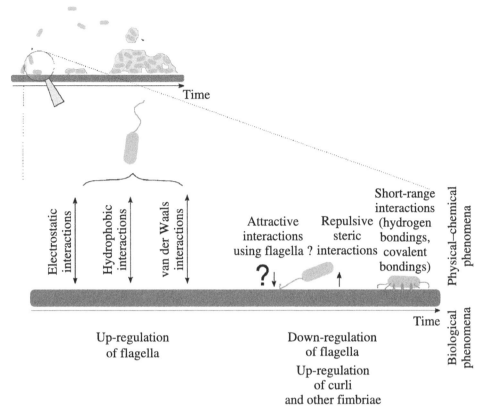

Figure 2. Scheme of the biofilm development. Details of the step 2 ("bacteria adhesion") in Fig. 1. Two separate stages are usually distinguished (the "reversible adhesion" and the "irreversible adhesion") involving different physical–chemical and chemical bacteria/surface interactions.

tion energy measured between *E. coli* bacteria and the surface of an atomic force microscope (AFM) silicon nitride tip [21]. A second limitation concerns the bacterial cell geometry, since bacteria are commonly approximated as spherical cells without filamentous-like appendages. Such appendages like flagella and pili may allow the bacterial cell to overstep the energetic barrier in the near surrounding of the surface [22]. To take into account their contribution in theoretical calculations has been the attempt by some authors [22, 23] but the generalization still remains unobvious. On the contrary, other molecules like lipopolysaccharides (LPS) may prevent a close contact with the surface due to sterical repulsion. Ong *et al.* [20] took into account this effect in the theoretical assessment of the bacteria/surface interaction forces and compared their results with the experimental forces measured by AFM for *E. coli* strains with different lengths of LPS. They observed that the presence and the length of LPS considerably affected the interaction force values, and demonstrated the possibility to improve the theoretical assessment by considering bacterial appendages. Finally, another limitation is that the DLVO theories ignore the capability of the cell to sense the surface ("surface sensing" [24]) and to adapt rapidly, by modifying the cell wall composition especially [25]. As an example, adhesive membrane appendages or receptors may be expressed, enhancing or speeding up the irreversible anchorage of the bacteria on the surface.

Therefore, due to the necessary simplifications of the bacterial features discussed above, the DLVO theories should be used to predict only the short period behavior corresponding to the arrival of the bacterium onto the surface, with particular care when applying bacteria species expressing flagella or/and pili.

The "Irreversible Adhesion"

This stage corresponds to the strengthening of the bacterial anchorage on the surface. Short-range forces are involved, like covalent and hydrogen bonding interactions. These interactions, stronger than those occurring in the reversible stage, are receptor/ligand-like bindings. They occur between bacterial outer cell wall and molecules present on the surface. In the case of biological surfaces (host tissues), bindings occur between biomolecules constituting the surface (proteins, carbohydrates) and specific receptors of the bacterial membrane, including appendages like flagella and pili (see Sections 4.1 and 4.2). These receptor/ligand bindings are known for their involvement in the bacterial infections occurring *in vivo* [26]. They are likely to also occur with artificial surfaces (biomaterials, model surfaces) covered with relevant biomolecules. Nevertheless, bacterial binding to immobilised biomolecules may not occur in each case, but depends considerably on the conformation, the orientation and the density of the biomolecules present on the surface.

Step 3: Proliferation and Synthesis of the Biofilm Matrix

After anchoring on the material surface, adherent bacteria proliferate and synthesize mostly the biofilm matrix. Proliferation is accomplished by cloning of the cells, leading to the formation of bacterial colonies. Bacteria motion ("twitching") and microcolonies motion ("swarming") at the solid surface accompany this process by enhancing the spreading of the colonies over the surface and, therefore, the area covered by bacteria [27, 28]. This movement ability depends on the presence of various appendages, like flagella and pili. In addition, the production of biosurfactants has been shown to improve the swarming-related spreading over the surface of some bacterial species (*Pseudomonas aeruginosa* — *P. aeruginosa* [29, 30], *Serratia liquefaciens* [29], *Bacillus subtilis* — *B. subtilis* [31]). Exogeneous surfactants were also reported to promote bacterial swarming by enhancing the surface wettability [31–33]. However, the importance of the surface wettability was also shown to be species and strain dependent [34].

Biofilm matrix consists for the large part of species-specific exopolysaccharides (EPS) [35]. Two well-known examples are alginic and colanic acids, produced by, respectively, *P. aeruginosa* [36] and *E. coli* [37]. They were found to participate in the 3-dimentional structure and cohesion of biofilm [17, 38–40]. Without EPS, bacteria are still able to form biofilms, however such biofilms are thinner and more tightly packed [17]. The nature and amount of produced EPS not only depends on the bacteria planktonic- or biofilm-associated state [41] but is also affected by environmental conditions [40]. Besides, other extracellular polymeric substances than EPS, including proteins and lipids and coming from the bacterial metabolism, are entrapped in biofilm, leading to specific chemical compositions of biofilms according to bacterial species and growth environment [36, 38, 40].

Step 4: Maturation and Detachment Events

Mature biofilms (Fig. 3) are complex bacterial communities with a specific mode of bacterial life [42]. During the last three decades, a large number of studies in this field have demonstrated the biofilm specificities. Among them is the 3-dimensional architecture that possesses pores and channels allowing bacterial access to nutrients and oxygen and protection against hostile surroundings [2]. Another essential biofilm feature is the specific gene expression of biofilm-associated bacteria compared to planktonic bacteria (for example, 38% of genes are differentially expressed in *E. coli* biofilm [25]), allowing cells to adapt to their surroundings. Typical examples are the modifications associated with flagella and curli gene expressions in *E. coli* mature biofilms, which has been related to the utility of curli for strengthening the biofilm structure and the reduction of flagella-induced motility in the

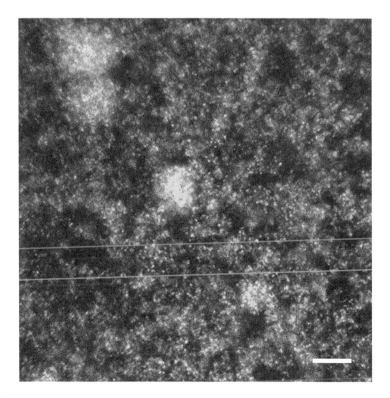

Figure 3. Image of *Escherichia coli* K12 biofilm observed by fluorescent confocal microscopy after Syto9® (Molecular Probes, USA) staining (scale bar 20 μm).

sessile mode of life [25]. Another example is the whole protein composition of the *P. aeruginosa* outer membrane which is significantly different (about 50%) in biofilms compared to planktonic cells [14] in agreement with the strategic role of the outer membrane in bacteria adaptation to surrounding stresses. These metabolic adaptations leading to the specific features of the bacterial mode of life associated with biofilms are allowed by the ability of the bacteria to communicate between each other using the so-called "Quorum Sensing" (QS) biochemical system. Based on the response to cell density, QS is used by bacteria to regulate a large number of physiological functions [43, 44]. In *P. aeruginosa*, for example, the LuxI/LuxR system based on the detection of specific concentrations of signalling acylated-homoserine lactones (HSL) like for all Gram-negative bacteria [43], has been implicated in the normal development of biofilms [40]. Production of wetting agents by *P. aeruginosa* and *Serratia* species [29], attachment of *Helicobacter pylori* to surfaces, or production of exopolysaccharides by *Vibrio cholerae* [45] are other examples of the QS involvement in bacterial adhesion and biofilm formation events. Also, QS is thought to have an important role in the detachment of biofilm parts occurring when biofilms become mature. Free bacteria are then released, able to further colonise other substrates. In *P. aeruginosa* biofilms, they were shown to be flagellated, recovering their motility properties [46] for the future dispersal. Other gene expression modifications under QS control, like cell death and autolysis for degrading the EPS matrix, have been recently assumed also to be involved in this stage [47].

Despite common characteristics, biofilms differ considerably between bacterial species and strains. The microbiological composition varies, for example, from mono-species composition for the model biofilms usually studied in the laboratory, to multi-species composition as observed in nature [48]. Mixed biofilms developing on urinary catheter or voice prostheses, for example, can also include yeast cells. Biofilm spatial structures also vary, from rough morphologies with the so-called "mushrooms" formed by *P. aeruginosa* to smooth architectures developed by some *E. coli* K12, depending, in particular, on the bacterial capacity to synthesize EPS. Chemical composition of the biofim matrix and growth kinetics are also highly affected by the bacterial species [49].

3. Role of the Material Properties

How characteristics of the material surface are able to affect biofilm formation is not known to the same extent according to the growth stage. A large number of published works have addressed the influence of material surface properties on bacterial adhesion. The roles of the surface topography and the surface chemistry described and discussed in the next two sections (Sections 3.1 and 3.2) concern mainly the first step of the biofilm development. On the contrary, only little is known about the material properties which may influence the mature biofilm features. A brief discussion on this topic is presented in Section 3.3.

3.1. Surface Topography

It is widely accepted that bacteria react to surface topographical features that are larger than bacterial cells. They adhere, for example, preferably to the bottom of crevices [50] rather than to the top. Thus, areas with visually higher degree of colonization are often observed on surface defects than on the major part of the surface, and have been reported by many authors either for bacteria [51–53] or yeast [54]. The fact that bacteria are preferentially observed on surfaces with features such as scratches can be easily verified (Fig. 4), although questions remain whether this is a function of preferential attachment or the end result of ineffective cleaning [55].

3.1.1. Bacterial Response to Microtopography

Bacteria response to micro-scale surface features is controversial. Some studies showed that the lowest R_a (average roughness parameter [56]) values resulted in the most hygienic surfaces [57–60], depending sometimes on bacterial species [61]. Others authors failed to demonstrate any clear relationship between R_a values and bacterial adhesion and/or cleanability [51, 62–65]. Hilbert *et al.* [66], for example, did not observe any influence of surface roughness ranging from 0.01 μm to 0.90 μm on bacterial attachment, colonization and removal. On the other hand, some authors proposed that the topographical surface features presenting the highest microbial retention [52, 53, 67] might have had dimensions similar to those of bacterial cells. For example, Flint *et al.* [52] and Taylor *et al.* [68] showed maximal retentions at $R_a = 0.9$ μm and $R_a = 1.24$ μm, suggesting that microbial cells were entrapped in the surface features. However Medilanski *et al.* [67] observed apparent contradictory behavior, since they showed a minimal level of bacterial adhesion on surfaces with scratches of 0.7 μm width ($R_a = 0.16$ μm).

Differences in bacterial behavior observed in the diverse studies presented above should be attributed partially to the variability in the bacterial species. In addition, contradictory results are often due to the high variability in the experimental methods. For example, microbial contamination cleaning processes vary a lot. Moreover, the detection of adherent bacteria is done sometimes by counting re-suspended bacteria after prior dislodgment from the surface, and sometimes by analyzing microscopic images recorded using optical or scanning electron microscopy. Other sources of variation are the differences in the surface chemistry of the materials used in the diverse studies even if materials are similar theoretically (stainless steel, for example). Small differences in material chemical compositions and surface finishes can lead to significant differences in the surface chemical properties. In some studies, in addition, surfaces are pre-conditioned with culture medium [66] or other solutions containing various biomolecules [68] which may influence bacterial attachment to the surface. Such differences in the material surface chemical properties and in the experimental procedures impede a direct comparison of the microbial retention results. Finally, the difficulty to propose a general view of the bacterial response to topographical features is due to the lack of surface features and bacterial colonization description. As well-known for eukaryotic cells [56], surface

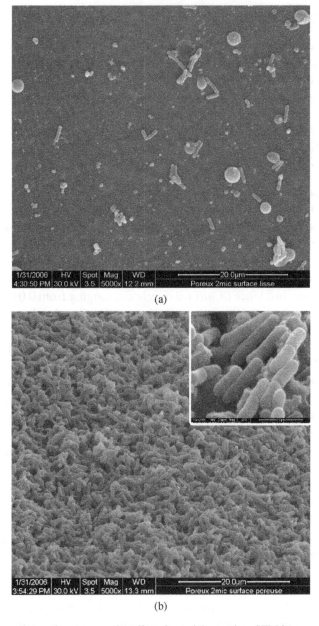

(a)

(b)

Figure 4. Evidence that surface topography affects bacterial retention. SEM images of *E. coli* colonisation on (a) smooth and (b) porous surfaces. Smooth and porous surfaces are two sides of a unique silica material sample. Both surfaces are chemically identical.

descriptions based only on R_a and R_q (root mean square amplitude roughness [56]) roughness parameters fail to provide topographical data relevant at the cell scales. More parameters are sometimes used [64, 65] but the morphology, organization and size of the topographical features are rarely taken into account [51, 67]. Shapes and orientations of the topographical features are, to our knowledge, never described quantitatively. Moreover, bacterial colonization is often given as an average on the whole sample, without determination of bacteria position and orientation. Consequently, correlation between surface topography and bacterial colonization at the surface feature scale cannot be established.

Bacterial attachment to microtopographical features like grooves and other crevices has been attributed to a protection strategy against shear forces and to the increase in the convection transport [50, 52, 53, 55, 69]. The attachment of bacteria in and on surface features should also be due to the enhancement of cell-surface contact area, thereby allowing an increase in the binding energy [53]. However, the energy used by microbial cell when distorting may justify, in part, that bacteria do not always localize and align with topographical features and seem in many cases not to react to surface roughness and features. Optimal size characteristics may, therefore, exist at the microscale. Shape, profile and orientation of these features should also play an important role in the quality of the binding, as Edwards and Rutenberg [53] already proposed and demonstrated theoretically. In addition, cell shape, membrane rigidity, and also bacterial ability to form cell chains and to involve membrane structures allowing mobility, sensing and anchorage to the surface, may affect bacterial response to the topographic features. Flagella, pili and other fimbriae (see Section 4.1), in particular, may enhance the capability of bacteria to create contact with the surface. Some authors have proposed that bacteria response to nanometer scale roughness may be mediated by fimbriae-like structures [70]. In addition, Scheuerman *et al.* [55] and Medilanski *et al.* [67] showed that microstructured surfaces were less colonized from non-motile mutant bacteria than from the wild strains. This result suggests that flagella may be involved in the capability of bacteria to move into the grooves or crevices and/or to recognize the topographical features. However, no study has addressed this topic experimentally up to now, even if some evidence exists that flagella-related motility is of high importance for bacteria to reach the surface [71]. Swarming- and twitching-like bacterial displacements may also be affected by topographical features which may slow down the spreading of the bacterial colonization onto the surface. However, swarming *versus* surface topography has not yet been addressed in the literature.

Finally, and although many of the authors cited above agree that microbial retention is locally affected by the surface defects, bacterial response to microtopography is still largely unknown. Especially, the molecular mechanisms regulating the bacterial response to microtopography have not yet been elucidated. As a consequence, effects at the nanoscale are even less well understood and only a few papers have addressed the effect of nanotopography on bacteria response.

3.1.2. Bacterial Response to Nanotopography

So far only a few studies have concentrated on the influence of nanostructured surfaces on bacteria adhesion. However, some results that have shown opposite behaviors of human osteoblasts and *Staphylococcus epidermidis* (*S. epidermidis*)

[72] or E. coli [73] on nano-structured materials are particularly interesting since they have demonstrated that materials could be specifically designed for promoting cells function while reducing bacterial colonization. Moreover, knowing that eukaryotic cells are able to sense nanotopographical features and given the difference in size between eukaryotic cells and bacteria, it could be expected that topographical features will also have an effect on bacteria attachment and retention. However, bacteria cells are much less deformable than eukaryotic cells and maintain their shape when attaching to a surface. Therefore, the interactions between bacteria and topographical surface features may be obstructed, thus limiting the possibility for bacteria to sense them. In light of this, it is unclear whether bacteria can react to nanoscale surface features. Nevertheless, flagella, pili and other fimbriae that can express bacteria depending on the species and strains (Section 4.1) may interact with the topographical features, allowing sensing by bacteria of topographical surface properties at the nanoscale. The impact not only on global bacterial adhesion but also on localization, orientation and metabolism of bacteria must, therefore, be tested with particular care.

Some studies have already started this delicate work and have shown evidence of the ability of bacteria to sense topographical nanofeatures. Using multiple regression analysis, Bruinsma et al. [74] showed that initial deposition rate of bacteria was statistically different on unworn lenses ($R_a = 4$ nm) compared to overworn lenses ($R_a = 10$ nm). Bakker et al. [70] reported a positive relation between roughness and adherent bacteria by multiple linear regression analysis on pre-conditioned polyurethane-coated glass plates. In these studies, however, the complex surface properties due to the commercial origin of the materials tested, and multiple other chemical changes occurring during the wear of the material, render it difficult to contend that the effect observed was only due to the surface topography. Furthermore, neither the distribution of adherent bacteria nor the size and morphology of the topographical features were described.

Whitehead et al. [57], Campoccia et al. [75] and Diaz et al. [76] addressed the question of the size and the morphology of the surface features by using model surfaces structured at the nanoscale. Whitehead et al. [57] studied the primary colonization (1 h) of several microorganisms (3 bacteria and 1 yeast) on titanium surfaces presenting pits of 200 nm, 500 nm, 1 μm and 2 μm diameter and 200 nm, 500 nm, 750 nm and 1 μm depth, resulting in R_a values ranging from 45 nm to 220 nm. Pit-to-pit distances that varied for the same surface type and between the different surface types were not specified. Aside from the general increase in adherent bacteria number with surface roughness, they observed that the number of adherent bacteria specifically retained on the pits did not follow the same trend, since surfaces with 500 nm diameter pits showed a different behavior than the other surfaces regarding the number of bacteria localized on or in the topographical features. The number of bacteria adhered was the smallest on the 500 nm diameter pits while it increased with the pit size for the other feature sizes (200 nm, 1 μm and 2 μm diameter). This effect, that the authors did not discuss in their paper, could be related to the size effect discussed by Edwards and Rutenberg on the basis of their theoretical calculations [53]: the 500 nm pits presented a low contact area for bacteria measuring around 1 μm diameter, while, on the other hand, the other pit sizes may result in more energetically favorable situations. Moreover, pit shape was different on the 200 nm and 500 nm diameter pits surfaces and the 1 μm and 2 μm diameter pits surfaces. This might have led to the mixing of the pit size effect with the pit shape effect for influencing the bacterial retention on the surfaces.

Campoccia *et al.* [75] also studied bacterial adhesion on nanostructured surfaces. Structured and reference surfaces were made of poly(ethylene terephthalate). Structured surfaces consisted of nanocylinders of 160 nm height and 110 nm diameter, with a spacing of 220 nm, while reference surfaces had a roughness of 1.0 ± 0.2 nm and were considered to be smooth. Using *Staphylococcus aureus* (*S. aureus*) and static cultures, the authors did not observe any differences in the amounts of bacteria adhered on the structured surfaces compared to the smooth ones. However, the lack of surface chemistry characterization means that it is not known whether both structured and smooth surfaces were chemically identical. Any differences in the surface chemistry might have influenced the bacterial adhesion and, therefore, might have masked the effect of the topography. Moreover, the characterization of bacterial colonization was performed using a global technique (here chemiluminometry) which does not allow the determination of modifications in bacteria localization and orientation.

The work performed by Diaz *et al.* [76] addressed the question of bacteria localization and orientation in nanoscale features, by imaging *Pseudomonas fluorescens* adhered on nanostructured model surfaces by atomic force microscopy (AFM). Three types of gold surfaces were used: smooth, random nanometer size structures (amplitude from 4 nm to 20 nm), and grooves 750 nm wide and 120 nm deep. The results showed a clear alignment of the cells with the grooves while no orientation was visible on the random nanometer-structured surfaces. However, AFM imaging was performed in air on dried substrates, which likely results in modification of the native distribution and orientation of bacteria. This process, which is suggested by several micrographs and is related to the dewetting of the bacteria suspension film initially present on the surface, may explain the retention of the bacteria in the grooves. Nevertheless, the absence of bacterial orientation on random nanometer-structured surfaces even after drying in air suggests that these nano-topographic features failed to impact the behavior of the bacterial species.

In a recent work, Puckett *et al.* [77] studied the influence of nanofeature properties (organization and shape) on bacterial adhesion. The numbers of total and viable bacteria attached to unmodified and chemically nanostructured titanium foil surfaces were observed by fluorescence microscopy and compared to each other. Nanostructure was provided by electron-beam evaporation to obtain "nanorough" surfaces and anodization with hydrofluoric acid, resulting in nanotextured or nanotubular surfaces depending on the etching time and concentration. This work showed significant differences depending on the surface type: nanorough surfaces were less colonized than the unmodified Ti surfaces, while the nanotextured and nanotubular surfaces were significantly more colonized. On the other hand, the viable-to-total bacteria ratio was significantly reduced on nanotextured and nanotubular surfaces compared to the unmodified and nanorough surfaces. Nanofeatures present on surfaces, therefore, seem to affect bacterial adhesion differently depending on their organization and shape. However, analysis of the surface energy and contact angles as well as chemical analysis of the surfaces using XPS showed significant differences between the different preparation methods (incorporation of fluorine and differences in the oxide content) preventing any direct comparison of these surfaces with the conventional and electron-beam evaporation ones.

In addition to the small number of studies conducted on the influence of nanotopographic surface features on bacteria adhesion and retention, the sizes of the features studied remain mostly larger than 100 nm. Wiencek and Fletcher [78] worked at the molecular scale and were interested in both chemistry and topography. They studied the effect of topography by using methyl- and hydroxyl-terminated self-assembled monolayers (SAMs) of different lengths. Bacteria adhesion was determined by optical imaging in a dynamic mode. Using these surfaces, they concluded that the nanometer-scale topography could influence bacterial adhesion as well as detachment. However, changes in adhesion were not observed with certain mixed chain-length methyl-terminated SAMs, suggesting a coupling between chemical and topographical effects on bacterial attachment, perhaps due to the polyethylene chains exposed by the lengthiest molecules. Their apparently contradictory results underlie the difficulty to decouple, here as well as in many other studies, the respective influences of the surface chemistry and the surface topography.

Recently, a very interesting study addressed the influence of very small topographical changes on the bacterial adhesion characteristics [79]. Mitik-Dineva *et al.* [79] compared not only the number of adherent bacteria, but also the size and shape of the adherent cells and the amount of EPS produced on non-etched and etched glass surfaces. Surprisingly, although etching resulted in changes in the R_a (and R_q), R_{max} and R_z of only 1 nm, 8 nm and 10 nm respectively, it affected the various parameters of the bacterial attachment for all the five bacteria species tested. For each species, EPS deposits on the surfaces were higher on etched (smoothest) substrates. However, the size and shape of cells were affected differently according to species. Some species became larger on the etched surfaces without significant shape modification, while others appeared smaller on the etched surfaces with a less elongated shape than on the non-etched surfaces. Furthermore, although it is commonly expected that increases in surface roughness should result in increase of bacterial attachment, this study demonstrated clearly an increase of the attached bacteria number resulting from a decrease of the topographical features due to the etching. Coupling of topographical and chemical surface properties influences is expected since smooth substrates were obtained by chemical etching using hydrofluoric acid. However, based on X-ray photoelectron and X-ray fluorescence spectroscopy analyses, the authors concluded that chemical etching did not change the chemical composition of the glass surfaces. Finally, the most important finding of this study is that bacteria may be far more susceptible to nanoscale surface roughness than was previously thought.

Finally, mechanisms of bacterial response to topographical surface features have not yet been elucidated. At the nanoscale, flagella-, pili- or other fimbriae-like membrane structures might be involved in the sensing of the topographical features by bacteria thus improving their ability to bind to these features. Mitik-Dineva *et al.* [79] proposed also that bacteria attaching successfully to smooth surfaces with features at the nanoscale might use EPS for improving their attachment.

3.2. Surface Chemistry

Although the coupling with the topographical characteristics (see Section 3.1) often prevents to conclude definitely about the effects of given chemical parameters or functionalities, ample evidence exists in the literature that chemical properties of material surfaces play an essential role in the bacteria adhesion to surfaces. Aside from specific functional groups, the surface hydrophobicity/hydrophilicity, the surface energy and the surface charge appear to be the chemical factors most likely to influence the bacteria adhesion.

3.2.1. Hydrophobicity

It is commonly accepted that the substrate hydrophilic and hydrophobic characters play an important role in bacteria adhesion. However, several contradictory trends have been reported. An and Friedman [60] maintained that hydrophilic materials might be more resistant to bacterial adhesion than hydrophobic ones. Other studies confirmed this assertion [80]. For example, Tegoulia and Cooper [81] observed that *S. aureus* adhesion to SAMs decreased according to the methyl > -carboxyl > hydroxyl termination order. Cerca *et al.* [82] also reported that adhesion occurred to a greater extent on hydrophobic surfaces that on hydrophilic ones for all the nine *S. epidermidis* strains tested. In this study, bacterial cell hydrophobicity seemed not to have significant influence on bacterial adhesion.

However, other studies rather demonstrated that bacteria with hydrophobic surfaces prefer hydrophobic materials while bacteria with hydrophilic surfaces prefer hydrophilic materials. Using multiple linear regression analysis, Bakker *et al.* [70] showed that three marine bacteria adhered on pre-conditioned polyurethane films according to this rule. Nevertheless, this study showed also that bacteria with more hydrophobic surface character adhered to a greater extent than hydrophilic bacteria, as already stated by other authors [83]. Recently, in an interesting study, Boks *et al.* [84] reported that similar number of *S. epidermidis* bacteria adhered on both hydrophilic and hydrophobic surfaces. However, mode and dynamics of bacterial adsorption and desorption observed in a flow chamber were significantly different according to substrate for all the four strains tested. Adsorption and desorption occurred significantly more on hydrophilic surfaces than on hydrophobic ones. In addition immobilization and detachment onto and from the surface were preceded by bacterial sliding movements to greater extent on hydrophilic surfaces than on hydrophobic ones. These differences in bacterial attachment and detachment on hydrophilic and hydrophobic materials, in agreement with previous results [85, 86], are consistent with the hypothesis that hydrophobic surfaces are more favourable to bacterial adhesion than hydrophilic surfaces. This behaviour was attributed to repulsive and attractive acid–base interactions between bacteria and hydrophilic and hydrophobic surfaces, respectively. In this study, hydrophobic and hydrophilic characters of the bacterial strains were not specified, preventing any discussion about their role in these bacteria/material interactions. Nevertheless, these results demonstrate that relationships between bacterial adhesion and hydrophobic/hydrophilic character of the substrate are complex.

It is important to note that the material surface properties can be modified by exogenous surfactants produced by bacteria. This should result in significant modification of the initial hydrophobic/hydrophilic character of the material surface. Several authors reported an increase of the surface wettability resulting from the presence in the surroundings of biosurfactants produced by bacteria and leading to the enhancement of bacteria swarming on the surface [31, 32].

3.2.2. Surface Free Energy

Some authors studied the relations between bacterial adhesion and material physical–chemical properties using the surface free energy parameter for characterizing the substrate. For example, Tsibouklis *et al.* [87] observed significant inhibition of bacterial adhesion on low surface free energy polymer surfaces, while others reported windows in surface free energies within which adhesion was minimal [88]. On the other hand, Speranza *et al.* [89] showed an important role of the acid/base interactions in *E. coli* adhesion to polymer surfaces. Subramani *et al.* [90] also determined stronger correlations of bacterial adhesion to filtration membrane with acid–base component of surface free energy than with Lifshitz–van der Waals component. Relationships between bacterial adhesion and surface free energy appear not to be obvious, likely, in part, because studies rarely considered also surface properties of the bacterial cell wall.

3.2.3. Surface Charge

Except for a few bacteria species, bacterial cell wall has a negatively charged surface at the physiological pH ($\sim$7), even if absolute values can vary from a few to several tenths of mV [91, 92]. Therefore, attractive interactions are expected to occur between bacterial cells and positively charged surfaces, while repulsive interactions should occur when the surface is negatively charged. Such behaviour has been shown by Roberts [93] who specified that mineral composition of the substrate also determined bacterial adhesion when surfaces were negatively charged, i.e., unfavourable to bacterial adhesion. Terada *et al.* [94] found also that electrostatic interaction is the most decisive factor for bacterial adhesion when it works as an attractive force, i.e., on positively charged surfaces. Also, results obtained by Li and Logan [95] indicated that surface electrostatic charge was important in adhesion, since ionic strength affected bacterial adhesion. However, no significant correlation was found between bacterial adhesion and surface charge. Thus, bacterial behaviour on charged surfaces is not so simple. Gottenbos *et al.* [96] demonstrated that surface charge effect depends on the growth duration on the surface and, also, varies with bacteria species. They reported that bacterial attachment was higher on positively charged surfaces for the four species tested (*E. coli, P. aeruginosa, S. epidermidis* and *S. aureus*). However, after attachment, bacteria grew faster on moderately negatively charged than on positively charged surfaces. After 7 h of incubation, colonisation on negatively charged surfaces was even higher for two of the strains (*P. aeruginosa* and *S. aureus*). These behaviours were attributed to the probable decrease of cell viability, due to strong binding between surface and bacteria. Thus, elongation of adhered bacteria needed for cell proliferation might be inhibited in part and bacterial cell membrane might even disrupt. On the other hand, Komarony *et al.* [97] reported surprising results, stating that *E. coli* and *S. aureus* attached better on surfaces (unmodified silicon wafers and 11-mercapto-1-undecanol grafted surfaces) with similar zeta-potential to cell surface zeta-potential. In this study, neither hydrophilic character nor roughness alone could be related to bacterial retention differences. However, surface charge, roughness and medium ionic strength might have a combined influence to lead to this surprising result.

3.2.4. Coupling Hydrophobicity/Surface Charge/Medium Properties

Contradictory results previously reported must be mostly attributed to the coupling of several influences, coming from the surface hydrophobicity, from the surface charge and from the surface roughness. This was highlighted by the work of Bruisma *et al.* [98] who showed by multiple regression analysis that separate influences of these different factors combined to provide the observed experimental result. Properties of the surrounding medium including pH and ionic strength may also play a role, modifying apparent surface properties of both bacteria and substrate material. Typical example is given by Gaboriaud *et al.* [99] who reported significant changes in bacterial adhesion according to pH and ionic strength of the medium. This behaviour was attributed to modifications of mechanical and structural characteristics of the bacterial envelope, rather than to sole contribution of electrostatic forces. Camesano and Logan [100] also reported high influence of pH on the interactions between surface and bacteria measured by AFM, while ionic strength of the medium affected only slightly the interaction force intensity.

3.2.5. Surface Functionalities

Chemical functionalities can lead to specific binding between a bacterium and material surface through receptor/ligand-like interactions. The functional groups involved in such interactions are mostly parts of biomolecules immobilized on the surface. Their role in bacterial adhesion not only to host tissue but also to an inert material is important. Their implication in bacteria/material interactions is reported and discussed in Section 5. Chemical functionalities that are unable to create specific receptor/ligand-like interactions are also known to affect bacterial adhesion in a particular way. In this case, only their physical–chemical properties are involved in the interactions. Poly(ethylene glycol) (PEG) group which is well-known for its anti-adhesive properties regarding bacteria is one of them. A recent work of Krsko *et al.* [101] using PEG hydrogel patterned surfaces illustrates the cell-repulsive properties of PEG. These properties are attributed to a stable interfacial water layer preventing direct contact between bacteria and surface [102]. Such functionalities are often proposed for developing biomaterial coatings leading to bacterial adhesion prevention [103].

In this section, the focus was on the initial adhesion of bacteria to surfaces. Nevertheless, an influence of surface properties like hydrophobicity and charge on the further development of biofilm should be expected. So far, this question has been rarely addressed. It is discussed in Section 3.3.

3.3. Role of the Material Surface Properties in the Mature Biofilm Characteristics

Only little is known about the influence of the material surface properties on the development and the characteristics of the mature biofilm. Nevertheless, some evidence exists that the surface should affect the evolution of the biofilm and its composition. We demonstrated recently [104] that physical–chemical properties of the substrate are able to affect the growth kinetics of biofilms and the structure of newly formed biofilms. We observed differences in the kinetics of development and morphology of *E. coli* biofilms on hydrophobic and hydrophilic surfaces, but with similar numbers of adherent bacteria on the different substrates in the initial step of adhesion [104]. Similarly, Patel *et al.* [105] obtained similar adhesion rate of *S. epidermidis* on two hydrophobic polyurethane samples (unspecified water contact angle values) in the first adhesion step, but significantly different rates after 12 h of incubation. Gottenbos *et al.* [106] reported also that growth of *S. epidermidis* and *P. aeruginosa* was affected by physical–chemical surface properties to a greater extent than the initial adhesion. In light of this, long-term bacteria/surface interactions should be more intensively addressed in the future. Topographical features influence should be particularly studied since the mature biofilm cohesion and stability may be affected by the irregularities of the underlying surface.

Characteristics of mature biofilm like its microbiological composition and the chemical composition of the extracellular polymeric substances may depend on the initial properties of the material surface. Obviously, since bacteria interact differently with surfaces according to both the surface properties and the bacterial cell wall properties which themselves depend on the bacterial species (Section 4), material surface properties can affect indirectly the composition of bacterial species and strain present in the mature biofilm. Consequently, chemical composition of the extracellular polymeric substances may also be indirectly influenced by the material surface properties depending on the bacteria present in the biofilm. In addition, some additional chemical components in the biofilm may be generated by the bacteria-induced degradation of the substrate, thus promoting the release and entrapment of material surface components in the biofilm matrix. To our knowledge, however, published studies supporting these hypotheses are rare, except concerning microbiologically induced corrosion which was shown to release chemical components of the substrate in the biofilm matrix [107].

4. Role of the Bacterial Cell Wall Properties and Features

Bacteria cell wall properties are obviously crucial for the development of bacteria adhesion with material surfaces. The specific composition of the cell membrane is responsible for ligand/receptor-like binding between the cell and the surface. Aside from its obvious role in the physical–chemical properties at the molecular level, the bacterium cell wall governs also the physical–chemical properties of the overall cell, i.e., at the macroscopic scale. In the next section (Section 4.1), bacterial cell wall constituents like appendages and other receptors able to be involved in the interactions with materials are presented before addressing their involvement in the bacteria adhesion to surfaces (Section 4.2).

4.1. "Bacterial Tools" for Adhesion (Fimbriae, Lectins and Other Receptors)

Bacteria possess at their surface diverse macromolecules that can be used by bacteria to interact with their surroundings, in particular for bacterial adhesion and further biofilm formation. They allow cells to adhere to a surface and also to adhere to one another. Mostly, these so-called adhesin molecules are composed of proteins, such as lectins, flagella, pili and other fimbriae. Nevertheless, bacterial surface carbohydrates are also likely to play a role in bacteria-to-surface and bacteria-to-bacteria interactions.

4.1.1. Pili

Pili are non-flagellar filamentous protein appendages anchored in the bacterial cell membrane. They are named both "pili" or "fimbriae" from the latin terms for "hair" and "fiber" and can be distinguished in several groups (4 in Gram-negative bacteria; 2 in Gram-positive bacteria). They usually measure less than 10 nm in diameter and are a few micrometers long, sometimes only a few hundreds nanometers. They are composed of so-called pilin subunits, which vary according to bacterial species and are assembled in homo- or hetero-polymers [26]. Pili are considered to be the most prevalent organelles for the adhesion to inert surfaces and for the attachment to host tissues. According to bacterial species and pili types, they have various functions in surface colonisation.

Type 1 pili are of particular significance in mediating adhesion of *E. coli* to eukaryotic cells, since they are able to recognize mannose (see also Section 4.1.3 on lectins) [108]. Pratt and Kolter [16] showed also their involvement in the *E. coli* initial attachment to several inert surfaces. Besides Otto *et al.* [109] proposed that type 1 pili play a role in the surface-sensing mechanism of *E. coli* activating the adaptation of the bacteria outer membrane protein composition [110]. In the specific case of pyelonephritic *E. coli* responsible for urinary tract infections with involvement of kidneys, another type of pili, the P pili, have been implicated. Its adhesin (papG) is similar to the type 1 pili adhesin located on the pili tip and recognizes glycan receptors on host cells [111]. Type IV pili are another type of fimbriae, expressed in various Gram-negative bacteria species, which have been shown to mediate the interactions with host cells through an adhesin function. Type IV pili have several attributes, including DNA binding [112] and a crucial role in twitching motility through elongation and retraction movements allowing bacterial displacement over the surface [113]. This pili-mediated motility was shown, for example, by *Legionella pneumophila* [114] and *P. aeruginosa* [115] and is thought to be involved also in other bacteria [116]. Other fimbriae, the curli, are amyloid-like fibers (Fig. 5) and have been identified in many *E. coli* and *Salmonella* species. They play an important role in attachment not only to host tissues [117] but also to inert surfaces [118–120]. They were also shown to bind many human proteins like fibronectin [121, 122], inducing bacteria internalisation by cells (entry of bacteria into host cells) [123]. They play also a crucial role in the 3-dimensional architecture of the biofilm by enhancing the cell-to-cell interactions [17, 117]. Besides these few examples, a number of other pili-like structures have been described in Gram-positive as well as in Gram-negative bacteria, often associated with serious infections. Their number is too large to be all reported here. An excellent recent review gives more details about pili and their assembly mechanisms [26].

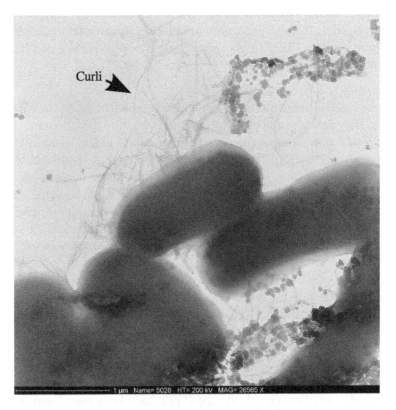

Figure 5. *E. coli* K12 curli observed by Transmission Electron Microscopy.

4.1.2. Flagella

Flagella are thicker (some tens of nanometers in diameter) filamentous protein structures present in single or multiple arrangements. They are well known for their role in the locomotion of planktonic bacteria [124]. They have also often been implicated in bacterial swarming on solid surfaces [125]. Besides, their involvement in the first step of bacterial adhesion has been proposed, which has been supported by several experimental works [126, 127]. For example, *P. aeruginosa* has been reported to adhere to inert surfaces in less extent when its unique flagellum is lacking [128]. However, the specific role of flagellum has not yet been elucidated even if it has been proposed that flagella may overcome the repulsive forces between the cell and the surface, thus increasing the probability of bacterium-to-surface contact [16]. In addition, flagella have also been shown to be used by bacteria for their adhesion to host cells [129, 130], involving the specific recognition of molecules like oligosaccharides present at the surface of host tissues [131]. However, this role of adhesin seems not to be the rule since authors reported that *E. coli* strains with and without flagella were equally able to develop biofilms on inert surfaces [17, 132].

4.1.3. Lectins

Lectins are proteins able to bind specifically carbohydrate moieties present in polysaccharides. Depending on the species and strain, lectins are located in the bacterial cell wall or on the tip of pili, which is, for example, the case of the mannose-specific adhesin of *E. coli* (FimH) [26]. In general, lectins are an important class of adhesins that are involved in many infections of animal tissues. They have been described in numbers of pathogenic bacteria [133]. In *P. aeruginosa*, for example, they are an essential tool of bacterial cells for binding to the carbohydrate residues present at the surface of the host tissue (lungs of cystic fibrosis patients, for example) before further host tissue colonisation [131]. They are also essential for uropathogenic *E. coli* to bind to the bladder epithelium cells [134].

4.1.4. Other Proteins

Other non-fimbriae proteins, known as autotransporters, have been reported for their adhesin function. Antigen 43 (Ag43), for example, has been shown to mediate bacterial autoaggregation through Ag43–Ag43 interactions [135].

4.1.5. Polysaccharides

Aside from proteins, some polysaccharides play an adhesive role in bacterial adhesion. This is the case of the lipopolysaccharides (LPS) which are carbohydrate macromolecules present in large numbers in the outer membrane of Gram-negative bacteria. Modifications in the structure of LPS have been shown to significantly affect the rate of bacteria deposition and further attachment [136, 137]. AFM force measurements supported this assertion, showing that intact LPS were necessary to obtain strong adhesion between bacteria and surfaces [21]. However, whether LPS impacted bacteria-to-surface contact was not clearly elucidated. The role of the charges conferred on the whole bacterium by the various LPS structures has been implicated, but no correlation with the length of LPS was shown. In addition, LPS are expected to result in steric repulsion when bacterium approaches the surface, in contradiction with the favourable role of intact LPS observed experimentally. Aside from their role in the adhesion to inert surfaces, LPS are involved in specific interactions with host tissue. In *E. coli* K12, for example, some oligosaccharide sites of LPS are lectin-binding receptors, able to recognize exogenous lectins specific themselves for different carbohydrate moieties present at bacterial surfaces and in exopolysaccharides of the biofilm matrix (D-galactose and L-fucose of the colanic acid, for example) [138]. Through exogenous lectins, they participate in the linkage between bacteria and the biofilm matrix. In general, lectin-polysaccharide recognition is also thought to constitute the molecular basis for the cell–cell aggregation.

Capsular polysaccharides that are not part of bacterial cells but constitute the largest part of the biofilm polymeric matrix have been also implicated in the adhesion of bacteria to surfaces. For example, Quintero and Weiner [139] showed that the attachment of *Hyphomonas* spp. was inhibited when capsular polysaccharide synthesis was lacking. However, the favourable role of capsular polysaccharides in bacterial attachment is not the rule since Hanna *et al.* [140] demonstrated that capsular colanic acid of uropathogenic *E. coli* blocks the interactions between bacteria and inert surfaces. Other authors proposed that exopolysaccharides might be rather critical for the formation and the cohesion of the 3-dimensional structure of biofilms [39].

All bacterial species or strains do not possess all the structures previously described. In addition, the role of these adhesins varies from one species to another. Their expression depends on the growth state of bacteria and is affected by the surrounding conditions. As addressed in the first part of this article (Section 2, step 4 "maturation and detachment"), a large number of adhesin- and fimbriae-related genes are especially modified during the development of biofilm.

4.2. Non-specific and Specific Adhesion

As suggested by the ability of bacterial surface structures to bind to host tissues or/and to attach to inert surfaces, two types of adhesion can be distinguished. The first one, named "non-specific adhesion", involves physical–chemical interactions as described typically by the DLVO theory and its extensions. It is mediated by macroscopic surface properties like surface free energy and surface charge and occurs at large and intermediate separation distances between cell and material. The second one known as "specific adhesion" corresponds to specific recognition between ligand and receptor (one of them located on the bacterial surface structure). It involves microscopic surface properties of cell and material, and corresponds to short-range interactions. Busscher and coworkers [141, 142] discussed early this difference in scale ranges in which each adhesion type occurs. Although they argued that non-specific and specific adhesion types are only different expressions of the same phenomenon involving basically the same physicochemical forces, they highlighted also that this artificial distinction was suited for addressing the specificities of both types of adhesion to inert surfaces as well as to biological surfaces.

Non-specific and specific adhesion types are usually addressed separately in the literature. Non-specific adhesion is mostly addressed in the materials science field, with typical applications to food-contact surfaces, drinking water pipelines and biomaterial surfaces. Non-specific adhesion was already studied intensively. Two factors of influence are usually considered, material surface properties (see Section 3) and cell surface physical–chemical properties which is the subject of this section (Section 4.2). As highlighted by Busscher *et al.* [141], these overall surface properties of bacteria are the result of several factors including the absence or presence of adhesins like those considered in Section 4.1.

Specific adhesion is studied by the research community interested in human and animal tissue infections and related immune responses. A number of studies address this topic which is not the aim of the present article. However, the necessity to consider the specific adhesion not only to host tissue but also in case of cell-to-material contact has been recently recognized in the research community. In the field of biomaterials science, biological fluid components and eukaryotic cells are present around the biomaterial, besides bacterial cells. Therefore, they are likely to create more complex interactions between bacteria and biomaterial than only non-specific adhesion bindings. For example, proteins adsorbed onto the material may lead to specific bindings with some bacterial surface structures. Another reason for considering specific adhesion in the biomaterials field is the development of new biomaterials with the aim to promote adequate integration in mammalian tissues. Proteins, peptides and other biomolecules are used for this purpose, incorporated in the material or in its coating. Unfortunately, they will not only activate expected responses from the animal cells, but also are likely to be recognized by specific receptors of the bacteria membrane. The risk of biofilm formation and further infections may, therefore, be increased.

4.2.1. Non-specific Adhesion

The chemical composition of the cell membrane including pili, flagella, LPS, etc., participates in the non-specific interactions between bacteria and inert surfaces through the associated physical–chemical properties conferred on the overall bacteria. Interactions with the substrate involve Lifshitz–van der Waals, electrostatic and Lewis acid–base forces as well as, obviously, hydrophobic, steric and other short-range forces (Fig. 2). In this frame work, DLVO theories are often used for describing the adhesion of an entire bacterium. Despite good qualitative agreements, theories have often failed to correlate quantitatively with experimental results [21]. Nevertheless, several authors demonstrated the importance of global physical–chemical characteristics of cell outer membrane in the first stage of cell-to-surface contact, especially concerning cell surface hydrophobicity and charge. However contradictory results are usual, leading to the impossibility to propose common trends. For example, Flemming *et al.* [143] studied the adhesion of *P. aeruginosa* wild type and LPS-deficient mutant strains to glass and polished stainless steel. They showed dependence between rate of adhesion and cell surface hydrophobicity and charge: strains with stronger hydrophobic and anionic characters adhered in significantly greater amounts than the other strains. On the contrary, using LPS-deficient mutants, Walker *et al.* [137] observed that the deposition rate was the highest for the least negatively charged *E. coli* strain, while the most negatively charged strain exhibited the lowest deposition rate. Besides, Haznedaroglu *et al.* [144] observed changes in both the retention of two *E. coli* isolates and their cell surface characteristics according to the starvation conditions. However, no common correlation was found between retention and physical–chemical characteristics for the two isolates. Even, they observed opposite relations between retention and hydrophobicity: the most adhesive bacteria were found to be the most hydrophobic for one isolate but the least hydrophobic for the second one. These apparent contradictions in the experimental relations between physical–chemical cell wall characteristics and rate of adhesion must be attributed, in part, to the differences among studies concerning the properties of both bacteria and material used. The second main reason is the large range of the growth conditions and experimental set-ups used to address this topic. For example, different transport dynamics or surface conditioning films may lead to different rates of adhesion. Haznedaroglu *et al.* [144] stated that sugar and protein productions were observed at the beginning of the starvation treatment, simultaneously with distinct adhesion behaviours. Other surrounding factors like nutrient access or growth medium pH may also lead to differences between studies, since bacterial features and, consequently, physical–chemical characteristics of the whole cell depend on these factors (see Section 4.3).

Fibre-like membrane features (see Section 4.1) alone are considered to play also a crucial role in the first cell-to-surface contact when bacteria are just getting closer to the surface. These macromolecules are supposed to overstep the repelling forces near the material surface (for example, van der Waals ones) allowing, therefore, bacterial adhesion to surface. Some experimental evidences exist that pili and flagella are essential for many bacteria species to attach to inert surfaces. Dramatic lack of *E. coli* bacterial adhesion was shown for flagella-deficient [16] and curli-deficient [17, 119] strains. Abu-Lail and Camesano [21] showed also the importance of LPS in the first interactions between *E. coli* and the AFM tip. Nevertheless, Burks *et al.* [136] observed that adhesion was not consistently correlated with the length of the LPS molecules, even when important sub-structures such as the O-antigen, the inner/outer core polysaccharide and the ketodeoxyoctonate (KDO) were lacking. Otto *et al.* [109] proposed that type 1 pili sensed the surface before any physiological adaptation of the adherent *E. coli*. However, depending on the strain, *E. coli* do not need all these organelles to activate attachment to the surface and develop further biofilms. *E. coli* K12 [17] were shown not to require flagella for initiating adhesion and forming biofilms. Moreover, Walker *et al.* [137] did not observe any consistent relationship between bacterial deposition rate and LPS length for three different *E. coli* strains.

4.2.2. Specific Adhesion

Pili, lectins and other membrane structures (see Section 4.1) are used by bacteria for a number of specific interactions involving parts of these macromolecules, known as adhesins. For example, type 1 pili possess a mannose-binding receptor (FimH) located on their tips and allowing bladder infections [26]. In general, such lectins bind to biofilm exopolysaccharides and carbohydrates moieties of cell membrane [145]. The interaction of bacteria with exogenous lectins is also possible through lectin receptors located on bacterial LPS and is able to specifically recognize lectins present in the bacteria environment [138]. Curli of *E. coli* were shown to bind to a large range of human proteins, including fibronectin, fibrinogen and plasminogen, using two selective protein-binding peptides [146]. As previously evoked, type IV pili are able to bind DNA for further DNA uptake [112].

On the other hand, biomolecules adsorb naturally onto biomaterials as soon as they are implanted into the human body. Furthermore, biomaterial surface coatings including whole or fragments of some of these pili-target biomolecules constitute current approaches. Therefore, studying interactions between bacteria (and their organelles) and biomolecules immobilised on inert surfaces appears now crucial. In particular, model surfaces presenting well-defined biomolecules conformations and densities should provide essential data about the important factors directing specific bacteria/biomolecule bindings. The most frequent bacteria-to-biomolecule interactions are addressed below.

4.2.2.1. Bacteria/Carbohydrate Interactions. Studies addressing mechanisms of bacteria-to-immobilized carbohydrate interactions are usually coupled with the development of glycoarrays created for detecting pathogens [147] for identifying novel carbohydrate-binding proteins [148] or potential inhibitors of bacterial adhesion [149]. Most of them concern the recognition of immobilized mannose residues by *E. coli* K12 and their type 1 pili [147]. In particular, the forces required to break the interactions between *E. coli* and surfaces presenting immobilized mannose groups were characterised and quantified [150]. Moreover, various structures and accessibility of immobilized mannose residues were tested [151]. Significant differences were related to the nature of mannosides and to how oligomannoses were branched, suggesting synergistic bindings of the bacterium with several mannose moieties. The nature of the linker used for the mannoside grafting (propyl- or tri(ethylene glycol)thiol) was shown to influence bacteria adhesion as well. However, no effect of the immobilised mannoside density was observed irrespective of the static or dynamic culture conditions. This work highlights that the specific adhesion is dependent not only on the nature but also on the structure and the accessibility of the immobilised biomolecules. It demonstrates also the difficulty to distinguish the effects of immobilised molecules from those of the underlying surface.

4.2.2.2. Bacteria/Protein Interactions. A particular interest appeared recently in the mechanisms of bacteria/immobilised protein interactions, demonstrated by the

rapid increase in the number of publications focusing on this topic. Specific interactions of bacteria with fibronectin, fibrinogen, collagen, laminin and bovine serum
albumine (BSA) have been intensively studied. It is now well known that BSA inhibits the adhesion of *S. aureus*, *S. epidermidis* and *E. coli* [152–154] while collagen
can promote or inhibit adhesion depending on bacterial species (respectively *S. aureus* and *E. coli*) [155]. Fibrinogen was shown to increase *S. aureus*, *S. epidermidis*
and *E. coli* retention to surface [152, 156] and *Streptococcus mutans* (*S. mutans*)
binds laminin specifically [157].

A particular example is given by fibronectin, whose interactions with bacteria are
the subject of a number of studies. Fibronectin is known to be specifically recognized by bacteria like *S. aureus*, *S. epidermidis* and *E. coli* [158, 159], involving
the N-terminal region of the protein [153]. For a few bacteria species like *S. aureus*
binding sites have already been identified in the bacterial membrane. A region of
E. coli curli has also been proposed to bind fibronectin specifically [146]. Binding
strength between bacteria (*S. aureus*) and immobilised fibronectin has been characterised by several authors [154, 155, 160]. In particular, Xu *et al.* [154] studied
the role of fibronectin-specific surface proteins of *S. aureus* in the bacteria adhesion to fibronectin-coated surfaces and characterised the strength of the binding.
Due to the essential role of the RGD (arginine-glycine-aspartic acid) fibronectin
site in eukaryotic cell adhesion to the extracellular matrix, immobilised RGD peptides are more and more used for coating biomaterials. Therefore, their properties
regarding bacterial adhesion have been studied. Several studies were conducted on
various bacteria species (*S. epidermidis*, *S. aureus*, *S. mutans* and *P. aeruginosa*)
using adsorbed [103, 161] or grafted [162] RGD-modified polymers on titanium.
They all showed that the chemical modification of the substrate did not affect the
original adhesion properties of the substrate. In addition, Chua *et al.* [163] failed
to observe any bacteria adhesion differences irrespective of the RGD peptide density grafted on polyelectrolyte multilayers. However, due to the lack of studies using
well-organised surfaces with various densities of grafted RGD, the question remains
about the possibility that peptide conformation or its accessibility may influence
the binding to bacteria. Controlled immobilisation of larger domains should also
be considered, which would provide relevant information about the role of the biomolecule accessibility to bacteria. An interesting work of Jarvis and Bryers [159]
demonstrated the influence of the immobilised fibronectin orientation in the recognition by bacteria while Hull *et al.* [164] proposed that some bacteria species need
multiple adjacent fibronectin molecules, showing their N-terminal groups, for successful binding to fibronectin.

4.2.2.3. Bacteria/Lectin Interactions.   Studies addressing specific interactions between immobilised lectin and bacteria are rare. Bundy and Fenselau [165] captured
bacteria on Concanavalin A-coated surfaces. However, the work was done for further bacteria identification using MALDI-MS analysis and the lectin surfaces used
in this study were not exploited for a better understanding of lectin/bacteria interactions. Also, Koshi *et al.* [166] developed interesting lectin arrays, however these

were intended for carbohydrate detection and were never used for understanding of mechanisms.

4.2.2.4. Bacteria/DNA Interactions. Studies addressing specific binding between immobilised DNA and bacteria are rare as well. Despite the important development of DNA arrays, model surfaces are not commonly used for characterising interactions between DNA and bacteria. Yet, some evidences exist that bacterial surface structure is able to recognize DNA. In Section 4.1, we reported that type IV pili of *P. aeruginosa* are known to bind DNA [112]. Panja *et al.* [167] demonstrated also interactions between plasmid DNA and *E. coli* LPS. Besides, *E. coli* adhesion tests carried out in our group showed more expression of curli by bacteria adhered on DNA terminated SAMs in comparison with glass substrates, without any differences in the numbers of adherent bacteria [168]. However, mechanisms leading to such bacterial response have not been yet understood.

This short overview illustrates the large range of interactions potentially involved in bacterial adhesion. Interactions between bacteria and biomolecules may also vary considerably depending on the biomolecule domains exposed for the binding. Since biomolecules immobilized by grafting may expose other domains than soluble proteins [169] or proteins simply adsorbed onto a surface, the necessity of addressing the biomolecule/bacteria interactions using well-defined model surfaces is crucial.

4.3. Influence of the Growth and Environmental Conditions

Cell membrane properties and features are highly dependent on the conditions of bacterial growth. Growth conditions might induce changes both in the biochemical composition of the membrane and in its physical–chemical characteristics. These changes result from bacterial physiological state, nutrient access, or environmental physical–chemical factors. They are obviously able to impact the bacteria/material interface.

4.3.1. Effect of Growth Conditions on Cell Membrane Fatty Acid Composition

Fatty acids of bacterial membranes constitute the scaffold of the cell wall. They were shown to be modified by environmental factors. For example, changes were related to the growth temperature [170, 171], to the growth medium pH [172, 173] or to the carbon source [174]. Besides, correlation between the physiological state of bacteria or their growth phase and the fatty acid composition of their membrane was already reported [170].

4.3.2. Effect of Growth Conditions on Cell Membrane Physical–Chemical Properties

Depending on the membrane composition changes, physical and physical–chemical properties of cell wall might be modified. In general, surface charge, hydrophobicity and electron donor and acceptor characteristics of bacteria were shown by many authors to vary according to growth medium composition, temperature and pH [175, 176]. For example, Kannenberg and Carlson [177] reported that reduced oxygen concentration induced structural modifications in LPS which caused a switch from predominantly hydrophilic to predominantly hydrophobic molecular forms. Increased hydrophobicity of LPS was positively correlated with an increase in the surface hydrophobicity of the whole cell. Besides, elasticity of the bacterium was also shown to vary depending on the polarity of the solvent added to the bacterial suspension [178]. In the same study, AFM measurements suggested that the conformation of cell surface biopolymers can be affected by these different growth conditions.

4.3.3. Effect of Growth Conditions on Adhesion to Surface

Since bacterial cell wall is the location of bacteria/material interface, bacterial growth conditions are expected to impact the interactions between bacteria and surface. This was reported by many authors. Walker [179] compared growth kinetics on surfaces and surface characteristics of bacteria grown in nutrient-rich (Lysogeny broth) and -poor (basal salts medium) media. He concluded that attachment efficiency increased with the level of nutrient presence. He related this behaviour to modifications of bacteria membrane and bacteria size. Consistent with it, Haznedaroglu *et al.* [144] observed that starved *E. coli* adhered less on quartz sand than non-starved bacteria, however without affecting cell viability. Besides, Walker *et al.* [180] studied the impact of bacterial growth phase on the adhesion kinetics of *E. coli*. They observed that stationary-phase bacteria adhered to a higher extent to quartz surfaces than mid-exponential phase bacteria. They attributed this behaviour to the heterogeneous distribution of charged functional groups on the bacterial surface, more pronounced on stationary cells than on the mid-exponential ones.

4.3.4. Effect of Experimental Procedures

Briandet *et al.* [181] demonstrated that the procedure used for bacterial growth prior to adhesion study was likely to modify the results of adhesion assays. The study was focused on the storage conditions (4°C and −80°C) and on the pre-culture procedure done prior to adhesion assays. The results of adhesion assays varied with storage and cultivation procedure as well as with bacterial strain, allowing to conclude that rigorous and reproducible experimental procedures were necessary.

An important factor for changes in bacteria/material interactions is the growth medium ionic strength. Many authors have demonstrated its significant influence on the first adhesion step to inert surfaces. Burks *et al.* [136] reported that collision efficiency of three different *E. coli* strains with glass bead surfaces in column bioadhesion tests was increased with increasing ionic strength. Walker [179] observed also that adhesion efficiency increased with ionic strength whatever the growth medium. Since material and bacteria surfaces were negatively charged (see Section 4.1), this behaviour was attributed to a decrease of the electrostatic double-layer repulsion. Using AFM force measurements in different ionic strength and pH conditions, Camesano and Logan [100] reported only a little effect of ionic strength on the measured interaction forces, while the effect of pH was much more pronounced. They attributed the lack of ionic strength effect to the use of organic salt buffers in comparison with other authors who used inorganic salts for their experiments. These contradictory results highlight again the difficulty to compare studies, due to the large variations in experimental conditions and procedures used.

5. Influence of Proteins Adsorbed on the Surface

Materials implanted into the body are immediately modified by the adsorption of proteins and other molecules from plasma or saliva, before cell and bacteria adhere. Consequently, considerable changes in the surface chemical properties occur: physical–chemical properties are modified; new chemical functions specific for bacterial adhesion may appear; and, on the contrary, specific functions initially grafted on the surface may be hidden. Therefore, adsorption of proteins may reduce or promote bacterial adhesion, depending on material, bacterial species and protein adsorbed. In this section, some examples of the studies showing the important role played by plasma and salivary biomolecules adsorption in bacteria adhesion onto surfaces is reported. Some details about the impact of individual proteins are given in Section 4.2.

Saliva, plasma and serum were reported by many authors to usually reduce bacteria adhesion. This was demonstrated by Lima *et al.* [182] for *S. mutans* on different materials (titanium, zirconia ceramic, hydroxyapatite) previously coated with saliva or/and serum and by Patel *et al.* [105] for *S. epidermidis* after 2 h of incubation on serum-coated surfaces. Carlen *et al.* [183] observed this trend with other species but reported marked reductions on plasma-coated hydroxyapatite than on saliva-coated ones. However, some other bacteria species like *Actinomyces naeslundii* were shown not to be affected by saliva or serum coating [182]. Moreover, the inhibition effect caused by these coatings seems to be restricted to the adhesion step of biofilm formation since the impact of the serum coating was opposite for incubation times up to 24 h [105]. Thick biofilms developed on serum-coated surfaces while only monolayers of bacteria were observed on non-coated materials.

Besides proteins, other biomolecules are also able to highly modify how bacteria sense and develop onto a surface. Biosurfactants produced by some bacteria species and known to play important roles in bacterial swarming and in biofilm 3D-architecture (see Section 2) were shown to also inhibit the adhesion of other species to stainless steel [184]. They were, therefore, proposed for treating metallic surfaces, aiming to prevent adhesion of pathogenic bacteria.

These few examples highlight (i) the high importance of considering the adsorbed protein layers for well-characterising and well-understanding the response of bacteria to surface properties, (ii) the necessity of studying the impact of adsorbed layers on bacterial adhesion and on biofilm formation since these two effects may be opposite, and (iii) the impossibility of generalizing the already shown adsorbed-layer effects for all bacterial species.

6. Competition with the Eucaryotic Cells ("the Race for the Surface")

When biomaterials come in contact with biological media, proteins and other molecules from the environment adsorb first at the surface. Then, eukaryotic cells and bacteria are transported through sedimentation and Brownian movements and compete for the surface, responding to surface properties with their own features (see the article of K. Anselme *et al.* in this same issue for specification about eukaryotic cell/material surface interface). The expression "the race for the surface" was introduced by Gristina *et al.* [9] for describing this competition. Although its role in a biomaterial implant is crucial due to the risk to develop infections when bacteria win, this competition was until recently ignored. For example, He *et al.* [155] compared the adhesion of fibroblast cells and *E. coli* and *S. aureus* on RGD-modified dopamine-grafted glass coverslips, showing the possibility to develop surfaces favourable to cell adhesion but neutral for bacterial adhesion. Shi *et al.* [162] analysed mouse osteoblast cell and *S. epidermidis* and *S. aureus* bacterial responses to RGD-grafted surfaces, using in addition antibacterial chitosan as the substrate. This work demonstrated the possibility of improving cell adhesion and preventing bacterial adhesion simultaneously. Recently we published a study aiming to compare cell and bacterial behaviours in response to the same topographically and chemically patterned model surfaces [73]. The opposite responses of bacteria and cells to surface chemistry and topography demonstrated the possibility to create cell-selective surfaces, not only by grafting active molecules but also by playing with the innate material surface characteristics.

Studies previously mentioned addressed competition between bacteria and cells in parallel. Bacteria and cells were inoculated on the same substrates in separate cultures. Recently, Subbiahdoss *et al.* [185] developed a valuable experimental methodology for more directly studying the competition. They first developed a modified culture medium by combining classical bacterial and cell culture media, aiming to optimised the growth of both cells (U2OS osteosarcoma-derived cells) and bacteria (*S. epidermidis*) in the same medium. Adhesion assays were performed in static and dynamic conditions in two steps, first the bacterial adhesion and then the adhesion of cells. Their results confirmed the impact of previous bacterial colonisation on further adhesion of osteoblastic cells. Both number and area of spreading cells were clearly affected by the presence of bacteria. Interestingly, spreading and survival rate of cells in presence of bacteria coating were greater under flow compared to under static conditions. This behaviour was attributed to flow-induced removal of the majority of bacterial endotoxins produced, which was further confirmed experimentally. We, however, note that the cellular model used in this work is derived from bone tumor and notably has a growth capacity much higher than that of healthy cells. We think that this work should be the beginning of new experimental approaches for *in vitro* evaluation of bi-functionalized biomaterial coatings efficiency using pertinent cell models like pathological bacteria strains *versus* normal cells.

7. Conclusion

The overview presented here demonstrates that a large range of factors influence bacteria/material interface. Aside from the necessity to take all these factors into account when studying bacterial adhesion and further biofilm development, these factors constitute potential targets for capturing bacteria or controlling biofilm formation. Antiadhesive surfaces obtained by grafting of molecules like PEG have been already exploited for preventing bacteria adhesion. Specific bacterial cell wall receptors have also been targeted with relevant biomolecules grafted onto surfaces for capturing bacteria in carbohydrate arrays. However, controlling bacterial adhesion and further biofilm development remains a challenge due especially to the complexity of the phenomena possibly occurring simultaneously. We believe, therefore, that:

(i) Studies should now focus on well-defined systems using well-controlled immobilisation of molecules and thorough characterisation of the surface. In this frame, the materials and surface science will play essential roles by providing new grafting methods and surface characterisation techniques (see the article by Ponche *et al.* in this same issue).

(ii) Analysis of bacteria adherent to surfaces should allow not only specific quantification but also localization of bacteria. This can be achieved obviously by confocal microscopy and thorough statistical analysis. Nevertheless, we are also convinced that the standardisation of the observation and analysis procedures used in the fundamental studies would allow significant improvements in the fundamental knowledge by allowing relevant comparison between the results obtained by the large number of groups working on this subject all around the world.

References

1. J. W. Costerton, P. S. Stewart and E. P. Greenberg, *Science* **284**, 1318 (1999).
2. G. G. Anderson and G. A. O'Toole, *Current Topics in Microbiology and Immunology* **322**, 85 (2008).
3. W. M. Dunne, *Clinical Microbiology Reviews* **15**, 155 (2002).
4. R. M. Donlan and J. W. Costerton, *Clinical Microbiology Reviews* **15**, 167 (2002).
5. N. B. Hallam, J. R. West, C. F. Forster and J. Simms, *Water Res.* **35**, 4063 (2001).
6. L. V. Poulsen, *Lebensm.-Wiss. u.-Technol.* **32**, 321 (1999).
7. P. A. Zaini, L. De La Fuente, H. C. Hoch and T. J. Burr, *FEMS Microbiol. Lett.* **295**, 129 (2009).
8. M. McCann, B. Gilmore and S. Gorman, *J. Pharm. Pharmacol.* **60**, 1551 (2008).
9. A. Gristina, P. Naylor and Q. Myrvik, *Med. Prog. Technol.* **14**, 205 (1989).
10. G. Chen and K. A. Strevett, *J. Colloid Interface Sci.* **261**, 283 (2003).
11. M. Hermansson, *Colloids Surfaces, B* **14**, 105 (1999).
12. P. Becker, W. Hufnagle, G. Peters and M. Herrmann, *Appl. Environ. Microbiol.* **67**, 2958 (2001).
13. C. Prigent-Combaret, E. Brombacher, O. Vidal, A. Ambert, P. Lejeune, P. Landini and C. Dorel, *J. Bacteriol.* **183**, 7213 (2001).

14. D. Seyer, P. Cosette, A. Siroy, E. Dé, C. Lenz, H. Vaudry, L. Coquet and T. Jouenne, *Biofilms* **2**, 27 (2005).
15. C. J. P. Boonaert, Y. F. Dufrêne and P. G. Rouxhet, in: *Encyclopedia Environmental Microbiology*, G. Bitton (Ed.), pp. 113–132. Wiley, New York (2002).
16. L. A. Pratt and R. Kolter, *Mol. Microbiol.* **30**, 285 (1998).
17. C. Prigent-Combaret, G. Prensier, T. T. Le Thi, O. Vidal, P. Lejeune and C. Dorel, *Environm. Microbiol.* **2**, 450 (2000).
18. J. Israelchvili, *Intermolecular and Surfaces Forces.* Academic Press, London (1991).
19. C. J. van Oss, *Colloids Surfaces, B* **54**, 2 (2007).
20. Y. L. Ong, A. Razatos, G. Georgiou and M. M. Sharma, *Langmuir* **15**, 2719 (1999).
21. N. I. Abu-Lail and T. A. Camesano, *Environm. Sci. Technol.* **37**, 2173 (2003).
22. J. Sjollema, H. C. Van der Mei, H. M. W. Uyen and H. J. Busscher, *J. Adhesion Sci. Technol.* **4**, 765 (1990).
23. J. M. Meinders, H. C. van der Mei and H. J. Busscher, *J. Colloid Interface Sci.* **176**, 329 (1995).
24. P. Lejeune, *Trends Microbiol.* **11**, 179 (2003).
25. C. Prigent-Combaret, O. Vidal, C. Dorel and P. Lejeune, *J. Bacteriol.* **181**, 5993 (1999).
26. T. Proft and E. Baker, *Cellular and Molecular Life Sciences (CMLS)* **66**, 613 (2009).
27. D. Kaiser, *Current Biology* **17**, R561 (2007).
28. R. M. Harshey, *Annual Review in Microbiology* **57**, 249 (2003).
29. R. Daniels, J. Vanderleyden and J. Michiels, *FEMS Microbiology Reviews* **28**, 261 (2004).
30. S. J. Pamp and T. Tolker-Nielsen, *J. Bacteriol.* **189**, 2531 (2007).
31. V. Leclère, R. Marti, M. Béchet, P. Fickers and P. Jacques, *Arch. Microbiol.* **186**, 475 (2006).
32. C. Niu, J. D. Graves, F. O. Mokuolu, S. E. Gilbert and E. S. Gilbert, *J. Microbiological Methods* **62**, 129 (2005).
33. A. Toguchi, M. Siano, M. Burkart and R. M. Harshey, *J. Bacteriol.* **182**, 6308 (2000).
34. I. H. Pratt-Terpstra, A. H. Weerkamp and H. J. Busscher, *Curr. Microbiol.* **16**, 311 (1988).
35. S. S. Branda, A. Vik, L. Friedman and R. Kolter, *Trends Microbiol.* **13**, 20 (2005).
36. C. Ryder, M. Byrd and D. J. Wozniak, *Current Opinion in Microbiology* **10**, 644 (2007).
37. G. Stevenson, K. Andrianopoulos, M. Hobbs and P. Reeves, *J. Bacteriol.* **178**, 4885 (1996).
38. I. W. Sutherland, *Trends Microbiol.* **9**, 222 (2001).
39. P. N. Danese, L. A. Pratt and R. Kolter, *J. Bacteriol.* **182**, 3593 (2000).
40. M. R. Parsek and T. Tolker-Nielsen, *Current Opinion in Microbiology* **11**, 560 (2008).
41. J. Kives, B. Orgaz and C. SanJosé, *Colloids and Surfaces B: Biointerfaces, BioMicroWorld 2005 — A Collection of Papers Presented at the 1st International Conference on Environmental, Industrial and Applied Microbiology* **52**, 123 (2006).
42. J. W. Costerton, Z. Lewandowski, D. E. Caldwell, D. R. Korber and H. M. Lappin-Scott, *Annual Review in Microbiology* **49**, 711 (1995).
43. B. L. Bassler, *Current Opinion in Microbiology* **2**, 582 (1999).
44. S. C. Winans, *Trends Microbiol.* **6**, 382 (1998).
45. M. R. Parsek and E. P. Greenberg, *Trends Microbiol.* **13**, 27 (2005).
46. K. Sauer, M. C. Cullen, A. H. Rickard, L. A. H. Zeef, D. G. Davies and P. Gilbert, *J. Bacteriol.* **186**, 7312 (2004).
47. L. Ma, M. Conover, H. Lu, M. R. Parsek, K. Bayles and D. J. Wozniak, *PLOS Pathogens* **5**, 1 (2009).
48. S. Wuertz, S. Okabe and M. Hausner, *Water Sci. Technol.* **49**, 327 (2004).
49. P. Stoodley, K. Sauer, D. G. Davies and J. W. Costerton, *Annu. Rev. Microbiol.* **56**, 187 (2002).
50. W. G. Characklis, *Biotechnol. Bioeng.* **23**, 1923 (1981).

51. K. A. Whitehead and J. Verran, *International Biodeterioration & Biodegradation* **60**, 74 (2007).
52. S. H. Flint, J. D. Brooks and P. J. Bremer, *J. Food Eng* **43**, 235 (2000).
53. K. J. Edwards and A. D. Rutenberg, *Chem. Geol.* **180**, 19 (2001).
54. J. Verran and C. J. Maryan, *J. Prosthetic Dentistry* **77**, 535 (1997).
55. T. R. Scheuerman, A. K. Camper and M. A. Hamilton, *J. Colloid Interface Sci.* **208**, 23 (1998).
56. K. Anselme and M. Bigerelle, *Biomaterials* **27**, 1187 (2006).
57. K. A. Whitehead, J. Colligon and J. Verran, *Colloids Surfaces, B* **41**, 129 (2005).
58. S. E. Tebbs, A. Sawyer and T. S. J. Elliott, *British J. Anaesthesia* **72**, 587 (1994).
59. M.-N. Leclercq-Perlat and M. Lalande, *J. Food Eng* **23**, 501 (1994).
60. Y. H. An and R. J. Friedman, *J. Biomed. Mater. Res.* **43**, 338 (1998).
61. L. M. Barnes, M. F. Lo, M. R. Adams and A. H. L. Chamberlain, *Appl. Environ. Microbiol.* **65**, 4543 (1999).
62. N. Hosoya, K. Honda, F. Iino and T. Arai, *J. Dentistry* **31**, 543 (2003).
63. J. Verran, D. L. Rowe and R. D. Boyd, *J. Food Protection* **64**, 1183 (2001).
64. C. Faille, J.-M. Membre, J.-P. Tissier, M.-N. Bellon-Fontaine, B. Carpentier, M.-A. Laroche and T. Benezech, *Biofouling* **15**, 261 (2000).
65. L. Boulange-Petermann, J. Rault and M. N. Bellon-Fontaine, *Biofouling* **11**, 201 (1997).
66. L. R. Hilbert, D. Bagge-Ravn, J. Kold and L. Gram, *International Biodeterioration & Biodegradation* **52**, 175 (2003).
67. E. Medilanski, K. Kaufmann, L. Y. Wick, O. Wanner and H. Harms, *Biofouling* **18**, 193 (2002).
68. R. L. Taylor, J. Verran, G. C. Lees and A. J. P. Ward, *J. Mater. Sci.: Mater. Med.* **9**, 17 (1998).
69. K. A. Whitehead and J. Verran, *Trans IChemE, Part C, Food and Bioproducts Processing* **84**, 253 (2006).
70. D. P. Bakker, H. J. Busscher, J. van Zanten, J. de Vries, J. W. Klijnstra and H. C. van der Mei, *Microbiology* **150**, 1779 (2004).
71. A. Houry, R. Briandet, S. Aymerich and M. Gohar, *Microbiology*, doi:10.1099/mic.0.034827 (2009).
72. G. Colon, B. C. Ward and T. J. Webster, *J. Biomed. Mater. Res. A* **78A**, 595 (2006).
73. L. Ploux, K. Anselme, A. Dirani, A. Ponche, O. Soppera and V. Roucoules, *Langmuir* **25**, 8161 (2009).
74. G. M. Bruinsma, M. Rustema-Abbing, J. de Vries, B. Stegenga, H. C. van der Mei, M. L. van der Linden, J. M. M. Hooymans and H. J. Busscher, *Investigative Ophthalmology and Visual Science* **43**, 3646 (2002).
75. D. Campoccia, L. Montanaro, H. Agheli, D. S. Sutherland, V. Pirini, M. E. Donati and C. R. Arciola, *Intl J. Artificial Organs* **29**, 622 (2006).
76. C. Diaz, P. L. Schilardi, R. C. Salvarezza and M. Fernandez Lorenzo de Mele, *Langmuir* **23**, 11206 (2007).
77. S. D. Puckett, E. Taylor, T. Raimondo and T. J. Webster, *Biomaterials* **31**, 706 (2010).
78. K. M. Wiencek and M. Fletcher, *Biofouling* **11**, 293 (1997).
79. N. Mitik-Dineva, J. Wang, V. K. Truong, P. R. Stoddart, F. Malherbe, R. J. Crawford and E. P. Ivanova, *Biofouling* **25**, 621 (2009).
80. K. M. Wiencek and M. Fletcher, *J. Bacteriol.* **177**, 1959 (1995).
81. V. A. Tegoulia and S. L. Cooper, *Colloids Surfaces, B* **24**, 217 (2002).
82. N. Cerca, G. B. Pier, M. Vilanova, R. Oliveira and J. Azeredo, *Res. Microbiol.* **156**, 506 (2005).
83. M. C. van Loosdrecht, J. Lyklema, W. Norde, G. Schraa and A. J. Zehnder, *Appl. Environ. Microbiol.* **53**, 1893 (1987).

84. N. P. Boks, H. J. Kaper, W. Norde, H. C. van der Mei and H. J. Busscher, *J. Colloid Interface Sci.* **331**, 60 (2009).
85. R. Bos, H. C. van der Mei, J. Gold and H. J. Busscher, *FEMS Microbiol. Lett.* **189**, 311 (2000).
86. A. M. Gallardo-Moreno, M. L. Gonzalez-Martin, J. M. Bruque and C. Perez-Giraldo, *Colloids Surfaces, A* **249**, 99 (2004).
87. J. Tsibouklis, M. Stone, A. A. Thorpe, P. Graham, V. Peters, R. Heerlien, J. R. Smith, K. L. Green and T. G. Nevell, *Biomaterials* **20**, 1229 (1999).
88. J. Genzer and K. Efimenko, *Biofouling* **22**, 339 (2006).
89. G. Speranza, G. Gottardi, C. Pederzolli, L. Lunelli, R. Canteri, L. Pasquardini, E. Carli, A. Lui, D. Maniglio, M. Brugnara and M. Anderle, *Biomaterials* **25**, 2029 (2004).
90. A. Subramani, X. Huang and E. M. V. Hoek, *J. Colloid Interface Sci.* **336**, 13 (2009).
91. V. Vadillo-Rodriguez, H. J. Busscher and H. C. van der Mei, in: *Biofilms in Medicine, Industry and Environmental Biotechnology*, P. Lens, A. P. Moran, T. Mahony, P. Stoodley and V. O'Flaherty (Eds), pp. 6–15. IWA Publishing, London (2003).
92. J. Li and L. A. McLandsborough, *Intl J. Food Microbiol.* **53**, 185 (1999).
93. J. A. Roberts, *Chem. Geol.* **212**, 313 (2004).
94. A. Terada, A. Yuasa, S. Tsuneda, A. Hirata, A. Katakai and M. Tamada, *Colloids Surfaces, B* **43**, 99 (2005).
95. B. Li and B. E. Logan, *Colloids Surfaces, B* **36**, 81 (2004).
96. B. Gottenbos, D. W. Grijpma, H. C. van der Mei, J. Feijen and H. J. Busscher, *J. Antimicrob. Chemother.* **48**, 7 (2001).
97. A. Komaromy, R. I. Boysen, H. Zhang, I. McKinnon, F. Fulga, M. T. W. Hearn and D. V. Nicolau, *Microelectron. Eng* **86**, 1431 (2010).
98. G. M. Bruinsma, M. Rustema-Abbing, J. de Vries, H. J. Busscher, M. L. van der Linden, J. M. M. Hooymans and H. C. van der Mei, *Biomaterials* **24**, 1663 (2003).
99. F. Gaboriaud, E. Dague, S. Bailet, F. Jorand, J. Duval and F. Thomas, *Colloids Surfaces, B* **52**, 108 (2006).
100. T. A. Camesano and B. E. Logan, *Environm. Sci. Technol.* **34**, 3354 (2000).
101. P. Krsko, J. B. Kaplan and M. Libera, *Acta Biomaterialia* **5**, 589 (2009).
102. M. Katsikogianni and Y. F. Missirlis, *European Cells and Materials* **8**, 37 (2004).
103. L. G. Harris, S. Tosatti, M. Wieland, M. Textor and R. G. Richards, *Biomaterials* **25**, 4135 (2004).
104. L. Ploux, S. Beckendorff, M. Nardin and S. Neunlist, *Colloids Surfaces, B* **57**, 174 (2007).
105. J. D. Patel, M. Ebert, R. Ward and J. M. Anderson, *J. Biomed. Mater. Res.* **80A**, 742 (2007).
106. B. Gottenbos, H. C. van der Mei and H. J. Busscher, *J. Biomed. Mater. Res.* **50**, 208 (2000).
107. F. Feugeas, G. Ehret and A. Cornet, *J. Trace Microprobe Techniques* **19**, 375 (2001).
108. T. J. Wiles, R. R. Kulesus and M. A. Mulvey, *Experimental and Molecular Pathology Special Issue: Molecular Pathology and Molecular Diagnostics* **85**, 11 (2008).
109. K. Otto, J. Norbeck, T. Larsson, K.-A. Karlsson and M. Hermansson, *J. Bacteriol.* **183**, 2445 (2001).
110. K. Otto and M. Hermansson, *J. Bacteriol.* **186**, 226 (2004).
111. L. Hansson, P. Wallbrandt, J.-O. Andersson, M. Bystrom, A. Backman, A. Carlstein, K. Enquist, H. Lonn, C. Otter and M. Stromqvist, *Biochimica Biophysica Acta (BBA) — General Subjects* **1244**, 377 (1995).
112. E. J. van Schaik, C. L. Giltner, G. F. Audette, D. W. Keizer, D. L. Bautista, C. M. Slupsky, B. D. Sykes and R. T. Irvin, *J. Bacteriol.* **187**, 1455 (2005).
113. B. Maier, L. Potter, M. So, H. S. Seifert and M. P. Sheetz, *Proc. Natl Acad. Sci. USA* **99**, 16012 (2002).

114. D. A. Coil and J. Anné, *FEMS Microbiol. Lett.* **293**, 271 (2009).
115. I. Alarcon, D. J. Evans and M. J. Fleiszig, *Investigative Ophthalmology and Visual Science* **50**, 2237 (2009).
116. J. S. Mattick, *Annual Review in Microbiology* **56**, 289 (2002).
117. T. Kikuchi, Y. Mizunoe, A. Takade, S. Naito and S.-i. Yoshida, *Microbiol. Immunol.* **49**, 875 (2005).
118. O. Vidal, R. Longin, C. Prigent-Combaret, C. Dorel, M. Hooreman and P. Lejeune, *J. Bacteriol.* **180**, 2442 (1998).
119. D. M. Pawar, M. L. Rossman and J. Chen, *J. Appl. Microbiol.* **99**, 418 (2005).
120. J. W. Austin, G. Sanders, W. W. Kay and S. K. Collinson, *FEMS Microbiol. Lett.* **162**, 295 (1998).
121. S. K. Collinson, P. C. Doig, J. L. Doran, S. Clouthier, T. J. Trust and W. W. Kay, *J. Bacteriol.* **175**, 12 (1993).
122. A. Olsen, A. Jonsson and S. Normark, *Nature* **338**, 652 (1989).
123. U. Gophna, T. A. Oelschlaeger, J. Hacker and E. Z. Ron, *FEMS Microbiol. Lett.* **212**, 55 (2002).
124. H. Berg and R. A. Anderson, *Nature* **245**, 380 (1973).
125. N. Verstraeten, K. Braeken, B. Debkumari, M. Fauvart, J. Fransaer, J. Vermant and J. Michiels, *Trends Microbiol.* **16**, 496 (2008).
126. G. A. O'Toole and R. Kolter, *Mol. Microbiol.* **30**, 295 (1998).
127. S. M. Kirov, *FEMS Microbiol. Lett.* **224**, 151 (2003).
128. K. Sauer, A. K. Camper, G. D. Ehrlich, J. W. Costerton and D. G. Davies, *J. Bacteriol.* **184**, 1140 (2002).
129. K. Roy, G. M. Hilliard, D. J. Hamilton, J. Luo, M. M. Ostmann and J. M. Fleckenstein, *Nature* **457**, 594 (2009).
130. R. M. La Ragione, W. A. Cooley and M. J. Woodward, *J. Med. Microbiol.* **49**, 327 (2000).
131. A. Imberty, M. Wimmerova, C. Sabin and E. P. Mitchell, in: *Protein-Carbohydrate Interactions in Infections Disease*, C. Hewley (Ed.), pp. 30–48. Royal Society of Chemistry, Cambridge (2006).
132. A. Reisner, J. A. Haagensen, M. A. Schembri, E. L. Zechner and S. Molin, *Mol. Microbiol.* **48**, 933 (2003).
133. I. Ofek, D. L. Hasty and R. J. Doyle, *Bacterial Adhesion to Animal Cells and Tissues*. ASM Press, Washington, DC (2003).
134. I. Ofek, D. L. Hasty, S. N. Abraham and N. Sharon, *Adv. Exp. Med. Biol.* **485**, 183 (2000).
135. R. Van Houdt and C. W. Michiels, *Res. Microbiol.* **156**, 626 (2005).
136. G. A. Burks, S. B. Velegol, E. Paramonova, B. E. Lindenmuth, J. D. Feick and B. E. Logan, *Langmuir* **19**, 2366 (2003).
137. S. L. Walker, J. A. Redman and M. Elimelech, *Langmuir* **20**, 7736 (2004).
138. S. Stoitsova, R. Ivanova and I. Dimova, *J. Basic Microbiol.* **44**, 296 (2004).
139. E. Quintero and R. M. Weiner, *Appl. Environ. Microbiol.* **61**, 1897 (1995).
140. A. Hanna, M. Berg, V. Stout and A. Razatos, *Appl. Environ. Microbiol.* **69**, 4474 (2003).
141. H. J. Busscher, M. M. Cowan and H. C. van der Mei, *FEMS Microbiol. Lett.* **88**, 199 (1992).
142. H. J. Busscher and A. H. Weerkamp, *FEMS Microbiology Reviews* **46**, 165 (1987).
143. C. A. Flemming, R. J. Palmer Jr, A. A. Arrage, H. C. van der Mei and D. C. White, *Biofouling* **13**, 213 (1998).
144. B. Z. Haznedaroglu, C. H. Bolster and S. L. Walker, *Water Res.* **42**, 1547 (2008).
145. A. Imberty, E. P. Mitchell and M. Wimmerova, *Curr. Opin. Struct. Biol.* **15**, 525 (2005).
146. A. Olsen, H. Herwald, M. Wikström, K. Persson, E. Mattson and L. Björck, *J. Biological Chem.* **277**, 34568 (2002).

147. M. D. Disney and P. H. Seeberger, *Chemistry & Biology* **11**, 1701 (2004).
148. E. W. Adams, D. M. Ratner, H. R. Bokesch, J. B. McMahon, B. R. O'Keefe and P. H. Seeberger, *Chemistry & Biology* **11**, 875 (2004).
149. X. Qian, S. J. Metallo, I. S. Choi, H. Wu, M. N. Liang and G. M. Whitesides, *Anal. Chem.* **74**, 1805 (2002).
150. M. N. Liang, S. P. Smith, S. J. Metallo, I. S. Choi, M. Prentiss and G. M. Whitesides, *Proc. Natl Acad. Sci. USA* **97**, 13092 (2000).
151. K. A. Barth, G. Coullerez, L. M. Nilsson, R. Castelli, P. H. Seeberger, V. Vogel and M. Textor, *Adv. Functional Mater.* **18**, 1459 (2008).
152. G. W. Charville, E. M. Hetrick, C. B. Geer and M. H. Schoenfisch, *Biomaterials* **29**, 4039 (2008).
153. Y. Liu, J. Strauss and T. A. Camesano, *Biomaterials* **29**, 4374 (2008).
154. C.-P. Xu, N. P. Boks, J. de Vries, H. J. Kaper, W. Norde, H. J. Busscher and H. C. van der Mei, *Appl. Environ. Microbiol.* **74**, 7522 (2008).
155. T. He, Z. L. Shi, N. Fang, K. G. Neoh, E. T. Kang and V. Chan, *Biomaterials* **30**, 317 (2009).
156. C. Tedjo, K. G. Neoh, E. T. Kang, N. Fang and V. Chan, *J. Biomed. Mater. Res.* **82A**, 479 (2007).
157. H. J. Busscher, B. van de Belt-Gritter, R. J. B. Dijkstra, W. Norde, F. C. Petersen, A. A. Scheie and H. C. van der Mei, *J. Bacteriol.* **189**, 2988 (2007).
158. A. Olsen, M. J. Wick, M. Morgelin and L. Bjorck, *Infect. Immun.* **66**, 944 (1998).
159. R. A. Jarvis and J. D. Bryers, *J. Biomed. Mater. Res.* **75A**, 41 (2005).
160. K. H. Simpson, M. G. Bowden, S. J. Peacock, M. Arya, M. Höök and B. Anvari, *Biomolecular Eng* **21**, 105 (2004).
161. R. R. Maddikeri, S. Tosatti, M. Schuler, S. Chessari, M. Textor, R. G. Richards and L. G. Harris, *J. Biomed. Mater. Res.* **84A**, 425 (2008).
162. Z. Shi, G. Neoh, E. T. Kang, C. Poh and W. Wang, *J. Biomed. Mater. Res.* **86A**, 865 (2008).
163. P.-H. Chua, K.-G. Neoh, E.-T. Kang and W. Wang, *Biomaterials* **29**, 1412 (2008).
164. J. R. Hull, G. S. Tamura and D. G. Castner, *Acta Biomaterialia* **4**, 504 (2008).
165. J. Bundy and C. Fenselau, *Anal. Chem.* **71**, 1460 (1999).
166. Y. Koshi, E. Nakata, H. Yamane and I. Hamachi, *J. Am. Chem. Soc.* **128**, 10413 (2006).
167. S. Panja, P. Aich, B. Jana and T. Basu, *Biomacromolecules* **9**, 2501 (2008).
168. N. Cottenye, F. Teixeira Jr, A. Ponche, Reiter, K. Anselme, W. Meier, L. Ploux and C. Vebert-Nardin, *Macromolecular Bioscience* **8**, 1161 (2008).
169. K. M. Holgers and A. Ljungh, *Biomaterials* **20**, 1319 (1999).
170. F. Dubois-Brissonnet, C. Malgrange, L. Guérin-Méchin, B. Heyd and J. Y. Leveau, *Intl J. Food Microbiol.* **55**, 79 (2000).
171. A. Álvarez-Ordóñez, A. Fernández, M. López, R. Arenas and A. Bernardo, *Intl J. Food Microbiol.* **123**, 212 (2008).
172. E. M. Fozo, J. K. Kajfasz and Quivey, *FEMS Microbiol. Lett.* **238**, 291 (2004).
173. L. Hua, Z. WenYing, W. Hua, L. ZhongChao and W. AiLian, *Food and Bioproducts Processing* **87**, 56 (2009).
174. A. D. Syakti, N. Mazzella, F. Torre, M. Acquaviva, M. Gilewicz, M. Guiliano, J.-C. Bertrand and P. Doumenq, *Res. Microbiol.* **157**, 479 (2006).
175. R. Briandet, T. Meylheuc, C. Maher and M.-N. Bellon-Fontaine, *Appl. Environ. Microbiol.* **65**, 5328 (1999).
176. M. Naïtali, F. Dubois-Brissonnet, G. Cuvelier and M.-N. Bellon-Fontaine, *Intl J. Food Microbiol.* **130**, 101 (2009).
177. E. L. Kannenberg and R. W. Carlson, *Mol. Microbiol.* **39**, 379 (2001).
178. N. I. Abu-Lail and T. A. Camesano, *Colloids Surfaces, B* **51**, 62 (2001).

179. S. L. Walker, *Colloids Surfaces, B* **45**, 181 (2005).
180. S. L. Walker, J. E. Hill, J. A. Redman and M. Elimelech, *Appl. Environ. Microbiol.* **71**, 3093 (2005).
181. R. Briandet, V. Leriche, B. Carpentier and M.-N. Bellon-Fontaine, *J. Food Protection* **62**, 994 (1999).
182. E. M. C. X. Lima, H. Koo, A. M. V. Smith, P. L. Rosalen and A. A. Del Bel Cury, *Clinical Oral Implants Research* **19**, 780 (2008).
183. A. Carlen, S. G. Rüdiger, I. Loggner and J. Olsson, *Oral Microbiology Immunology* **18**, 203 (2003).
184. T. Meylheuc, C. Methivier, M. Renault, J.-M. Herry, C.-M. Pradier and M. N. Bellon-Fontaine, *Colloids Surfaces, B* **52**, 128 (2006).
185. G. Subbiahdoss, R. Kuijer, D. W. Grijpma, H. C. van der Mei and H. J. Busscher, *Acta Biomaterialia* **5**, 1399 (2009).

Role of Proteins and Water in the Initial Attachment of Mammalian Cells to Biomedical Surfaces: A Review

Purnendu Parhi [a], **Avantika Golas** [b] and **Erwin A. Vogler** [a,b,*]

[a] Department of Materials Science and Engineering, The Pennsylvania State University, University Park, PA 16802, USA
[b] Department of Bioengineering, The Pennsylvania State University, University Park, PA 16802, USA

Abstract
Anchorage-dependent mammalian cells are typically grown *in vitro* on hydrophilic glass and plastic substrata in a medium supplemented with 5–20% v/v blood-serum proteins. Inoculated single cells gravitate from suspension to within close proximity of substrata surfaces whereupon initial contact and attachment occurs followed by progressive cell adhesion, spreading, and ultimately proliferation. A critical examination of the role of proteins and water in the initial attachment phase concludes that the cell attachment phase is not mediated by biological recognition of surface-adsorbed ligands by cell membrane receptors as frequently depicted in various textbook explanations of cell adhesion. This conclusion is based on extensive experimental evidence showing that blood proteins do not adsorb on hydrophilic surfaces that are most conducive to cell growth but do adsorb on hydrophobic surfaces that are not conducive to cell growth. As a consequence, the conventional idea that initial cell attachment is mediated by various adhesin factors adsorbed from serum-protein solutions is viewed as untenable. Rather, it is concluded that the initial contact-and-attachment of cells to hydrophilic surfaces is controlled by physicochemical interactions unrelated to biological recognition. The general physics of these interactions is known but an adequate descriptive theory that can be tested against experimentally measured cell adhesion kinetics has yet to be developed. The role of these physicochemical interactions in stimulating biological machinery within cells to fully adhere and proliferate on surfaces of biotechnical interest is unknown but is of great significance to the science underlying various biomedical and biotechnical applications of materials.

Keywords
Water, bioadhesion, protein adsorption

1. Introduction

It is common knowledge that most, if not all, anchorage-dependent mammalian cells attach and adhere to, and subsequently proliferate on, hydrophilic (water-wettable surfaces) much more efficiently than on hydrophobic (less water wettable)

---

* To whom correspondence should be addressed. E-mail: EAV3@PSU.EDU

Surface and Interfacial Aspects of Cell Adhesion
© Koninklijke Brill NV, Leiden, 2010

counterparts [1, 2]. Indeed, commercial production of sterile disposable tissue cultureware is largely based on use of surface treatments of plastics used in this industry (typically polystyrene) [3] to render inherently hydrophobic surfaces more hydrophilic and thus useful as substrata for the culture of mammalian cells *in vitro*. Surface treatment technology arose from research of the early 1970's, stimulated by Rappaport's pioneering studies [4], among others. Soon after, a variety of surface synthesis strategies were explored, ranging from use of liquid-phase chemical oxidants [5, 6] to the application of gas-discharge treatments [7] that set the precedence for widespread application of modern gas discharge (plasma) technology in biomedicine [8] and biotechnology [9–11].

Although broad generalizations are always dangerous in the face of profound biological complexity (and a very broad literature base), it seems that dependence of mammalian cell attachment efficiency on substratum hydrophilicity (from serum-protein containing media) is a general phenomenon. Otherwise, surface treatment of commercial tissue culture grade polystyrene would be substantially unnecessary. We [12–15] and others (see, for example, [16–18]) have found that cell attachment efficiency pivots around a narrow range of water wettability measured in terms of basal medium (aqueous buffer) contact angles θ or buffer adhesion tension $\tau^\circ \equiv \gamma_{lv}^\circ \cos\theta$ (where γ_{lv}° is the liquid–vapor interfacial tension of basal medium) [3, 12, 19].

Attachment efficiency is here defined as the maximum, steady-state cell adhesion (measured as the percent of a cell inoculum attached, % I_{max}) estimated from the plateau of attachment kinetics curves like that illustrated in Fig. 1 [20]. Attachment efficiency to surfaces exhibiting $\theta < 65°$ (defining hydrophilic as used in [19, 21]) is thus found to be much greater than cell attachment to surfaces exhibiting $\theta > 65°$ (defining hydrophobic) [12, 22, 23], as illustrated in Fig. 2 for human fetal osteoblast (hFOB bone) cells [15, 16]. Similar attachment efficiency response to substratum surface energy has been measured for a variety of soft tissue cells as well (see Fig. 3, for example) [13, 14, 21]. Thus, attachment efficiency exhibits a pronounced hydrophobic/hydrophilic contrast that pivots around $\theta = 65°$, sharply increasing from low levels on hydrophobic surfaces to maximal values on hydrophilic surfaces characteristic of the cell type and fluid phase composition (e.g., serum content). A strikingly similar hydrophobic/hydrophilic contrast arises in a variety of biological responses to artificial materials — bacterial adhesion, blood coagulation, and protein adsorption — which we have correlated with the putative structure and reactivity of water at surfaces [12, 13].

Limited available information seems to suggest that maximal attachment efficiency cannot be significantly increased by simply increasing hydrophilicity substantially beyond the 'pivot point'. Rather, increasing substratum wettability to $\theta < 35°$ seems neither to measurably increase % I_{max} nor increase proliferative potential of adherent cells (see Fig. 2, for example) [12, 14, 24]. On the other hand, anchorage-dependent cells in contact with substratum more hydrophobic than the pivot point typically perish or persist in a rounded shape. If and when these

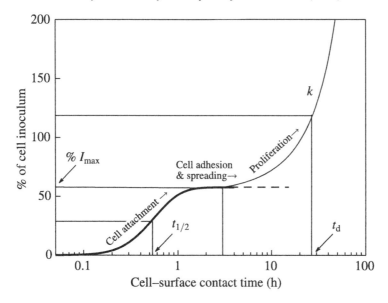

Figure 1. Schematic illustration of mammalian cell adhesion and proliferation kinetics identifying quantitative parameters that can be extracted from measurement of the number of cells attached (expressed here as percentage of viable cell inoculum; % I) with time. % I_{max} is the maximum percentage of a cell inoculum at steady state that adheres to a surface from a sessile cell suspension and $t_{1/2}$ measures half-time to % I_{max}. The proliferation rate (k) and cell number doubling time (t_d) are a measure of attached cell viability (adapted from [14, 15]). Sigmoidal attachment-kinetics is found to be typical of both hydrophilic and hydrophobic substrata materials, with % I_{max} generally higher for hydrophilic surfaces than hydrophobic surfaces [20]. % I_{max} and k are found to scale with substratum surface energy for certain hard- and soft-tissue cell types [12–15, 20] (see Fig. 2).

physically-compromised cells finally do spread on poorly-adhesive hydrophobic substratum, attached cells retain a spindle-shaped morphology for much longer than the same cells on hydrophilic surfaces, exhibiting few cell–substratum points of contact [14].

Interestingly, adhesion to hydrophobic surfaces can be forced by reducing suspending medium interfacial tension with low-concentrations of non-ionic surfactants in a manner that parallels manipulation of fluid-phase interfacial tension with added proteins [26, 27] (proteins exhibit 'biosurfactant' properties, reducing basal medium interfacial tension from $\gamma_{lv}^o \sim 72$ mN/m to $\gamma_{lv}' = 55$–60 mN/m at 5–20% v/v serum concentrations; where the "o" and prime superscripts differentiate basal and surfactant/protein containing media, respectively; see also [3]). It is further found that presence of serum proteins in suspending medium is not an absolute requirement for cell attachment [16, 25–31], although longer-term cell viability and proliferation is critically dependent on the presence of proteins.

The above observations draw attention to a critical role that water and proteins play in the initial phases of cell adhesion, at least as it occurs in the widespread application of commercial cultureware. The experimentally observed hydrophilic/hydrophobic contrast in cell attachment efficiency, ability to attach cells in

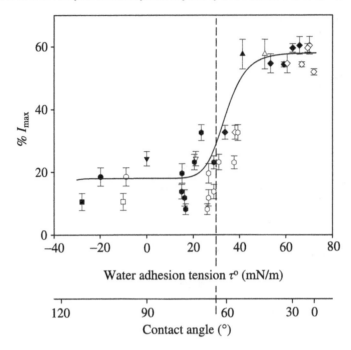

Figure 2. Correlation of % I_{max} (see Fig. 1) with substratum surface energy for human fetal osteoblast (hFOB) cells. Surface energy is measured by water adhesion tension $\tau^0 = \gamma_{lv}^0 \cos\theta$, where $\gamma_{lv}^0 = 72.8$ mN/m at 20°C for pure water and θ is the angle subtended by a droplet of basal medium (no protein) on the surface under study (advancing θ = filled symbols, receding θ = open symbols; adapted from [14, 15]). Error bar represents standard deviation ($N \geqslant 3$) determined for the same surface type. Trend-line through advancing and receding data is guide to the eye; ▲ = TCPS (tissue-culture grade polystyrene); ▼ = BGPS (bacteriological-grade polystyrene); ● = silanized glass; ■ = silanized quartz; ◆ = PTPS (plasma-treated polystyrene); ◕ = biodegradable polymers of PLGA 5/5 (M_n = 80 k), PLGA 7/3 (M_n = 96 k), PLA (M_n = 160 k), PCL (M_n = 80 k), PLCL 7/3 (M_n = 82 k), PLGCL 2.5/2.5/5 (M_n = 60 k), PLGCL 3.5/3.5/3 (M_n = 54 k). M_n = number-average molecular weight by GPC. PLGA = poly(lactide-*co*-glycolide); PLCL = poly(lactide-*co*-caprolactone); PLGCL = poly(lactide-*co*-glycolide-*co*-caprolactone). See [15] for details on materials preparation and characterization.

the absence of serum proteins, and ability to manipulate cell attachment efficiency using ordinary surfactants are all signatures of the effect of water wetting (interfacial tensions) on the cell adhesion process. The seemingly absolute requirement for proteins in maintaining cell viability after attachment to cell culture substrata emphasizes a critical biological role in the latter stages of bioadhesion but does not directly address the role in initial cell attachment.

The purpose of this paper is to critically review the role of water and proteins in the attachment of mammalian cells to cell culture labware that is frequently underemphasized or overlooked altogether in discussion of bioadhesion. The widely-held view that cell attachment is necessarily mediated by adsorbed proteins is skeptically examined, concluding that the interactions between cell and substratum that dom-

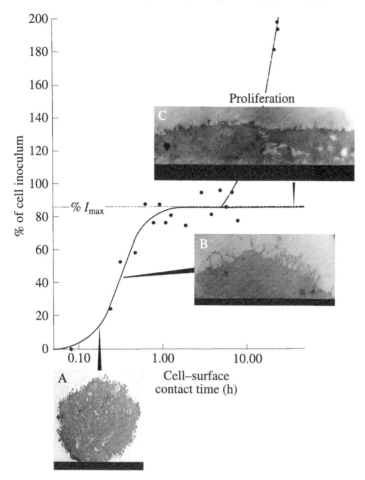

Figure 3. Attachment rate curve for Madin Darby Canine Kidney (MDCK, epitheliod) cells contact-ing, attaching and adhering to tissue culture grade polystyrene (TCPS) from 10% FBS-containing media (data taken from [20]). Notice the sigmoidal increase in attached cell number (expressed as % of inoculum, compare to Fig. 1). Insets are low-resolution cross-sectional views of osmium-stained attached cells at various times obtained by transmission electron microscopy (TEM, see [26] for exper-imental details; images have been processed to increase contrast and the planar TCPS substratum has been colored black to emphasize location relative to attached cells). Initial cell attachment (Panel A, MDCK in contact with TCPS for 20 min, 18 576× magnification) occurs through the agency of cellu-lar extensions (pseudopodia, see [172] and citations therein) at only a few points along portion of the cell membrane nearest the surface (compare to Fig. 4). These initially-attached cells resist removal by 3× buffer washes. Attached cells undergo progressive cell adhesion and spreading (Panel B, MDCK in contact with TCPS for 40 min, 18 500× magnification). Fully-spread cells (Panel C, MDCK in contact with TCPS for 24 h 18 500× magnification) deposit a thick extra-cellular matrix layer and proliferate over the entire unoccupied surface of the cell culture substratum.

inate the initial phase of cell attachment are physicochemical in nature [1, 3, 23] and do not involve biological recognition by cells of surface-adsorbed proteins as frequently depicted in the literature (see, for example, [32–34] and Fig. 4).

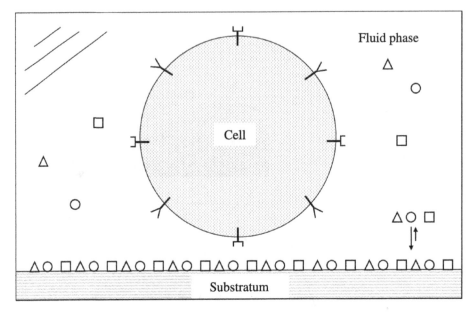

Figure 4. Highly-schematic diagram of cell attachment to a surface mediated by biological recognition (adapted from figures appearing in [33, 34, 40], compare to Figs 3 and 5). Ligands (such as fibronectin, vitronectin, and fibrinogen) adsorbed from serum-containing medium (represented by circles, squares, and triangles) are thought to undergo specific interactions with cell membrane receptors forming bonds that hold cells to the surface. Cells complete adhesion to the substratum by progressive flattening and production of extra-cellular matrix prior to proliferation (see Figs 1, 3).

2. Phases of Cell Adhesion and Proliferation

Adhesion of anchorage-dependent mammalian cells to a substratum surface has been historically viewed as occurring in four major sequential steps that precede proliferation: protein adsorption (or lack thereof), cell–substratum contact, cell–substratum attachment, and cell adhesion/spreading (Fig. 3) [1, 25, 26, 35]. Cells that have contacted the surface and formed an attachment (that can resist removal by buffer rinsing, for example) have a rounded morphology similar to cells remaining in suspension at the same cell–substratum contact time, with only a small portion of the cell membrane in apparent contact with the surface (see also Fig. 6(A)). Gradually, this contact area increases as cells flatten into the adherent state. Finally, cells spread on a layer of extracellular matrix (ECM) produced by adherent cells. Under the right conditions, certain kinds of cells can be grown into a multilayer tissue with cells enmeshed in ECM (see, for examples drawn from authors' laboratory, [14, 36, 37]).

Individual steps of this overall process appear to be linked together in a chain of cause-and-effect such that cell proliferation is found to be generally compromised on surfaces to which cells do not efficiently attach-and-adhere, as compared to proliferation on surfaces to which cells efficiently attach-and-adhere (see [15] for an example specific to osteoblast cells). Hence, cytocompatibility *in vitro* follows

David Williams' 'four components of biocompatibility' paradigm [39, 40] wherein acute molecular events transpiring at surfaces mediate subsequent events. It is for this reason that water and protein interactions with materials used in tissue culture are of both fundamental and practical interest.

Contact, attachment and proliferation are time-dependent processes best studied by measuring cell adhesion/proliferation kinetics. Attachment kinetics generally follows a sigmoidal trend like that shown in Fig. 1. These kinetic profiles quantify half-time to maximal attachment $t_{1/2}$ and doubling time t_d (see Fig. 1 annotations). Poor cell–substratum compatibility is marked by low maximal attachment % I_{max} and long $t_{1/2}$. Cell–substratum compatibility is frequently assessed using end-point assays (instead of kinetic profiles) in which determinants such as cell number, phenotype, and/or viability are measured after some arbitrary length of cell–substratum contact time. For interpretation of end-point assays, one should bear in mind that cell attachment, adhesion, and proliferation are inherently kinetic processes. Hence, an outcome measured at any particular time t is a composite of preceding steps, each with increasing biological complexity. As a consequence, counting the number of cells attached to a substratum after cells have begun to spread and divide is not a good measure of cell attachment efficiency because this total number also depends on adhesion/spreading and proliferation rates.

This article is specifically focused on the initial attachment step that ultimately moderates cell adhesion and proliferation through the aforementioned chain of cause-and-effect. The rudimentary question that comes to mind is: how do cells gravitating from suspension to within close proximity of a substratum surface form the *first points of physical attachment* to the surface that hold the cell in place until cells have time to progressively increase cell–substratum contact area (adhere). Specifically, we focus on the effect of substratum wettability (surface energy) on the efficiency of cell attachment and the relationship to adsorbed protein.

Each of the sequential steps of mammalian cell adhesion mentioned above can be further broken down to identify fundamental interactions that lead initially-suspended cells to gravitate to within 50 nm or so of a surface whereupon physical, biochemical, and biological forces apparently conspire to close the cell–surface distance gap [14]. Biologists and physical scientists emphasize different fundamental interactions. These interactions are discussed separately below in terms of biological, colloid science, and thermodynamic perspectives.

2.1. The Biological View

Among the many research papers and reviews articulating the cell biologist's perspective on cell adhesion, perhaps Grinnell's 1978 review [35] has been the most enduring. Content of this classic has become the conceptual outline of events involved in the adhesion of mammalian cells to surfaces now familiar to biologists, elaborated by research of the three decades that have followed Grinnell's review (see [14] and citations therein). Biologists quite naturally contemplate a cell in all its glorious biological complexity, envisioning cell attachment to substrata surfaces as occurring by the mating of various membrane-bound receptors to ligands adsorbed onto substrata surfaces in preceding protein adsorption events [33, 40] (see Fig. 4). Various signaling cascades are thereby potentiated that control gene expression. Up-regulated biochemical machinery is directed to the task of synthesizing and secreting extra-cellular matrix (ECM) proteins that promote firm adhesion of relatively loosely-attached cells. Physics possibly involved in the biological description of cell adhesion is largely ignored in this view.

2.1.1. Water, Protein Adsorption and the Biological View

That little or no mention of water is made in the biologist's paradigm might be easily understood — or even justified given the emphasis of purely cellular events — were it not for the proposed critical step of making the receptor-ligand bonds that constitute specific recognition. For this whole scheme to work, serum protein derived ligands (a.k.a. adhesins such as fibrinogen, fibronectin, and vitronectin) must adsorb from suspending aqueous medium onto the hydrophilic surfaces that are most conducive to cell adhesion. Apparently, however, proteins do not adsorb on hydrophilic surfaces from aqueous cell culture media (see Section 3), militating against the idea that biological recognition is solely responsible for the initial attachment of cells to surfaces. *This is not to say that specific recognition cannot or does not occur* under conditions purposely engineered to promote such specific recognition, such as surface-immobilized RGD (or similar) amino-acid sequences or various adhesins (see [41–44] as examples drawn from many). Rather, lack of protein adsorption on hydrophilic surfaces calls into question the generality of specific recognition as the primary driver of cell attachment from serum containing medium to hydrophilic tissue culture labware.

The general picture of protein adsorption emerging from data outlined in Section 3 has protein expelled from solution by what amounts to be the hydrophobic effect [12, 45] with a modest free energy gain per gram or mole of protein. Protein partitions into the near (vicinal) surface region of a solution-contacting adsorbent (a.k.a. interphase), necessarily displacing water within this interphase because two objects cannot occupy the same space at the same time. Displacement of interphase water has an energetic cost that depends on substratum surface energy (water wettability). For hydrophobic adsorbents, these energetic costs are less than the energy gain from expelling protein from solution. As adsorbent hydrophilicity increases, however, energy costs related to displacing interphase water increase until a balance occurs near the pivot point mentioned in Section 1. Free energy of adsorption decreases to zero and protein adsorption commensurately decreases to zero near the pivot point [46, 47].

Energetics of protein adsorption on hydrophilic surfaces does not mean that there is a significant depletion of protein within the interphase relative to bulk solution concentration. Rather, protein concentration within the interphase of hydrophilic surfaces is equal, or nearly equal, to the solution concentration. However, protein does not become anchored or bound to hydrophilic surfaces through protein–substratum interactions because displacement of interphase water is energetically prohibitive (unless the surface exhibits ion-exchange properties; see further Section 3). Thus, experimental evidence shows that blood proteins do not adsorb on commercial hydrophilic cell culture substrata that support efficient cell adhesion. Conversely, experimental evidence clearly shows that blood proteins do efficiently adsorb on hydrophobic surfaces that do not support efficient cell adhesion. A straightforward reconciliation of these lines of evidence is that initial stages of cell attachment to surface are not mediated/moderated by adsorbed protein ligands involved in biological recognition of the kind cartooned in Fig. 4. More on this topic is presented in Section 2.4.

2.2. The Colloid Science View

Colloid science views cell attachment through the lens of DLVO theory of colloid stability (see [1, 3] and citations therein). The living cell, so carefully described by the biologist, is reduced to a colloidal object (a particle). This particle has a net electric charge and Hamaker constant that control electrostatic and dispersion (van der Waals) interactions, respectively, between cell and substratum. Likewise, the substratum is treated as a planar 'collector' with a net electric charge and Hamaker constant. No doubt obscure to the biologist, and perhaps surface chemist as well, these simplifications seek justification on the basis that only average physical properties need to be considered in close approach to a surface — in the same sense that only mass of a human body needs to be considered in determining weight due to gravitational attraction to Earth, not the detailed structure of various tissues comprising the body.

DLVO theory further has it that competition between (typically) repulsive electrostatic and attractive dispersion interactions creates a net attractive potential between cells and substratum that brings suspended cells close enough to a surface whereupon physical adhesion can occur. At cell–substratum separations of 10 nm or so, it is argued that a tumbling cell is too distant from the substratum surface for registration between ligands and receptors or points-of-charge to occur, leading to a smeared-out average set of physical properties. This force-over-distance argument is silent about what occurs once close contacts are made and, to this extent, is non-competitive with the biological view (see Section 2.1).

2.2.1. Water, Protein Adsorption and the Colloid View

Simplified applications of DLVO theory to cell adhesion typically treat the suspending aqueous phase as a uniform dielectric medium lacking structure or graded properties throughout the system, up to the point of contact with the substratum. In this regard, water is not so much ignored as it is simplified into a pervasive ether through which interactions occur [48]. There are good reasons to be skeptical that water is uniform in properties at close proximity to a surface — to say nothing of patch-wise variations in particle and collector surface chemistry that could greatly complicate dielectric constant profiles near a real surface.

Indeed, evidence from decades of research suggests that structure and reactivity of water in close proximity to a surface (vicinal water) differs substantially from bulk solution properties [12, 13, 19]. This effect is thought to arise from the extensively hydrogen-bonded (self-associated) nature of water (molecular simulations suggest that approximately 75% of water molecules are transiently bonded to three or four nearest neighbors at room temperature [49]). Changes in hydrogen bonding caused by the imposed presence of a macroscopic surface propagate over some distance into bulk solution through this pervasive 'flickering' hydrogen-bond network. The extent of hydrogen bonding can be thereby locally increased or decreased relative to that occurring in bulk solution in a way that depends on the water-contacting surface chemistry [13, 19]. Significant unresolved questions are (i) how far into bulk solution this perturbation of self-association extends from a water contacting surface? (ii) How this perturbation depends on surface chemistry? and (iii) how vicinal water properties influence events such as adsorption and adhesion? [19]. But there seems little doubt that water is *definitively not* a continuum fluid without structure, uniform in chemistry and dielectric properties [50], as usually assumed in ordinary application of DLVO theory.

Assumptions of uniform dielectric properties are all the more dubious if protein adsorption is involved. Experiment shows that, at 5–20% v/v serum concentrations typically used in the culture of cells *in vitro* (~2.5–10 mg/ml), protein adsorption from cell culture media completely saturates hydrophobic surfaces, yielding interphase concentrations most probably of the order of 300 mg/ml for a substratum such as untreated, hydrophobic bacteriological-grade polystyrene [51]. This adsorbed protein apparently resides in a multilayer interphase [12, 51, 52]. Blood serum consists of about 1000 different proteins with concentrations varying over more than three orders of magnitude [53–55]. Among the proteins frequently identified as adhesion factors (adhesins), fibronectin (460 kDa) and vitronectin (65 and 75 kDa forms) are quite dilute in serum only about 300 μg/ml, or roughly 15–60 μg/ml in serum-supplemented media (a few hundred picomoles/ml at most). The most abundant adhesin, fibrinogen, at *plasma* concentrations ranging between 2–4 mg/ml is depleted by coagulation of plasma in the production of serum. Thus, serum is dominantly comprised (on both weight and molar concentration bases) of proteins other than adhesins (such as albumin and immunoglobulin). Clearly, the solution environment encountered by a cell in close approach to such an interphase must be very much different than experienced in solution away from the hydrophobic substratum. Perhaps this resembles the circumstance envisioned in Fig. 5, which is quite unlike the flat plane typically considered by DLVO theory and very different from diagrams like Fig. 4 that popularly appear in some rudimentary texts [32–34]. Accordingly, adhesins must be co-adsorbed with a plethora of other blood proteins. It is only reasonable to suspect that many (or all) of these adsorbed adhesins would be inaccessible to a cell near the surface. If the jumble of adsorbed proteins diagramed in Fig. 5 is relatively loosely bound, as suggested by interfacial rheology (see [56] and citations therein), then it seems reasonable to suggest that cells would have difficulty 'gripping' the substratum, by whatever means used by a cell to grip

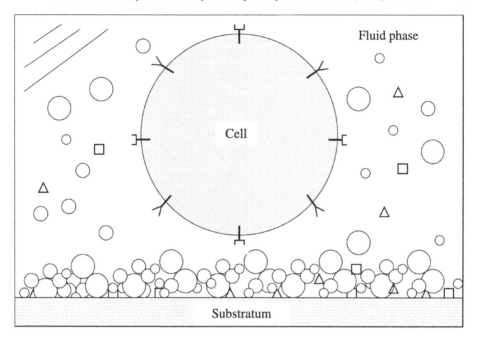

Figure 5. Highly-schematic diagram of a cell attaching to a hydrophobic surface that has adsorbed a multitude of different-sized blood proteins (represented by different size circles) from serum at much greater solution concentrations than adhesion ligands (squares and triangles). The interfacial region (a.k.a. interphase) concentrates proteins by adsorption. Competition among serum proteins on the basis of solution concentration would significantly dilute the presence of adhesion ligands at the surface, perhaps explaining why mammalian cell attachment to hydrophobic surfaces is low.

the surface. Perhaps this qualitatively explains the observed low cell attachment to hydrophobic surfaces from serum-containing media.

All of the above taken into consideration strongly suggests that the interfacial environment encountered by a cell approaching a hydrophilic surface immersed in a serum-containing medium must be very much different than that of a hydrophobic substratum immersed in the same medium. Proteins within the proximal fluid phase would be more-or-less distributed as in bulk solution because protein adsorption does not occur on hydrophilic surfaces. Likewise, ion distribution would be similar to bulk solution because, in buffered media with >0.1 M salt concentrations, the Debye length is a nanometer or less. Indeed, the hydrophilic surface may be quite similar to that contemplated by DLVO theory, although there is still good reason to suspect that interfacial water properties are different than bulk water [12]. However, the interphase environment may actually appear to an attaching cell, thus it seems evident that the suspending aqueous phase cannot be treated as a uniform dielectric medium in-depth. No doubt this greatly complicates practical application of DLVO theory to real cell adhesion problems. Indeed, the success that DLVO theory has enjoyed in predicting at least general trends in cell adhesion may well depend on the fact that proteins do not adsorb on the hydrophilic surfaces typically used in cell culture.

2.3. The Thermodynamic View

Thermodynamics approaches cell adhesion from the standpoint of energetics involved in forming and destroying interfaces between cell, solution, and substratum. Figure 6(A–D) shows progressive levels of abstraction of the cell attachment process. Panel A shows an MDCK cell that has formed initial attachment to tissue culture grade polystyrene (see Fig. 3 legend for details). Panel B schematizes the arrangement of a small portion of a cell extension (pseudopodia) contacting the substratum surface, juxtaposing the cell membrane and surface. Of course, Panel B is purely hypothetical and, among other important details, does not consider nanoscale roughness of the substratum surface. However, the speculation of Panel B invites contemplation of the molecular-scale arrangement of the lipid bilayer with the substratum and the nature of communication between cell and surface that might arise from such a physical interaction.

Panel C further abstracts the cell of Panel A into uniform sphere (in cross section) with a uniform cell–surface (cs) interface that is reversibly formed by attaching to the surface. This latter process is at the expense of cell–liquid (cl) and surface–liquid (sl) interfaces that are destroyed when the cell forms a physical contact with the surface (as in Panel B). The cell need not actually move macroscopic distances to make and break this (sc) interface. Nor is it necessary to conclude that the cell is uniform over the entire surface as suggested in Panel C because thermodynamics models only that portion of a cell that actually forms an interface. This latter aspect of the thermodynamic model is made clearer in Panel D which reduces the cell–surface contact region into unit interfacial areas, each with a characteristic interfacial energy. The net free energy involved in cell adhesion is construed in this simplified Dupre'-like analysis to be the energy gained by forming a cell–substratum interface (γ_{cs}) at the expense of breaking cell–liquid (γ_{cl}) and surface–liquid (γ_{sl}) interfaces (all other factors such as temperature and pressure are presumed constant).

Probably more obscure to the biologist than even the colloid view (see Section 2.2), abstractions of the real cell adhesion process diagrammed in Fig. 6 seek justification on the basis that interfacial energetics subsumes all contributions to the adhesion process — biological, chemical, and physical — into an exchange of interfacial energies. Details at the microscopic level need not be considered, including mating of receptors with ligands [3]. Although, this interaction-in-contact argument is non-competitive with either the biological or colloid viewpoint (see Sections 2.1 and 2.2), it is silent about molecular details involved in adhesion and how the cell actually gets close enough to the surface to form adhesion contact [3, 14, 26]. There are many subtleties embedded in thermodynamic arguments, not the least of which perhaps is that cell substratum and cell–liquid interfacial tensions are not easily

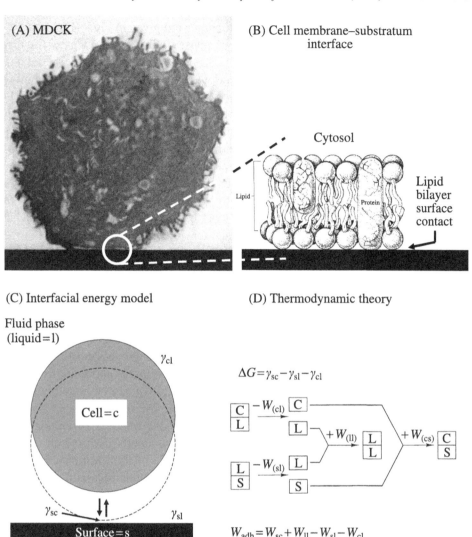

Figure 6. Various stages of abstraction of the cell attachment process. Panel A is expanded from Fig. 4. Panel B contemplates the nature of the interface between the surface and the cell membrane of an attaching cell. Panel C abstracts the cell of Panel A into a spherical object reversibly forming an attachment with a surface indicating cell–surface (cs), cell–liquid (cl), and surface–liquid (sl) interfaces and associated interfacial tensions γ. Panel D interprets the cell adhesion process sketched in Panel C as the net energy of interface formation and destruction, equivalently formulated in terms of either interfacial tensions or work of interface formation and destruction (see text for discussion).

measured for use in free energy of adhesion (ΔG) calculations that might then be compared to cell adhesion in test of the applicability of thermodynamics [3, 25, 26].

2.3.1. Water, Protein Adsorption and the Thermodynamic View

A beauty of classical thermodynamics is that molecular details are implicit in macroscopic thermodynamic properties and need not be directly considered. As such, the Dupre' calculation of free energy of adhesion given in Fig. 6(D) implicitly includes all chemical aspects of the aqueous phase, including any water-structure and protein-adsorption effects (see Section 2.2.1) that may occur. If interfacial tensions can be measured or estimated in some way, then ΔG takes all contributing factors into account [25, 26]. Of the three perspectives on cell adhesion discussed to this point, thermodynamic view is the only one that accounts for the role of water in the adhesion process, even if in an indirect way.

The role of water in cell adhesion can be made still more explicit, without loss of generality, by decomposing the simple Dupre' relationship into work of adhesion (W_{adh}) components, as shown in the lower portion of Fig. 6(D). Here, work of forming and destroying interfaces ($-\Delta G$ by convention) is considered instead of interfacial tensions because the concept of work expended or gained has somewhat more intuitive physical appeal than interfacial tensions. The first step shown in the lower portion of Fig. 6(D) separates unit areas of cell–liquid and substratum–liquid interfaces, expending work components W_{cl} and W_{sl}, respectively, in the process. In the second step, these separated interfaces are re-formed into cell–substratum and liquid–liquid interfaces, generating work components W_{cs} and W_{ll}. The total work W_{adh} is the sum of separate work contributions.

Among the work contributions illustrated in Fig. 6(D), W_{ll} is cell-adhesion most notable because it is a strong contributor to moving the process forward. Physically, W_{ll} is the work of water (aqueous phase) cohesion that is substantially due to the self-associated (hydrogen-bonded) nature of water. W_{ll} can be unambiguously calculated from the liquid interfacial tension γ'_{lv} from the relation $W_{ll} = 2\gamma'_{lv}$. As already mentioned in Section 1, the prime superscript is used to differentiate interfacial tension of serum-containing medium most frequently used in cell culture from the interfacial tension of the basal medium $\gamma^0_{lv} \approx 72$ mN/m from which culture medium is prepared ($\gamma'_{lv} \approx 60$ mN/m for 5–20% serum-containing medium [3, 25, 26, 47]). In fact, W_{ll} is so large in comparison to other work components that the term $[W_{ll} - W_{sl}] \geqslant 0$ for the full range of possible substratum surface energies up to fully water-wettable because $W_{sl} \equiv \gamma'_{lv} + \gamma'_{lv} \cos\theta' < 2\gamma'_{lv}$ for all $\theta' > 0$. Furthermore, $[W_{ll} - W_{sl}]$ is *largest* for hydrophobic surfaces to which *cells do not adhere* ($W_{adh} \to 0$). Thus, hydrophobic surfaces are non-adhesive even though $[W_{ll} - W_{sl}]$ contributes strongly to the cell adhesion process. Conversely hydrophilic surfaces are adhesive even though $[W_{ll} - W_{sl}]$ contributes substantially less to the cell adhesion process.

Evidently, then, the term $[W_{cs} - W_{cl}]$ counterbalances $[W_{ll} - W_{sl}]$ at hydrophobic substrata, driving $W_{adh} \to 0$. Given that W_{cl} is a (typically) unknown positive value, it must be concluded that W_{cs} is relatively small at hydrophobic substrata such that $[W_{cs} - W_{cl}] < 0$ in a way that compensates for $[W_{ll} - W_{sl}] \geqslant 0$. In other words, cells fail to form strong attachment to hydrophobic surfaces that adsorb serum protein from solution. This conclusion is consistent with the inference drawn from the discussion of Fig. 5 in Section 2.2.1 but inconsistent with the standard idea that cell adhesion is mediated by protein ligands adsorbed from solution.

2.4. Summary in Dialogue

The situation outlined in Sections 2.1–2.3 brings to mind the premise behind an old joke that has three very different individuals (ethnicity, race, creed...) forced to interact in a humorous circumstance (bar, desert island, place of worship...) whereupon an argument ensues with comic outcome. Applied to the current (not so comic) circumstance, the three individuals might be a (young and vibrant) biologist, a (chastened) colloid scientist, and a (crusty old bearded) thermodynamicist.

The biologist emphatically asserts that experimental evidence supporting receptor-ligand mediation of cell adhesion is unequivocal. The colloid scientist counters by pointing out that much of this experimental evidence is based on highly-contrived experiments in which known ligands are purposely immobilized on surfaces and thus not necessarily predictive of cell adhesion from serum-containing medium. The colloid scientist hastens to add that the biologist's theory is, at best, incomplete because it does not explain how cells overcome electrostatic repulsion encountered at a surface that would otherwise prevent the close cell–surface contacts required for bonding to occur between surface-adsorbed ligands and membrane-bound receptors. The biologist argues back that the colloid scientist's theory is also incomplete because it does not specify how cells actually adhere once the cell is in close proximity of the surface.

The thermodynamicist joins the argument by admonishing that neither biologist nor colloid scientist takes into account the obvious experimental facts that cells fail to efficiently adhere to protein adsorbent hydrophobic surfaces. Yet these same cells adhere efficiently from the same suspending solution onto hydrophilic surfaces that adsorb much less, arguably no (Section 3), protein from solution. According to the thermodynamicist, the biologist is right for reasons he cannot square with the reality that protein ligands do not adsorb on hydrophilic surfaces and the colloid scientist is right about electrostatics but cannot properly construct a theory that applies to the real-world bioadhesion process.

The biologist advances an alternative interpretative paradigm based on the idea that various adhesins involved in cell adhesion have special adsorption properties that are not shared by all other serum proteins. These asserted special properties give rise to chemically-specific (acid–base type) interactions with hydrophilic surfaces that promote adsorption in an orientation conducive to interaction with cell membrane receptors. The thermodynamicist is happy with but one aspect of this suggestion: that serum proteins do not adsorb on hydrophilic surfaces. Otherwise, he explains, competitive adsorption with about 1000 other proteins that make up the serum proteome [53, 55] would surely dilute out putatively specifically-adsorbed adhesin factors. The colloid scientist further notes that for the biologist's idea to work, additional supporting propositions must be made: (1) adhesins do not adsorb on hydrophobic surfaces, as other blood proteins are known to do; or (2) adhesins adsorb on hydrophobic surfaces in an inactive orientation that is not conducive to interaction with cell-membrane receptors; or (3) adhesins adsorbed on hydrophobic surfaces are effectively blocked by the coadsorption of a plethora of serum proteins as suggested by the thermodynamicist for the hydrophilic case; or (4) some combination of the aforementioned three.

But where is the experimental evidence for such a proposal built on the idea of exceptional adsorption behavior? The thermodynamicist adds that a nearly identical patchwork of arguments has been shown to be untenable in explication of the sharp hydrophobic/hydrophilic contrast activation of blood plasma coagulation [57]. How could this argument work for bioadhesion if it does not work for coagulation?

Unabashed, the biologist adds one more conspiring eventuality to keep the biological recognition idea alive — that biologically significant surface concentrations are too low to be detected under conditions relevant to conventional cell culture. Aforementioned experiments with surface-immobilized ligands show that ligands receptor mediated cell adhesion can occur, and, therefore, must occur in other situations. Both the colloid scientist and thermodynamicist rebuke the biologist for using an argument that tenuously survives only by the logical impossibility of proving something not... proving that some phenomenon does not occur just below ever-receding detection limits. The colloid scientist adds that the feasibility of this argument is further strained by the fact that low surface concentrations of adsorbed ligands is inconsistent with numbers of attachment points that would be required to hold cells to a surface. Experiment shows that MDCK cell attachment efficiency, for example, does not vary as a function of inoculum concentration ranging between 10^2 cells/cm^2 to 10^5 cells/cm^2 (10^5 cells/cm^2 is approximately monolayer surface coverage of attached round cells) [20]. If, in fact, adhesion factors are adsorbed at undetectably low surface concentrations, then sparsely distributed ligands would surely be titrated by increasing cell numbers arriving from suspension. Such a circumstance would lead to a sharp decrease in cell attachment at some inoculum concentration less than that required to fill the surface with cells at jammed packing density. But this is not observed.

Frustrated by attacks on two fronts, the biologist turns the table on his colleagues by asserting the point of contention combatively: if cell adhesion is not mediated by biological recognition, then how does cell adhesion actually occur? The response from both colloid scientist and thermodynamicist is unanimous — initial cell attachment is a physical, not biological, process. There are but these two choices they say biological or physical. The thermodynamicist invokes the law of parsimony (Occam's razor) requiring the minimal interpretation that fits experimental data. Rather than propose special adhesion/adsorption properties or biological activity at vanishingly small surface concentrations, the parsimonious conclusion is that initial stages of cell adhesion are substantially, if not entirely, mediated by purely physicochemical interactions between cell and surface and not by biological recognition.

Not good enough for the biologist, who continues to insist that the colloid scientist and thermodynamicist specify the physical forces involved and how these forces are propagated through aqueous medium in a way that ultimately leads to cell adhesion. And why is thermodynamics such a useful tool for what is quite apparently a kinetic problem [3, 20, 25, 58–61]? If not biological recognition, then what signals newly-attached cells to turn on the biological machinery that leads to increased adhesion, production of ECM, and ultimately proliferation? If not biological recognition, how does a cell sense the difference between hydrophilic and hydrophobic substrata? If not biological recognition, what physical forces can induce cells to adhere, spread, and proliferate on surfaces and why have these physical forces not yet been elaborated?

2.5. The Problem with Theories

Figure 7 coarsely categorizes different theoretical approaches that have been taken to the problem of bioadhesion, along with some early (but not necessarily first) literature citations that, in the authors' view, are archetypes for work that was to follow along the same theme. In the early years, say 1960's through mid-1980's, there was enthusiasm that bioadhesion (including bacteria adhesion) could be substantially understood using colloid science, surface chemical, or surface thermodynamic theories. As early as 1924, Mudd and Mudd were applying surface thermodynamics to bacteria adhesion [62, 63]. Over time, a number of imaginative physicochemical theories were developed, each using distinct perspectives outlined in the preceding sections to explain how bioadhesion occurred (see, for example, [25, 26, 64–72]). Although some of these theories have been useful in separating and weighing the relative importance of various material properties (such as charge, wettability, surface density of cell-binding ligands, etc.), it seems clear now that cell–substratum interactions ranging from cell–surface contact through proliferation to chronic cell–material interaction are far too complex to be meaningfully embraced by relatively simple (tractable) mathematical models [14].

Also in these early years, experiments revealed a strong dependence of cell adhesion/proliferation on substratum surface chemistry, giving rise to the expectation that cell adhesion could be understood using surface chemical principles.

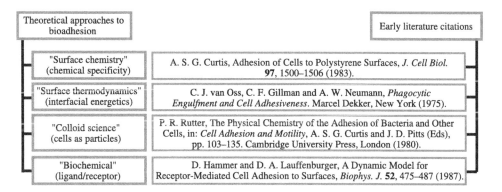

Figure 7. Theoretical approaches to bioadhesion include categories listed on the left hand side represented by archetypical literature examples listed on the right hand side (adapted from [14]). Each of these theories contemplates only a facet of a multifaceted bioadhesion process. A challenge is to integrate the various views into an overarching theory that can be linked with biological aspects of cell adhesion.

Evidence mounted supporting the idea that a particular surface functional group — hydroxyl or carboxyl, for example — was particularly stimulatory to cell adhesion and proliferation [11, 73–76] over other functional groups. However, it has proven difficult in subsequent research to clearly separate cause-and-effect in the cell adhesion/proliferation process, especially in the ubiquitous presence of proteins, and by doing so, unambiguously separate surface chemistry from all other influences (such as surface energy/water wettability) [14, 77]. The most general rule connecting material properties with cell–substratum compatibility emerging from decades of focused research is that anchorage-dependent mammalian cells strongly favor hydrophilic surfaces [1–3, 12, 13, 20, 25, 26], as already mentioned in Section 1. No doubt surface chemistry and wettability are inextricably convolved properties because it is the hydrogen bonding of water to surface functional groups that most profoundly influences wettability (see, for example, [3, 12, 19]). Wetting and surface chemistry are not separate factors, as is sometimes asserted in the cell adhesion literature [78]. Indeed, surface chemistry is responsible for wetting properties.

Interest in, or even remembrance of, physicochemical and surface science theories faded very quickly after the identification of the pantheon of adhesins biologists now use to qualitatively describe cell adhesion [79–81]. Physical scientists, such as Hammer and colleagues [82, 83] mentioned in Fig. 7 attempted to develop quantitative adhesion theories based on biological recognition. Such theories effectively ignore physicochemical interactions between cells, water, and surfaces; even though it is self-evident that biology, biochemistry, and physics are simultaneously operative at some level (see [84] and citations therein for more recent theoretical developments).

A significant problem encountered in the development of bioadhesion theories is that cell attachment, adhesion, and proliferation span broad temporal and spatial ranges. The different theoretical approaches to cell adhesion captured in Fig. 7 effectively study different facets of a multi-faceted cell adhesion problem that most likely become more-or-less important at various stages of cell adhesion. For example, there can be little doubt that surface chemistry, colloid forces, and surface thermodynamics are more important earlier in the process than later when biology strongly influences production of extracellular matrix and cell proliferation. Thus, an unsolved problem in cell adhesion is to integrate these separated temporal views in a way that establishes how preceding stages of cell–surface interaction influence succeeding stages, thereby developing a theory that spans a useful range of the bioadhesion process. Pursuing the example mentioned above a bit further to illustrate this latter point, it is evident that colloid science considers only forces between cell and surface that occur in close proximity — but not contact — whereas surface thermodynamics contemplates the energetics of interface formation and destruction commensurate with intimate cell–surface contact [3]. Thus, colloidal principles might speak volumes about the forces that bring cells to within a few nanometers of a surface but colloid science is silent about the adhesion process itself. Conversely, surface thermodynamics might address cell–surface adhesivity but says nothing about getting the cell close enough to the surface to actually form a cell–surface interface. A connecting theory is required to bring these parts of the problem together and explain how the surface contact step can influence final adhesion. Modified colloid and surface thermodynamic theories might indeed build such a bridge [25, 26, 84], but the span between the physics and biology appears much, much broader. Worse, it is not yet apparent what kind of information can fill this physics–biology gap or how closure can be accomplished in terms that relate materials properties to cell–substratum compatibility.

As Hammer and Tirrell put it so well in [82] "...specific recognition between reactive biomolecules and receptors occurs against a backdrop of polymeric and long-range nonspecific forces". An important question is: in what way and over what time frame does biology, biochemistry, and physics conspire in the bioadhesion process? And in what way does this conspiracy change with changing substratum surface properties? Until this fundamental aspect of bioadhesion is resolved, the debate captured in Section 2.4 will continue to be a stalemate. The conclusion drawn from this review that initial attachment of cells to hydrophilic substrata is not mediated by biological recognition should greatly simplify formulation of an overarching theory of cell adhesion, at least for this restricted set of materials such as commercial tissue culture substrata.

3. Protein Adsorption on Substrata Used in Cell Culture

There can be little doubt that protein adsorption is among the most contentious and important open issues in biomaterials surface science. Contentious aspects include the reversibility/irreversibility of protein adsorption, mechanism of the so-called Vroman effect [32, 85–106], capacity of proteins to adsorb in multilayers [46, 51, 52, 56, 100, 107–120], energetics of protein adsorption [46, 56, 106, 109–111, 121–125], and the applicability of thermodynamic/computational models [107]. The importance of protein adsorption lies in the widely-held view that protein adsorption is among the first steps in the biological response to artificial materials [32, 40, 109, 128–130], especially including cell adhesion according to the biological view discussed in Sections 2.1 and 2.4.

Given this very brief introduction to a problem that has received intense scrutiny for at least the last five decades, it should be of little surprise that the relationship between protein adsorption and bioadhesion is likewise a matter of some considerable debate. As a means of illustrating how this debate can manifest itself, we critically examine Hoffman and Ratner's intrepid attempt to succinctly summarize complex relationships among bioadhesion, substratum water wettability, and protein adsorption appearing in [129]: "*It is generally acknowledged that surfaces that strongly adsorb proteins will generally bind cells, and that surfaces that resist protein adsorption will also resist cell adhesion. It is also generally recognized that hydrophilic surfaces are more likely to resist protein adsorption, and that hydrophobic surfaces usually will adsorb a monolayer of tightly adsorbed protein. Exceptions to these generalizations exist, but, overall, they are accurate statements".*

It is quite apparent that the first part of Hoffman and Ratner's statement *". . .It is generally acknowledged that surfaces that strongly adsorb proteins will generally bind cells. . ."* is not exclusively true. Inspection of Fig. 2 shows that hydrophobic surfaces (as defined by the pivot point mentioned in Section 1) are not conducive to the adhesion of mammalian cells from serum-containing media, even though numerous studies have shown that proteins do indeed adsorb on hydrophobic materials (consistent with Hoffman and Ratner's subsequent statement that *". . .hydrophobic surfaces usually will adsorb. . . protein. . .")*. Thus, we conclude that protein adsorbent surfaces are not necessarily bioadhesive, at least as the term bioadhesive applies to mammalian cells and substrata used in conventional tissue culture.

The second part of Hoffman and Ratner's statement *". . .surfaces that resist protein adsorption will also resist cell adhesion. . ."* is sufficiently vague with respect to the kinds of surfaces contemplated and the notion of "resistant to protein adsorption" that it is difficult to accept as a useful generalization. Certain products of modern surface engineering, such as hydrophilic self-assembled monolayers (SAMs) prepared from oligo(ethylene glycol) terminated thiols [130–137] have indeed been shown not to adsorb proteins (a.k.a. "protein resistant"), and it is apparently true that cells do not efficiently attach to such surfaces [134, 138]. But these water-swollen, hydrogel-like materials are certainly not representative of hydrophilic materials such as gas-discharge-treated polystyrene widely used in mammalian cell culture. This second statement is furthermore difficult to rationalize with the experimental data of Fig. 2, especially when taken in combination with the third statement that *". . .It is also generally recognized that hydrophilic surfaces are more likely to resist protein adsorption. . ."*. Data of Fig. 2 clearly show that mammalian cell adhesion is most efficient to hydrophilic surfaces, which are claimed to resist protein adsorption by the third statement. But efficient cell adhesion to surfaces that resist protein adsorption is in conflict with the second statement that anticipates cell adhesion to protein adsorbing substrata. Indeed, if it were not true that hydrophilic surfaces resisted protein adsorption, then wholesale adsorption from serum protein-containing solutions would surely swamp adsorption of adhesins thought to mediate mammalian cell adhesion (as discussed in Section 2.4). Finally, if hydrophilic surfaces resist protein adsorption, which indeed we find to be the case, then how is it that adhesins adsorb on hydrophilic surfaces to mediate cell adhesion through biological recognition?

Our studies of protein adsorption briefly reviewed in the following sections strongly support Hoffman and Ratner's ideas that *". . .hydrophilic surfaces are more likely to resist protein adsorption. . ."* and that *". . .hydrophobic surfaces usually will adsorb. . . protein. . ."*. However, we find that protein does not necessarily adsorb on hydrophobic surfaces as *". . .a monolayer of tightly adsorbed protein. . ."*. Instead, we find that proteins can adsorb on hydrophobic surfaces in multiple layers, depending on protein size (molecular weight), solution concentration, and adsorbent surface area. Furthermore, the free energy of adsorption is found to be quite modest, only a few multiples of thermal energy ($\Delta G^{\mathrm{o}}_{\mathrm{ads}} \sim -5RT$ for hydrophobic surfaces, with monotonically decreasing $\Delta G^{\mathrm{o}}_{\mathrm{ads}}$ with increasing adsorbent hydrophilicity — hardly adsorption that might be reasonably termed 'tight').

All taken together, we cannot agree with Hoffman and Ratner that "...*these generalizations... are accurate statements*" and take this opportunity to offer a revision made specific to the culture of mammalian cells *in vitro*: "*Hydrophobic surfaces to which protein readily adsorbs will generally resist mammalian cell adhesion whereas hydrophilic surfaces to which protein does not adsorb promote mammalian cell adhesion. Exceptions to this general rule might include hydrogel and hydrogel-like surfaces that* ABsorb *significant amounts of water and/or surfaces bearing strong Lewis Acid/Base functionalities exhibiting ion-exchange properties. Initial stages of mammalian cell adhesion to hydrophilic substrata from serum-containing media are not mediated by biological recognition involving surface-adsorbed ligands for cell receptors*".

3.1. Brief Review of Protein Adsorption

A thoroughgoing review of the protein adsorption literature is well beyond the intended scope of this paper. However, for the purpose of providing some technical support for opinions rendered herein, we briefly summarize results of an extended survey of blood protein adsorption obtained in the authors' laboratories. This compilation should not be accepted as a substitute for a broader review of work appearing in disparate literature sources (such as [128]) nor inclusive of all possible combinations of proteins and adsorbents. However, a significant effort has been made in our studies to include diverse blood proteins spanning three decades in molecular weight and adsorbent surfaces incrementally sampling the full range of observable water wettability bearing a broad range of surface chemistries.

Before engaging in the details of subsequent sections, it is worthwhile to discuss and define some nomenclature. We categorize all physicochemical events leading to an excess (positive or negative) accumulation of a particular solute (e.g., an adhesin), over and above bulk solution concentration, at the interface between a solid surface (or air) and aqueous solution as adsorption of that solute [3, 139]. Descriptors such as binding, charge interactions, directed assembly, ion-exchange, and the like are nothing more than specific ways surface active solutes can absorb on a surface [140] and are not different processes than adsorption [12]. Adsorption can be detected and quantified by a great number of different sensitive techniques [3, 12, 141]. Perhaps the simplest of these methods are contact angle and wetting methods capable of detecting minute traces of organic substances adsorbed on water-wettable (hydrophilic) surfaces [3, 142]. This richness in analytical information is actually a burden in the interpretation of protein adsorption because of the ensuing difficulty in comparing results on a consistent basis [12]. In particular, those analytical methods involving the removal of a surface from protein solution and/or rinsing of the surface to remove bulk solution (e.g., dip-rinse-measure protocols sometimes used in ellipsometry and radiometry) are, in our view, very likely to introduce artifacts that significantly compromise results (see, for example, [142]). Protein bound to a surface after destruction of the hydrated interface in equilibrium with protein solution falls outside of the definition of adsorption as applied herein. Work briefly reviewed in the following sections was performed using either tensiometric (contact angles and wetting) or solution-depletion (protein mass remaining in solution after contact with an adsorbent) methods that do not significantly perturb adsorption dynamics or the interface region.

3.1.1. Hydrophobic Surfaces

As discussed in Section 3, there is little, if any, disagreement in the current literature that blood proteins adsorb on hydrophobic surfaces immersed in aqueous protein solutions. How much protein adsorbs as a function of solution concentration, strength of protein/surface interaction, reversibility of adsorption, and adsorption into multilayers all remain contentious issues. But the fact that protein adsorbs on hydrophobic surfaces seems incontrovertible. Our work has focused on obtaining mass and energy balances of adsorption on hydrophobic surfaces with the goal of resolving some of the remaining open issues. Table 1 condenses this work into a few key findings.

Comparison of adsorption energetics to hydrophobic solid surfaces (solid–liquid interface, sl) to adsorption at the hydrophobic liquid–air interface (liquid–vapor, lv) leads us to the conclusion that the basic mechanism of adsorption to (sl) and (lv) interfaces is basically the same (row 1, Table 1) [109, 110, 122, 124, 125]. Viewed in retrospect, this is no particular surprise given that physical interaction between hydrophobic surfaces and proteins is limited to dispersion (Lifshitz–van der Waals) forces which must be substantially similar at hydrophobic (sl) and (lv) interfaces. At the time we carried out these studies, however, there was expectation within the biomaterials community that the molecularly-smooth and maximally hydrophobic (lv) interface was somehow not representative of hydrophobic solid interfaces. Data show that this expectation is unwarranted [125].

It was a surprise, however, that adsorption of diverse proteins on hydrophobic surfaces was remarkably similar (row 2, Table 1). The significance of this observation is that it suggests that the structural variability that confers profoundly different bioactivity does not greatly affect interaction energetics in water that drives adsorption on the hydrophobic interface [109–111, 121, 122, 124]. Again at the time of these studies, it was widely anticipated that protein adsorption was complex at the molecular scale, critically dependent on structural differences among proteins. Detailed examination of interfacial energetics and mass balance of adsorption does certainly reveal differences among various proteins, but these differences seem smaller than might otherwise have been expected considering the diversity in proteins studied. Adsorption avidity was quantified by partition coefficients that measure the ratio of interfacial and bulk solution phase w/v (mg/ml) concentrations W_I and W_B, respectively, where the partition coefficient $P \equiv W_I/W_B$. For a broad variety of proteins, it was found that $45 < P < 520$, meaning that the hydrophobic

Table 1.

Summary of principal results on protein adsorption on hydrophobic surfaces

Experimental observation	Interpretation
Protein adsorption on the hydrophobic buffer-air and buffer-SAM surface was essentially identical.	Mechanism underlying protein adsorption on hydrophobic liquid–vapor and solid–liquid interfaces from aqueous-buffer solution is the same.
Adsorption energetics was surprisingly similar among proteins with partition coefficients $10 < P < 10^3$.	Biochemical diversity among proteins is unrelated to the interaction energetics with water (amphiphilicity) that drives adsorption, corroborating earlier studies [143]: $$\Delta G^o_{\substack{ads \\ phobic}} = -RT \ln P \approx -5RT,$$ corroborating estimates from hydrophobic chromatography [118].
Protein adsorption energetics followed 'Traube-rule-like' progression in MW that is a signature of the hydrophobic effect.	Molar variability in protein amphilicity is achieved by aggregating greater mass of similar amphiphilic character (blocks of amino acids), as opposed to accumulating greater amphilicity with MW.
A quasi-thermodynamic theory predicated on interfacial packing of hydrated spheroids rationalized interfacial energetics of protein adsorption.	Predicts and explains Traube-rule-like ordering and observed similarity in interfacial energetics of adsorption among proteins. Hydrophobic effect dominates protein adsorption. Proteins with $MW > 125$ kDa adsorb in multiple layers.
Interfacial shear rheology of blood proteins adsorbed on the hydrophobic buffer-air surface exhibited concentration-dependent viscoelastic properties.	Adsorbed protein forms an organized, shear-sensitive network consistent with multiple layers for proteins with $MW > 125$ kDa.

surface region was concentrated 45–520 fold over bulk concentration. The perspective arising from these measurements is that of a *very concentrated proteinaceous interface region with very different chemical properties than bulk solution* [3, 12, 13] (see Fig. 5 and discussion thereof in Section 2.2.1).

Partition coefficients in the range $45 < P < 520$ quantify considerable protein-to-protein variation in adsorption properties, but this variation in P translates into a narrow range of apparent free energy of adsorption $\Delta G^o_{\substack{ads \\ phobic}}$ falling between $-6RT < (\Delta G^o_{\substack{ads \\ phobic}} = -RT \ln P) < -4RT$ (where the 'phobic' subscript emphasizes outcomes for protein adsorption on hydrophobic surfaces). This apparent free energy of adsorption $\Delta G^o_{\substack{ads \\ phobic}}$ is thus surprisingly low: only about $-5RT$ units, plus or minus 20% or so. Again viewed in retrospect, low $\Delta G^o_{\substack{ads \\ phobic}}$ might well have been anticipated because it had long been recognized that proteins were weak 'bio-surfactants' [3, 143], reducing air/water interfacial tensions by only about 20 mN/m [109, 110, 121] (comparable to ordinary aliphatic soaps but much less than the >50 mN/m reductions obtained by many synthetic surfactants [140]). But at the time of these studies, there was a general expectation within the biomaterials community that proteins bind irreversibly or at least 'very tightly' (implying a large P-value) to hydrophobic surfaces (see Section 3, for example). Indeed, a small fraction of the total protein adsorbed on hydrophobic solid surfaces does seem to resist buffer rinsing [142, 144] and perhaps this fraction may be fairly considered as 'tightly bound' (although it is not clear how effective buffer rinsing is at the interfacial scale). But the majority of protein adsorbed on hydrophobic surfaces appears to be reversibly adsorbed [122, 124, 125], readily displaced by different proteins in adsorption competition experiments [52, 106]. Low $\Delta G^o_{ads \atop phobic}$ and weak biosurfactancy are manifestations of the fact that water is indeed a good solvent for proteins [145], readily dissolving mg/ml concentrations of proteins exceeding 1000 kDa [53]. Thus we conclude that *protein adsorption out of aqueous solution is not energetically favorable by a large amount.*

This survey of adsorption of blood proteins with molecular weight (size) varying over three decades on hydrophobic surfaces further revealed a systematic pattern evocative of the so-called 'Traube rule' of surfactant adsorption. It was found that blood proteins adsorbed as a 'homologous series in size' [111, 122] (row 3, Table 1). A quasi-thermodynamic model predicated on the packing of hydrated spheroids at a surface seems to adequately explain general experimental trends (row 4, Table 1) and achieves a level of agreement between mass and energetic balances for protein adsorption that is rarely reported in the protein adsorption literature [48, 53]. This same 'volumetric interpretation' of protein adsorption predicts and accommodates adsorption of proteins in multilayers [46, 51, 52, 107, 108].

The above findings and considerations have led us to propose that the free energy of protein adsorption to any surface, ΔG^o_{ads}, consists of three essential components [46, 51]:

(i) The free energy (gain) due to the hydrophobic effect operating on proteins, $\Delta G^o_{phobic \atop effect}$.

(ii) The free energy (cost) due to vicinal water displacement (surface dehydration), $\Delta G^o_{dehydration}$.

(iii) The free energy (gain) due to protein–protein and protein–surface interactions $\Delta G^o_{interaction}$.

The first component (i) is the hydrophobic effect that expels protein from solution to recover hydrogen bonds among water molecules otherwise separated by proteins in solution. The hydrophobic effect is approximately constant (on a weight basis) for all globular proteins because the partial specific volume v^o of proteins falls within a conserved range of $0.70 \leqslant v^o \leqslant 0.75$ cm^3/g protein [146] (see [147–152] for basic information regarding spherical dimensions and molecular packing of proteins). In other words, the number of hydrogen bonds recovered by water by expulsion of protein is approximately constant per gram protein because protein volume/gram is nearly constant. Hence it happens that $\Delta G^o_{\text{phobic}\atop\text{effect}}$ is approximately constant per gram protein, for purified proteins and mixtures alike [122, 124]. The second component (ii) is the energetic cost of displacing vicinal water by adsorbing protein, which presumably increases with increasing adsorbent surface water wettability ($\Delta G^o_{\text{dehydration}} > 0$). For a particular surface with particular water wettability, however, $\Delta G^o_{\text{dehydration}}$ is approximately constant per gram protein for the same reasons $\Delta G^o_{\text{phobic}\atop\text{effect}}$ is approximately constant per gram protein. Likewise, the third component (iii) $\Delta G^o_{\text{interaction}}$ must depend in some way on the chemistry of the adsorbent surface and would include any nanoscopically localized interactions between protein and surface. We infer from interfacial rheology [56] that $|\Delta G^o_{\text{phobic}\atop\text{effect}}| \geqslant |\Delta G^o_{\text{interaction}}|$ (where "| |" denotes absolute value) because blood proteins adsorbed on the hydrophobic buffer/air interface are shear sensitive (row 5, Table 1; see also [153] and citations therein). And, as mentioned above in reference to interfacial energetics, $\Delta G^o_{\text{ads}\atop\text{phobic}}$ is only of the order of a few RT units, so $\Delta G^o_{\text{interaction}}$ cannot be larger. That is to say, low viscoelasticity and low $\Delta G^o_{\text{ads}\atop\text{phobic}}$ implies relatively weak intermolecular interactions among adsorbed protein molecules and hydrophobic surfaces. Weak intermolecular interactions are, of course, consistent with dispersion (van der Waals) forces that dominate hydrophobic surface reactions with solvent and solute molecules alike [19, 45]. All of the above taken together suggests that the physical chemistry of protein adsorption from purified aqueous solution follows the basic rule $\Delta G^o_{\text{ads}} = (\Delta G^o_{\text{phobic}\atop\text{effect}} + \Delta G^o_{\text{dehydration}} + \Delta G^o_{\text{interaction}}) \approx (G^o_{\text{phobic}\atop\text{effect}} + \Delta G^o_{\text{dehydration}})$, where the approximation excludes surfaces that happen to bear specialized immobilization chemistries, specific biochemical ligands for the protein(s) in question, or strong Lewis acid/base groups exhibiting ion-exchange type chemistries [107].

Thus *it is apparent that water substantially controls or dominates protein adsorption on hydrophobic surfaces*. Given that efficiency of cell adhesion increases sharply near the $\theta = 65°$ pivot point where protein adsorption decreases below detection limits and that water controls protein adsorption through the $\Delta G^o_{\text{dehydration}}$ term of the three-component free energy rule, it seems reasonable to conclude that water controls cell attachment. Cell attachment to hydrated surfaces is apparently favorable at least partly because hydrophilic surfaces are not blocked by a jumble of adsorbed protein.

3.1.2. Hydrophilic Surfaces

As discussed in the Section 3, protein adsorption on hydrophilic surfaces (as defined herein) is generally regarded as much less efficient than on hydrophobic counterparts [85, 129, 154–164]. Ascertaining exactly how much protein adsorbs on hydrophilic surfaces from the literature is complicated by the notorious lack of a consensus definition among investigators using hydrophobic/hydrophilic terminology [19, 21, 165], failure to systematically categorize hydrophilic materials according to material class (hydrogel, Lewis acid/base strength of surface functional groups, etc.) [57], and the wide variety of analytical methods applied that defy meaningful comparison of results [12] (modern methods include atomic force microscopy [166], ellipsometry [167], internal reflection spectroscopy [104], surface plasmon resonance [168], and tensiometry [122] to name but a few from an extensive list of surface science tools). Moreover, different investigators use different proteins (with different molecular weights and biological activities) at quite different concentrations, sometimes metered in units of weight/volume and sometimes in units of mole/volume. Either measure is, of course, entirely acceptable unless adsorption behavior of two proteins with very different molecular weights are directly compared on the same basis — the same weight of two such proteins contains different moles and the same molarity contains different masses of protein. It is not at all obvious *a priori* how protein adsorption should be scaled in a way that accounts for the significant variation in protein size.

Our approach to understanding protein adsorption on hydrophilic materials has been to measure and compare full adsorption isotherms for a group of test proteins varying over three decades of molecular weight adsorbing on surfaces with incrementally increasing hydrophilicity over the approximate range $0° \leqslant \theta \leqslant 120°$ [46, 47, 125]. In very brief summary, we found that adsorption capacity systematically decreased with increasing adsorbent hydrophilicity and fell below detection limits near the pivot point mentioned in Section 1. Adsorption energetics was entirely consistent with mass measurements, confirming the decreasing adsorption to increasingly hydrophilic surfaces trend. This trend appears to hold for adsorption from complex proteinaceous mixtures such as plasma and serum (derived from various mammalian species) as well as purified proteins [47] (see following section). Interpreted according to the three-component free energy rule mentioned in Section 3.1.1, decreasing protein adsorption on increasingly hydrophilic surfaces occurs because of the increasing energy cost of dehydrating the adsorbent surface. The free energy of adsorption ΔG°_{ads} monotonically decreases to detection limits near the $\theta = 65°$ pivot. Again, surfaces bearing ion-exchange (charged) functionalities are exceptional in that these wettable surfaces adsorb protein. But even in these exceptional cases, $\Delta G^{\circ}_{dehydration}$ appears to control adsorbent capacity [107]. No such ion-exchange-type interaction has been observed between proteins and hydrophilic surfaces bearing 'ordinary' oxidized-type chemistries (carboxyl, hydroxyl, ether, etc.).

Thus we conclude that proteins do not adsorb on hydrophilic surfaces as defined herein by the $\theta \sim 65°$ pivot point. These observations do not prove, however, that some amount of protein falling just below detection limits (usually of the order of 0.1 mg/m^2) adsorbs on hydrophilic surfaces. However, as discussed in Section 2.4, it is not logically possible to prove that something does not exist just above or below a detection horizon. Rather, we argue that the trend of decreasing protein adsorption on increasingly hydrophilic adsorbents due to increasing $\Delta G^\text{o}_\text{dehydration}$ effectively guarantees no adsorption on surfaces at some definite wettability.

3.1.3. Adsorption of Serum Proteins

Measurement of proteins adsorption from complex mixtures such as blood plasma and serum requires special treatment because these mixtures are chemically undefined. Quantitative measure of the amount of any single protein from such mixtures is certainly technically challenging. However, it is relatively straightforward to measure adsorption energetics using contact angle and wetting methods that do not directly rely on knowledge of fluid phase chemical composition [47, 121, 124, 169, 170]. Using these measures of interfacial energetics, we find that adsorption of purified plasma proteins cannot be confidently distinguished from that of plasma or serum [121, 124]. This finding is completely consistent with the observation that biochemical diversity among proteins does not substantially affect interfacial energetics of adsorption mentioned earlier. We further find that adsorption from serum or plasma decreases to below limits of detection near the pivot point [47] in a manner that is identical to purified blood proteins [125]. We thus conclude from interfacial energetics that blood proteins do not adsorb on hydrophilic surfaces from serum or plasma.

Solution depletion is an unambiguous measure of adsorption that is essentially free from experimental artifacts that might perturb adsorption dynamics. The basic idea behind the depletion method is to measure the concentration of protein in solution before and after contact with adsorbent particles. Concentration may be expressed in any convenient unit. For chemically-defined solutions of purified proteins, concentration in mass (or moles) per volume (e.g., mg/ml) allows the amount of protein adsorbed to be directly calculated from measured solution depletion $D \equiv (W^\text{o}_\text{B} - W_\text{B})$, where W^o_B and W_B are w/v (molar) concentrations before and after contact with adsorbent, respectively (adsorbed mass or moles $= DV_\text{B}$, where V_B is solution volume). Any number of methods can be used to quantify W^o_B and W_B, such as electrophoresis [46, 51, 52, 107, 108], radiometry [144], or non-specific dye uptake of proteins (so-called Bradford assay) [171]. In particular, the Bradford assay is useful for measuring adsorption from protein mixtures such as serum in culture media because it does not require protein separation. As applied in the depletion method of measuring adsorption from serum-containing culture media, W^o_B and W_B can be expressed as dilution (e.g., v/v or w/w in % serum). In this case, D is interpreted as change in solution concentration due to adsorption but not absolute mass (or moles) adsorbed.

Results shown in Fig. 8 were obtained using the Bradford assay for measuring fetal bovine serum (FBS) proteins adsorbed from Dulbecco's Modified Eagle basal medium (DMEM) solution on either silanized (hydrophobic = solid triangles in Panel c) or fully water wettable (hydrophilic = open circles in Panel c) glass particles (hydrophilic 425–600 μm diameter glass beads with 0.25 m^2/g BET surface area obtained from Sigma Aldrich were prepared by oxidation in hot 30% H$_2$O$_2$/concentrated H$_2$SO$_4$ followed by 3× sequential washes in 18 MΩ de-ionized water, isopropanol, and chloroform before air-plasma treatment of a single layer of treated particles in a 15 mm Pyrex glass petri dish for 10 min at 100 W in a Herrick plasma cleaner. So-prepared particles were optionally silanized by 1.5 h reaction with 5% octadecyltrichlorosilane in chloroform followed by 3× rinsing in chloroform and annealing in a vacuum oven at 110°C for 12 h. Similar treatment of class coverslip witness samples revealed that clean glass was fully water wettable (advancing contact angle $\theta = 0°$) whereas $\theta \sim 110°$ on silanized surfaces). Figure 8(a, b) compare curvilinear calibration curves relating color intensity at 595 nm to bovine serum albumin (BSA) solution concentration in mg/ml or FBS solution concentration in v/v %, respectively, showing that the Bradford assay can be used for either purified protein mixtures or chemically undefined milieu such as serum. Figure 8(c) is a solution-depletion assay of FBS protein adsorption on hydrophobic and hydrophilic particles obtained using the calibration curve of Fig. 8(b). Inspection of Fig. 8(c) reveals that FBS proteins clearly adsorbed on hydrophobic adsorbent particles but adsorption of FBS proteins on hydrophilic glass adsorbent particles was not detectable above estimated limits of detection (LOD, double headed arrow annotation on the lower right corner of Fig. 8(c)). These results are completely consistent

---

Figure 8. Measurement of serum protein adsorption on silanized (hydrophobic, filled triangles in Panel (c)) or fully water wettable glass particles from fetal bovine serum (FBS) containing Dulbecco's Modified Eagle Medium (DMEM) solutions using a dye-binding assay. Panel (a) is a curvilinear calibration curve obtained for the purified protein bovine serum albumin (BSA) relating dye color intensity at 595 nm to BSA concentration (mg/ml) for comparison to the calibration curve of Panel (b) corresponding to FBS (v/v % in DMEM). Panel (c) is the result of the depletion method of measuring FBS protein adsorption on hydrophobic (filled triangles) and hydrophilic (open circles) glass adsorbent particles (see Section 3.1.3 for detailed interpretation). Estimated limit of detection (LOD) is indicated by the double-arrow annotation at the lower right of Panel (c). Note that FBS adsorption on hydrophobic particles was clearly measurable whereas no FBS adsorption on hydrophilic particles could be detected. We estimate that the ~250 cm^2 hydrophobic particle surface area saturates at about 13% FBS. FBS (Gibco) solutions were prepared by serial dilution of a 25% v/v FBS stock solution in DMEM. 30 μl of each dilution was mixed with 100 mg of 425–600 μm diameter glass particles in a 1 ml microtube (VWR) by one minute centrifugation (40 RPM in a Hettick microtube fixed rotor centrifuge) and resuspension by pipette aspiration. After 60 min equilibration, 80 μl of protein solution was removed from the microtube, mixed with 160 μl Bradford reagent (Coomassie Blue G-250, Sigma) and diluted to a total volume of 1 ml with phosphate buffer saline (PBS). Adsorption was measured at 595 nm with a Beckman Coulter DU 800 UV-Visible spectrophotometer in a polystyrene cuvette.

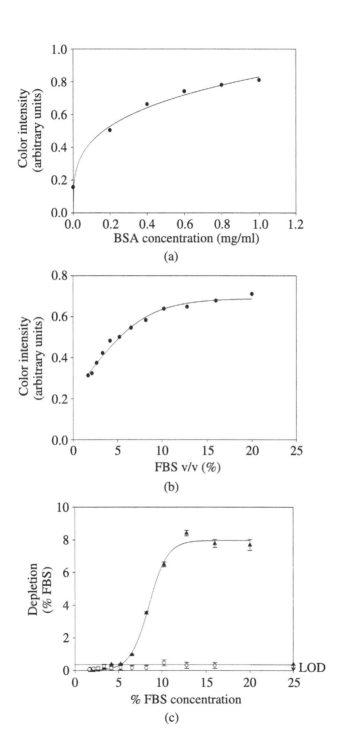

(a)

(b)

(c)

with studies of adsorption of single proteins briefly reviewed in preceding sections in that FBS adsorption on hydrophilic surfaces could not be detected.

Furthermore, general trends observed in the adsorption of FBS proteins on hydrophobic particles are very similar to that of a variety of purified blood proteins on different surfaces. In particular, increasing adsorption as a function of solution concentration to a plateau level whereupon increasing solution concentration does not lead to additional adsorption is characteristic of the adsorption of all proteins we have studied by depletion on different surfaces [46, 51, 52, 107, 108]. We estimate from data of Fig. 8(c) that surface saturation occurs at about 13% v/v FBS. Thus we conclude that the hydrophobic adsorbent surface area ~250 cm^2 (100 mg glass particle adsorbent at 0.25 m^2/g BET surface area) immersed in $V_B = 30$ μl protein solution cannot accommodate additional adsorbed protein [46, 51] beyond 13% v/v FBS. Based on this, we further estimate that saturation of hydrophobic surfaces by FBS-containing solutions occurs at about $(13/100)(30 \times 10^{-3}$ ml$/0.025$ m$^2) = 1.56 \times 10^{-5}$ (ml serum/cm^2) or about 16 nl serum/cm^2. Adsorbent capacity for serum appears very low because serum actually contains about 50 mg/ml protein. This high concentration of protein, coupled with the fact that protein adsorption is not an efficient process (see Section 3), leads to low apparent adsorbent capacity when expressed as volume of serum per area. To put this in perspective, typical cell culture uses about 5 ml medium per 25 cm^2 surface area at 5–20% v/v serum, which converts to 10 to 40 μl serum/cm^2. In other words, typical cell culture medium contains greater than 1000× more protein than required to completely saturate hydrophobic substrata. If it is supposed, for the continued sake of argument, that serum proteins adsorb on hydrophilic surfaces in the same manner as hydrophobic surfaces (which is clearly not the case), then serum-containing medium would contain much more protein than required to completely saturate the cell culture substratum. Therefore, the purely hypothetical situation of protein adsorption on hydrophilic surfaces (a hypothesis that is not supported by experiment) would lead to a circumstance more like that depicted in Fig. 5 than depicted in Fig. 4. Clearly, low abundance adhesins in serum-containing solutions would require exceptional adsorption characteristics to specifically adsorb on hydrophilic surfaces in the manner of Fig. 4 without becoming swamped by much higher concentration blood proteins as in Fig. 5 (see also Section 2.2.1).

All taken together, we conclude that serum proteins, including adhesins, do not adsorb on surfaces more hydrophilic than the $\theta = 65°$ pivot point. That is to say, serum proteins do not adsorb on hydrophilic labware widely used in tissue culture. And if serum proteins did adsorb on hydrophilic surfaces like those used in routine tissue culture, then adhesins in serum would almost assuredly be overwhelmed by high-abundance blood proteins such as albumin and IgG. Arguing that adhesins have special adsorption properties not shared by other blood proteins runs afoul of the experimental fact that protein adsorption on hydrophilic surfaces cannot be detected using unambiguous solution-depletion method or tensiometry.

4. Conclusion

Extensive studies of blood protein adsorption on a wide variety of materials reveal that serum proteins do not adsorb on hydrophilic surfaces but do adsorb on hydrophobic counterparts. It is well known that most anchorage-dependent mammalian cells attach and adhere to hydrophilic substrata widely used in the art of tissue culture from serum-containing medium but these same cells do not attach and adhere to hydrophobic substrata. Thus, it is concluded that adsorbed proteins do not mediate initial attachment of mammalian cells *in vitro*. This conclusion stands in sharp contrast to the prevailing idea that cells attach to tissue culture substrata by the interaction of surface-adsorbed ligands to cell membrane bound receptors.

The behavior of water at surfaces controls protein adsorption. Protein adsorption controls the adhesion of mammalian cells to substrata widely used in the art of tissue culture, but not in the manner contemplated by biological recognition theory. Thus it can be concluded that water ultimately controls cell adhesion. We propose that a general rule applicable to anchorage-dependent mammalian cells is: *Hydrophobic surfaces on which proteins readily adsorb will generally resist mammalian cell adhesion whereas hydrophilic surfaces on which proteins do not adsorb promote mammalian cell adhesion. Exceptions might include hydrogel and hydrogel-like surfaces that* ABsorb *significant amounts of water and/or surfaces bearing strong Lewis Acid/Base functionalities exhibiting ion-exchange properties. Initial stages of mammalian cell adhesion to hydrophilic substrata from serum-containing media are not mediated by biological recognition involving surface-adsorbed ligands for cell receptors.*

Acknowledgements

PP appreciates support by the American Chemical Society Petroleum Research Fund grant #44523-AC5. AG and EV acknowledge support from the National Institute of Health grant PHS 2R01HL069965. Authors appreciate additional support from the Materials Research Institute and Departments of Materials Science and Engineering and Bioengineering of the Pennsylvania State University.

References

1. D. Barngrover, in: *Mammalian Cell Technology*, W. G. Thilly (Ed.), p. 131. Butterworths, Boston (1986).
2. T. A. Horbett and L. A. Klumb, in: *Interfacial Phenomena and Bioproducts*, J. L. Brash and P. W. Wojciechowski (Eds), p. 351. Marcel Dekker, New York (1996).
3. E. A. Vogler, in: *Wettability*, J. C. Berg (Ed.), p. 184. Marcel Dekker, New York (1993).
4. C. Rappaport, in: *The Chemistry of Biosurfaces*, M. L. Hair (Ed.), Vol. 2, p. 449. Marcel Dekker, New York (1972).
5. T. Matsuda and M. H. Litt, *J. Polym. Sci. Polym. Chem. Edition* **12**, 489 (1974).
6. H. G. Klemperer and P. Knox, *Lab. Practice* **26**, 179 (1977).
7. R. W. Benedict and M. C. Williams, *Biomater. Med. Art. Org.* **7**, 477 (1979).

8. B. O. Aronsson, J. Lausmaa and B. Kasemo, *J. Biomed. Mater. Res.* **35**, 49 (1997).

9. R. D'Agostino, P. Favia and F. Fracassi, *Plasma Processing of Polymers*. Kluwer Academic Publishers, Dordrecht (1997).

10. H. Yasuda, *Plasma Polymerization*. Academic Press, Orlando, FL (1985).

11. W. S. Ramsey, W. Hertl, E. D. Nowlan and N. J. Binkowski, *In Vitro* **20**, 802 (1984).

12. E. A. Vogler, *Adv. Colloid Interface Sci.* **74**, 69 (1998).

13. E. A. Vogler, *J. Biomater. Sci. Polym. Edn* **10**, 1015 (1999).

14. X. Liu, J. Y. Lim, H. J. Donahue, R. Dhurjati, A. M. Mastro and E. A. Vogler, *Biomaterials* **28**, 4535 (2007).

15. J. Y. Lim, X. Liu, E. A. Vogler and H. J. Donahue, *J. Biomed. Mater. Res. A* **68**, 504 (2004).

16. K. Nakazawa, Y. Izumi and R. Mori, *Acta Biomaterialia* **5**, 613 (2009).

17. E. M. Harnett, J. Alderman and T. Wood, *Colloids Surfaces B* **55**, 90 (2007).

18. K. Kirchhof and T. Groth, *Clinical Hemorheology Microcirculation* **39**, 247 (2008).

19. E. A. Vogler, in: *Water in Biomaterials Surface Science*, M. Morra (Ed.), p. 269. Wiley, New York (2001).

20. E. A. Vogler and R. W. Bussian, *J. Biomed. Mater. Res.* **21**, 1197 (1987).

21. E. A. Vogler, in: *Water in Biomaterials Surface Science*, M. Morra (Ed.), p. 150. Wiley, New York (2001).

22. J. H. Lee, J. W. Lee, G. Khang and H. B. Lee, *Biomaterials* **18**, 351 (1997).

23. K. Anselme, *Biomaterials* **21**, 667 (2000).

24. K. Webb, V. Hlady and P. A. Tresco, *J. Biomed. Mater. Res.* **41**, 422 (1998).

25. E. A. Vogler, *Biophys. J.* **53**, 759 (1988).

26. E. A. Vogler, *Colloids Surfaces* **42**, 233 (1989).

27. M. D. M. Evans and J. G. Steele, *Expl. Cell Res.* **233**, 88 (1997).

28. S. R. Benhabbour, H. Sheardown and A. Adronov, *Biomaterials* **29**, 4177 (2008).

29. J. H. Lee, Y. M. Ju, W. K. Lee, K. D. Park and Y. H. Kim, *J. Biomed. Mater. Res.* **40**, 314 (1998).

30. J. H. Lee, G. Khang, J. W. Lee and H. B. Lee, *J. Biomed. Mater. Res.* **40**, 180 (1998).

31. J. H. Lee and H. B. Lee, *J. Biomed. Mater. Res.* **41**, 304 (1998).

32. T. Horbett, in: *Biomaterials: Interfacial Phenomena and Applications*, S. L. Cooper, N. A. Peppas, A. S. Hoffman and B. D. Ratner (Eds), ACS Symp. Ser. No. 199, p. 234. Am. Chem. Soc., Washington, DC (1982).

33. F. J. Schoen and R. N. Mitchell, in: *Biomaterials Science: an Introduction to Materials in Medicine*, B. Ratner and A. Hoffman (Eds), p. 260. Elsevier Academic Press, San Diego, CA (2004).

34. K. C. Dee, D. A. Puleo and R. Bizios, *An Introduction to Tissue–Biomaterial Interactions*. Wiley, Hoboken, NJ (2002).

35. F. Grinnell, in: *International Review of Cytology*, G. H. Bourne, J. F. Danielli and K. W. Jeon (Eds), Vol. 53, p. 67. Academic Press, New York (1978).

36. R. Dhurjati, X. Liu, C. V. Gay, A. M. Mastro and E. A. Vogler, *Tissue Engineering* **12**, 3045 (2006).

37. R. Dhurjati, V. Krishnan, L. A. Shuman, A. M. Mastro and E. A. Vogler, *Clin. Expl. Metastasis* **25**, 741 (2008).

38. D. F. Williams, in: *Handbook of Biomaterial Properties*, J. Black and G. Hastings (Eds), p. 481. Chapman and Hall, London (1998).

39. D. F. Williams, *Biomaterials* **29**, 2941 (2008).

40. T. A. Horbett, in: *Biomaterials Science: an Introduction to Materials in Medicine*, B. Ratner and A. Hoffman (Eds), p. 237. Elsevier Academic Press, San Diego, CA (2004).

41. Y. Hirano, Y. Kando, T. Hayashi, K. Goto and A. Nakajima, *J. Biomed. Mater. Res.* **25**, 1523 (1991).
42. C. Roberts, C. S. Chen, M. Mrksich, V. Martichonok, D. E. Ingber and G. M. Whitesides, *J. Am. Chem. Soc.* **120**, 6548 (1998).
43. J. A. Neff, K. D. Caldwell and P. A. Tresco, *J. Biomed. Mater. Res.* **40**, 511 (1998).
44. A. Rezania and K. E. Healy, *J. Biomed. Mater. Res.* **52**, 595 (2000).
45. V. V. Yaminsky and E. A. Vogler, *Current Opinion Colloid Interface Sci.* **6**, 342 (2001).
46. H. Noh and E. A. Vogler, *Biomaterials* **27**, 5801 (2006).
47. E. A. Vogler, D. A. Martin, D. B. Montgomery, J. Graper and H. W. Sugg, *Langmuir* **9**, 497 (1993).
48. B. W. Ninham, in: *Biophysics of Water*, F. Franks and S. Mathias (Eds), p. 105. Wiley, Chichester, UK (1982).
49. G. W. Robinson, S.-B. Zhu, S. Singh and M. W. Evans, *Water in Biology, Chemistry, and Physics.* World Scientific, Singapore (1996).
50. E. A. Vogler, in: *Water in Biomaterials Surface Science*, M. Morra (Ed.), p. 4. Wiley, New York (2001).
51. H. Noh and E. A. Vogler, *Biomaterials* **27**, 5780 (2006).
52. H. Noh and E. A. Vogler, *Biomaterials* **28**, 405 (2007).
53. N. L. Anderson and N. G. Anderson, *Molecular & Cellular Proteomics* **1**, 845 (2002).
54. F. W. Putnam, in: *The Plasma Proteins: Structure, Function & Genetic Control*, F. W. Putnam (Ed.), Vol. 1, p. 58. Academic Press, New York (1975).
55. N. L. Anderson, M. Polanski, R. Pieper, T. Gatlin, R. S. Tirumalai, T. P. Conrads, T. D. Veenstra, J. N. Adkins, J. G. Pounds, R. Fagan and A. Lobley, *Molecular & Cellular Proteomics* **3**, 311 (2004).
56. F. Ariola, A. Krishnan and E. A. Vogler, *Biomaterials* **27**, 3404 (2006).
57. E. A. Vogler and C. A. Siedlecki, *Biomaterials* **30**, 1857 (2009).
58. E. Ruckenstein, A. Marmur and S. R. Rakower, *Thrombos. Haemostas.* **36**, 334 (1976).
59. E. Ruckenstein, *J. Theor. Biol.* **51**, 429 (1975).
60. E. Ruckenstein, A. Marmur and W. N. Gill, *J. Theor. Biol.* **58**, 439 (1976).
61. R. Srinivasan and E. Ruckenstein, *J. Colloid Interface Sci.* **79**, 390 (1981).
62. S. Mudd and E. B. H. Mudd, *J. Expl. Medicine* **40**, 647 (1924).
63. S. Mudd and E. Mudd, *J. Expl. Medicine* **40**, 633 (1924).
64. B. A. Pethica, *Expl. Cell Res.* **8**, 123 (1961).
65. P. R. Rutter, in: *Cell Adhesion and Motility*, A. S. G. Curtis and J. D. Pitts (Eds), p. 103. Cambridge University Press, London (1980).
66. D. F. Gerson, in: *Physicochemical Aspects of Polymer Surfaces*, K. L. Mittal (Ed.), Vol. 1, p. 229. Plenum Press, New York (1983).
67. P. Bongrand, C. Capo and R. Depieds, *Progr. Surface Sci.* **12**, 217 (1982).
68. B. A. Pethica, in: *Microbial Adhesion to Surfaces*, R. C. W. Berkeley, J. M. Lynch, J. Melling, P. R. Rutter and B. Vincent (Eds), p. 19. Ellis Horwood, London (1983).
69. G. Bell, M. Dembo and P. Bongrand, *Biophys. J.* **45**, 1051 (1984).
70. D. C. Torney, M. Dembo and G. I. Bell, *Biophys. J.* **49**, 501 (1986).
71. P. J. Facchini, A. W. Neumann and F. DiCosmo, *Appl. Microbiol. Biotechnol.* **29**, 346 (1988).
72. W. Norde and J. Lyklema, *Colloids Surfaces* **38**, 1 (1989).
73. A. S. G. Curtis, J. V. Forrester, C. McInnes and F. Lawrie, *J. Cell Biol.* **97**, 1500 (1983).
74. A. Curtis and C. Wilkinson, *Studia Biophysica* **127**, 75 (1988).
75. N. F. Owens, D. Gingell and A. Trommler, *J. Cell Sci.* **91**, 269 (1988).

76. S. Margel, E. A. Vogler, L. Firment, T. Watt, S. Haynie and D. Y. Sogah, *J. Biomed. Mater. Res.* **27**, 1463 (1993).

77. E. A. Vogler, *J. Electron. Spectr. Rel. Phenom.* **81**, 237 (1996).

78. Y. Arima and H. Iwata, *Biomaterials* **28**, 3074 (2007).

79. J. P. Duguid, *J. General Microbiol.* **21**, 271 (1959).

80. J. P. Duguid and D. C. Old, in: *Bacterial Adherence*, E. H. Beachey (Ed.), p. 185. Chapman and Hall, London (1980).

81. E. H. Beachey, *J. Infectious Diseases* **143**, 325 (1981).

82. D. A. Hammer and M. Tirrell, *Annu. Rev. Mater. Sci.* **26**, 651 (1996).

83. D. Hammer and D. A. Lauffenburger, *Biophys. J.* **52**, 475 (1987).

84. T. Y. Yang and M. H. Zaman, *J. Chem. Phys.* **126**, 206A (2007).

85. J. Brash and D. Lyman, in: *The Chemistry of Biosurfaces*, J. L. Brash (Ed.), p. 177. Marcel Dekker, New York (1971).

86. L. Vroman, *Bull. N. Y. Acad. Med.* **48**, 302 (1972).

87. L. Vroman, A. L. Adams, M. Klings and G. Fischer, in: *Applied Chemistry at Protein Interfaces: Adv. in Chem. Series No. 145*, p. 255. American Chemical Society, Washington, DC (1975).

88. S. L. Cooper, N. A. Peppas, A. S. Hoffman and B. D. Ratner (Eds), *Protein Adsorption on Biomaterials*, p. 539. Am. Chem. Soc., Washington, DC (1982).

89. J. Brash and P. T. Hove, *Thrombos. Haemostas.* **51**, 326 (1984).

90. H. G. W. Lensen, D. Bargeman, P. Bergveld, C. A. Smolders and J. Feijen, *J. Colloid Interface Sci.* **99**, 1 (1984).

91. L. Vroman and A. Adams, *J. Colloid Interface Sci.* **111**, 391 (1986).

92. P. Wojciechowski, P. T. Hove and J. L. Brash, *J. Colloid Interface Sci.* **111**, 455 (1986).

93. H. Elwing, A. Askendal and I. Lundstrom, *J. Biomed. Mater. Res.* **21**, 1023 (1987).

94. H. Shirahama, J. Lyklema and W. Norde, *J. Colloid Interface Sci.* **139**, 177 (1990).

95. E. F. Leonard and L. Vroman, *J. Biomater. Sci. Polym. Edn* **3**, 95 (1991).

96. M. Wahlgren and T. Arnebrant, *Trends Biotech.* **9**, 201 (1991).

97. P. Wojciechowski and J. L. Brash, *J. Biomater. Sci. Polym. Edn* **2**, 203 (1991).

98. J. L. Brash and P. T. Hove, *J. Biomater. Sci. Polym. Edn* **4**, 591 (1993).

99. L. Vroman, *J. Biomater. Sci. Polym. Edn* **6**, 223 (1994).

100. P. M. Claesson, E. Blomberg, J. C. Froberg, T. Nylander and T. Arnebrant, *Adv. Colloid Interface Sci.* **57**, 161 (1995).

101. J. C. Lin and S. L. Cooper, *J. Colloid Interface Sci.* **182**, 315 (1996).

102. J. H. Lee and H. B. Lee, *J. Biomed. Mater. Res.* **41**, 304 (1998).

103. H. Derand and M. Malmsten, in: *Biopolymers at Interfaces*, M. Malmsten (Ed.), Surfactant Science Series, Vol. 75, p. 393. Marcel Dekker, New York (1998).

104. M. Malmsten (Ed.), *Biopolymers at Interfaces*. Marcel Dekker, New York (1998).

105. S.-Y. Jung, S.-M. Lim, F. Albertorio, G. Kim, M. C. Gurau, R. D. Yang, M. A. Holden and P. S. Cremer, *J. Am. Chem. Soc.* **125**, 12782 (2003).

106. A. Krishnan, C. A. Siedlecki and E. A. Vogler, *Langmuir* **20**, 5071 (2004).

107. H. Noh and E. A. Vogler, *Biomaterials* **29**, 2033 (2008).

108. N. Barnthip, H. Noh, E. Leibner and E. A. Vogler, *Biomaterials* **29**, 3062 (2008).

109. A. Krishnan, J. Sturgeon, C. A. Siedlecki and E. A. Vogler, *J. Biomed. Mater. Res. A* **68**, 544 (2004).

110. A. Krishnan, Y.-H. Liu, P. Cha, D. Allara and E. A. Vogler, *J. Biomed. Mater. Res. A* **75**, 445 (2005).

111. A. Krishnan, C. Siedlecki and E. A. Vogler, *Langmuir* **19**, 10342 (2003).

112. S. Lee and E. Ruckenstein, *J. Colloid Interface Sci.* **125**, 365 (1988).
113. J. Jeon, R. Superline and S. Raghavan, *Appl. Spectroscopy* **46**, 1644 (1992).
114. D. E. Graham and M. C. Phillips, *J. Colloid Interface Sci.* **70**, 415 (1979).
115. J. A. D. Feijter, J. Benhamins and F. A. Veer, *Biopolymers* **17**, 1759 (1978).
116. E. Brynda, N. Cepalova and M. Stol, *J. Biomed. Mater. Res.* **18**, 685 (1984).
117. B. Lassen and M. Malmsten, *J. Colloid Interface Sci.* **180**, 339 (1996).
118. W.-Y. Chen, H.-M. Huang, C.-C. Lin, F.-Y. Lin and Y.-C. Chan, *Langmuir* **19**, 9395 (2003).
119. C. Zhou, J.-M. Friedt, A. Angelova, K.-H. Choi, W. Laureyn, F. Frederix, L. A. Francis, A. Campitelli, Y. Engelborghs and G. Borghs, *Langmuir* **20**, 5870 (2004).
120. C. Beverung, C. Radke and H. Blanch, *Biophys. Chem.* **81**, 59 (1999).
121. A. Krishnan, A. Wilson, J. Sturgeon, C. A. Siedlecki and E. A. Vogler, *Biomaterials* **26**, 3445 (2005).
122. A. Krishnan, Y.-H. Liu, P. Cha, D. Allara and E. A. Vogler, *J. Royal Soc. Interface* **3**, 283 (2006).
123. A. Krishnan, Y.-H. Liu, P. Cha, R. Woodward, D. Allara and E. A. Vogler, *Colloids Surfaces: Biointerfaces* **43**, 95 (2005).
124. A. Krishnan, P. Cha, Y.-H. Liu, D. Allara and E. A. Vogler, *Biomaterials* **27**, 3187 (2006).
125. P. Cha, A. Krishnan, V. F. Fiore and E. A. Vogler, *Langmuir* **24**, 2553 (2008).
126. T. A. Horbett, *Cardiovac. Pathol.* **2**, 137S (1993).
127. T. A. Horbett, in: *Biomaterials Science: an Introduction to Materials in Medicine*, B. D. Ratner (Ed.), p. 133. Academic Press, San Diego, CA (1996).
128. J. D. Andrade and V. Hlady, *Adv. Polym. Sci.* **79**, 3 (1986).
129. A. S. Hoffman and B. D. Ratner, in: *Biomaterials Science: an Introduction to Materials in Medicine*, B. Ratner and A. Hoffman (Eds). Elsevier Academic Press, San Diego, CA (2004).
130. K. Feldman, G. Hahneer, N. D. Spencer, P. Harder and M. Grunze, *J. Am. Chem. Soc.* **121**, 10134 (1999).
131. R. G. Chapman, E. Ostuni, S. Takayama, R. E. Holmlin, L. Yan and G. M. Whitesides, *J. Am. Chem. Soc.* **122**, 8303 (2000).
132. H. Kitano, K. Ichikawa, M. Ide, M. Fukuda and W. Mizuno, *Langmuir* **17**, 1889 (2001).
133. P. Harder, M. Grunze, R. Dahint, G. M. Whitesides and P. E. Laibinis, *J. Phys. Chem. B* **102**, 426 (1998).
134. L. Deng, M. Mrksich and G. M. Whitesides, *J. Am. Chem. Soc.* **118**, 5136 (1996).
135. E. Ostuni, R. G. Chapman, R. E. Holmiln, S. Takayama and G. M. Whitesides, *Langmuir* **17**, 5605 (2001).
136. S. J. Sofia and E. W. Merrill, in: *Poly(ethylene glycol) Chemistry: Biological Applications*, J. M. Harris and S. Zalipsky (Eds), p. 342. American Chemical Society, Washington, DC (1997).
137. K. L. Prime and G. M. Whitesides, *Science* **252**, 1164 (1991).
138. G. P. Lopez, B. D. Ratner, C. D. Tidwell, C. L. Haycox, R. J. Rapoza and T. A. Horbett, *J. Biomed. Mater. Res.* **26**, 415 (1992).
139. R. Aveyard and D. A. Haydon, *An Introduction to the Principles of Surface Chemistry*. Cambridge University Press, London (1973).
140. M. J. Rosen, *Surfactants and Interfacial Phenomena*. Wiley, New York (1978).
141. K. S. Birdi (Ed.), *Handbook of Surface and Colloid Chemistry*. CRC Press, Boca Raton, FL (1997).
142. E. A. Vogler, J. C. Graper, H. W. Sugg, L. M. Lander and W. J. Brittain, *J. Biomed. Mater. Res.* **29**, 1017 (1995).
143. B. C. Tripp, J. J. Magda and J. D. Andrade, *J. Colloid Interface Sci.* **173**, 16 (1995).

144. E. S. Leibner, N. Barnthip, W. Chen, C. R. Baumrucker, J. V. Badding, M. Pishko and E. A. Vogler, *Acta Biomaterialia* **5**, 1389 (2009).

145. R. H. Pain, in: *Biophysics of Water*, F. Franks and S. Mathias (Eds), p. 3. Wiley, Chichester, UK (1982).

146. T. V. Chalikian and K. J. Breslauer, *Biopolymers* **39**, 619 (1996).

147. F. M. Richards, *Ann. Rev. Biophys. Bioeng.* **6**, 151 (1977).

148. C. Chothia, *Nature* **254**, 304 (1975).

149. S. Miller, A. M. Lesk, J. Janins and C. Chothia, *Nature* **328**, 834 (1987).

150. S. Miller, J. Janin, A. M. Lesk and C. Chothia, *J. Mol. Biol.* **196**, 641 (1987).

151. J. Tsai, R. Taylor, C. Chothia and M. Gerstein, *J. Mol. Biol.* **290**, 253 (1999).

152. M. Gerstein and C. Chothia, *Proc. Natl. Acad. Sci. USA* **93**, 10167 (1996).

153. J. Benjamins, J. Lyklema and E. H. Lucassen-Reynders, *Langmuir* **22**, 6181 (2006).

154. G. Sagvolden, I. Glaever and J. Feder, *Langmuir* **14**, 5984 (1998).

155. D. L. Elbert and J. A. Hubbell, *Ann. Rev. Mater. Sci.* **26**, 365 (1996).

156. V. Hlady, *Appl. Spectroscopy* **45**, 246 (1991).

157. H. Uyen, J. Schakenraad, J. Sjollema, J. Noordmans, W. L. Jongebloed, I. Stokross and H. J. Busscher, *J. Biomed. Mater. Res.* **24**, 1599 (1990).

158. W. Norde and J. Lyklema, *J. Biomater. Sci. Polym. Edn* **2**, 183 (1991).

159. W. Norde, *Adv. Colloid Interface Sci.* **25**, 267 (1986).

160. W. Norde and J. Lyklema, *J. Colloid Interface Sci.* **66**, 257 (1978).

161. W. Norde and J. Lyklema, *J. Colloid Interface Sci.* **66**, 266 (1978).

162. W. Norde and J. Lyklema, *J. Colloid Interface Sci.* **66**, 277 (1978).

163. W. Norde and J. Lyklema, *J. Colloid Interface Sci.* **66**, 285 (1978).

164. W. Norde and J. Lyklema, *J. Colloid Interface Sci.* **66**, 295 (1978).

165. L. Gao and T. J. McCarthy, *Langmuir* **24**, 9183 (2008).

166. A. Sethuraman, M. Han, R. S. Kane and G. Belfort, *Langmuir* **20**, 7779 (2004).

167. P. Tengvall, I. Lundstrom and B. Liedberg, *Biomaterials* **19**, 407 (1998).

168. R. J. Green, M. C. Davies, C. J. Roberts and S. J. B. Tendler, *Biomaterials* **20**, 385 (1999).

169. E. A. Vogler, *Langmuir* **8**, 2005 (1992).

170. E. A. Vogler, *Langmuir* **8**, 2013 (1992).

171. M. M. Bradford, *Anal. Biochem.* **72**, 248 (1976).

172. J. Condeelis, *Annu. Rev. Cell Biol.* **9**, 411 (1993).

Part 2
Methods to Study Cell Adhesion

Applications of Micro- and Nano-technology to Study Cell Adhesion to Material Surfaces

Franz Bruckert * and Marianne Weidenhaupt

Laboratoire des Matériaux et du Génie Physique, UMR CNRS 5628, Grenoble Institute of
Technology, Minatec, 3 parvis Louis Néel, BP 257, 38016 Grenoble Cedex 1, France

Abstract

Micro- and nano-technologies provide new tools to study and control cell adhesion. These technologies
are well suited to reconstitute the cell's natural environment and mimic its biochemical, morphological
and mechanical peculiarities. 'Smart' material surfaces and coatings, which allow controlling the binding
and release of specific macromolecules and/or cells, are being developed at a high pace. Moreover, the
application of geometrical constraints at the micro- or nano-scale reveals some of the physico-chemical
principles underlying molecular and cellular organization.

These technologies, combined with the growing knowledge in molecular biology and the spatial and temporal resolutions given by the various microscopy techniques, are going to boost our understanding of cell
physiology. The availability of well-defined multicellular assemblies opens new ways to test and analyze
cells, either in a cluster, or in assemblies mimicking tissue organization. These techniques help bridging the
gap between molecular biology and tissue or organism physiology.

First, we list the main chemical and physical parameters in the cell micro-environment that influence its
survival, proliferation, differentiation or migration. Then we review some examples where micro- and nano-
technologies are used to control cell spreading and adhesion in different ways: (1) *via* the distance between
adhesive molecules, (2) *via* the geometry of adhesive zones, (3) *via* surfaces with switchable adhesiveness,
or (4) *via* three-dimensional coatings used as reservoir of active molecules.

Keywords

Living tissues, cell adhesion, cell differentiation, cell motility, mechanosensitivity, nanosciences, excitable
surfaces

1. Cell Microenvironment: an Overview

In the living tissues of pluricellular organisms, cells are attached to each other either
directly by molecules embedded in the plasma membrane, or through the extracellular matrix, a mixture of polymers secreted by cells, onto which they adhere. Blood
vessels are a good example of such complex organized structures. They are made

---

* To whom correspondence should be addressed. Tel.: (33) 4 56 52 93 21; e-mail:
Franz.Bruckert@grenoble-inp.fr

Surface and Interfacial Aspects of Cell Adhesion
© Koninklijke Brill NV, Leiden, 2010

of three layers. First, in contact with the blood flow, a monolayer of endothelial cells provides an impermeable barrier to most molecules and cells, but allows the transport of specific molecules (e.g., glucose) by transcytosis and the transmigration of specific cells (e.g., monocytes) between the lumen and the vessel wall. In this very thin layer (a few μm), cells are attached to each other by adherens junctions. Second, layers of smooth muscle cells, organized around the endothelial cell monolayer, control the vessel diameter in response to mechanical stimuli (blood pressure, blood flow shear stress). These are relayed by chemical mediators secreted by the endothelial layer. For example vasodilation and vasoconstriction are regulated by nitric oxide and angiotensin, respectively. Depending on the vessel size, the smooth muscle cell layer thickness varies from 50 μm to several mm. Third, an external layer made of elastic fibers and fibroblast cells ensures mechanical stiffness. In large vessels, this layer may be a few millimeters thick and also contains nutritive vessels and nerves that stimulate smooth muscle cells. A specific 80 nm thick extracellular matrix layer, called the basal membrane, separates the endothelial cells from smooth muscle cells [1]. Another specific extracellular matrix layer separates the smooth muscle cell layer from the external layer. This example shows that in living tissues, several differentiated cell types co-exist in oriented structures organized at different scales, from about 10 nm to several mm.

Despite its apparent stability, the structure of living tissues is dynamic. Cells continuously synthesize and degrade the extracellular matrix [2], divide themselves and die in response to precise molecular clues [3]. These mechanisms allow living tissue growth, physiological adaptation and repair. Cell signalling may occur directly, by cell–cell contact, by secretion of growth factors, differentiation factors, chemo-attractants (molecules that influence cell proliferation, differentiation state, or directed movement), or indirectly by the secretion of extracellular matrix and guidance molecules. Many proteins constituting the extracellular matrix indeed contain molecular motives analogous to growth factors or differentiation factors (e.g., fibulins [4]). Therefore, a specific interaction between cells and the extracellular matrix is necessary for cell survival, differentiation or proliferation. In addition, hydrated, charged polymers contained in the extracellular matrix bind many growth and differentiation factors with various specificities by hydrogen bonds and electrostatic interactions. These immobilized molecules may come in contact with cell explorative structures such as filopodia, where they can stimulate, in an oriented manner, intracellular signalling pathways. Conversely, cells produce enzymes that degrade specific molecules in the extracellular matrix (e.g., matrix metalloproteinase [5]). The mixture of macromolecules and cells composing living tissues react therefore together in an interdependent manner. Taking again blood vessels as an example, cells may proliferate axially or radially, which either extends or widens the vessel [6].

In living tissues, stem cells play a special role. Contrary to differentiated cells which undergo a limited number of symmetrical divisions, they may proliferate without limit, in specific conditions. They generate differentiated cells by asymmetric division: one daughter cell keeping stem cell properties while the other differentiates. Stem cells maintain the steady-state number of differentiated cells and are, therefore, essential for tissue development, renewal and repair. A well-known example of stem cells are blood stem cells, which are localized in the bone marrow, generating erythrocytes, lymphocytes, monocytes and platelets in response to specific growth and differentiation factors. For blood vessels, stem cells reside in the wall around the endothelial cell layer (pericytes). During angiogenesis, endothelial cells are also able to degrade the basement membrane and together reorganize into new vessels. Recent progress has shown that stem cells exist in most tissues, and their stimulation could be a target for future regenerative medicine [7]. An alternative challenge is to integrate them in biomaterials to reconstitute tissues, or tissue components.

Thanks to the progress made in molecular and cellular biology, the use of mutants, the visualization of fluorescently labelled molecular structures and numerous molecular identification techniques (transcriptomics and proteomics), a good knowledge of genes involved in cell differentiation and tissue biogenesis is nowadays available. For instance, in blood vessels, VE-cadherin and the type 2 VEGF receptor are specific for endothelial cells, caldesmon and β-actin for smooth muscle cells, and collagen I for fibroblasts. Similarly, signalling pathways triggered by growth or differentiation factors, chemoattractants and adhesion molecules have been thoroughly studied. Generally speaking, the binding of extracellular molecules to membrane-bound receptors triggers some conformational changes that allow the cooperative assembly of intracellular molecules at the plasma membrane, resulting in ion flow regulation, protein phosphorylation and second messenger biosynthesis (outside-in signalling). These transient molecular modifications assure the relay of the signal within the cell and they may amplify it. They have a direct, short-term effect on macromolecules already present in the cell (primary response, ms to min) or an indirect, long-term effect by acting on transcription factors that translocate to the nucleus and control protein biosynthesis (delayed response, 20 min to days). Cells are also able to change the number and conformation of adhesion molecules (e.g., integrins) by membrane trafficking and protein phosphorylation (inside-out signalling). In living tissues, many of these signals prevent programmed cell death (apoptosis) and control symmetrical or asymmetrical cell division (see below). It is therefore essential to understand and to be able to reconstitute the proper molecular environment of cells to ensure their survival, differentiation state and physiological function. Micro- and nano-technologies are well suited to conceive appropriate environments for different cell types in a very reproducible manner.

Although significant progress has been made in the identification of relevant genes and macromolecules involved in cell physiology, an integrated picture of the cell molecular mechanisms is still lacking. Even for relatively simple cell structures mediating attachment to the extracellular matrix such as focal adhesions, the hierarchical assembly and disassembly of these multi-molecular structures and the cross-talk between different signalling pathways are largely unknown [8]. Moreover, focal adhesions, stable adhesion structures often observed in cells anchored in the extracellular matrix, and podosomes, i.e., dynamic adhesion structures involved in cell migration through tissues, share the same set of structural proteins: integrins, talin, paxillin, vinculin, actin, etc., but are controlled by different phosphorylation signals (FAK *versus* Src, [9]). The assembly of adherens junctions is modulated similarly. Cadherins connect cells together, and their assembly is stabilized by their connection to the actin cytoskeleton *via* catenins. All these various cell structures consist of the same basic elements that ensure plasma membrane binding to the outside (e.g., integrins, cadherins), plasma membrane binding to the actin cytoskeleton (talin, paxillin), localized actin polymerization catalysis (ERM proteins) and mechanical force application and mechanosensitivity (myosins).

Many studies have shown that mechanical forces can trigger biochemical signalling at cell–cell or cell–extracellular matrix contacts. The effect of blood flow on endothelial cell morphology is one of the earliest documented examples [10]. In adherens junctions, an increase of β-catenin tyrosine phosphorylation in response to fluid shear stress regulates the linkage between VE-cadherin and the actin cytoskeleton and is involved in their redistribution [11]. The molecular mechanisms of mechanotransduction are unknown. It is possible that specific proteins such as those forming ion channels are sensitive to mechanical stimuli [12, 13] or that local mechanical forces control the association and dissociation rates within cooperative molecular assemblies [14]. Recently, Whitehead *et al.* [15] reported that heterozygous expression of a mutant form of APC (Adenomatous Polyposis Coli) protein renders colon tissue especially sensitive to mechanical stimuli. Compression of the colon tissue increases the expression of c-myc and twist-1 transcription factors by activating the Wnt/β-catenin pathway. Translocation of β-catenin from adherens junctions to the nucleus in response to mechanical forces exerted during embryogenesis has also been reported in drosophila and is dependent on a specific Src activation [16]. Altogether, this shows that cell–cell adherens junctions are mechanosensitive. Furthermore, focal adhesions are also sensitive to applied mechanical forces [17, 18]. Since ICAPs (integrin cytoplasmic domain associated proteins) shuttle between activated integrins and the nucleus, where they stimulate cell proliferation [19], long term effects of mechanical forces acting on cells embedded in the extracellular matrix can be anticipated [20]. These examples show that for practical applications, the stiffness of materials should be taken into consideration, in addition to surface functionalization. This has nicely been illustrated by studies using thin elastic films of different stiffness [21–24]. From a theoretical point of view, force application to deformable structures results in the localization of mechanical stress in rather sharp zones. In the case of the interaction between a living cell and a soft material, forces are concentrated at the border of cell–surface contact areas [25–27]. This explains why the distribution of mechanical forces provides orienting clues for the cells.

In summary, cell adhesion to material surfaces depends on many parameters. First of all, the presence and the location of specific extracellular matrix molecules, chemoattractants and contact with other cells will be paramount. Adsorption of macromolecules and surface functionalization are therefore essential. On a long term, remodelling of the extracellular matrix, secretion or storage of growth and differentiation factors and proper material stiffness will be determinant. A complete characterization of material properties is thus necessary. Mastering these processes is crucial for a good integration of substituting biomedical materials and for the compatibility between medical implants and living tissues. It is also important for research in biology, since eukaryote cells are often grown on material surfaces. Furthermore, as the interaction between cells and materials extends over different scales, from nm (typical size of macromolecules) to several μm (cell geometry), micro- and nano-technologies are, therefore, well suited to engineer material surfaces for biological use, in order to provide cells with precise and well-characterized conditions.

2. Examples of Applications of Nano-technologies to Cell Adhesion

In this section, we will review several advances in materials science and miniaturization technologies that open new ways to tackle biological or medical questions.

2.1. Surface Activation of Adhesion Proteins

The precise spatial arrangement of integrin molecules and actin microfilaments in adhesive zones is not known, but it is likely to be ordered, because of geometrical constraints and the limited diffusion range of biochemical processes. An illustration of precise integrin clustering during formation of cell adhesive structures is provided by the effect of surface nanostructuring on cell spreading and motility. Using block copolymer nanolithography, Spatz and coworkers engineered material surfaces covered with microarrays of small gold particles exposing cyclic RGDfK motives (an $\alpha_V\beta_3$ integrin ligand) separated by protein-repellent PEGylated surfaces. The distance between adjacent gold particles was precisely defined by the size of the polymer coating of the gold particles [28]. Their small size (8 nm) ensured that only one integrin head domain (9 nm) could bind only a single gold particle. Cells were strikingly sensitive to the distance between adjacent gold particles. Spreading was similar on uniformly distributed RGDfK surfaces and on gold particles exposing RGDfK motives separated by 25 or 58 nm, but they did not spread efficiently on nanoparticles separated by 73 nm or 108 nm [29]. This effect was observed with REF-52 fibroblasts, 3T3 fibroblasts, MC3T3 osteoblasts or B16 melanocytes, all expressing $\alpha_V\beta_3$ integrins. The limitation in cell spreading could be due either to the increased separation distance between adjacent gold particles or the global decreased RGDfK density. In order to address this issue, two elegant experiments were done. First, combining block copolymer nanolithography and electron-beam lithography, Arnold *et al.* [30] prepared squared surfaces with side lengths ranging from 100 nm up to 3 μm, separated by the same distance, and containing ordered gold particles separated by 58 nm. The distance between adjacent squares was equal to their side length, and the global RGDfK density was thus equivalent to that of ordered gold particles separated by 116 nm, an experimental condition where cells did not spread. On all nano-patterned squares, cells were able to spread showing that the local RGDfK density, but not the global one, drove adhesive structure formation. Second, using homopolymer polystyrene to prevent the ordered hexagonal packing of gold particles, Huang *et al.* [31] prepared disordered arrays of RGDfK-functionalized gold particles, keeping the average interparticle distance constant from 58 to 100 nm. In contrast to ordered arrays, MC3T3-E1 osteoblasts were almost insensitive to RGDfK density when spreading on disordered arrays. In such disordered arrays, small zones indeed exist, where a few RGDfK integrin ligands are separated by less than 58 nm, the critical distance for the formation of adhesive structures. Altogether, this shows that integrin clustering depends on the underlying order of RGDfK ligand arrangement on the surface (Fig. 1(a)). The formation of integrin clusters is evidenced by the increase of the force necessary to detach cells, which also critically depends on integrin binding site spacing [32]. At the molecular level, this critical distance corresponds to the cooperative assembly of adjacent integrin molecules linked to the actin cytoskeleton. It should

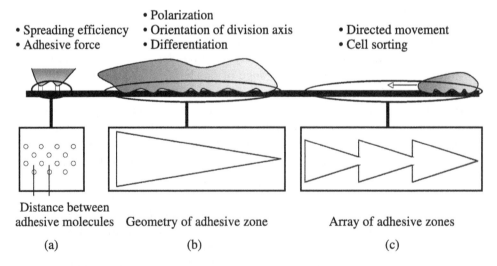

Figure 1. Tailoring adhesive surfaces to control cell adhesion, morphology and physiology. (a) The distance between adhesive molecules controls the formation and mechanical properties of adhesive zones. (b) The geometry of adhesive zones controls cell polarization, the orientation of the cell division axis and cell differentiation. (c) Array of adhesive zones can be used to direct cell movement and to sort cell types.

be noted that this 50–70 nm distance, much larger than integrin or actin monomer sizes (9 and 5 nm, respectively), is indicative of the formation of extensive molecular assemblies.

Besides cell adhesion studies, it should be noted that nanopatterning technologies could have other applications in biology, especially because specific biomolecules can be attached to gold particles [33]. For instance, this could be used to reproduce *in vitro* the geometry of large identical molecular assemblies: photosynthetic or respiratory complexes, vesicular coats, intermediate filaments.... Using nanobeads of different sizes and different surface chemistries could even make multiple-patterning possible in the near future. In this section, we have seen that patterning at the nanometer scale reveals interesting features of molecular adhesion mechanisms. In the next section, we will show that patterning at the micrometer scale is an important new tool to study cell internal organization.

2.2. Tailoring Adhesive Surfaces to Control Cell Geometry

Micropatterning technology has been used to study cell internal organization, which can be controlled by adhesion geometry at the microscale. Early studies by Whitesides, Ingber and Chen showed that cell fate (survival, differentiation or programmed cell death) critically depends on the available adhesive area [34, 35]. In addition, the shape of the adhesive zone strongly influences the dynamics of the actin and microtubule cytoskeletons [36, 37] and the internal distribution of the cell organelles [38, 39] (Fig. 1(b)). The rationale is that the adhesive zone geometry controls the distribution of mechanical forces and the release of signalling molecules at adhesive contacts within the cell. Because of mechanical equilibrium, adhesion structures localize in regions of maximal stress. The adhesive contact distribution drives cell polarity, resulting in directional cell movement [40–42] and orienting the cell division axis [43, 44] (Fig. 1(b) and 1(c)). Furthermore, adhesive patterns can be applied to multicellular assemblies, and therefore used to study intercellular interactions [45, 46] or to separate cell types in a population [42, 47].

The key aspects of this technology that should attract the attention of biologists are (i) the relative ease to prepare micropatterns using photolithography or imprinting techniques, (ii) the versatility of pattern design, and (iii) the fact that uniform cell size, shape and orientation greatly facilitate cell image analysis, in combination with fluorescence microscopy [44]. In molecular cell biology, this technique opens the way to mutant library screens with much more complex — and better controlled — environmental situations than ever envisioned. For instance, polarized cell arrays can be combined with parallel multiple optical trap technology to simultaneously stimulate all cells at the same location [48]. In tissue engineering, it opens the way to reproduce the precise geometry of complex cell assemblies within the tissue, and to study the physiological relevance of geometrical parameters in cell signalling [46].

2.3. Smart Surfaces to Control Cell Adhesion

Material surfaces can be engineered not only to selectively control cell adhesion in a consistent manner, but also to switch from a non-adhesive to an adhesive state. A range of surfaces have been developed, whose hydrophobicity can be controlled either electrically, electrochemically, thermally, or photoactively [49] (Fig. 2). Surface hydrophobicity is an interesting parameter to modulate cell adhesion because most proteins, including extracellular matrix ones, bind more strongly to hydrophobic surfaces than to hydrophilic ones. However, large physico-chemical changes are necessary to significantly modify protein adsorption and alter cell adhesion. In addition, caution should be taken when using physical forces since living cells are very sensitive to their environment. Electrowetting, for instance, requires large electrical voltages to be effective in physiologically relevant solutions, which may trigger electrophysiological responses. In the same way, strong UV illumination is necessary for photo-induced wetting, which is harmful to cells. As a consequence, these techniques have not yet been employed to control cell adhesion. Electrochemical and thermal switchings are more cell-friendly techniques and several researchers have already demonstrated promising applications.

Electrochemical switching (Fig. 2(a) and 2(b)) can be achieved in different ways. One possibility is to change the redox state of a molecule grafted onto the material surface. The resulting surface voltage change exerts repulsive or attractive forces on adsorbed or covalently bound molecules, which drives a conformational change. Wang et al. [50] tethered bipyridinium molecules through an alkylated linker to an electrode and showed that redox modification of the bipyrinidium group bent the linker towards the surface, exposing the most hydrophobic part of the molecule. A reversible, but modest, surface energy change accompanies voltage application. Another approach is to release or bind biomolecules from or to the surface. Yeo et al. [51] prepared a self-assembled monolayer presenting an RGD peptide linked to an O-silyl hydroquinone group. Applying an electrical potential to the substrate oxidized the hydroquinone and released the RGD group, resulting in the detachment of cells attached to the RGD moiety. Subsequent treatment of the surface with diene-tagged RGD peptides restored cell adhesion. They used this dynamic surface to confine 3T3 fibroblasts to adhesive patterns and release them over the entire surface. Thiol chemistry on gold surfaces can also be used to drive reductive electrochemical desorption of self-assembled monolayers. Mali et al. [52] demonstrated that proteins could be patterned on addressable gold electrodes and selectively released from them. Inversely, Tang et al. [53] coated an indium tin oxide (ITO) microelectrode array with a protein-resistant poly(L-lysine)-g-poly(ethylene glycol) graft copolymer. Application of a positive electrical potential resulted in localized polymer desorption, due to the positively charged poly-L-lysine moiety and freed the ITO surface for subsequent protein binding. It should be noted that these techniques are relatively slow, since tens of seconds are required to fully remove adsorbed

Before stimulation   After stimulation

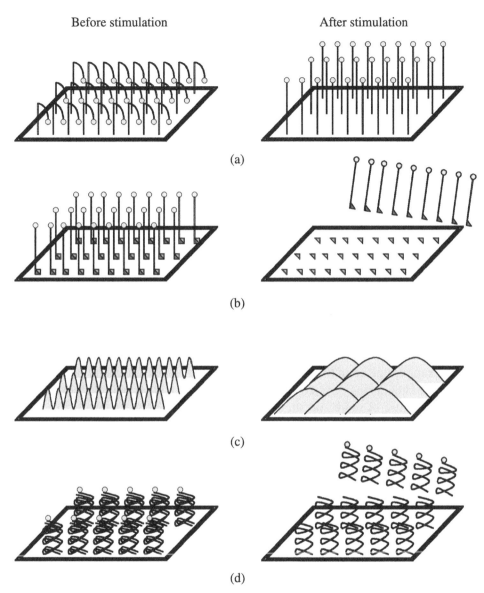

(a)

(b)

(c)

(d)

Figure 2. Smart surfaces to control cell adhesion. (a, b): Electrochemical switching; (c, d): thermal switching. In (a) and (c), a conformational change modifies the surface properties of the material, whereas in (b) and (d), this is achieved by desorption of a functional group.

molecules from the electrodes. State of the art electrochemical switching is, therefore, well applicable to cells that spread or move rather slowly.

Thermal switching is based on hydrogels that are coated over the surface and exhibit a transition between a collapsed and a swollen structure at a lower critical solution temperature (LCST) (Fig. 2(c) and 2(d)). An example of such a thermoresponsive polymer is poly(N-isopropylacrylamide) or PNIPAAm, whose LCST is in the range of 32–35°C. Cells adhere, spread and grow well on PNIPAAm hydrogels at 37°C, since the dehydrated polymer surface is hydrophobic which allows strong extracellular matrix protein binding. Reducing the temperature by about 3–5°C makes the surface hydrophilic and swelling exerts large mechanical forces, which induce detachment of a cell layer including an intact extracellular matrix. This lift-off technique ensures only limited cellular damage and is now widely used for tissue engineering [54], and multiple cell layers can even be generated. As an example of a medical application, corneal epithelial cell layers grown on PNIPAAm dishes can be directly implanted on the patient's cornea [55], or co-cultures of hepatocytes and endothelial cells can be obtained that mimic liver tissue organization [56].

Recently, oligonucleotide–peptide and oligonucleotide–protein conjugates have been synthesized [57]. These molecules are interesting in the field of cell adhesion because their oligonucleotide moiety is able to reversibly bind a complementary DNA sequence. Michael *et al.* [58] recently grafted oligonucleotides onto titanium surfaces and hybridized them to complementary oligonucleotide–GRGDSP conjugates. This surface was then able to bind osteoblasts due to the RGD-containing motif. Desorption of the cells should, in principle, be obtained by local denaturation above the melting temperature of the oligonucleotide strands.

The examples given in this section show different ways to directly control cell–surface adhesion, using physico-chemical means. We will show in the next section how to make use of more specific biochemical or cellular mechanisms to control cell adhesion.

2.4. Polymers, Hydrogels and Multilayer Polyelectrolyte Film Coatings

In the presence of living cells, material surfaces are exposed to a changing environment and anti-biofouling coatings, currently available, do not last for more than a few days in cell culture or inside the human body. Active surfaces are therefore needed that may release or capture molecules of interest. Polymers are of special interest because their extended and flexible structure accommodates large conformational changes, which allows large binding capacity compared to two-dimensional surface immobilization techniques. Furthermore, protein activity can often be preserved in these hydrated molecules. Attaching polymers on materials, therefore, extends the volume in which cells interact with the underlying material and allows a better control of bound molecules.

The surface modification of blood contacting materials provides a nice illustration of the application of polymer versatility. An antithrombin–heparin complex covalently linked to a protein-repellent poly(ethylene oxide) polymer efficiently catalyzes the inhibition of either thrombin or factor Xa by blood antithrombin, preventing activation of the clotting cascade [59]. Similarly, a composite molecule made of a lysine aminoacid moiety exposing its ε-amino group attached to a poly(ethylene glycol) polymer both prevents fibrinogen binding and adsorbs plasminogen and tissue plasminogen activator, resulting in fibrin clot lysis [60]. These coatings strongly reduce platelet adhesion and may greatly improve the hemocompatibility of cardiovascular devices. Active surfaces are, therefore, an efficient way to control cell adhesion.

Polyelectrolyte Multilayer Films are assembled by the alternate adsorption of polyanions and polycations [61]. They can be prepared from a large variety of natural or synthetic polymers. Linearly growing films maintain a stratified structure, but not exponentially growing ones in which polyelectrolytes can diffuse [62]. These coatings are interesting to control cell adhesion because in addition to surface functionalization with adhesion receptor ligands, their mechanical stiffness can be modulated by crosslinking [63, 64]. As previously explained, adherent cells exert large forces on material surfaces and soft materials thus prevent cell spreading. Cell adhesion patterns could, therefore, be engineered using photo-crosslinking to adjust the coating stiffness over the surface [65]. The development of thermosensitive films or their electrochemical destabilization may also provide temporal control of cell adhesion. Another interest of polyelectrolyte multilayer films is the possibility to fill them with bioactive molecules, such as growth factors [66] or differentiation factors [67] and thus create a material-associated reservoir for cell stimulation.

The increasing complexity of polyelectrolyte multilayer film technology is similar to actual trends in hydrogel technology [68]. Tissue engineering scaffolds also release bioactive molecules such as growth factors and cytokines [69–71]. The main difference between hydrogels and polyelectrolyte multilayer films is that hydrogels additionally provide multiple scale porosity, allowing therefore cells to grow within the material and even angiogenesis to occur [72, 73].

3. Future Directions in Nanobiosciences to Study Cell–Material Adhesion

The examples given show that nanosciences can be used both to study and to influence cell–surface adhesion. The temporal and spatial control of cell adhesion is the focus of many current studies. In addition, there is a trend to extend the cell–material interaction zone towards the third dimension. Simple surface functionalization is progressively replaced by multi-layers of molecules, each fulfilling a specific function.

This increased complexity requires the development of new experimental tools and conceptual models to monitor protein conformational changes and homologous and heterologous interactions. Similarly, studying cellular organization in three dimensions requires powerful microscopy techniques and analytical methods. The presence of macromolecules in a reduced effective volume should enhance cooperative effects in their interaction, which needs to be adequately monitored. Theoretically, the increased complexity should drive more studies to model the consequences of 3D confinement on macromolecule interactions: diffusion in a crowded environment, electrostatic and electrochemical effects under the Debye length and kinetic and thermodynamic control of reaction specificity.

4. Conclusion

In conclusion, these emerging technologies open the way to new biological discoveries at different scales, from internal molecular assemblies to cell–cell interactions and tissue morphogenesis. By using materials to reduce the variations associated with cell shape or concentrations, modelling signalling pathways becomes easier and can better explain experimental results.

References

1. M. R. Hayden, J. R. Sowers and S. C. Tyagi, *Cardiovasc. Diabetol.* **4**, 9 (2005).
2. H. C. Blair, M. Zaidi and P. H. Schlesinger, *Biochem. J.* **364**, 329 (2002).
3. B. Pucci, M. Kasten and A. Giordano, *Neoplasia* **2**, 291 (2000).
4. R. Timpl, T. Sasaki, G. Kostka and M. L. Chu, *Nature Rev. Mol. Cell Biol.* **4**, 479 (2003).
5. P. Van Lint and C. Libert, *J. Leukoc. Biol.* **82**, 1375 (2007).
6. E. A. Jones, F. le Noble and A. Eichmann, *Physiology (Bethesda)* **21**, 388 (2006).
7. H. Nakagami, N. Nakagawa, Y. Takeya, K. Kashiwagi, C. Ishida, S. Hayashi, M. Aoki, K. Matsumoto, T. Nakamura, T. Ogihara and R. Morishita, *Hypertension* **48**, 112 (2006).
8. E. Zamir and B. Geiger, *J. Cell Sci.* **114**, 3583 (2001).
9. S. Linder and M. Aepfelbacher, *Trends Cell Biol.* **13**, 376 (2003).
10. S. Noria, D. B. Cowan, A. I. Gotlieb and B. L. Langille, *Circ. Res.* **85**, 504 (1999).
11. J. A. Ukropec, M. K. Hollinger and M. J. Woolkalis, *Exp. Cell Res.* **273**, 240 (2002).
12. O. P. Hamill and B. Martinac, *Physiol. Rev.* **81**, 685 (2001).
13. P. Delmas, *Cell* **118**, 145 (2004).
14. A. Zumdieck, R. Voituriez, J. Prost and J. F. Joanny, *Faraday Discuss.* **139**, 369 (2008).
15. J. Whitehead, D. Vignjevic, C. Futterer, E. Beaurepaire, S. Robine and E. Farge, *HFSP J.* **2**, 286 (2008).
16. N. Desprat, W. Supatto, P. A. Pouille, E. Beaurepaire and E. Farge, *Dev. Cell* **15**, 470 (2008).
17. D. Riveline, E. Zamir, N. Q. Balaban, U. S. Schwarz, T. Ishizaki, S. Narumiya, Z. Kam, B. Geiger and A. D. Bershadsky, *J. Cell Biol.* **153**, 1175 (2001).
18. N. Q. Balaban, U. S. Schwarz, D. Riveline, P. Goichberg, G. Tzur, I. Sabanay, D. Mahalu, S. Safran, A. Bershadsky, L. Addadi and B. Geiger, *Nature Cell Biol.* **3**, 466 (2001).
19. H. N. Fournier, S. Dupe-Manet, D. Bouvard, F. Luton, S. Degani, M. R. Block, S. F. Retta and C. Albiges-Rizo, *Mol. Biol. Cell* **16**, 1859 (2005).
20. A. Millon-Fremillon, D. Bouvard, A. Grichine, S. Manet-Dupe, M. R. Block and C. Albiges-Rizo, *J. Cell Biol.* **180**, 427 (2008).
21. H. B. Wang, M. Dembo and Y. L. Wang, *Am. J. Physiol. Cell Physiol.* **279**, C1345 (2000).
22. D. E. Discher, P. Janmey and Y. L. Wang, *Science* **310**, 1139 (2005).
23. K. Ren, T. Crouzier, C. Roy and C. Picart, *Adv. Funct. Mater.* **18**, 1378 (2008).
24. A. Schneider, G. Francius, R. Obeid, P. Schwinte, J. Hemmerle, B. Frisch, P. Schaaf, J. C. Voegel, B. Senger and C. Picart, *Langmuir* **22**, 1193 (2006).

25. E. Decave, D. Garrivier, Y. Brechet, F. Bruckert and B. Fourcade, *Phys. Rev. Lett.* **89**, 108101 (2002).

26. A. Nicolas, B. Geiger and S. A. Safran, *Proc. Natl Acad. Sci. USA* **101**, 12520 (2004).

27. F. Chamaraux, S. Fache, F. Bruckert and B. Fourcade, *Phys. Rev. Lett.* **94**, 158102 (2005).

28. J. Blummel, N. Perschmann, D. Aydin, J. Drinjakovic, T. Surrey, M. Lopez-Garcia, H. Kessler and J. P. Spatz, *Biomaterials* **28**, 4739 (2007).

29. E. A. Cavalcanti-Adam, T. Volberg, A. Micoulet, H. Kessler, B. Geiger and J. P. Spatz, *Biophys. J.* **92**, 2964 (2007).

30. M. Arnold, M. Schwieder, J. Blummel, E. A. Cavalcanti-Adam, M. Lopez-Garcia, H. Kessler, B. Geiger and J. P. Spatz, *Soft Matter* **5**, 72 (2009).

31. J. Huang, S. V. Grater, F. Corbellini, S. Rinck, E. Bock, R. Kemkemer, H. Kessler, J. Ding and J. P. Spatz, *Nano Lett.* **9**, 1111 (2009).

32. C. Selhuber-Unkel, M. Lopez-Garcia, H. Kessler and J. P. Spatz, *Biophys. J.* **95**, 5424 (2008).

33. D. Aydin, M. Schwieder, I. Louban, S. Knoppe, J. Ulmer, T. L. Haas, H. Walczak and J. P. Spatz, *Small* **5**, 1014 (2009).

34. C. S. Chen, M. Mrksich, S. Huang, G. M. Whitesides and D. E. Ingber, *Science* **276**, 1425 (1997).

35. R. McBeath, D. M. Pirone, C. M. Nelson, K. Bhadriraju and C. S. Chen, *Dev. Cell* **6**, 483 (2004).

36. K. K. Parker, A. L. Brock, C. Brangwynne, R. J. Mannix, N. Wang, E. Ostuni, N. A. Geisse, J. C. Adams, G. M. Whitesides and D. E. Ingber, *FASEB J.* **16**, 1195 (2002).

37. M. Thery, V. Racine, A. Pepin, M. Piel, Y. Chen, J. B. Sibarita and M. Bornens, *Nature Cell Biol.* **7**, 947 (2005).

38. M. Thery, V. Racine, M. Piel, A. Pepin, A. Dimitrov, Y. Chen, J. B. Sibarita and M. Bornens, *Proc. Natl Acad. Sci. USA* **103**, 19771 (2006).

39. M. Thery, A. Pepin, E. Dressaire, Y. Chen and M. Bornens, *Cell Motil Cytoskeleton* **63**, 341 (2006).

40. J. James, E. D. Goluch, H. Hu, C. Liu and M. Mrksich, *Cell Motil Cytoskeleton* **65**, 841 (2008).

41. X. Jiang, D. A. Bruzewicz, A. P. Wong, M. Piel and G. M. Whitesides, *Proc. Natl Acad. Sci. USA* **102**, 975 (2005).

42. G. Mahmud, C. J. Campbell, K. J. M. Bishop, Y. A. Komarova, O. Chaga, S. Soh, S. Huda, K. Kandere-Grzybowska and B. A. Grzybowski, *Nature Physics* (2009).

43. M. Thery and M. Bornens, *Curr. Opin. Cell Biol.* **18**, 648 (2006).

44. M. Thery, A. Jimenez-Dalmaroni, V. Racine, M. Bornens and F. Julicher, *Nature* **447**, 493 (2007).

45. R. A. Desai, L. Gao, S. Raghavan, W. F. Liu and C. S. Chen, *J. Cell Sci.* **122**, 905 (2009).

46. C. M. Nelson, M. M. Vanduijn, J. L. Inman, D. A. Fletcher and M. J. Bissell, *Science* **314**, 298 (2006).

47. Y. Roupioz, N. Berthet-Duroure, T. Leichle, J. B. Pourciel, P. Mailley, S. Cortes, M. B. Villiers, P. N. Marche, T. Livache and L. Nicu, *Small* **5**, 1493 (2009).

48. C. O. Mejean, A. W. Schaefer, E. A. Millman, P. Forscher and E. R. Dufresne, *Opt. Express* **17**, 6209 (2009).

49. S. L. Gras, T. Mahmud, G. Rosengarten, A. Mitchell and K. Kalantar-zadeh, *Chem. Phys. Chem.* **8**, 2036 (2007).

50. X. Wang, A. B. Kharitonov, E. Katz and I. Willner, *Chem. Commun.*, 1542 (2003).

51. W. S. Yeo, M. N. Yousaf and M. Mrksich, *J. Am. Chem. Soc.* **125**, 14994 (2003).

52. P. Mali, N. Bhattacharjee and P. C. Searson, *Nano Lett.* **6**, 1250 (2006).

53. C. S. Tang, M. Dusseiller, S. Makohliso, M. Heuschkel, S. Sharma, B. Keller and J. Vörös, *Anal. Chem.* **78**, 711 (2006).

54. Y. Tsuda, A. Kikuchi, M. Yamato, A. Nakao, Y. Sakurai, M. Umezu and T. Okano, *Biomaterials* **26**, 1885 (2005).
55. M. Nitschke, S. Gramm, T. Gotze, M. Valtink, J. Drichel, B. Voit, K. Engelmann and C. Werner, *J. Biomed. Mater. Res. A* **80**, 1003 (2007).
56. M. Hirose, M. Yamato, O. H. Kwon, M. Harimoto, A. Kushida, T. Shimizu, A. Kikuchi and T. Okano, *Yonsei Med. J.* **41**, 803 (2000).
57. K. Kaihatsu and D. R. Corey, *Methods Mol. Biol.* **283**, 207 (2004).
58. J. Michael, L. Schonzart, I. Israel, R. Beutner, D. Scharnweber, H. Worch, U. Hempel and B. Schwenzer, *Bioconjug. Chem.* **20**, 710 (2009).
59. Y. J. Du, J. L. Brash, G. McClung, L. R. Berry, P. Klement and A. K. Chan, *J. Biomed. Mater. Res. A* **80**, 216 (2007).
60. D. Li, H. Chen, W. Glenn McClung and J. L. Brash, *Acta Biomater.* **5**, 1864 (2009).
61. E. Leguen, A. Chassepot, G. Decher, P. Schaaf, J. C. Voegel and N. Jessel, *Biomol. Eng.* **24**, 33 (2007).
62. C. Picart, J. Mutterer, L. Richert, Y. Luo, G. D. Prestwich, P. Schaaf, J. C. Voegel and P. Lavalle, *Proc. Natl Acad. Sci. USA* **99**, 12531 (2002).
63. A. Schneider, C. Vodouhe, L. Richert, G. Francius, E. Le Guen, P. Schaaf, J. C. Voegel, B. Frisch and C. Picart, *Biomacromolecules* **8**, 139 (2007).
64. L. Richert, A. J. Engler, D. E. Discher and C. Picart, *Biomacromolecules* **5**, 1908 (2004).
65. C. Pozos Vazquez, T. Boudou, V. Dulong, C. Nicolas, C. Picart and K. Glinel, *Langmuir* **25**, 3556 (2009).
66. L. Ma, J. Zhou, C. Gao and J. Shen, *J. Biomed. Mater. Res. B Appl. Biomater.* **83**, 285 (2007).
67. T. Crouzier, K. Ren, C. Nicolas, C. Roy and C. Picart, *Small* **5**, 598 (2009).
68. S. Chaterji, I. K. Kwon and K. Park, *Prog. Polym. Sci.* **32**, 1083 (2007).
69. H. Liu, H. Fan, Y. Cui, Y. Chen, K. Yao and J. C. Goh, *Biomacromolecules* **8**, 1446 (2007).
70. F. G. Rocha, C. A. Sundback, N. J. Krebs, J. K. Leach, D. J. Mooney, S. W. Ashley, J. P. Vacanti and E. E. Whang, *Biomaterials* **29**, 2884 (2008).
71. Y. Liu, S. Cai, X. Z. Shu, J. Shelby and G. D. Prestwich, *Wound Repair Regen.* **15**, 245 (2007).
72. Y. Du, E. Lo, S. Ali and A. Khademhosseini, *Proc. Natl Acad. Sci. USA* **105**, 9522 (2008).
73. E. De Giglio, S. Cometa, M. A. Ricci, A. Zizzi, D. Cafagna, S. Manzotti, L. Sabbatini and M. Mattioli-Belmonte, *Acta Biomater.* **6**, 282 (2010).

Label-Free and Non-invasive Biosensor Cellular Assays for Cell Adhesion

Ye Fang [*]

Biochemical Technologies, Science and Technology Division, Corning Inc., Corning, NY 14831, USA

Abstract
Cells rely on their matrix adhesion complexes to sense multiple environmental cues including extracellular matrix components, adhesive ligand density, surface rigidity and molecular and mechanical perturbations, thus governing their functions and behavior. Label-free and non-invasive biosensors offer generic, integrated, multi-parametric and kinetic measures of cell adhesion and its associated cellular processes. Various label-free biosensor cellular assays are reviewed, and their uses for cell adhesion are discussed.

Keywords
Cell adhesion, biosensor, optical biosensor, extracellular matrix, dynamic mass redistribution

1. Introduction

Cells in multicellular organisms are in contact with each other and with extracellular matrix (ECM). The adhesion of cells to each other or to the ECM is a complex and dynamic process involving physical and chemical interactions and biological signaling processes. As a consequence, cell adhesion is responsible for a wide range of normal and aberrant cellular activities [1–3]. This includes the regulation of cell growth and differentiation during development, and the modulation of the migration of immune cells to sites of infection, the invasion and metastasis of tumor cells, and angiogenesis during wound healing. Changes in cell adhesion have profound effects in many disease states through processes such as angiogenesis, apoptosis and inflammation and, therefore, can be the defining event in many diseases, including cancer, osteoporosis and arthritis.

Given the important roles of cell adhesion in regulating cellular behavior, cellular assays that are capable of quantitative characterization of cell adhesion and its kinetics are highly desirable. In recent years, a wide range of label-free biosensors have been developed and employed for whole cell sensing, including cell adhesion [4–6]. Unlike current techniques for characterizing cell adhesion which are often

---

[*] E-mail: fangy2@corning.com

Surface and Interfacial Aspects of Cell Adhesion
© Koninklijke Brill NV, Leiden, 2010

destructive and require manipulations including cell engineering and labels [7, 8], label-free biosensor cellular assays offer generic, integrated, multi-parametric and kinetic measures of cell adhesion [9]. Here, recent developments in label-free biosensor technologies that are used in whole cell sensing are reviewed, and their uses for studying cell adhesion are discussed.

2. Label-Free Biosensors for Cellular Assays

A label-free biosensor for cellular assays is a device consisting of three components — a biological component (i.e., live cells), a detector element and a transducer. The transducer is used to convert a molecular recognition event or a stimulus-induced cellular response occurred within the biological component into a quantifiable signal, as detected by the detector element. Depending on the type of transducer used, label-free biosensors for whole cell sensing can be largely classified into three categories: acoustic, electrical and optical biosensors (Fig. 1). These biosensors differ greatly in operating principle; as a result, these biosensors measure different aspects of cellular responses discussed in detail below.

2.1. Acoustic Biosensors

Acoustic biosensors such as quartz crystal resonators utilize acoustic waves to characterize cellular responses [10, 11]. The acoustic waves are generally generated and received using piezoelectric sensors (Fig. 1(a)). An acoustic biosensor is often designed to operate in a resonant-type sensor configuration. In a typical setup, thin quartz discs are sandwiched between two gold electrodes. Application of an alternating current (AC) signal across electrodes leads to the excitation and oscillation of the crystal, which acts as a sensitive oscillator circuit. The output sensor signals are the resonance frequency and motional resistance. The resonance frequency is largely a linear function of total mass of adsorbed material when the material is rigid. Under liquid environments, the acoustic sensor response is sensitive not only to the mass of bound molecules, but also to changes in viscoelastic properties and charge of the molecular complexes formed or of live cells [12]. By measuring the resonance frequency and the motional resistance of cells adhered on the crystals, cellular processes including cell adhesion [10, 13–15] and cytotoxicity [16, 17] can be studied in real time. Since live cells are sensitive to changes in electric potential or field as well as to mechanical stress, the use of the AC signal and its resultant mechanical motion of the biosensor system may influence cellular behavior. Furthermore, the sensitivity of the acoustic biosensor signals to mass density, viscoelastic property and charge distribution of cells could complicate data interpretation, particularly for receptor signaling. In addition, acoustic biosensors generally have low throughput.

2.2. Electrical Biosensors

Electrical biosensors employ impedance to characterize cellular responses, including cell adhesion [18, 19]. In a typical setup, live cells are brought in contact with

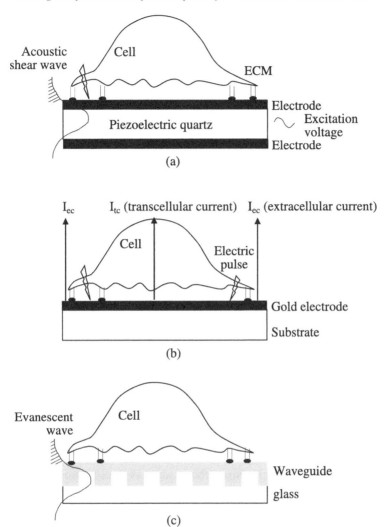

Figure 1. Principles of three types of biosensors for whole cell sensing. (a) An acoustic biosensor for monitoring the viscoelasticity of live cells. Piezoelectric quartz is sandwiched between two electrodes. Live cells are brought in contact with the electrode embedded surface. A surface bound acoustic shear wave is generated by high frequency piezoelectric pulses. (b) An electric biosensor for monitoring the ionic environment surrounding the biosensor and the cells. Cells are brought in contact with the surface of a biosensor having arrayed gold microelectrodes. Flows of both extracellular (I_{ec}) and transcellular (I_{tc}) currents are measured, and a low AC voltage at variable frequencies is applied to the cell layer. (c) A RWG biosensor for monitoring the dynamic mass redistribution in live cells. Cells are brought in contact with the surface of a biosensor consisting of a glass substrate, a waveguide thin film within which a grating structure is embedded. A surface bound electromagnetic wave is generated by illuminating the biosensor with light.

a biosensor surface in which an integrated electrode array is embedded (Fig. 1(b)). Sinusoidal voltages that are swept through a range of frequencies in continuous wave mode are used to generate electric fields between the electrodes, which are impeded by the presence of cells. The electric voltage waves are generated on site using an integrated electric circuit; and the electrical current through the circuit is followed with time. The resultant impedance is a measure of changes in the electrical conductivity or permeability of the cell layer. The cellular plasma membrane acts as an insulating agent forcing the current to flow between or beneath the cells, leading to quite robust changes in impedance. Recent developments in sophisticated algorithm allow the separation of the flows of extracellular and transcellular currents [20]. Impedance-based measurements have been applied to study a wide range of cellular events, including cell adhesion and spreading [21], cell micromotion [22], cell morphological changes [23], cell death [24, 25] and cell signaling [26–29]. Similar to the acoustic biosensors, the use of the electric voltage waves may lead to somewhat unexpected perturbations in cells, which, in turn, could influence the cellular responses upon stimulation. Moreover, since the impedance is largely a function of ionic environment surrounding or across the cell layer, the electrical biosensor is largely sensitive to and, thus, biased towards, changes in cellular morphology and ionic redistribution. In addition, the throughput is often low to medium for the electrical biosensors.

2.3. Optical Biosensors

Optical biosensors primarily employ a surface-bound electromagnetic wave to characterize cellular responses [4, 5, 30]. The surface-bound waves can be achieved on thin metal films using either light excited surface plasmons (surface plasmon resonance, SPR) [31, 32] or on dielectric substrates using diffraction grating coupled waveguide mode resonances (resonance waveguide grating, RWG) [33]. For SPR, the readout is the resonance angle at which a minimal in intensity of reflected light occurs. Similarly, for RWG biosensor, the readout is the resonance angle or wavelength at which a maximum incoupling efficiency is achieved. The resonance angle or wavelength is a function of the local refractive index at or near the sensor surface [33]. Unlike SPR, which is limited to only a few flow channels for assaying at the same time, RWG biosensors are amenable for both high throughput screening (HTS) and cellular assays, due to recent advancements in instrumentation and assays [4, 5, 34]. In a typical RWG, the cells are directly placed into a well of a microtiter plate in which a biosensor consisting of a material with high refractive index is embedded (Fig. 1(c)). Local changes in the refractive index lead to a dynamic mass redistribution (DMR) signal of live cells upon stimulation [9]. These biosensors have been used to study diverse cellular processes, including receptor biology [35–38], ligand pharmacology [39–41] and cell adhesion [6, 9, 42–44]. Since most optical biosensors use long-wavelength light for illumination, these biosensors are believed to be, or close to be, truly non-invasive. Furthermore, these refractive index-sensitive biosensors largely measure DMR in cells upon stimulation; and the resultant DMR signal is mostly associated with actin remodeling, cell adhesion changes and protein trafficking [35–37]. However, since most optical biosensors have limited detection volume due to the short penetration depth of the evanescent wave, the cellular responses measured are often biased towards cellular changes within the bottom portion of cells [9].

3. Label-Free Biosensor Cellular Assays for Cell–Substrate Interactions

The adhesion of live cells to a surface is crucial to the development of biomaterials for tissue engineering and of novel cell culture vessels, as well as to the fundamental understanding of cell biology. Label-free biosensors are well-suited for quantitative assessment of cell–substrate interactions, partly because of their capability of being non-invasive, multi-parametric and kinetic for measurements [9], and partly because of the suitability of the biosensor surfaces to diverse chemical or biochemical modifications [6]. These surface modifications permit a systematic study of cell adhesion as a function of specific extracellular matrix components [45, 46], adhesive ligand density and surface rigidity [47], as well as chemical [48, 49] and mechanical interventions [50]. Thus, it is no surprise that there are many reports describing the use of these biosensors for studying cell adhesion. Here only a few examples are selected to illustrate such applications.

Cell-to-substrate adhesion is an active and dynamic process. To study cell adhesion process with a label-free biosensor, the biosensor surface is often coated with a specific material, and pre-incubated with a medium solution until it reaches an equilibrium state to achieve a steady baseline. Afterwards, cells suspended in the same medium are introduced. The biosensor response is monitored over time. These measurements offer unique views how cells interact with the surface, and how the biosensor surface chemistry governs cell adhesion and functions.

Planar optical waveguides, such as RWG biosensors, offer an ideal substratum for cells to reside on. The waveguide material itself can be the same as the coatings of medical implants (e.g., the oxides of niobium, tantalum and titanium), or can be similar to polymeric materials used for cell culture. The surfaces can be further modified both chemically and morphologically. The substrates are often transparent, enabling simultaneous examinations with light and fluorescence microscopes. For example, Ramsden *et al.* utilized optical waveguide light-mode spectroscopy (OWLS) to study the kinetics of adhesion and spreading of animal cells [42]. Results showed that the adhesion of live cells to the surfaces is sensitive to compositions of both the surface and the cell medium, but follows a typical biphasic process: an initial passive sedimentation phase followed by an active adhesion phase (i.e., spreading). Similar results have been obtained for diverse cell types using OWLS or other types of biosensors [6, 9, 12, 13, 15, 17, 21, 43–45, 51–53]. Interestingly, depending on type of cells and surface chemistry, the biosensor signals of cell adhesion could differ greatly in their fine features, such as kinetics and transition time for the occurrence of the active adhesion phase [6, 51].

The abovementioned OWLS measurements used a RWG biosensor with a relatively short penetration depth ($\sim$200 nm). Using a nanoporous silica support with a refractive index of 1.2 to replace glass, a so-called symmetry waveguide biosensor was achieved [52]. Glass is the most commonly used waveguide support for RWG biosensors, but it has a high refractive index of ~1.5. The symmetry waveguide biosensor supports two different modes and, thus, gives rise to tunable and relatively long penetration depths, enabling multi-depth screening of cell adhesion. Such symmetry RWG biosensor has been shown to be valuable in real-time monitoring the temporal and spatial evolution of cell adhesion process, leading to useful information about refractive index inhomogeneities within the cell layer perpendicular to the biosensor surface.

The ability of cells to recognize, interact and respond to environmental signals, including ECM components, is central to many biological processes including inflammation and organogenesis. Of high interest is to screen drug molecules that intervene the cell adhesion process [5, 17]. Hong et al. [45] utilized an acoustic biosensor to study chemicals to intervene cell–surface adhesion interactions. Results showed that for bovine aortic endothelial cells, soluble RGD peptides significantly retarded the rate of cell adhesion to gelatin coated biosensor surfaces, while heparinase III reduced the long-term cell adhesion. RGD peptides are known to bind to integrins — a class of cell surface receptors responsible for cell–ECM interactions. Heparinase III is an enzyme that digests heparin sulfate selectively from the membrane surface, thus causing degradation of cell surface glycocalyx. Wegener et al. [10] used impedance readout to monitor cell adhesion to different ECM-coated electrode surfaces. Results showed that epithelial MDCK cells displayed surface-dependent kinetic profiles during adhesion, and laminin-mediated cell adhesion primarily through a glycolipid, the so-called Forssman antigen, while fibronectin interaction with integrin receptors initiated a signaling cascade that involved both adhesion and spreading.

Label-free optical biosensors are capable of probing cell biology with a multiparametric and kinetic readout, termed DMR signals. Recently, we developed a HTS-compatible RWG biosensor for whole cell sensing [9]. This biosensor can detect both horizontal and vertical DMR signals in live cells, due to its ability of online monitoring cell responses with multiple optical output parameters. When changes in shape of the resonance spectrum are mainly attributed to stimulation-modulated inhomogeneous redistribution of cellular contents parallel to the sensor surface, shifts in resonance angle or wavelength reflect primarily the vertical DMR signals. Optical signatures were obtained and used to characterize several cellular processes including cell adhesion and spreading, detachment by trypsinization, and signaling through either epidermal growth factor receptor (EGFR) or bradykinin B_2 receptor. For human skin cancerous cell line A431, its adhesion to the biosensor surface leads to both vertical and horizontal mass redistributions within the biosensor sensing volume; cells interact with the surface through multiple steps, each with its own characteristics; and the presence of vincristine not only suppresses the responses, but also reduces the kinetics of cell spreading (Fig. 2). Vincristine is a plant alkaloid that inhibits microtubule assembly by binding to tubulin.

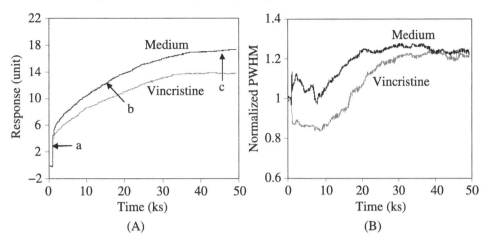

Figure 2. Multi-parametric real-time monitoring of the adhesion and spreading of A431 cells in the absence and presence of vincristine using the shift in the incident angle (A) or the PWHM (Peak-Width-at-Half-Maximum) of the transverse magnetic or p-polarized TM_0 mode (B) with a RWG biosensor. Cells were suspended in culture medium containing 5% fetal bovine serum in the absence and presence of vincristine at room temperature (25°C), and applied to different wells of a 96-well Corning® Epic® biosensor micorplate. The biosensors in the microplate were pre-incubated with the same medium for at least 1 h to reach a steady baseline before cell addition. An angular interrogation imaging system was used to monitor the biosensor response. In the absence of vincristine, the real time kinetics of the shift in the incident angle followed three major phases: an immediate and rapid increased signal due to the increased bulk index, the immobilization of serum proteins onto the sensor surface, and the sedimentation of cells and subsequent contact of cells with the surface (a), followed by a prolonged increased signal due to spreading (b) until it reaches a saturated level (c). The baseline-normalized PWHM of the simultaneously obtained resonance peak was also found to be dynamic and exhibited fine characteristics. These results indicated that the cells interact with the surface through multiple steps, each with its own characteristics. Reproduced with permission from Ref. [9].

The specific targeting of immune cells to sites of infection is crucial to human physiology. Central to these processes is the interaction of cells with adhesion molecules involved in migration and invasion. Thus, it is important to study the interaction of cells with proteins or antibodies for understanding the molecular mechanisms of cell functions including migration and invasion, and identifying specific molecular recognition codes for the homing of cells to different organs. Label-free biosensor cell attachment assays have been developed and applied to identify specific antigen positive cells, including carcinoembryonic antigen (CEA) expressing cells [54]. CEA is one of the most useful human tumor markers.

4. Label-Free Biosensor Cellular Assays for Cell–Cell Interactions

In the human immune defense system, the immune cells are capable of homing to a specific anatomic location from the blood, sticking to the surface of blood vessels, penetrating the extracellular matrix surrounding the vessel, and then targeting the damaged tissues, resulting in the characteristic curing of these tissues. Critical to these processes is the interaction between an effector cell and a target cell, and the effector cell-mediated cytotoxicity of target cells [55, 56]. Cell-mediated killing requires cell-to-cell contact, which is mediated by the pairwise recognition between multiple receptors presented on the surfaces of effector and target cells. To study cell–cell interactions with a label-free biosensor, the target cells are brought in contact with the biosensor surface, primarily through cell culture. Once a monolayer of target cells is formed, the cell medium is often replaced with an assay buffer. Once a steady baseline is established within the biosensor detection system, the effector cells suspended in the same assay buffer are introduced, primarily *via* a liquid handling device, to the biosensor having the target cell layer. The biosensor response is monitored over time. For example, Glamann and Hansen [57] utilized an electrical biosensor, termed as real-time cell electrical sensing (RT-CES), to detect the interactions between natural killer (NK) cells in suspension and adherent breast cancer cells MCF7 cultured on the electrode biosensor surface. Results showed that NK cells caused apoptosis of MCF7 cells, and the NK cell-mediated killing was dependent on the NK cell-to-target cell ratio. The NK cells are considered the major cytotoxic effector cells for innate immunity that can recognize and kill malignantly transformed and infected cells.

5. Label-Free Biosensor Cellular Assays for Cell Signaling Involving Adhesion

Cell adhesion is a dynamic remodeling process of live cells upon stimulation with environmental cues. The cell adhesion network contains many classes of signaling molecules, including adaptor proteins, cytoskeletal proteins, actin-binding proteins, kinases, protein phosphatases, transmembrane receptors and adhesion proteins [58]. Many of these adhesion components can be switched on and off by signaling elements. Label-free biosensor cellular assays are well-suited for studying cell signaling, particularly cellular pathways that lead to the alteration of cell adhesion. To study cell signaling with a label-free biosensor, a cell line expressing a target receptor is brought in contact with the biosensor surface, primarily through cell culture. Once a monolayer of cells is formed, the cell medium is often replaced with an assay buffer. Once a steady baseline is established within the biosensor detection system, a solution having an activator that turns on the target receptor is introduced, primarily *via* a liquid handling device, to the biosensor having the cultured cells. The biosensor response is monitored over time. For example, EGFR is a cell surface receptor and plays important roles in cell migration, adhesion, and proliferation, as well as in innate immune response in human skin. Mutations that lead to EGFR overexpression or overactivity have been associated with a number of cancers, including lung, colon and breast cancers [59]. Recently we had employed the RWG biosensor technology, termed as MRCAT [9, 35], to systematically study systems cell biology of EGFR in A431 cells. In conjunction with conventional cell biology and chemical biology approaches, we had found that EGFR signaling was strongly dependent on the cellular status (i.e., proliferating or quiescent states), and the DMR signal of quiescent A431 cells upon stimulation with EGF required EGFR tyrosine kinase activity, actin polymerization and dynamin, and proceeded to cell detachment mainly through the MEK pathway.

6. Concluding Remarks

Cell adhesion is a fundamental process for many biotechnologies ranging from cell culture, bioprocess, tissue engineering to drug discovery. Cell adhesion is also central to human physiology and pathophysiology. Label-free biosensors provide a generic, multi-parametric, and kinetic measure of cell adhesion and its associated cell signaling pathways. Their ability to be non-invasive, label-free and generic permits the study of diverse arrays of cellular processes including cell adhesion and proliferation. With recent advancements in biosensor instrumentation such as Epic$^{®}$ system, high throughput screening becomes a reality for both biochemical interaction analysis and the cellular activity assessments of drug molecules. The robustness of biosensor output signals, partly due to the significant amplification of cells as delicate signaling machineries and partly due to the improvements in sensitivity of new-generation biosensors, allows for large scale applications such as drug discovery. Surface modifications and patterning can provide novel dimensions to control and fine-tune cell adhesion process and, thus, cell functions. The next generation biosensor systems will have higher sensitivity and spatial resolution [60], and smarter assay designs due to our ever-evolving understanding of cell adhesion [61, 62], thus, will provide new perspectives in cell biology including cell adhesion.

Acknowledgement

This work was partially supported by the National Institutes of Health Grant 5U54MH084691.

References

1. M. A. Schwartz and M. H. Ginsberg, *Nature Cell Biol.* **4**, E65 (2002).
2. A. R. Ramjaun and K. Hodivala-Oilke, *Intl. J. Biochem. Cell Biol.* **41**, 521 (2009).
3. R. Zaidel-Bar, S. Itzkovitz, A. Ma'ayan, R. Iyengar and B. Geiger, *Nature Cell Biol.* **9**, 858 (2007).
4. Y. Fang, *Assay Drug Dev. Technol.* **4**, 583 (2006).
5. Y. Fang, A. G. Frutos and R. Verkleeren, *Comb. Chem. High Throughput Screen.* **11**, 357 (2008).
6. J. J. Ramsden and R. Horvath, *J. Receptors Signal Transduction* **29**, 211 (2009).
7. C. Cozens-Roberts, J. A. Quinn and D. A. Lauffenburger, *Biophys. J.* **58**, 857 (1990).
8. G. Wrobel, M. Holler, S. Ingebrandt, S. Dieluweit, F. Sommerhage, H. P. Bochem and A. Offen-hausser, *J. Roy. Soc. Interface* **5**, 213 (2008).
9. Y. Fang, A. M. Ferrie, N. H. Fontaine, J. Mauro and J. Balakrishnan, *Biophys. J.* **91**, 1925 (2006).
10. J. Wegener, C. R. Keese and I. Giaever, *Exp. Cell Res.* **259**, 158 (2000).

11. A. Alessandrini, M. A. Croce, R. Tiozzo and P. Facci, *Appl. Phys. Lett.* **88**, 083905 (2006).
12. J. Wegener, J. Seebach, A. Janshott and H.-J. Galla, *Biophys. J.* **78**, 2821 (2000).
13. D. Le Guillou-Buffello, R. Bareille, M. Gindre, A. Sewing, P. Laugier and J. Amédée, *Tissue Eng. — Part A* **14**, 1445 (2008).
14. F. Li, J. H.-C. Wang and Q.-M. Wang, *Sensors Actuators B* **128**, 399 (2008).
15. J. Li, C. Thielemann, U. Reuning and D. Johannsmann, *Biosensors Bioelectronics* **20**, 1333 (2005).
16. L. Tan, X. Jia, X. Jiang, Y. Zhang, H. Tang, S. Yao and Q. Xie, *Biosensors Bioelectronics* **24**, 2268 (2009).
17. T. S. Hug, *Assay Drug Dev. Technol.* **1**, 479 (2003).
18. C. A. Thomas Jr., P. A. Springer, G. E. Loeb, Y. Berwald-Netter and L. M. Okum, *Exp. Cell Res.* **74**, 61 (1972).
19. I. Giaever and C. R. Keese, *Proc. Natl. Acad. Sci. USA* **81**, 3761 (1984).
20. G. J. Ciambrone, V. F. Liu, D. C. Lin, R. P. McGuinness, G. K. Leung and S. Pitchford, *J. Biomol. Screen.* **9**, 467 (2004).
21. C. Tiruppathi, A. B. Malik, P. J. Del Vecchio, C. R. Keese and I. Giaever, *Proc. Natl. Acad. Sci. USA* **89**, 7919 (1992).
22. I. Giaever and C. R. Keese, *Proc. Natl. Acad. Sci. USA* **88**, 7896 (1991).
23. I. Giaever and C. R. Keese, *Nature* **366**, 591 (1993).
24. J. Zhu, X. Wang, X. Xu and Y. A. Abassi, *J. Immunol. Methods* **309**, 25 (2006).
25. C. Xiao, B. Lachance, G. Sunahara and J. H. Luong, *Anal. Chem.* **74**, 5748 (2002).
26. N. Yu, J. M. Atienza, J. Bernard, S. Blanc, J. Zhu, X. Wang, X. Xu and Y. A. Abassi, *Anal. Chem.* **78**, 35 (2006).
27. M. F. Peters and C. W. Scott, *J. Biomol. Screen.* **14**, 246 (2009).
28. B. Xi, N. Yu, X. Wang, X. Xu and Y. A. Abassi, *Biotechnol. J.* **3**, 484 (2008).
29. E. Verdonk, K. Johnson, R. McGuiness, G. Leung, Y.-W. Chen, H. R. Tang, J. M. Michelotti and V. F. Liu, *Assay Drug Dev. Technol.* **4**, 609 (2006).
30. Y. Yanase, H. Suzuki, T. Tsutsui, T. Hiragun, Y. Kameyoshi and M. Hide, *Biosensors Bioelectronics* **22**, 1081 (2007).
31. R. H. Ritchie, *Phys. Rev.* **106**, 874–881 (1957).
32. M. A. Cooper, *Nature Rev. Drug Discov.* **1**, 515 (2002).
33. K. Tiefenthaler and W. Lukosz, *J. Opt. Soc. Am. B* **6**, 209 (1989).
34. E. Tran and Y. Fang, *J. Biomol. Screen.* **13**, 975 (2008).
35. Y. Fang, A. M. Ferrie, N. H. Fontaine and P. K. Yuen, *Anal. Chem.* **77**, 5720 (2005).
36. Y. Fang, G. Li and J. Peng, *FEBS Lett.* **579**, 6365 (2005).
37. Y. Fang, G. Li and A. M. Ferrie, *J. Pharmacol. Toxicol. Methods* **55**, 314 (2007).
38. R. Schroeder, N. Merten, J. M. Mathiesen, L. Martini, A. K. Letunic, F. Krop, A. Blaukat, Y. Fang, E. Tran, T. Ulven, C. Drewke, J. Whistler, L. Pardo, J. Gomeza and E. Kostenis, *J. Biol. Chem.* **284**, 1324 (2009).
39. Y. Fang and A. M. Ferrie, *FEBS Lett.* **582**, 558 (2008).
40. P. H. Lee, A. Gao, C. van Staden, J. Ly, J. Salon, A. Xu, Y. Fang and R. Verkleeren, *Assay Drug Dev. Technol.* **6**, 83 (2008).
41. J. Antony, K. Kellershohn, M. Mohr-Andrä, A. Kebig, S. Prilla, M. Muth, E. Heller, T. Disingrini, C. Dallanoce, S. Bertoni, J. Schrobang, C. Trankle, E. Kostenis, A. Christopoulos, H.-D. Höltje, E. Barocelli, M. De Amici, U. Holzgrabe and K. Mohr, *FASEB J.* **23**, 442 (2009).
42. J. J. Ramsden, S. Y. Li, E. Heinzle and J. E. Prenosil, *Biotechnol. Bioeng.* **43**, 939 (1994).

43. R. Horvath, K. Cottier, H. C. Pedersen and J. J. Ramsden, *Biosensors Bioelectronics* **24**, 805 (2008).
44. T. S. Hug, J. E. Prenosil and M. Morbidelli, *Biosensors Bioelectronics* **16**, 865 (2001).
45. S. Hong, E. Ergezen, R. Lec and K. A. Barbee, *Biomaterials* **27**, 5813 (2006).
46. J. Li, C. Thielemann, U. Reuning and D. Johannsmann, *Biosensors Bioelectronics* **20**, 1333 (2005).
47. E. A. Scott, M. D. Nichols, L. H. Cordova, B. J. George, Y.-S. Jun and D. L. Elbert, *Biomaterials* **29**, 4481 (2008).
48. Q. Liu, J. Yu, L. Xiao, J. C. O. Tang, Y. Zhang, P. Wang and M. Yang, *Biosensors Bioelectronics* **24**, 1305 (2009).
49. S. J. Braunhut, D. McIntosh, E. Vorotnikova, T. Zhou and K. A. Marx, *Assay Drug Dev. Technol.* **3**, 77 (2005).
50. M. S. Jenkins, K. C. Y. Wong, O. Chhit, J. F. Bertram, R. J. Young and N. Subaschandar, *Biotechnol. Bioeng.* **88**, 392 (2004).
51. A. Aref, R. Horvath, J. McColl and J. J. Ramsden, *J. Biomedical Optics* **14**, 010501 (2009).
52. R. Horvath, K. Cottier, H. C. Pedersen and J. J. Ramsden, *Biosensors Bioelectronics* **24**, 799 (2008).
53. I. Giaever and C. R. Keese, *Jpn. J. Appl. Phys.* **47** (Part 2), i–v (2008).
54. M. Zourob, S. Elwary, X. Fan, S. Mohr and N. J. Goddard, *Methods in Molecular Biology: Biosensors and Biodetection* **503**, 89 (2009).
55. L. L. Lanier, *Annu. Rev. Immunol.* **23**, 225 (2005).
56. J. Lieberman, *Nat. Rev. Immunol.* **3**, 361 (2003).
57. J. Glamann and A. J. Hansen, *Assay Drug Dev. Technol.* **4**, 555 (2006).
58. R. Zaidel-Bar, S. Itzkovitz, A. Ma'ayan, R. Iyengar and B. Geiger, *Nature Cell Biol.* **9**, 858 (2007).
59. H. Zhang, A. Berezov, Q. Wang, G. Zhang, J. Drebin, R. Murali and M. I. Greene, *J. Clin. Invest.* **117**, 2051 (2007).
60. F. Vollmer and S. Arnold, *Nature Methods* **5**, 591 (2008).
61. D. E. Discher, P. Janmey and Y.-L. Wang, *Science* **310**, 1139 (2005).
62. J. T. Groves, *Signal Transduction Knowledge Environment* **301**, e45 (2005).

Cell Adhesion Monitoring Using Substrate-Integrated Sensors

Andreas Janshoff[a], **Angelika Kunze**[a], **Stefanie Michaelis**[b], **Vanessa Heitmann**[b], **Bjoern Reiss**[b] **and Joachim Wegener**[b,*]

[a] Institut für Physikalische Chemie, Universitaet Goettingen, Tammannstr. 6, 37077 Goettingen, Germany
[b] Institut fuer Analytische Chemie, Chemo- & Biosensorik, Universitaet Regensburg, Universitaetsstr. 31, 93053 Regensburg, Germany

Abstract
Adhesion of mammalian cells to *in vitro* surfaces is an area of active research and it attracts considerable interest from various scientific disciplines, most notably from medical technology and biotechnology. One important issue in the context of cell–surface adhesion is the time course of attachment and spreading upon surfaces that are decorated with proteins to make them cytocompatible. This article reviews two emerging non-microscopic techniques capable of monitoring the adhesion process label-free and in real-time. Both approaches, electric cell–substrate impedance sensing (ECIS) and the quartz crystal microbalance (QCM), are based on substrate-integrated transducers that transduce cellular adhesion into an electrical signal. A short introduction of both techniques is followed by a set of examples that illustrate the performance of these sensors, their individual merits and limitations. In order to analyze the integral and complex signals of both sensors in contact with mammalian cells in more detail, we also studied their individual readouts during the adsorption of liposomes with well-defined structure and chemical composition.

Keywords
Electric cell–substrate impedance sensing (ECIS), quartz crystal microbalance (QCM), cell adhesion, cell spreading, cell-matrix interactions, cell–surface junction, liposome adsorption

1. Adhesion of Animal Cells to *in Vitro* Surfaces: Why Bother?

The interactions of living cells with *in vitro* surfaces play a key role in a growing range of biomedical and biotechnological applications that aim to anchor cells tightly to inorganic substrates *in* or *ex vivo*. But also the opposite situation is sometimes of importance when all kinds of efforts are undertaken to keep cells from adhering upon such surfaces. In both cases suitable sensors and devices are required to study the adhesion of cells to such substrates. In this context, the term *cell*

* To whom correspondence should be addressed. Tel.: +941-943-4546; Fax: +941-943-4491; e-mail: Joachim.Wegener@chemie.uni-regensburg.de

Surface and Interfacial Aspects of Cell Adhesion
© Koninklijke Brill NV, Leiden, 2010

generally encompasses both eukaryotic as well as prokaryotic organisms but this article will only deal with sensors to detect the adhesion of *animal or human cells* to *in vitro* surfaces. Throughout this manuscript we will use the terms biomaterial surface, material surface, technical surface or *in vitro* surface as synonyms for the surfaces of man-made materials that are used for implants or other devices that are brought into contact with living mammalian cells.

Medical technology is one of the most important areas in which the interactions between animal or even human cells and biomaterial surfaces is of critical importance, in particular when it comes to designing endoprostheses or implants. When an implant is placed inside a living organism in order to fulfill structural or functional tasks, its biocompatibility is an unconditional prerequisite [1]. Very often the term biocompatibility means integration of the device into the target tissue and settling of the tissue-specific cells on its surface without creating a foreign-body response. But in many cases this is just a minimum requirement and the proper functionality of the implant requires strong and mechanically stable adhesion of the cells to the surface. Well-known examples of such implants are polymer tubes that are used to create bypasses, for instance, around plugged coronary arteries. In order to make the polymer surface blood compatible, endothelial cells (i.e. the cells that line the native blood vessels *in vivo*) are grown on the inner surface of the tubing, providing a vascular surface similar to the one inside the native vessels [2, 3]. Since the endothelial cells are exposed to significant shear forces by the circulating blood stream, their adhesion to the inner wall of the tubing is crucial and has to withstand considerable mechanical stress. In the exciting development of neuroprostheses, the requirements with respect to the cell-material contact zone are even more challenging as a real functional interfacing of the cells with electrodes or semiconductor devices is required. Besides mechanical stability, the cell–surface junction has to allow for a sensitive bi-directional transfer of electrical signals between the cells and the *in vitro* transducer [4, 5]. Animal cells interfaced with semiconductor devices have nowadays become emerging tools for drug and cytotoxicity screening *in vitro* [6].

Another less medical but more biotechnological example that demonstrates the importance of tailor-made cell–substrate adhesion and proper sensing devices to study it is the design of bioreactors in which animal cells are used for the large-scale production of proteins or fine chemicals. To improve the space–time-yield of such processes, the producer cells are frequently grown on the surface of micro-carrier beads that provides a higher ratio of available growth surface to reactor volume. However, the cells have to withstand the shear forces associated with the flow-through of medium that provides oxygen and nutrients. Thus, the anchorage of the cells to the surface of the carrier beads is decisive for the productivity of the bioreactor [7] and needs to be studied to find the optimum process conditions.

The interactions of cells with *in vitro* surfaces are considered and studied in the above-mentioned fields of applied science where they are important for the functionality of a certain process or device. Whenever cells are cultured *in vitro* they will encounter *in vitro* surfaces, attach to them *via* cell–surface interactions and may even require substrate anchorage for their survival. It is well known that so-called *anchorage-dependent cells* will die if they cannot find a suitable place to adhere to and express proper cell–surface interactions [7]. But the surfaces have to fulfill certain conditions in order to be accepted or tolerated as an adhesion site for the cell. It is important to recognize that in many cell types certain signal transduction cascades are triggered upon attachment and spreading on a particular surface and this, in turn, may alter their differentiation or functional properties. Thus, the appropriate *in vitro* handling of stem cells, that typically show an undifferentiated phenotype but may follow certain wanted or unwanted differentiation pathways if they encounter *in vitro* surfaces, requires detailed knowledge about the molecular architecture of their cell–surface adhesion sites [8].

All the above-mentioned examples demonstrate that cell–surface adhesion plays a role in many fields of applied and fundamental research. A major problem that might have hampered research progress is the fact that the contact zone between cell and surface is not easy to access experimentally since it is buried between the cell body on the one side and the substratum on the other. Modern and extremely powerful techniques like scanning force microscopy (SFM), scanning electron microscopy (SEM) or other scanning probe techniques can provide very detailed images of the upper cell surface but they do not have access to the interface between lower cell membrane and the substratum to which the cell is attached. Of course, the cell bodies can be removed in order to study the molecular composition on the surface but it is an inherent problem whether the removal of the cell body has changed the interface with respect to its chemical and structural composition. To overcome this limitation, specialized analytical techniques and sensors have been developed that are tailor-made to study the contact area between cells and biomaterial surfaces. In this article we will focus on two emerging techniques that are based on physical sensors that are an integral part of the growth surface and that have proven to be very versatile and sensitive to monitor cell adhesion.

2. Key Events during Adhesion of Animal Cells to *in Vitro* Surfaces

Attachment and spreading of cells on *in vitro* surfaces is an enormously complex process. It requires the presence of adhesion-promoting proteins that are pre-immobilized on the surface of interest providing binding sites for specific receptor molecules expressed on the cell surface. Animal cells are unable to adhere to bare surfaces that are not decorated with pre-adsorbed proteins. The adhesion-promoting proteins are either pre-deposited from a protein solution before cells are seeded or the adhesive molecules are synthesized and secreted by the cells themselves. In particular, the latter option is often overlooked when cells attach and spread on an *in vitro* surface in a protein-free environment. Given this critical dependence of cell adhesion on the presence of adhesion-promoting proteins on the surface, it is obvious that the suitability of the surface to allow for protein adsorption is a prerequisite for being *cell compatible*. Here, wettability and surface topography are generally considered as the most relevant surface parameters for protein adsorption [9]. Neglecting the differences observed for individual proteins, there seems to be a general trend that the amount of protein that adsorbs on a given surface is higher when the surface is hydrophobic. However, when encountering a hydrophobic surface the protein may experience unfolding accompanied by a loss of its adhesive properties. On a hydrophilic surface the amount of protein that adsorbs is generally smaller but the protein retains its biological folding and will be recognized by the cell–surface receptors. Thus, this very general concept implies that *in vitro* surfaces should be hydrophilic in nature in order to be well suited for cell adhesion. Protein adsorption on a suitable surface generally occurs instantaneously. Spatz and collaborators recently demonstrated that the cells need suitable binding sites on the surface within certain distances in order to nucleate the formation of stable adhesion sites. Distances between individual surface binding sites of more than 73 nm do not allow for stable cell adhesion, while distances of less than 58 nm provided mechanically stable focal adhesion sites [10]. *Focal adhesion* is a term used in cell biology to describe the accumulation of individual, molecular cell-to-surface interactions in certain spots or foci on the surface which are particularly stable.

When a cell suspension is allowed to settle on an *in vitro* surface that is decorated with adhesive proteins, the initially suspended and mostly spherical cells first have to get close to the surface by sedimentation. The time necessary for a cell to reach the surface can be calculated from Stoke's law. For a spherical cell of 10 μm radius the sedimentation velocity can be estimated to be approximately 1 mm/min (assuming a cell density of 1.05 g/cm^3). Close to the surface, however, sedimentation becomes slower as the drainage of fluid between cell body and the flat surface requires additional time. Formation of first adhesive contacts between an almost spherical cell and the *in vitro* surface is generally considered as *attachment* or *adhesion*. Soon after first molecular contacts between substrate-immobilized proteins and the corresponding receptors on the cell surface have been established, the cells start to actively spread out accompanied by an extension of their contact area as illustrated in Fig. 1. Spreading out on the growth surface and areal extension of the cell-to-surface junction requires overcompensation of the cortical tension of the plasma membrane and the formation of new cell–surface contact sites along the periphery of the advancing cell body. The latter is a complex competition between specific cell–surface receptors and non-specific electrostatic interactions on the one hand with steric repulsion due to the necessary compression of extracellular material by the anchored cell body on the other [11]. It has been shown recently that the rate of cell spreading, s, is directly proportional to the ratio of energy necessary for cell adhesion E_{adh} and the cortical tension of the membrane σ_{Mem} [12]:

$$s \propto E_{adh}/\sigma_{Mem}. \tag{1}$$

Thus, reliable measurements of spreading kinetics for a given cell type but different biomaterial surfaces will provide the individual adhesion energies E_{adh} on a relative

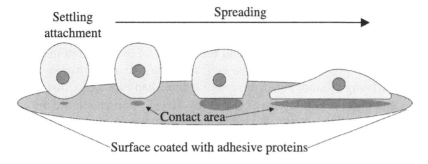

Figure 1. Changes in the three-dimensional shape of a mammalian cell during spreading on a cytocompatible culture substrate. The contact area between cell and surface increases continuously forming the cell-material interface.

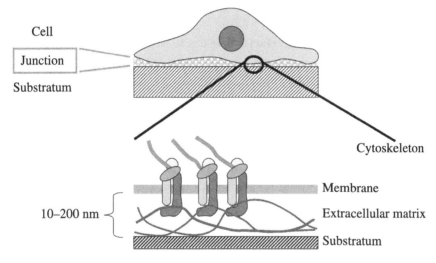

Figure 2. Schematic of the cell-material interface. Anchorage is provided by cell–surface receptors that bind specifically to components of the extracellular matrix on the substrate surface and are linked to the cytoskeleton on the intracellular side.

scale if one assumes that the cortical tension of the membrane is not affected by the nature of the surface, which is a reasonable assumption.

The geometrical distance between the substrate-facing cell membrane and the material's surface, after the cell has completely attached and spread, is a matter of debate. Figure 2 illustrates the situation at the cell–surface junction underneath a spread animal cell. Some authors report that in areas of closest adhesion between membrane and surface, so called *focal contacts,* the membrane gets as close as 5 nm to the surface [13]. Others report distances of up to 30 nm in these special regions. Areas in which the membrane is free of cell–surface receptors and, thus, is not involved in substrate adhesion can be as far as 200 nm away from the surface. However, in contrast to textbook presentations, Iwanaga *et al.* [14] did not find any correlation between the sites of *focal contacts* and sites of closest apposition of cell membrane and substrate. The *average* cell–substrate separation distance (averaged along the entire contact area) is generally reported to be between 25 and 200 nm, depending on the cell type and the coating of the substratum. The physical and chemical properties inside the cleft between cell and substrate are just on the verge of being characterized. It is well established nowadays that the narrow channel contains at least proteins and carbohydrates — which are the most prominent constituents of the extracellular material in almost any soft tissue — as well as water and salts. Very recent studies report that the specific conductivity within this small channel is indistinguishable from the bulk fluid that the cells are bathed in [15]. However, when vesicles or erythrocyte ghosts were attached to a surface, it was found that the conductivity in the remaining cleft was considerably reduced [16]. The reason for this change in ion mobility beneath the cells remains to be resolved.

3. ECIS and QCM: Two Interfacial Sensors to Monitor the Dynamics of Cell Adhesion

The number of experimental techniques capable of probing the contact area between adherent cells and their growth support is rather limited. All these techniques have in common that they were designed for one of the following objectives (i) to image the contact area between cells and substrate, (ii) to measure the distance between basal (i.e., substrate-facing) plasma membrane and surface, or (iii) to monitor changes in cell adhesion as a function of time. It is beyond the scope of this review to provide a comprehensive survey about all these available methods and to list their individual performances and limitations. The interested reader is referred to an article by one of us [17] and the references therein.

In this article, however, we focus on two unrelated, non-imaging techniques that are both label-free, non-invasive and capable of providing the dynamic aspect of cell adhesion in real-time. In the literature they are referred to as *Electric Cell-Substrate Impedance Sensing* (ECIS) and the *Quartz Crystal Microbalance* (QCM) techniques. In both cases the cells are allowed to adhere directly to the surface of the sensor and the accompanying sensor response is quantified by means of non-invasive electrical measurements. Compared to invasive ultra-structural or label-dependent techniques, the common strength of both approaches is to provide a dynamic time-resolved readout that allows to resolve even subtle details of the adhesion kinetics. In the following sections both approaches will be introduced briefly before their performance with respect to monitoring cell adhesion is discussed and compared.

3.1. A Brief Introduction to the ECIS Sensor Device

The idea of electric cell–substrate impedance sensing (ECIS) was introduced by Ivar Giaever and Charles R. Keese, who were the first to grow mammalian cells directly on the surface of gold-film electrodes and to record and analyze the corresponding changes in the electrode's electrical impedance [18, 19]. In the meantime

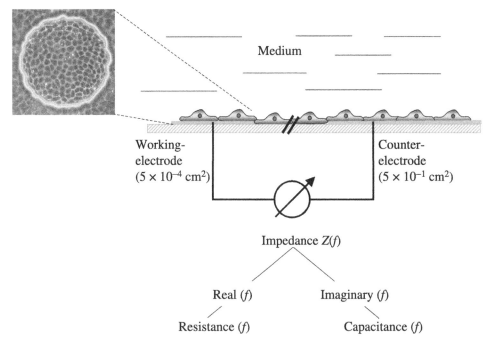

Figure 3. The principal of electric cell–substrate impedance sensing (ECIS). The cells under study are grown on the surface of gold-film electrodes deposited on the bottom of a cell culture vessel. The impedance measurement is dominated by the electrical properties of the smaller working electrode. The area of the working electrode is delineated by a circular opening ($d = 250$ µm) in a photoresist overlayer that insulates the rest of the deposited gold from the bulk electrolyte. Note that the size of the electrode in the magnified presentation is not drawn to scale with respect to cell size.

several other groups have followed this basic concept but used other electrode materials, geometries, or different recording setups [20–22].

The basic principle of the ECIS technique is sketched in Fig. 3. Two coplanar gold-film electrodes are deposited on the bottom of a polymer dish and the cells are allowed to settle, adhere and spread on these surfaces under ordinary cell culture conditions. The electrical connection between the two electrodes is provided by the culture medium, a buffered physiological salt solution of app. 15 ms/cm conductivity that contains all nutrients and growth factors that the cells require. The two electrodes differ with respect to their surface areas. By making the counter electrode 1000 times larger than the small working electrode, the impedance of the latter dominates the read-out of the entire circuit. Thus, the observed changes in electrical impedance can be clearly assigned to changes that occur at the small working electrode with negligible contributions due to the presence of cell bodies on the counter electrode or the electrical wiring of the setup. The size of the working electrode is one of the most critical parameters when the sensitivity of the measurement is considered [23]. For bigger working electrodes it becomes increasingly difficult to record cell-related changes in impedance as the resistance of the bulk electrolyte gradually masks the impedance of the cell-covered electrode. Most data available in literature have been recorded with circular electrodes of 250 μm diameter.

In ECIS, impedance data are usually recorded over a frequency range between 1–10^6 Hz. In terms of information content this broad frequency range can be subdivided into two regimes:

(a) For the major fraction of this frequency band (<10 kHz for working electrodes with 250 μm diameter, as used here) the cells behave essentially like insulating particles forcing the current to flow around the cell body on paracellular pathways. Current leaving a cell-covered electrode has to flow through the confined and narrow channels between the ventral plasma membrane and the electrode surface before it can escape through the paracellular shunt between adjacent cells into the bulk phase (Fig. 2). Since the current has to bypass the cell bodies it picks up impedance contributions from the cell–substrate adhesion zone as well as from the contact area between neighboring cells. Readings of the total impedance are thus sensitive to changes in cell–cell and cell–substrate contacts or cell shape in general. Recording the impedance at several frequencies in this regime together with theoretical modeling allows to assign the individual impedance contributions either to the cell–substrate or the cell–cell contact sites. Details of the modeling are not addressed here but the interested reader is referred to [24].

(b) For monitoring cell adhesion, high frequency readings (>10 kHz for working electrodes with 250 μm diameter, as used here) of the complex impedance are particularly useful [25]. At these frequencies, the current can capacitively couple through the cells passing the ventral and the dorsal membranes in the form of a displacement current. At these frequencies it does not flow around the cells to a significant extent.

For cell adhesion studies it is worthwhile to look deeper into measurements of the complex impedance. The impedance, in general, consists of a real (resistance, R_{total}) and an imaginary part (reactance, X_{total}). The former contains all resistive contributions of the system, whereas the latter includes all capacitive contributions. From equation (2)

$$C_{total} = 1/(2 \cdot \pi \cdot f \cdot X_{total}) \tag{2}$$

it is possible to calculate the equivalent capacitance of the whole system C_{total} which is the most useful parameter in terms of dynamic cell adhesion monitoring as will be detailed below. In equation (2) f denotes the AC frequency.

For a cell-free electrode the measured capacitance C_{total} at a given frequency is equal to the electrode capacitance:

$$C_{total} = C_{electrode}. \tag{3}$$

For a *cell-covered* electrode and frequencies above 10 kHz, C_{total} is comprised of the capacitance of the electrode (as before) and the capacitances of the two membranes (ventral and dorsal) which are now arranged in series to

$C_{\text{electrode}}$. According to Kirchhoff's laws about electrical circuitry the measured capacitance C_{total} for an electrode entirely covered with cells is

$$1/C_{\text{total}} = 1/C_{\text{cell-covered}} = 1/C_{\text{electrode}} + 1/C_{\text{membrane1}} + 1/C_{\text{membrane2}}. \quad (4)$$

During cell attachment and spreading on the electrode surface, the fraction of the electrode that is covered with cells — or the plasma membranes of the cells — increases with time. As the capacitance scales with the area the following relationship applies for all intermediate situations with a partly covered electrode:

$$C_{\text{total}} = (1 - x_{\text{coverage}}) \cdot C_{\text{cell-free}} + x_{\text{coverage}} \cdot C_{\text{cell-covered}} \quad (5)$$

with x_{coverage} as the ratio of cell-covered and total electrode area ($0 \leqslant x_{\text{coverage}} \leqslant 1$). Correspondingly, the total capacitance C_{total} changes linearly with the fraction of the electrode that is covered by a planar layer of two cell membranes and thus with the fraction of the electrode covered by spread cells.

In summary, measuring the total capacitance of the system is the most direct approach to monitor the coverage of the electrode surface as a function of time. As the capacitance is easy to record at this frequency, measurements of C_{total} provide an easy and accurate determination of the spreading kinetics which can be used to determine the adhesion energy as stated above. However, this very convenient linear relationship only applies as long as the frequency of the AC current is high enough such that all current traversing the cell layer takes the transcellular route across the plasma membranes.

It should be noted at this point that other authors [20] chose to interpret the measured impedance by using an equivalent circuit of an overall resistance (R_{p}) in parallel to an overall capacitance (C_{p}). Plotting this parallel capacitance C_{p} as a function of time is, however, not an equivalent way of monitoring cell spreading as the parallel capacitance responds in a more complicated way to electrode surface coverage at most frequencies.

3.1.1. ECIS-Based Cell Adhesion Monitoring

The considerations discussed in the preceding section indicated that measuring the total capacitance of the ECIS electrode C_{total} at a sampling frequency >10 kHz should provide a sensitive measure to follow the spreading of animal cells in real time. We have studied the suitability of C_{total} experimentally by performing different sets of experiments that all required a very detailed recording of spreading kinetics. As sampling frequency we have arbitrarily chosen 40 kHz but any frequency above the frequency threshold of 10 kHz works similarly well. From an experimental viewpoint it is important to stress that we used rather high cell densities in a more or less mono-disperse cell suspension for inoculation ($5 \times 10^5/\text{cm}^2$) to ensure that the recorded signal only reported on cell attachment and spreading without any contributions from cell proliferation. The number of cells seeded into the electrode containing wells was sufficient to form a confluent monolayer on the

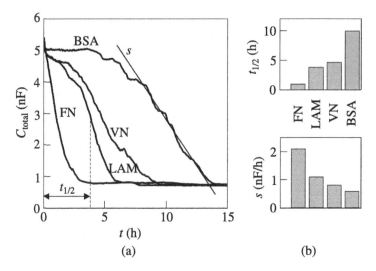

Figure 4. (a) Time courses of the capacitance C_{total} measured at a sampling frequency of 40 kHz when MDCK cells were seeded into ECIS arrays whose electrodes had been coated with fibronectin (FN), laminin (LAM), vitronectin (VN) or bovine serum albumin (BSA) in 100 µg/ml concentrations, respectively. The slope of each curve s between $C_{total} = 4$ nF and $C_{total} = 2$ nF — equivalent to the apparent rate of spreading — was extracted by linear regression as shown for the BSA-coated electrode. (b) Half-times $t_{1/2}$ and apparent spreading rates s as determined from the data shown in (a) [25].

surface without any need for cell division. Moreover, this approach ensured a homogeneous coverage of the well bottom so that no normalization for local cell density was necessary.

Figure 4(a) shows the time course of the electrode capacitance C_{total} at a sampling frequency of 40 kHz when equal numbers of suspended MDCK-II cells (Madin Darby canine kidney) attached and spread on four ECIS electrodes that had been coated with different protein layers prior to cell seeding [25]. Two parameters have been extracted to quantitatively compare the dynamics of cell spreading. The parameter $t_{1/2}$ denotes the time required for half-maximum spreading of the cells; whereas the parameter s stands for the apparent spreading rate (Fig. 4(b)), which is deduced from the slope of the curve at $t = t_{1/2}$ (Fig. 4(a)). It is noteworthy that according to equation (1), the spreading rate s is directly proportional to the adhesion energy E_{adh} of the cells for a particular surface.

The four traces in Fig. 4(a) show that MDCK cells apparently attach and spread much faster on surfaces coated with fibronectin (FN) compared to all other proteins used in this experiment. According to the parameters $t_{1/2}$ and s, the kinetics for cell adhesion to vitronectin (VN) and laminin (LAM) decorated electrodes is rather similar. For bovine serum albumin (BSA) coating it requires more than five hours before the cells even start to spread out significantly. The half-times $t_{1/2}$ of LAM, VN and BSA coatings clearly mirror this huge difference in spreading dynamics. Consistent with these findings BSA is considered to be a non-adhesive protein. The

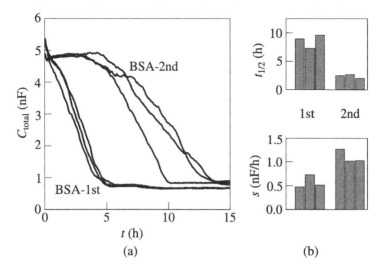

Figure 5. (a) Time course of the capacitance C_{total} measured at a sampling frequency of 40 kHz when MDCK cells were seeded into ECIS arrays whose electrodes had been coated with 100 µg/ml BSA (BSA-1st). After 24 h these cells were gently removed from the surface and a fresh cell suspension was seeded on the same electrodes (BSA-2nd). The figure shows triplicates for each condition. Serum-free medium was used in both cases. The figure shows (b) half-times $t_{1/2}$ and apparent spreading rates s as determined from the data shown in (a) [25].

apparent spreading rate s for BSA, however, is surprisingly close to the values for LAM and VN indicating at first sight that the cells eventually spread on this protein layer with similar kinetics — or in other words and according to equation (1) that the adhesion energy is apparently similar for these three coatings.

The experiment shown in Fig. 5 explains these data from a different viewpoint [25]. Here, equal numbers of MDCK cells were inoculated on three electrodes that were all pre-coated with BSA under identical conditions (1st inoculation, Fig. 5(a), BSA-1st). After 20 hours the cells were gently removed from the surface and a fresh cell suspension was inoculated on the identical electrodes that had been used in the preceding experiment (2nd inoculation, Fig. 5(a), BSA-2nd). Now the cells attach and spread much faster and the kinetic parameters of the second inoculation are similar to those that had been determined for a LAM coating (see Fig. 4). The explanation for these observations is that the cells that had been first inoculated on the BSA coated electrodes had synthesized adhesive proteins and secreted them on the electrode surface. Since the biosynthesis of these proteins requires time, it took roughly five hours before the cells started to spread on their self-made extracellular matrix. When the adherent cells were gently removed from the surface, their adhesive proteins were left behind and the cells that had been inoculated afterwards found a layer of adhesive proteins already on the surface. The spreading characteristics of the second inoculation indicate that the initially seeded cells might have secreted LAM onto the surface. Consistent with this experiment, it has been reported that MDCK cells synthesize and secrete LAM [26].

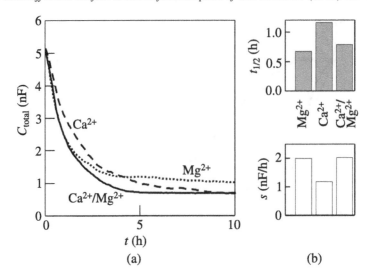

Figure 6. (a) Time course of the capacitance C_{total} measured at a sampling frequency of 40 kHz when MDCK cells were seeded into ECIS arrays whose electrodes had been coated with 100 µg/ml fibronectin. The cell suspension was prepared in Earles' balanced salt solution with Ca^{2+} and Mg^{2+} (solid); Mg^{2+} only (dotted), or Ca^{2+} (dashed) only. (b) Half-times $t_{1/2}$ and apparent spreading rates s as determined from the data shown in (a) [25].

The outstanding sensitivity of capacitance readings to monitor the substrate-anchorage of living cells could be nicely demonstrated in experiments, in which adhesion of MDCK cells to FN coated electrodes was studied in the presence of physiological concentrations of either Ca^{2+}, or Mg^{2+} or both. Binding capabilities of those cell–surface receptors that are responsible for FN binding depend on the presence of divalent cations. Some integrins show a selectivity for Ca^{2+} over Mg^{2+} or *vice versa* [27]. Figure 6 shows the time course of the electrode capacitance at 40 kHz when initially suspended MDCK cells attach and spread onto FN coated electrodes in the presence of either Ca^{2+}, Mg^{2+} or both. Apparently, MDCK cells attach and spread slightly faster onto a FN coating when Mg^{2+} is present in the culture fluid, either alone or in co-presence of Ca^{2+}. When Ca^{2+} is the only available divalent cation, spreading kinetics is slightly retarded. It is also apparent that the electrode capacitance drops to smaller values when Ca^{2+} is present in the fluid either alone or in co-presence of Mg^{2+} indicating that the cell morphology is slightly different in the presence or absence of Ca^{2+}.

3.1.2. Latex Spheres and Giant Liposomes as Simple Model Systems for Living Cells

In order to validate the performance of the ECIS device with well-defined chemical model systems we chose to study with exactly the same experimental setup the adhesion of (i) latex spheres (particle surface not chemically modified) of 3 µm diameter and (ii) liposomes (vesicles) that were prepared to have roughly cellular dimensions [28]. The latex spheres served as a model for hard particles that will

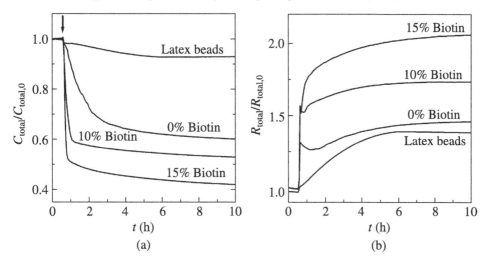

Figure 7. Time course of the capacitance C_{total} (a) and resistance R_{total} (b) measured at 40 kHz during adsorption of vesicles on the avidin-coated electrodes. Vesicles contain different concentrations of biotinylated lipid, 0%, 5%, 15%. For comparison, the time dependent adsorption of latex beads on uncoated electrodes is included. For better comparison both quantities are normalized to the starting values at time zero [28].

not spread but attach in spherical form. The giant liposomes were used to simulate cell spreading. In order to induce liposome attachment and spreading the electrodes were coated with the biotin-binding protein avidin while the liposome shell was doped with the corresponding biotin-labeled lipids. Comparing the liposome system with a cellular system, the biotinylated lipids represent the cell–surface receptors while the avidin coating of the electrodes simulates the deposition of adhesion-promoting proteins.

Figure 7 compares the time course of the measured capacitance C_{total} when giant vesicles doped with different mole fractions of biotinylated lipids as well as latex beads of 3 μm diameter were added to the bulk solution and were allowed to settle on an avidin-coated electrode [28]. The capacitance values have been normalized to the starting value at time zero. It is evident that the liposomes induce a considerably larger capacitance reduction compared to the latex beads. The reason for this is that the latex spheres behave like hard particles with only a very limited contact area with the electrode surface. Thus, the current flows exclusively around the particles even at 40 kHz. The more flexible giant liposomes, however, spread out to a certain degree on the electrode surface, forming an extended area of close contact with the electrode. The biotin-doped liposomes show faster adsorption kinetics compared to the biotin-free controls indicating that the specific biotin–avidin interaction affects the kinetics of cell adhesion and, thus, the adhesion energy E_{adh}. But not only the kinetics is different, we also observed significant differences in the final capacitance which ranged between 0.6 for the normalized capacitance for the undoped vesicles to 0.4 for the highest biotin doping of 15% (w/w). However, we know from preceding SFM studies that the liposomes show a stronger spreading and flattening on avidin-coated surfaces with increasing biotin content [29]. The contact area with the protein-coated substrate grows with increasing biotin content giving rise to a more pronounced capacitance decrease with the same number of vesicles on the surface. Thus, the magnitude of capacitance reduction is assumed to reflect the more sustained spreading of the liposomes with increasing biotin content. Figure 7(b) shows the time-dependent increase of the total resistance obtained at a sensing frequency of 40 kHz for the same model systems. Due to the presence of the dielectric structures close to the electrode surface the resistive portion of the impedance is increased as well. Moreover, with higher biotin concentration the resistance increase is more pronounced but rather low for the hard latex spheres. The resistance readout is, however, more difficult to interpret in terms of surface coverage since the initial jump in resistance is due to the unavoidable change in electrolyte composition that is associated with vesicle addition to the bathing fluid. Moreover, there is no simple correlation between surface coverage and magnitude of resistance increase since the latter is also affected by the constriction of current flow between adjacent liposomes.

Taken together, ECIS is one of the emerging techniques that can be applied in various modes to monitor the formation and modulation of cell–substrate interactions with high time-resolution compared to the time scale of the biological phenomenon under study. ECIS is not confined to electrodes made from gold but can also be applied to other conducting supports. Gold, however, is the best suited material due to its high electrical conductivity, chemical inertness and electrochemical characteristics. Unfortunately, the gold electrode cannot be coated with thin layers of other technical materials like, for instance, polymers without losing the capability for electrochemical measurements due to the presence of an insulating electrode coating. This limits the field of potential applications to electrode coatings that are conducting and do not interfere with the ECIS measurements.

3.2. A Brief Introduction to the QCM Technology

The second emerging technique to study the adhesion of cells to *in vitro* surfaces is the so-called *quartz crystal microbalance* (QCM) technique [30–32]. The QCM had already been well-known and established as an analytical tool to study adsorption phenomena at the solid–liquid interface when its potential to study cell–substrate adhesion was recognized. The approach is based on thin disks made from α-quartz that are sandwiched between two metal electrodes (Fig. 8). Due to the piezoelectric nature of α-quartz, any mechanical deformation of the crystal creates an electrical potential difference at the quartz surfaces and *vice versa*. Thus, mechanical oscillations of the crystal can be triggered, stabilized and recorded electrically. The mode of mechanical deformation (e.g., shear, torsion) in response to an electrical potential difference between the two surface electrodes depends on the crystallographic orientation by which the thin disk-shaped resonator has been cut out of a single crystal of α-quartz. For QCM purposes, only AT-cut resonators are used that

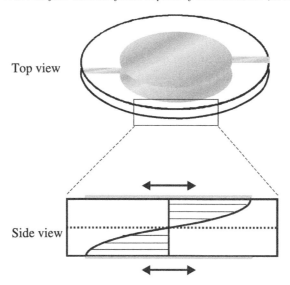

Figure 8. Top and side views on AT-cut thickness shear mode resonators as they are used in the QCM technique. The disk-shaped resonator is sandwiched between two gold-film electrodes that are used to drive the resonant oscillation and to read the resonance frequency. Under resonance conditions a standing acoustic wave is established between the crystal's surfaces that is sensitive to adsorption and desorption reactions occurring at the surface.

perform shear oscillations parallel to the surface with the maximum amplitude at the crystal faces (see Fig. 8). At resonance the mechanical shear displacement of the crystal responds very sensitively to the adsorption of a foreign material on the resonator surface. It was Sauerbrey who found already in 1961 [33] a linear correlation between the observed shift of the resonance frequency and the foreign mass that was deposited on the surface. Since frequency shifts of electrical oscillations can be measured very accurately, the device is capable of reporting the adsorption of sub-microgram quantities on the surface which gave it the name *quartz crystal microbalance*.

It was then recognized that the interactions of cells with the quartz surface also induced a shift in resonance frequency that was shown to be correlated with the degree of surface coverage [34–36]. Thus, time resolved measurements of the resonance frequency can be used to follow the attachment and spreading of cells on the quartz surface. As the fundamental resonance of the thickness shear mode (TSM) resonators that are used for QCM experiments is in the MHz regime, the time resolution of such measurements can be pushed down to milliseconds. Moreover, we and others have found that confluent monolayers of different cell types (i.e., 100% coverage) produce individual, cell-type specific shifts of the resonance frequency [37]. The structural reasons for these individual shifts in resonance frequency are not yet fully understood but may report on the individual molecular architecture of cell–substrate contacts, different adhesion mechanics, cell-type specific viscoelasticity, cell-type specific density of cell–substrate contacts per unit area or — and the most relevant — cell-type specific micro-mechanics of the membrane that is facing the substrate.

In contrast to ECIS, it is a unique and important feature of the QCM technique that the measurement is still possible when the quartz resonator is first coated with a thin layer of any material to be tested for biocompatibility. These thin material layers can be of metallic, polymeric or ceramic nature. The only limitation is that the pre-adsorbed material layer is rigid in nature, of limited thickness (generally below a few micrometers) and does not produce significant acoustic losses. In particular, the applicability to a wide variety of materials — after they had been coated on shear wave resonators — renders the QCM technique a universal and versatile sensor for cell-material interactions.

Several experimental issues are noteworthy for QCM-based cell adhesion studies:

(a) Due to energy trapping, the quartz resonator is only sensitive to changes of the surface load in those areas of the resonator that are covered with electrodes. We typically use quartz resonators with a diameter of 14 mm but the centro-symmetric electrodes on either side of the crystal are only 6 mm in diameter. Thus, similar to the ECIS setup the sensitive area inside this "wired Petri dish" is just a fraction of the total area that is available for cell adhesion.

(b) Even within the area that is covered by electrodes, the local sensitivity of the resonator is not uniform but falls off with increasing distance from the electrode center. Whereas the sensitivity is maximal in the center of the electrode, it fades to zero at the electrode edges. This well-known fact is important to recognize when cell adhesion experiments are conducted as it requires very homogeneous, single cell suspensions at the beginning of the experiment. Otherwise the readout may show significant scatter as it depends a great deal on where the cells adhere to the sensor [38].

(c) The Sauerbrey relationship, as it was introduced above, does not apply when cells are studied by means of QCM measurements. The cell bodies do not behave like a rigid mass layer but much more like a viscoelastic body [39, 40]. Thus, it is not valid to translate the observed frequency shift to biomass simply using the Sauerbrey equation and the integral mass sensitivity of the device. Ignoring this fact and applying the Sauerbrey relation anyway leads to a significant underestimation of the cell mass by QCM measurements [41].

(d) The QCM is often — and in most cases correctly — considered as a mass sensing device. However, in addition to what is discussed in (c) it is also important to realize that the QCM is only sensitive to the first monolayer of cells in direct contact with the resonator surface. Additional cells settling down on the first monolayer without direct contact with the sensor surface are not registered. The method is essentially blind to changes occurring beyond the first cell monolayer unless there is an indirect response of the cells to these activities or the surface-attached cells form very thin extensions with only a few hundred nanometers in height. If the resonator is loaded with pure water at room temperature, the decay length of the mechani-

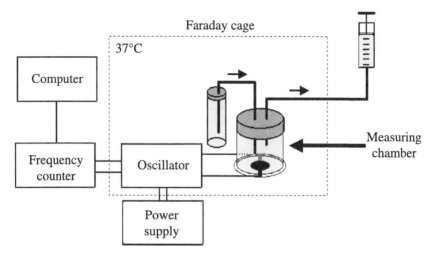

Figure 9. Experimental setup to monitor the time course of cell attachment and detachment by reading the resonance frequency of the quartz resonator that forms the bottom plate of the cell culture vessel. The measuring chamber is housed in a 37°C incubator.

cal shear wave amounts to 250 nm. This decay length is unknown for cell-covered resonators but we could show experimentally that the shear wave does not escape the cell bodies with considerable amplitude if the cells are several micrometers in height [42].

3.2.1. QCM-Based Cell Adhesion Monitoring

Figure 9 shows a schematic of the experimental setup that was used in our laboratory to measure the shift in resonance frequency during attachment and spreading of cells [37]. The quartz resonator with a fundamental resonance frequency of 5 MHz forms the bottom plate of a measuring chamber that holds approximately 0.5 ml of cell suspension. The oscillation at minimum impedance $Z_{\min}$ is stabilized by a feedback-control oscillator circuit[1] that is placed close to the crystal inside a temperature controlled Faraday cage (37°C). The oscillator circuit is driven by a 5 V power supply and the resonance frequency is determined by a commercially available frequency counter. The interested reader can find very different hardware approaches for QCM measurements in the literature that may even report two quantities: the resonance frequency and the quality factor of the oscillation which indicates viscous losses [43].

3.2.2. Time Course of Attachment and Spreading

Cells were seeded into the measuring chamber in a sterile flow hood. Immediately afterwards attachment and spreading of the cells was followed with time. Figure 10(a) compares the time-dependent shift in resonance frequency when increasing amounts of epithelial MDCK II cells are seeded into the chamber at time

---

[1] The oscillator circuit based on a Texas Instruments TTL chip was developed by A. Janshoff.

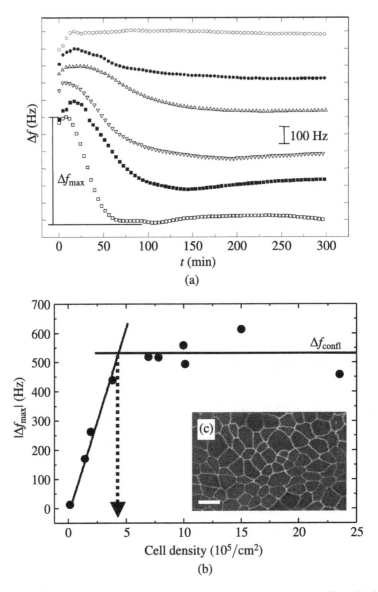

Figure 10. (a) Shift of the resonance frequency during attachment and spreading of initially suspended MDCK II cells. Each curve represents a different number of cells that were seeded at time zero. From the upper to the lower curve seeding densities were as follows (in cm^{-2}): open circles 0; filled circles 1.3×10^5; up triangles 1.8×10^5; down triangles 3.7×10^5; filled squares 7.7×10^5; open squares 1.5×10^6. Δf_{max} indicates the maximum frequency shift observed for a given seeding density. (b) Maximum frequency shift Δf_{max} as a function of the cell density that was seeded into the measuring chambers at time zero (see Fig. 10(a)). The intersection of the ascending and the horizontal lines corresponds to the number of cells on the surface (per unit area). Δf_{confl} denotes the frequency shift for a confluent cell monolayer. (c) Fluorescence micrograph of a confluent MDCK II cell layer after staining for a junctional protein exclusively localized at the cell border. The scale bar corresponds to 20 μm [36].

zero [36]. From the topmost to the bottommost curve in Fig. 10(a) the cell density in the measuring chamber was continuously increased from a cell-free control to a maximum of 1.5×10^6 cells/cm^2. Immediately after the start of the experiment there is a moderate increase of the resonance frequency by 50–100 Hz which is exclusively due to the warm-up of the medium inside the chamber to 37°C. After this transient maximum, the resonance frequency continuously decreases and it now reports the formation of cell–substrate adhesion sites and continuous progress in cell attachment and spreading. The time resolution of such measurements can be reduced well below one second so that even very subtle details of the cell adhesion kinetics can be monitored using QCM measurements. The slope of the curves is easily accessible and can be used to determine the adhesion energy for this cell type in contact with the biomaterial surface under study. The more cells are seeded the larger is the resulting shift in resonance frequency upon attachment and spreading.

When the maximum frequency shift $|\Delta f_{max}|$ for an individual experiment as shown in Fig. 10(a) is plotted against the number of cells seeded into the measuring chamber at time zero, we obtain a saturation type relationship that is presented in Fig. 10(b) [36]. The data in Fig. 10(b) can be interpreted as follows: as long as the density of seeded cells is small enough that all cells reaching the surface can find an adhesion site, an increase in the maximum resonance frequency shift $|\Delta f_{max}|$ is observed with increasing seeding density. This frequency shift is proportional to the fractional surface coverage as has been confirmed by others [34, 35]. However, when the number of seeded cells is increased further all adhesion sites on the surface are occupied and accordingly we do not find any further increase in $|\Delta f_{max}|$. This observation convincingly underlines that the QCM device is primarily sensitive to phenomena that occur at the quartz surface but does not report on cells that are beyond the first cell monolayer [42]. Based on these data it is hard to imagine that biological activities that occur at the apical surface of an established cell layer can be observed by QCM unless the cells are very thin. It is, however, reported that exocytotic events in adherent cells can be monitored by the QCM approach [44].

If only those cells in direct contact with the resonator surface contribute to the overall signal, we should be able to determine the cell density on the surface from measurements as the one shown in Fig. 10(b). To do so, we have chosen a two-case approach: (i) for low seeding densities the relationship is approximated by a straight line with positive slope that indicates a linear correlation between frequency shift and surface coverage; (ii) beyond a certain cell density the experimental adhesion curve is modeled by a horizontal line indicating that surplus cells, that do not find adhesion sites on the substrate, do not contribute to the measured QCM response. Accordingly, the interception between these two straight lines should mark the actual cell density on the surface. For MDCK cells (strain II), that were used in these experiments, we found the interception to be located at a seeding density of $(4.3 \pm 0.5) \times 10^5$ cells/cm^2 (arrow in Fig. 10(b)). For validation we have also determined the cell density in an entirely confluent monolayer microscopically after the cell borders had been stained by immuno-cytochemistry. Figure 10(c) shows a typical fluorescence micrograph that was used to determine the cell density. Images recorded by fluorescence microscopy revealed a cell density of $(5.5 \pm 0.3) \times 10^5$ cells/cm^2 on the surface which is slightly above the value extracted from QCM readings. However, microscopic experiments were conducted on cell monolayers that were allowed to grow to confluence for several days, while QCM experiments were limited to attachment and spreading within only 5 h. Since the cells tend to multiply to some degree even in a confluent monolayer before contact inhibition stops any further proliferation, it is not surprising to find somewhat higher cell densities in our microscopic control experiments. Repeating these kind of experiments with other cell types confirmed our conclusions. We found consistently that the number of cells on the surface was determined correctly from QCM readings. Interestingly, different cell types create individual shifts in resonance frequency when they adhere to the quartz surface. It is important to stress that these differences are not due to incomplete coverage of the quartz resonator but reflect individual differences in the contact mechanics or the mechanics of the substrate-facing membrane.

As already indicated above, *specific* molecular interactions of the receptor-ligand type as well as *non-specific* interactions contribute individually and at different times to the anchorage of cells to a given *in vitro* surface [11, 45, 46]. Compelling evidence has been collected that *specific* ligand–receptor interactions are more important for the final strength and the dynamic properties of the adhesion sites [45], whereas unspecific electrostatic or electrodynamic interactions are important during the first phase of the cell–surface encounter. Thus, the question arises whether the QCM response requires specific, receptor-mediated adhesion of the cells to the surface or the sole presence of the cell body close to the resonator surface. One strategy to answer this question is to block the specific interactions between cell–surface receptors and adhesive proteins on the substrate by adding short peptides to the culture fluid that compete with the adhesion-promoting proteins for the binding sites of the cell–surface receptors. When these soluble peptides are added to the cell suspension, they are expected to delay or entirely eliminate specific cellular interactions with substrate immobilized proteins.

In our experiments we used serum containing medium as culture fluid even though the chemical composition of serum is not precisely defined and may vary, to some degree, from batch to batch. Serum naturally contains the adhesive proteins vitronectin (VN) and fibronectin (FN), which adsorb instantaneously from solution to the surface. Both proteins, VN and FN, are recognized by cell–surface receptors *via* the same amino acid sequence Arg-Gly-Asp-Ser, or RGDS in one letter code. Thus, we studied the impact of soluble peptides with this amino acid sequence on the time course of cell attachment and spreading using the QCM approach. Figure 11 shows the outcome of two experiments in which either the penta-peptides Gly-Arg-Gly-Asp-Ser (GRGDS) or Ser-Asp-Gly-Arg-Gly (SDGRG) were added to the cell suspension in a concentration of 1 mM each [36]. The two penta-peptides GRGDS and SDGRG contain exactly the same amino acids but in reverse order. Thus, the two molecules carry the same charge density and would provide the same

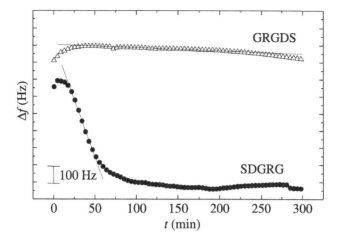

Figure 11. Time course of the resonance frequency during the attachment and spreading when similar numbers of MDCK II cells were seeded into the quartz dish in presence of the soluble peptides Gly-Arg-Gly-Asp-Ser (GRGDS) or Ser-Asp-Gly-Arg-Gly (SDGRG). The concentration of each peptide was 1 mM and the cell density was adjusted to 8×10^5 cm^{-2} [36].

perturbation to unspecific interactions — if at all. However, due to the reversal of the amino acid sequence only GRGDS has the correct sequence to interact specifically with the cell–surface receptors whereas SDGRG does not. The time course of the resonance frequency $\Delta f(t)$ as presented in Fig. 11 clearly demonstrates that in the presence of 1 mM GRGDS the resonance frequency does not indicate any cell adhesion to the resonator surface. In contrast, when SDGRG is added to the culture fluid, there is no difference compared to experiments in which no peptide is present at all (see Fig. 10(a)). Thus, when specific interactions between cell–surface receptors and substrate immobilized proteins are not allowed to form, we do not observe any measurable impact on QCM readings. The measurements clearly show that loose attachment of the cell bodies to the substrate does not produce any significant acoustic load and, moreover, that cells which are not capable of forming specific molecular interactions with substrate-immobilized proteins cannot be detected [36, 42]. We want to emphasize and repeat at this point that these kinds of cell adhesion measurements are not limited to adhesion to the bare or protein-decorated quartz resonator but are also possible with a thin film of any material that can be precoated on the resonator as long as it fulfills the requirements mentioned earlier in the text.

We also studied the situation at the interface between surface and cell membrane by means of reflection interference contrast microscopy (RICM) that has been used extensively to image the footprints of cells on transparent substrates. In RICM, the cell-covered coverslip is illuminated from below and the reflected light is used for imaging. The light can either be reflected at the glass–medium interface or at the adjacent medium–membrane interface. Interference between the two provides a contrast that codes and maps the distance between membrane and surface. In these experiments we seeded cells in serum-containing medium that was either supplemented with 1 mM RGDS (similar activity as GRGDS) or not. In both cases, the cells were allowed to attach and spread upon the glass surface for 200 min before we recorded RICM images of each sample. In the absence of RGDS the cells form typical cell–surface junctions with the substrate. The footprints indicate a spread morphology under these conditions. When RGDS is present the cells are hardly visible in the RICM image although they had settled on the surface. Based on the principles of RICM image formation and some experimental parameters of the microscope one can estimate that in the presence of 1 mM RGDS the lower cell membrane must be farther away from the substrate surface than 100 nm — probably significantly more — confirming the analysis of the QCM readings in the presence of these inhibitory peptides. Apparently, the presence of the cell bodies within this distance from the substrate surface and with only a very limited contact area — like a hard sphere on a flat surface — does not provide any significant acoustic load for the quartz resonator [17].

Thus, two main conclusions can be drawn from these experiments: (i) The QCM does only report on cells that are specifically anchored to the resonator surface. The method is blind to cells that just settle on the surface and attach only loosely. (ii) When specific cell–substrate interactions are absent, the cells stay away from the surface by more than 100 nm according to our RICM data. Theoretical considerations have previously indicated that cells may approach the surface as close as 5–10 nm just by unspecific attraction [47]. This is, however, neither confirmed by our optical measurements nor by the QCM data. We have already discussed above that the decay length of the mechanical oscillation in a QCM experiment in an aqueous environment at room temperature is approx. 250 nm. But apparently the loosely attached cell bodies (in presence of RGDS) are far enough away from the surface that the mechanical oscillation cannot sense them.

3.2.3. QCM Experiments with Well-Defined Model Systems

Due to the complex mechanical characteristics of living cells anchored to the resonator surface, which are inadequately described by the available micro-mechanical models, we studied chemically well-defined model systems to better understand what could be learned from QCM-based adhesion studies of animal cells [29, 48, 49]. As described for the ECIS experiments before, we used liposomes doped with varying amounts of biotinylated lipids to mimic the cell body (liposome) and its cell–surface receptors (biotin moieties). The adhesive proteins on the surface were modeled by a layer of pre-deposited avidin that provides binding sites for the biotin residues in the lipid shell. Thus, receptor density and protein concentration on the surface were under experimental control and could be adapted according to the experimental needs. In our initial studies we used large unilamellar vesicles made from dipalmitoylphosphatidylcholine (DPPC) doped with increasing molar ratios of dipalmitoylphosphatidylethanolamine (DPPE) carrying a biotin residue. The biotin was covalently attached to the lipid headgroup *via* a C_6 spacer.

In these experiments we used a special QCM setup that was originally described by Rodahl and coworkers [50] and is referred to as QCM-D. This device not only records the changes in resonance frequency Δf but also changes in the so-called dissipation factor D which is the inverse of the quality factor Q of the oscillation:

$$D = \frac{1}{Q} = \frac{Dissipated\ energy\ per\ cycle}{Stored\ energy\ per\ cycle}. \tag{6}$$

As expressed in equation (6), shifts in D reflect changes in energy dissipation of the shear oscillation. For dissipative systems the energy of the shear oscillation is transmitted into the material layer adsorbed on the quartz. Thus, measuring the change in energy dissipation becomes important whenever systems are studied that do not behave like a rigid mass. Only for homogeneous rigid mass films an experimentally observed frequency shift can be attributed unequivocally to mass deposition on the resonator surface according to the Sauerbrey relationship [33]. When the microviscosity or elasticity close to the quartz surface changes, the Sauerbrey equation no longer holds since these effects change the resonance frequency as well and are indistinguishable from simple mass deposition. Thus, viscous energy losses can make QCM measurements ambiguous and hard to interpret [51]. The device developed by Rodahl and coworkers [50] overcomes this problem by recording both the shift in resonance frequency as well as energy dissipation at the same time which makes data interpretation more robust and provides twice the information on the system under study.

When living animal cells were studied with this setup we typically found frequency shifts Δf between 50 and 500 Hz depending on the cell type. The cell-type specific change in dissipation factor ΔD ranged between 1 and 4×10^{-4} [48]. When we used undoped DPPC liposomes of 100 nm diameter that were allowed to settle on an avidin-coated resonator, we recorded frequency shifts in the order of 400–500 Hz, thus very similar to the readout for living cells even though these liposomes did not form specific molecular interactions with the surface-immobilized avidin. With respect to energy dissipation the liposomes, however, did not dissipate the same amount of energy as living cells did. For the undoped DPPC liposomes we observed an increase in energy dissipation in the order of 3×10^{-5} which is roughly an order of magnitude less than recorded for the substrate-anchored cells.

Adding biotin-labeled lipids in the liposome shell in order to allow for molecular recognition between liposome and surface bound protein led to a gradual reduction of both Δf and ΔD. As demonstrated in Fig. 12 there is a gradual drop in both parameters with increasing concentrations of biotin residues in the liposome shell. In other words, the more the ligand–receptor pairs were available the more the QCM response was reduced [48]. And this result does not depend on the size of the liposome. The data in figure 12 were recorded for large unilamellar liposomes with an average diameter of 100 nm but giant liposomes with diameters in the μm-range also showed a similar behavior. Even for these vesicles, that have roughly the size of a typical animal cell, we could not observe a similar energy dissipation as observed

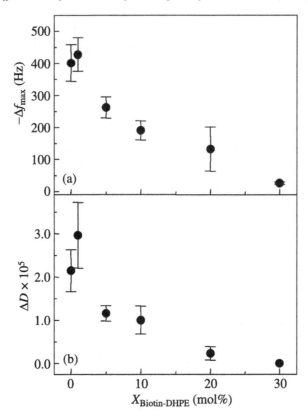

Figure 12. Summary of liposome adhesion studies as performed with a QCM-D setup described in the text. Panel (a) shows the maximum frequency shift Δf_{max} when the concentration of biotinylated lipids was gradually increased. Panel (b) summarizes the final shifts of the dissipation factor ΔD from the same experiments. Data points are averages of at least two independent experiments [48].

for adherent cells — independent of the presence or absence of specific molecular interactions.

The reason for this very unexpected behavior was revealed by scanning force microscopy [29]. With increasing biotin load in the liposome shell, the liposomes spread out on the surface. But eventually they rupture when the adhesion forces provided by the ligand–receptor interactions dominate over the intermolecular forces that keep the lipids together within the liposome shell. The ruptured liposomes eventually form a lipid double layer on the surface with the water-filled interior of the original liposome being emptied into the bulk phase. These lipid bilayers on the surface behave essentially like a rigid mass deposited on the surface so that the shifts in resonance frequency and dissipation decline. Kasemo and coworkers investigated this effect with similar model systems [52].

As a conclusion of this section, these experiments clearly revealed that unilamellar liposomes, as used here, are not a suitable model system to perform systematic studies of cell adhesion to *in vitro* surfaces by QCM — mostly due to the unavoidable rupture of the liposomes when surface attraction becomes too strong. Nevertheless, for intermediate biotin concentrations these experiments did show that an aqueous compartment surrounded by a lipid double layer was not sufficient to explain the high acoustic load that was exerted on the resonator by a confluent cell layer. There is more to it than just a membrane-confined fluid compartment close to the surface. As indicated several times throughout the last sections we have collected compelling evidence that the surface-facing membrane together with the cortical cytoskeleton and their micro-mechanical properties dominate the QCM readout. In other words, QCM-based adhesion studies provide a label-free and time-resolved view on the micromechanical changes at the substrate-facing lower membrane in addition to the time course of adhesion.

Acknowledgement

The authors would like to acknowledge generous financial support by the Kurt-Eberhard Bode Stiftung.

References

1. K. C. Dee, D. A. Puleo and R. Bizios, *Biocompatibility*. John Wiley & Sons, Chichester (2003).
2. P. Zilla, M. Deutsch and J. Meinhart, *Seminars Vascular Surg.* **12**, 52 (1999).
3. J. G. Meinhart, J. C. Schense, H. Schima, M. Gorlitzer, J. A. Hubbell, M. Deutsch and P. Zilla, *Tissue Eng.* **11**, 887 (2005).
4. T. Stieglitz, M. Schuettler and K. P. Koch, *IEEE Eng. Med. Biol. Mag.* **24**, 58 (2005).
5. P. Fromherz, *ChemPhysChem* **3**, 276 (2002).
6. T. Henning, M. Brischwein, W. Baumann, R. Ehret, I. Freund, R. Kammerer, M. Lehmann, A. Schwinde and B. Wolf, *Anticancer Drugs* **12**, 21 (2001).
7. I. Freshney, *Culture of Animal Cells: A Manual of Basic Techniques*. John Wiley & Sons, Chichester (2000).
8. T. Noll, N. Jelinek, S. Schmidt, M. Biselli and C. Wandrey, in: *Advances in Biochemical Engineering/Biotechnology*, T. Scheper (Ed.) **74**, p. 111. Springer Verlag (2002).
9. K. L. Prime and G. M. Whitesides, *Science* **252**, 1164 (1991).
10. M. Arnold, E. A. Cavalcanti-Adam, R. Glass, J. Blummel, W. Eck, M. Kantlehner, H. Kessler and J. P. Spatz, *ChemPhysChem* **5**, 383 (2004).
11. A. Pierres, A. M. Benoliel and P. Bongrand, *Eur. Cells Mater.* **3**, 31 (2002).
12. T. Frisch and O. Thoumine, *J. Biomech.* **35**, 1137 (2002).
13. A. S. G. Curtis, *Eur. Cells Mater.* **1**, 59 (2001).
14. Y. Iwanaga, D. Braun and P. Fromherz, *Eur. Biophys. J.* **30**, 17 (2001).
15. D. Braun and P. Fromherz, *Biophys. J.* **87**, 1351 (2004).
16. V. Kiessling, B. Müller and P. Fromherz, *Langmuir* **16**, 3517 (2000).
17. J. Wegener, in: *Encyclopedia of Biomedical Engineering*, M. Akay (Ed.) **6**, p. 1. Wiley & Sons, Hoboken, NJ (2006).
18. I. Giaever and C. R. Keese, *Nature* **366**, 591 (1993).
19. I. Giaever and C. R. Keese, *Proc. Natl Acad. Sci. USA* **81**, 3761 (1984).
20. R. Ehret, W. Baumann, M. Brischwein, A. Schwinde and B. Wolf, *Med. Biol. Eng. Comput.* **36**, 365 (1998).

21. L. Ceriotti, J. Ponti, P. Colpo, E. Sabbioni and F. Rossi, *Biosens. Bioelectron.* **22**, 3057 (2007).

22. D. Krinke, H. G. Jahnke, O. Panke and A. A. Robitzki, *Biosens. Bioelectron.* **24**, 2798 (2009).

23. I. Giaever and C. R. Keese, *Proc. Natl Acad. Sci. USA* **88**, 7896 (1991).

24. J. Wegener, in: *Nanotechnology*, H. Fuchs (Ed.) **6**, p. 325. VCH, Weinheim (2009).

25. J. Wegener, C. R. Keese and I. Giaever, *Exp. Cell. Res.* **259**, 158 (2000).

26. P. J. Salas, D. E. Vega-Salas and E. Rodriguez-Boulan, *J. Membr. Biol.* **98**, 223 (1987).

27. D. Kirchhofer, J. Grzesiak and M. D. Pierschbacher, *J. Biol. Chem.* **266**, 4471 (1991).

28. A. Sapper, B. Reiss, A. Janshoff and J. Wegener, *Langmuir* **22**, 676 (2006).

29. B. Pignataro, C. Steinem, H. J. Galla, H. Fuchs and A. Janshoff, *Biophys. J.* **78**, 487 (2000).

30. A. Janshoff and C. Steinem (Eds), *Piezoelectric Sensors*. Springer, Berlin (2007).

31. A. Janshoff, H.-J. Galla and C. Steinem, *Angew. Chemie Intern. Edition* **39**, 4004 (2000).

32. J. Wegener, A. Janshoff and C. Steinem, *Cell Biochem. Biophys.* **34**, 121 (2001).

33. G. Sauerbrey, *Z. Phys.* **155**, 206 (1959).

34. D. M. Gryte, M. D. Ward and W. S. Hu, *Biotechnol. Prog.* **9**, 105 (1993).

35. J. Redepenning, T. K. Schlesinger, E. J. Mechalke, D. A. Puleo and R. Bizios, *Anal. Chem.* **65**, 3378 (1993).

36. J. Wegener, A. Janshoff and H. J. Galla, *Eur. Biophys. J.* **28**, 26 (1998).

37. J. Wegener, S. Zink, P. Rosen and H. Galla, *Pflugers. Arch.* **437**, 925 (1999).

38. F. Josse, Y. Lee, S. J. Martin and R. W. Cernosek, *Anal. Chem.* **70**, 237 (1998).

39. A. Janshoff, J. Wegener, M. Sieber and H. J. Galla, *Eur. Biophys. J.* **25**, 93 (1996).

40. C. M. Marxer, M. C. Coen, T. Greber, U. F. Greber and L. Schlapbach, *Anal. Bioanal. Chem.* **377**, 578 (2003).

41. V. Heitmann, B. Reiss and J. Wegener, in: *Piezoelectric Sensors*, A. Janshoff and C. Steinem (Eds) **5**, p. 303. Springer, Berlin (2007).

42. J. Wegener, J. Seebach, A. Janshoff and H. J. Galla, *Biophys. J.* **78**, 2821 (2000).

43. M. Rodahl, F. Hook, C. Fredriksson, C. A. Keller, A. Krozer, P. Brzezinski, M. Voinova and B. Kasemo, *Faraday Discuss.* **107**, 229 (1997).

44. A. S. Cans, F. Hook, O. Shupliakov, A. G. Ewing, P. S. Eriksson, L. Brodin and O. Orwar, *Anal. Chem.* **73**, 5805 (2001).

45. P. Bongrand, *J. Dispersion. Sci. Technol.* **19**, 963 (1998).

46. G. I. Bell, M. Dembo and P. Bongrand, *Biophys. J.* **45**, 1051 (1984).

47. E. A. Vogler and Bussian, *J. Biomed. Mater. Res.* **21**, 1197 (1987).

48. B. Reiss, A. Janshoff, C. Steinem, J. Seebach and J. Wegener, *Langmuir* **19**, 1816 (2003).

49. E. Lüthgens, A. Herrig, K. Kastl, C. Steinem, B. Reiss, J. Wegener, B. Pignataro and A. Janshoff, *Measurement. Sci. Technol.* **14**, 1865 (2003).

50. M. Rodahl, F. Höök, A. Krozer, P. Brzezinski and B. Kasemo, *Rev. Sci. Instrum.* **66**, 3924 (1995).

51. M. V. Voinova, M. Jonson and B. Kasemo, *Biosens. Bioelectron.* **17**, 835 (2002).

52. E. Reimhult, M. Zach, F. Hook and B. Kasemo, *Langmuir* **22**, 3313 (2006).

Methods to Measure the Strength of Cell Adhesion to Substrates

Kevin V. Christ [a] and Kevin T. Turner [a,b,*]

[a] Materials Science Program, University of Wisconsin, Madison, WI 53706
[b] Department of Mechanical Engineering, University of Wisconsin, Madison, WI 53706

Abstract

Cell-substrate adhesion is a critical factor in the development of biomaterials for use in applications such as implantable devices and tissue engineering scaffolds. In addition, cell adhesion to the extracellular matrix is intertwined with a number of fundamental cell processes, and several diseases are characterized by cells with altered adhesion properties. While many approaches exist to characterize cell adhesion, only a fraction of the techniques provides quantitative measurements of the strength of adhesion by physically detaching cells through application of force or stress. In this review, the most commonly used techniques to measure the adhesion strength of cells adhered to substrates are summarized. These methods can be divided into three general categories: centrifugation, hydrodynamic shear and micromanipulation. For each method, the technique is described and its capabilities assessed. A comprehensive review of recent applications of the methods is given, and adhesion strength measurements performed using different techniques on fibroblasts, a commonly-studied cell, are compared. Finally, the strengths and drawbacks of the various techniques are discussed.

Keywords

Cell adhesion, adhesion strength, cell detachment, cell mechanics, biomaterials, hydrodynamic shear, cytodetachment, micropipette aspiration

Symbols

A: cell area (m^2)

D: characteristic flow chamber dimension (m)

D_h: hydraulic diameter (m)

F: force (N)

---

* To whom correspondence should be addressed. Department of Mechanical Engineering, University of Wisconsin, 1513 University Avenue, Madison, WI 53706-1572, USA. Tel.: 608-890-0913; Fax: 608-265-2316; e-mail: kturner@engr.wisc.edu

Surface and Interfacial Aspects of Cell Adhesion
© Koninklijke Brill NV, Leiden, 2010

g: gravitational acceleration (m/s^2)

h: chamber height (m)

k: stiffness constant (N/m)

ΔP: aspiration pressure (Pa)

Q: volumetric flow rate (m^3/s)

r: radial position (m)

R: characteristic radius (m)

RCF: relative centrifugal force (g)

Re: Reynolds number

u: fluid velocity (m/s)

U: average fluid velocity (m/s)

V_{cell}: cell volume (m^3)

w: chamber width (m)

x: displacement (m)

y: coordinate measured normal to channel wall (m)

δ: fluid boundary layer thickness (m)

μ: dynamic viscosity (kg/m s)

ρ: density of fluid (kg/m^3)

ρ_{cell}: density of the cell (kg/m^3)

ρ_{medium}: density of medium (kg/m^3)

σ: normal stress (Pa)

τ: shear stress (Pa)

ω: angular speed (rad/s)

1. Introduction

Cell adhesion is a broad area of study that is concerned with the manner in which cells interact and bind with one another and to the extracellular matrix (ECM). Cell–matrix and cell–cell adhesion are crucial mechanical phenomena in biology and are central in several basic aspects of cell function, including growth [1], differentiation [2] and migration [3]. Biological processes in which cell adhesion plays a major role include tissue development [4], wound healing [5] and immune response [6]. In addition, changes in cell adhesion accompany many pathological events, such as cancer metastasis [7] and the development of leukemia [8]. Finally, controlling cell adhesion is essential for the proper function of biomaterials [9, 10] and biosensors [11], and is central to the rapidly growing fields of tissue engineering [12] and stem cells [13].

The cell is an active system and can respond to mechanical stimuli by altering protein expression, geometry and motility [14]. It is well known that the cell processes mentioned above are influenced directly by the topography [15] and stiffness [16] of the surface to which the cells are adhered. Cell adhesion to the ECM occurs primarily *via* integrins, a family of transmembrane heterodimers that bond to external ligands (e.g., collagen, fibronectin, vitronectin) in the ECM [17]. Integrins cluster to form focal adhesions, which are small, discrete plaques that provide the primary mechanical junctions between the actin cytoskeleton and the matrix [18]. Focal adhesions are complex, multi-protein structures that not only provide a physical connection to the external surface but also provide routes for signaling [19]. Cell adhesion is a time-dependent process, as the adhesive components are not fully developed until several hours after the cell initially binds with the ECM [20]. During this time period, a significant level of adhesion strengthening occurs *via* cell spreading, integrin aggregation, and focal adhesion formation until a steady state is reached [21].

In this review, we provide a critical assessment of methods to measure the adhesion strength of cells to surfaces, many of which are functionalized with ECM protein coatings. Though most animal cells are capable of adhering to external surfaces, the ECM, or other cells, cells that form well-defined bonds with their surroundings and remain adhered in some fashion throughout their lifetimes (e.g., endothelial cells, fibroblasts, osteoblasts) are crucial in determining the mechanical properties of tissues. The adhesion properties of these cells are of fundamental importance in tissue engineering *via* scaffolds and the integration of structural biomedical implants, such as joint replacements, in the body. This review focuses on methods to detach and measure the adhesion strength of such cells that have been allowed to adhere and spread on two-dimensional substrates. Many of the techniques that will be discussed have also been used to characterize the adhesion of cells that travel primarily through the vasculature, such as leukocytes [22], and lymphocytes [23], but the adhesion and functionality of these cells are quite different and will not be discussed in detail here.

Common techniques to characterize cell–matrix adhesion include quantifying the spread area of the cell, analyzing focal adhesions, measuring cell migration, and simple wash assays. These techniques provide valuable information about adhesion and are relatively straightforward to perform, but they do not provide a quantitative measurement of the forces or stresses involved in cell attachment. There are two general classes of quantitative force assays that are used in investigating cell adhesion. The first, traction-force microscopy, allows the stresses induced below and around an adhered cell to be quantified by tracking the displacements of structures in the surface [24–26]. Traction force measurements have provided tremendous insight into how cells sense and respond to their mechanical and chemical environments; however, these techniques do not provide a direct measurement of the strength of the adhesive interaction between the cell and the matrix. The second class of force-based assays, which is the focus of this review, measures cell adhesion strength by quantifying the force or stress required to detach a cell from a substrate.

Adhesion strength assays can be divided into three general categories depending on the method of loading used to detach the cell: (1) centrifugal force, (2) hydrodynamic shear, and (3) single-cell micromanipulation. Centrifugal methods utilize common laboratory centrifuges to apply body forces normal to the surface with adhered cells in order to detach the cells. Hydrodynamic shear methods employ fluid flow chambers to apply shear stress on the cell surface. Common configurations of this type of assay include the spinning disk, radial flow chamber, and parallel plate flow chamber. Micromanipulation methods involve the use of a micropipette, atomic force microscope (AFM) probe, or similar instrument to apply a concentrated force to a single cell in order to detach it. These micromanipulation approaches are implemented in a range of configurations and can apply normal or shear stresses on the cell during detachment. All of these methods result in the measurement of either a force or stress that describes the strength of adhesion between the cell and the surface on which it is adhered. While there are other techniques, such as laser stress waves [27] and ultrasound [28], that have also been used to probe adhesion strength, these are much less widely employed than the principal assays identified above.

There are significant differences in the nature of the data that can be obtained from adhesion strength tests depending on whether the cells are tested individually or as part of a large population. In population studies, the fraction of cells remaining on a surface after global force or stress application is quantified, and from repeated trials with varying loads, an adhesion strength value at which 50% of the cells detach is typically determined. These studies are useful in measuring the effects of environmental variables, such as ECM protein type or concentration, on the average strength of cell adhesion. In contrast, single cell studies involve the detachment of cells one at a time, and thus allow characteristics of individual cells, such as spread area, phenotype, and focal adhesion arrangement, to be measured and correlated with adhesion strength for a particular cell. Centrifugation has been used exclusively to analyze large populations of cells, while micromanipulation approaches have only been used to study single cells. Fluid shear methods have been used to study populations or single cells.

The various adhesion strength measurement methods apply different types of loads to the cells. Fluid shear experiments apply a nearly uniform shear stress to the exposed surface of the cell, while centrifugation applies a normal force to the entire cell body. Micromanipulation techniques involve the interaction of a solid object with the cell and generally apply highly-localized forces. This force can be applied in varying directions, thus normal or shear force can be imparted on the cell with micromanipulation approaches. An understanding of the differences in force application is important in choosing an experimental method, as some cells may respond to certain stimuli differently than others — for example, endothelial cells are known to respond to shear flow by rearranging their geometry and adhesive components [29]. It is also important to recognize that the assays often apply these loads over different periods of time. As adhesion is known to be rate dependent (e.g., [30, 31]), the magnitude of the load as well as the duration of loading must be considered.

In this review, we provide a detailed description of the three classes of adhesion assays and discuss specific techniques and measurements that have been performed with them. For each method, we discuss the type of mechanical loading involved and provide the relevant equations that describe the stresses and forces. We then provide a comprehensive survey of contemporary works in which the method was used to measure the adhesion strength of cells seeded on 2-D substrates, which is often done in order to evaluate biomaterials and study the effects of chemical treatments on cells. Special attention is devoted to work in which fibroblast cell lines were examined. This common cell type has been investigated in many studies with similar surface chemistries, thus this facilitates comparisons of the different adhesion strength measurement techniques. Following the review of the techniques and measurements, we compare and contrast these methods and discuss the benefits and drawbacks of each. While there have been other review papers in the past decade that discuss cell adhesion strength measurement [20, 32, 33], the review here is different in that it covers many recent developments, including methods based on microfluidic devices, and provides a critical analysis of the mechanics of the assays and a comparison of numerous experimental results.

2. Cell Adhesion Measurement Techniques

2.1. Centrifuge Assay

Centrifuge adhesion strength assays are frequently used due to their simplicity and the wide availability of equipment. In the most common method, cells are seeded in a multiwell plate that is then placed in a swing-bucket centrifuge (Fig. 1). The plate is spun so that the bottom surface of the well faces normal to the axis of rotation; hence, the cells experience a body force acting in the direction normal to the bottom of the plate that pulls them away from the surface. To assess adhesion strength, the number of cells before and after application of load in the centrifuge is quantified. The fraction of cells that remains adhered after centrifugation can be determined by measuring the amount of radiation emitted from radiolabeled cells

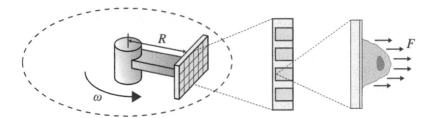

Figure 1. Schematic of a typical centrifuge assay. A multiwell plate is sealed and spun with the wells facing outwards. A body force, acting normal to the well bottom, is applied on the cells to cause detachment.

[34], quantifying the amount of cellular genetic material [35], or by using automated fluorescence analysis [36]. In many cases, the assay is used to assess the relative effect of treatments such as ECM protein type and concentration or the inhibition of a specific cellular function. Statistical comparisons of adherent fractions are performed to determine treatment significance. By using multiwell plates, many treatments in replicate can be assessed in a single experiment. In cases in which an adhesion strength value is determined, tests at multiple loads are performed and the force necessary to detach 50% of the cells from the surface is reported as the adhesion strength.

The load applied in the centrifuge assay is the body force exerted on each cell, which is given as

$$F = (\rho_{cell} - \rho_{medium}) \cdot V_{cell} \cdot \text{RCF}, \tag{1}$$

where ρ_{cell} is the density of the cell, ρ_{medium} is the density of the medium, V_{cell} is the cell volume, and RCF is the relative centrifugal force in units of g [34]. A typical value of ρ_{cell} for a cell is 1.07 g/cm^3, while 1.00 g/cm^3 is a typical value for ρ_{medium} [37]. The RCF is determined by the size and speed of the centrifuge and is calculated as:

$$\text{RCF} = \frac{R\omega^2}{g} \tag{2}$$

where R is the radius from the rotor center to the plate, ω is the angular speed in rad/s, and g is the gravitational acceleration. Typical RCF values are between $20g$ and $3000g$, with forces on individual cells ranging from 1 to 2000 pN (Fig. 2) [37, 38]. The average normal stress applied to the cell-substrate interface, ($\sigma = F/A$), depends on the cell's spread area, A, and the force applied on the cell, F, as shown in Fig. 2. The small forces and stresses that can be achieved with this technique restrict its use to weakly-adhered cells, which usually have incubation times of an hour or less. The times of load application in centrifuge tests typically range from 5 to 10 minutes [37, 38]. While the above type of loading is that most commonly used in centrifuge assays, alternate configurations that apply forces on the cell that act

Surface and Interfacial Aspects of Cell Adhesion (2010) 193–224

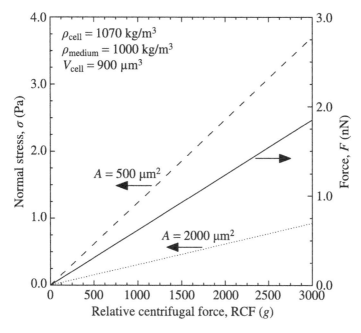

Figure 2. Normal stress and force applied to cells in the centrifuge assay as a function of the relative centrifugal force (RCF). The applied stress is a function of the cell area, A, and is shown for two cells of typical areas. The RCF values shown are attainable with standard laboratory centrifuges.

parallel to the surface and utilize ultracentrifuges with RCFs as high as $110\,000g$ (forces ∼ 100 nN) have also been reported [39, 40].

To illustrate the different implementations of the centrifuge assay, specifics of several studies that have employed this technique are summarized below. Lotz *et al.* analyzed the adhesion strengthening process of fibroblasts and glioma cells adhered to wells coated with fibronectin or tenascin using the centrifuge assay [41]. In their procedure, a modification of that presented in [34], a second multiwell plate was inverted and placed on top of the first in order to collect the cells detached during centrifugation. The cells were radioactively labeled, and the adherent fraction was determined using a gamma counter after removal from the wells. Radiolabeling has also been used to quantify the adherent fraction in a study of receptor-ligand bond characteristics that examined the adhesion strength of cancer cells to E-selectin [42].

Another common method for determining adherent fractions is to label cells with a fluorescent dye such as calcein-AM and use an automated plate reader to measure cell coverage before and after force application [36, 38]. This method was used by García and colleagues in a number of experiments investigating the effects of surface functionalization, mainly fibronectin on self-assembled monolayers (SAMs), on cell adhesion and integrin binding [43–48]. Reyes and García demonstrated the high-throughput capabilities of this method by analyzing the adhesion of HT-1080 cancer cells, NIH3T3 fibroblasts and MC3T3-E1 osteoblastic cells adhered in standard multiwell tissue-culture polystyrene (TCPS) plates coated with varying surface concentrations of collagen or fibronectin [37]. Incubation times varied between 30 min and 2 h, and two centrifugation speeds were used, corresponding to $45g$ and $182g$. Significant differences in adhesion were observed when comparing the various cell types and surface chemistries, and the 96-well plate format allowed a large number of surfaces to be examined in a single experiment. Using data from [37], we calculated the force required to detach 50% of the NIH3T3 cells used in one of these experiments; this is summarized in Table 1.

The final method that is commonly used to measure the adherent fraction of cells after loading is to rupture the detached and adhered cell samples after loading and measure the amount of released DNA from each sample. The CyQuant® assay from Invitrogen is one of several commercially-available DNA quantification assays and it was previously used to examine WT NR6 fibroblastoid cells adhered to RGD-functionalized poly(methyl methacrylate)/poly(ethylene oxide) (PMMA/PEO) substrates [35]. DNA quantification has also been used with a centrifuge assay to

Table 1.
Adhesion strength of fibroblasts measured using standard population assays

Method	Cell type	ECM conc. (ng/cm^2)	Incubation time (h)	Experiment duration	Applied load	Fraction adhered (%)	Ref.
Centrifugation*	NIH3T3	300	1	5 min	7 pN[†]	60	[37]
Spinning disk	NIH3T3	200	16	5 min	75 Pa	50	[21]
Radial flow	CCL 92	45	0.5	5 min	5 Pa	50	[73]
		320			11 Pa	50	
Parallel plate*	CCL 92	5	2	120 min	5 Pa	20	[82]
		10				85	
		5		2 min		90	
		10				100	
Parallel plate	CCL 92	182	0.5	2–10 s	6 Pa	60	[97]
			1		7 Pa	70	
			2		6 Pa	90	
Microfluidic channel	WT NR6	25**	0.5	12 min	400 Pa	5	[111]
					267 Pa	40	
					200 Pa	90	

Notes: Fibronectin was the ECM protein coating on the surface in all cases. Glass was the substrate in all cases except [37] in which a TCPS plate was used.

*ECM concentration was estimated from a plot in the respective reference.

[†] The cell volume was assumed to be 900 μm³ to calculate the load.

**ECM concentration was estimated by multiplying the reported volumetric concentrations by the channel height (25 μm).

characterize osteoblast adhesion to interpenetrating polymer networks grafted with peptides [49, 50].

2.2. Hydrodynamic Shear

Hydrodynamic shear flow assays are widely used and involve the application of well-controlled fluidic shear stresses to cells adhered on a substrate. The three most prevalent implementations of this concept are the spinning disk, radial flow chamber, and parallel plate flow chamber. Other techniques that rely on flow-induced stresses, such as jet impingement [51–53] and cone-and-plate shear chambers [54], are less widely used and will not be discussed here. In all hydrodynamic methods, the flow is generally maintained in the laminar regime (no turbulence) near the cell [55]. This allows the shear stress on the cell to be calculated based on the flow parameters through relatively simple relations. The limit of the laminar regime is determined by the Reynolds number,

$$Re = \frac{\rho U D}{\mu}, \tag{3}$$

where ρ is the mass density of the fluid, U is the average velocity, D is the characteristic dimension and μ is the fluid viscosity. When Re becomes sufficiently large (the exact threshold depends on the test geometry), the flow is no longer laminar.

The shear stress on a channel wall induced by flow of a Newtonian fluid is given by

$$\tau = \mu \frac{du(y)}{dy}\bigg|_{y=0}, \tag{4}$$

where $u(y)$ is the flow velocity and y is the direction normal to the wall. In fluid shear assays, the adhesion strength is typically defined as the wall shear stress that causes 50% of the initial population of adhered cells to detach. However, in some cases, cell adhesion strength may be measured directly as the parallel plate configuration is well-suited for direct observation of single cells using microscopy, provided that transparent materials are used in the chamber's construction.

2.2.1. Spinning Disk

The spinning disk apparatus utilizes shear stresses generated by spinning a disk with cells adhered to the surface inside a chamber that contains a buffer solution (Fig. 3). A detailed description of the device components, operation, and fluid dynamics has been presented in [56] and [57]. The shear stress applied on the cells is given as

$$\tau = 0.800r(\rho\mu\omega^3)^{1/2}, \tag{5}$$

where r is the radial distance from the center of the disk and ω is the angular speed of the disk (rad/s). A thin boundary layer forms near the surface of the disk during spinning and its thickness can be estimated by

$$\delta = 3.6\left(\frac{\mu}{\rho\omega}\right)^{1/2}. \tag{6}$$

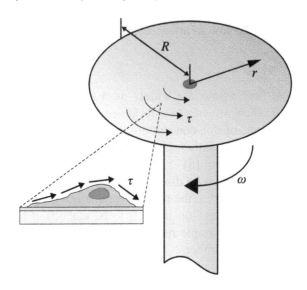

Figure 3. Schematic of spinning disk assay. The disk is spun inside a chamber filled with a buffer solution, resulting in a shear stress on the cells adhered on the surface of the disk. The shear stress increases linearly with the radial position on the disk.

Equation (5) is only valid for cases in which $R \gg \delta$, where R is the radius of the spinning disk. For this system, $\mathrm{Re} = \rho \omega R^2 \mu^{-1}$, and test conditions should be chosen such that Re is below the critical value of 1×10^5 [55]. A plot of the maximum achievable shear stress according to disk radius for a disk spinning in a buffer solution (PBS II, [56]) for $\mathrm{Re} = 100\,000$ is shown in Fig. 4. Note that shear stress decreases with disk radius in Fig. 4 because the Reynolds number is constant in this plot.

In the spinning disk method, cells are seeded on circular coverslips of glass or circular plates of another material of interest (typical dia. 10–50 mm) that are then fixed onto a disk that is spun at speeds ranging from 500 to 3000 RPM. A key advantage of the spinning disk method is that a shear stress that varies linearly with the radial position on the disk (equation (5)) is generated, thus a range of stresses are applied to the population of cells in a single experiment. Shear stresses upwards of 200 Pa have been applied [58, 59], so cells allowed to adhere for many hours that have high strengths can be detached. Experiments typically last no longer than 10 minutes [58, 60]. The adherent fractions of cells are generally quantified using microscopy by counting the number of cells before and after spinning using either a manual procedure [21] or automated image processing software [57].

This method has been used to examine a number of cell-substrate systems for specific applications. For example, it has been used to probe the strength of rat osteosarcoma cells on bioactive glass [57, 61, 62], human bone marrow cells on hydroxyapatite [60, 63] and MC3T3-E1 cells on RGD peptides on SAMs [64]. The method has also been employed in more fundamental experiments, including stud-

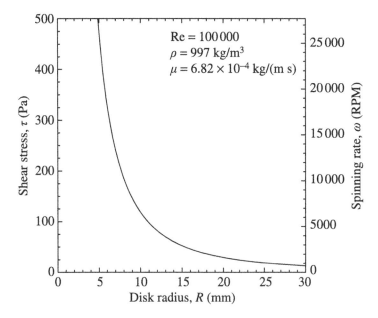

Figure 4. Maximum shear stress and corresponding spinning rate for the spinning disk assay as a function of disk radius (assuming $r = R$). For each value of disk radius, the spinning rate is selected such that the Reynolds number is 1×10^5. Thus, the corresponding stress is the maximum shear stress that can be applied in the assay in the laminar regime. The fluid properties assumed are that of the buffer solution (PBS II at 37°C) used in [56].

ies of the role of focal adhesion kinase, a protein central to focal adhesion function, in adhesion strength [59, 65]. Boettiger and colleagues investigated the adhesion strength response of both chick embryo fibroblasts [66] and human osteosarcoma cells [67] transformed with the oncogene v-src in order to study its effect on integrin function. HT-1080 cells were used to study the effects of cell exposure to insulin-like growth factor I (IGF-I) [68] and substrate surface charge [69] on adhesion strength. Finally, in a study on the adhesion strength of a specific integrin ($\alpha_5\beta_1$), García *et al.* measured both the short- and long-term adhesion strength of IMR-90 human fibroblasts adhered to fibronectin-coated glass [70].

In a unique study, Gallant and colleagues used the spinning disk to study NIH3T3 fibroblasts adhered to patterned circular fibronectin islands of different sizes [21, 58]. The circular islands ranged from approximately 3 to 314 μm^2 in area and the cells were allowed to adhere for times ranging from 15 min to 16 h. The results show adhesion strength correlates with cell area at small areas, but is independent of island size above areas of ~75 μm^2 [21]. The results of a control experiment in which fibroblasts were allowed to adhere to unpatterned fibronectin is detailed in Table 1 for comparison to other measurements.

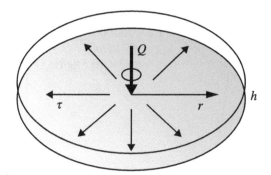

Figure 5. Schematic of radial flow chamber. Flow typically enters the circular chamber from the center and travels outwards, resulting in a shear stress on the cells adhered on the bottom surface. The shear stress in this configuration decreases nonlinearly with increasing radial distance.

2.2.2. Radial Flow Chamber

In contrast to the spinning disk, both the radial and parallel plate methods involve flowing fluid in a chamber over cells on a stationary substrate. Radial flow techniques direct the flow outwards from the center of a circular chamber over a substrate seeded with cells (Fig. 5). If the chamber is made of optically transparent materials, detachment events can be monitored *via* light microscopy. This technique has been used extensively by Goldstein and DiMilla [71–74] and by Healy and colleagues [75–79]. As was the case in the spinning disk method, the radial flow chamber applies a range of forces to a population of cells in a single experiment as a result of the experiment geometry. In the most common radial flow chamber configuration, the inlet flow is directed outwards from the center of the chamber. The mean velocity of the fluid, and hence the shear stress, decreases with increasing radial distance in a nonlinear fashion. The most commonly used expression to estimate the applied stress is [72]:

$$\tau = \left| \frac{3\mu Q}{\pi h^2 r} - \frac{3\rho Q^2}{70\pi^2 h r^3} \right|, \tag{7}$$

where Q is the flow rate and h is the chamber height. The first term of the equation represents the viscous wall shear stress generated by axisymmetric flow from a point source at the center of the chamber, and the second term, which decreases quickly with increasing radius, is a correction to account for inertial effects near the inlet. It is generally desirable to minimize inertial effects, thus data are typically recorded only from cells that reside sufficiently far from the inlet. For the inertial term in equation (7) to be less than 5% of the viscous term, measurements should only be taken from cells at locations where [72]

$$r > \left(\frac{2\rho Q h}{7\pi\mu} \right)^{1/2}. \tag{8}$$

The Reynolds number for the radial flow method is defined as $\text{Re} = Q\rho(\pi r\mu)^{-1}$, with a critical value of 2000 [72]. The smallest chamber height used in previous studies was 200 μm, and the maximum chamber radius was typically on the order of centimeters [71]. Accounting for the entry length condition (equation (8)) and limiting Re to 2000, the maximum shear stress that can be generated for chambers with typical heights is less than ∼20 Pa. Therefore, the method is generally restricted to studies in which the cells have been allowed to adhere for short times only. Typically, stress is applied to cells for less than 10 minutes and adherent fractions are determined by analysis of microscopy images.

Goldstein and DiMilla used the radial flow chamber in a number of studies related to the adhesion strength of murine fibroblasts adhered to fibronectin-coated SAMs on glass surfaces [71–74]. In one study, the cells were allowed to adhere for 30 min to surfaces with different fibronectin concentrations. The results showed that adhesion strength was relatively insensitive to fibronectin concentration up to 0.23 μg/cm^2 but then increased at a concentration of 0.32 μg/cm^2 [73]. An additional feature of this study was the transparent chamber that allowed visual tracking of cell deformation and strain during the experiments. Adhesion strength values from this experiment are summarized in Table 1.

The radial flow method is quite flexible and has also been used to look at a number of other cell types and materials. The adhesion of MC3T3-E1 cells to multilayer poly(allylamine hydrochloride) (PAH) heparin films was analyzed in order to evaluate biocompatibility of various film chemistries [80]. Other studies include the investigation of osteoblasts and osteoblast-like cells adhered to RGD peptides on quartz substrates [76, 78, 79], and endothelial cells on interpenetrating polymer networks [75]. Osteoblasts adhered to silane-treated quartz substrates were also studied [77]. Sordel *et al.* explored the adhesion strength of CHO cells on glass and silicon substrates with different surface modifications, including poly(dimethylsiloxane) (PDMS) coatings, oxygen plasma treatments, and deposited silicate layers [81].

2.2.3. Parallel Plate Flow Chamber

Parallel plate flow chambers (Fig. 6) have been used extensively to characterize cell adhesion strength. The flow in these systems is well-defined, and the chambers can be mounted on a microscope to allow direct observation of the cells during tests. The chambers are usually constructed by sandwiching a thin rubber gasket between two plates and placing a cell-seeded substrate inside. For long-term studies, heating elements can be included as well. The flow is often driven using hydrostatic pressure from a raised reservoir, with an automated pump for flow recirculation [82]. Pumps can be used independently to drive the flow as well [83]. For a given flow rate, the wall shear stress will be constant along the length of the channel beyond the entrance length, but it can vary across the width depending on the channel dimensions. Variation across the width of the channel can be minimized using low aspect ratio ($w > 20h$) cross sections [82].

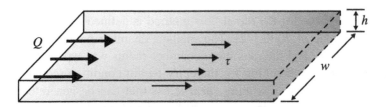

Figure 6. Schematic of parallel plate flow chamber. Flow enters at one end and travels through the channel, applying a shear stress to cells adhered on the bottom. The shear stress is constant along the length and highly uniform across the width for channels in which $w \gg h$.

The Reynolds number for this geometry is $\mathrm{Re} = (\rho Q D_{\mathrm{h}})(wh\mu)^{-1}$, where D_{h} is the hydraulic diameter of the flow channel that is defined as $D_{\mathrm{h}} = 2hw(h + w)^{-1}$. For channel flows, the flow is laminar up to $\mathrm{Re} \approx 2300$. The wall shear stress, which is typically taken as the stress applied on the cell, is

$$\tau = \frac{6\mu Q}{wh^2}, \tag{9}$$

when $h \ll w$. Typical channel heights range from 200 µm to 1 mm. Previous parallel plate flow studies of cells have involved a variety of incubation times, though the applied shear stresses were rarely greater than about 30 Pa. In most population type assays, particular areas of the flow chamber are imaged *via* microscopy a number of times throughout the experiment, and the adherent fractions as a function of time are then determined by counting the cells. While most studies involve cells adhered to ECM-coated surfaces such as glass or a rigid polymer, adhesion strengths of cells adhered to dentin [84] and cartilage [85] samples placed inside the flow chamber have also been reported.

Parallel plate flow chambers have been used to characterize human skin fibroblasts adhered to glass [83], Teflon [86] and other polymer substrates [87]. This same group has also used this method to examine endothelial cells adhered to glass without any ECM treatment [88]. In these studies, image sequences showing cell deformation under flow were recorded in addition to adhesion strength, which demonstrates the ability of the technique to acquire information on individual cells as well as the adhesion characteristics of the population. While the above studies primarily employed a constant flow rate, and hence a constant applied stress on the cells, pulsatile flow to probe the impact of time-varying stresses on fibroblast adhesion strength has also been employed in the parallel plate flow chamber [89].

The parallel plate system is very flexible and due to its simple construction allows a diverse range of surfaces to be studied. It has been used to examine the adhesion of a range of cell types to many different materials, including polyelectrolyte multilayer films [90], polyethylene films [91], poly-L-lactide (PLL) films [92], organosilane SAMs [93], and polyurethane [94]. Fibroblast adhesion to a variety of substrates, including the polymer surfaces discussed previously [86, 87], plasma-treated poly(L-lactic *co*-glycolic acid) (PLGA) and poly(L-lactic acid)

(PLLA) films [95, 96], and variously treated glass surfaces [82, 83, 87, 89, 97], has been investigated extensively using this method.

Truskey and Pirone used a parallel plate flow chamber to study fibroblasts adhered to fibronectin-coated glass surfaces [82]. The same system was used to examine the effect of silane treatments on fibroblast adhesion strength [97]. Incubation times in these experiments ranged from 30 minutes to 2 hours, and low shear stresses (<20 Pa) were applied over various loading durations, ranging from a few seconds to 2 hours, across the set of experiments. The relationship between cell spreading and adhesion strength was explored, and membrane fragments were fluorescently labeled in order to gain insight into the detachment mechanisms. This study showed evidence of a relationship between cell area and adhesion strength. Furthermore, it illustrated that cells can detach by peeling from the substrate, by receptor pull-out and membrane rupture, or by a combination of both. The results of these experiments are summarized in Table 1. The same flow system was also used to analyze the effect of glass silanization on fibronectin adsorption and endothelial cell adhesion strength [98].

Variously modified forms of the basic parallel plate flow chamber have been employed to achieve different loading conditions in the assay. The most common variation is a chamber in which the height varies along the length of the channel, thus allowing the generation of a range of stresses along the length. This assay was first demonstrated and used to study endothelial cell adhesion [99, 100]. Subsequently, it has been used in a number of studies exploring the adhesion strength of biotinylated endothelial cells adhered to glass with fibronectin or RGD peptide functionalization, notably [101–103]. Recently, chambers with tapered height have been employed to examine the effects of trypsin exposure on cell adhesion strength [104]. In contrast to varying the channel height, the width of the channel has also been varied to achieve a range of shear stresses in a single experiment [105]. This particular geometry has been used to examine hepatocytes adhered to polystyrene substrates coated with Matrigel, which is a solution that consists of several proteins and other biological factors [106]. Endothelial cells adhered to surfaces functionalized with artificial ECM proteins were also studied in this fashion [107].

A key limitation of conventional parallel plate assays is their inability to generate shear stresses large enough to detach well-spread cells while maintaining laminar flow; thus, they are generally limited to short-term adhesion studies, as typical chambers are capable of producing maximum shear stresses on the order of only 10 Pa under laminar conditions. As discussed above, the flow generally ceases to be laminar beyond Re $\approx$ 2300. In order to produce larger stresses on cells in hydrodynamic flow experiments, the wall shear stress must be increased while ensuring that the flow remains laminar. One method of achieving this is to reduce the channel height, as demonstrated in Fig. 7.

In recent years, cell adhesion has been investigated using flow in rectangular microfluidic channels to apply shear stresses to cells in the same fashion as the tra-

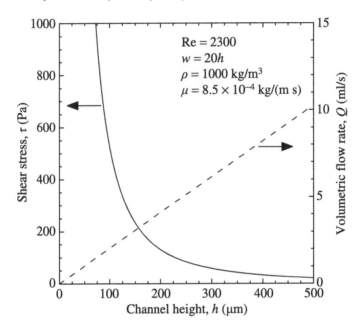

Figure 7. Maximum achievable shear stress and corresponding volumetric flow rate in a parallel plate flow chamber as a function of channel height, assuming flow in the laminar regime (Re = 2300) and $w = 20h$. The fluid properties assumed are that of the growth medium solution used in [97].

ditional parallel plate flow chambers discussed above. These devices are typically constructed from optically transparent PDMS bonded to glass using the soft lithography rapid prototyping process that allows many nearly-identical devices to be manufactured in a short amount of time [108]. The heights of these devices are frequently less than 100 μm, and as a result, shear stresses of several hundred Pa can be generated under laminar flow conditions. One drawback of the use of PDMS is that due to its low elastic modulus, it can deform under the pressure that develops in the flow and cause the channel dimensions to change [109]. In the case of large pressures, this deformation can be severe enough to require that it be accounted for in calculating the shear stresses [109, 110].

Lu *et al.* used a series of microfluidic devices to study the adhesion strength of WT NR6 fibroblasts adhered to fibronectin-coated glass surfaces [111]. The devices had channels with heights of 25 μm and were capable of generating wall shear stresses up to 640 Pa. Three different devices were used: one with four identically-sized channels with different ECM concentrations, one with four channels of different widths that yields a different shear stress in each channel, and a single channel with several branches that allows perfusion for long-term incubation. The results of a selection of these experiments are summarized in Table 1.

Christ *et al.* used a rectangular microchannel to analyze the adhesion strength of single NIH3T3 fibroblasts that were allowed to adhere for 24 hours to collagen and fibronectin coatings on glass [110]. In this work, single cells were imaged throughout the detachment process, and the relationship between adhesion strength and cell geometry was investigated. It was found that a statistically significant relationship exists between cell adhesion strength and cell area, and also between adhesion strength and cell circularity. In addition, it was observed that while some cells quickly peeled from the surface, others experienced significant membrane deformation or rupture prior to detachment. These findings highlight the heterogeneity of the adhesion characteristics in spread cells and suggest further investigations into the relationship between the mechanics at the cell-substrate interface and adhesion strength.

Microchannel assays are increasingly being employed and several recent reports illustrate their flexibility. For example, a device consisting of eight parallel channels has been used to assess the effect of varying collagen and fibronectin concentrations on adhesion strength of endothelial cells [112]. In another case, microchannel assays were used to examine the adhesion of various cell types on surfaces with various coatings, including collagen, glutaraldehyde, and silane [113]. Finally, Kwon *et al.* used a microfluidic shear device consisting of four parallel channels with different surface topography patterns to separate cancer cells mixed in a population of healthy cells based on adhesion strength [114].

2.3. Micromanipulation

While population studies have been successful in determining general adhesion characteristics of cells across a broad range of experimental conditions, single-cell adhesion studies can provide greater insight into the actual mechanical processes associated with adhesion strength and resistance to detachment. The key advantage of these methods is the precise control over the magnitude and location of the force being applied to the cell under scrutiny. However, these methods can be time-consuming and often require specialized equipment and a high level of expertise. Two of the most common micromanipulation approaches, cytodetachment and micropipette aspiration, are discussed in detail below. The main distinction between these two methods is the manner in which the force applied to the cell is quantified; in cytodetachment the force is measured by monitoring the elastic deflection of a probe, while in micropipette studies the force in generally determined based on the aspiration pressure. A number of other single-cell approaches, including microplates [115], optical tweezers [116] and magnetic tweezers [117], have also been used to examine adhesion at the single-cell level but are not reviewed in detail here.

2.3.1. Cytodetachment

A common micromanipulation technique is the use of an atomic force microscope (AFM) cantilever or other stiff probe to physically detach individual cells in an open medium environment such as a Petri dish (Fig. 8). The probe and the substrate to which the cell is adhered are typically mounted on separate linear stages, and one is translated at a specified rate such that cell is loaded and eventually detached.

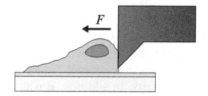

Figure 8. Schematic of a typical cytodetachment test. A micromanipulator, such as an AFM probe, is pushed against the cell and applies a tangential force that detaches the cell. The force is generally measured by monitoring the deflection of the probe. The direction of the loading and stress induced on the cell can be varied by adjusting the orientation of the probe with respect to the cell.

The force is generally quantified by measuring the elastic deflection of the probe used to apply force to the cell. When a commercial AFM is used, the deflection of the probe (i.e., cantilever) is measured from a laser signal that is reflected off the probe and monitored using a photodetector. In cases in which the probe and stage are self-constructed, images of probe deflection are often acquired optically during the cell detachment process and processed to determine the deflection [118, 119]. The force is calculated from the measured deflection of the probe at detachment, x, and the stiffness of the probe, k:

$$F = kx. \tag{10}$$

Determining detachment force in this particular manner requires rigorous calibration or modeling to determine the probe stiffness. The loading rate can be varied over a wide range with most stages, but rates employed in previous cell adhesion strength studies were typically 1–20 μm/s [120, 121]. Cytodetachment techniques can apply and measure a range of forces, and cell adhesion strength measurements ranging from 1 nN to 1 μN have been reported [118, 122].

Various commercial AFM or similar systems that are custom-built, with slightly different configurations, have been employed in previous adhesion studies. Sagvolden et al. developed an AFM-like experimental system designed to manipulate individual cells and used it to measure the adhesion strength of cervical carcinoma cells on fibronectin- and laminin-coated polystyrene surfaces [123]. An AFM cantilever was used as the probe, and the cell was translated against it at a rate of 2.5 μm/s by displacing the substrate. The cantilever deflection was measured by a fiber optic sensor, and hundreds of cells were tested. In other work, a commercial AFM system was used to detach NIH3T3 cells from SAMs on gold-coated microelectrodes [124]. Su and colleagues [122] used an AFM probe mounted on a specially-constructed manifold to detach cells. Images were recorded over the course of the experiments, and the probe deflection was determined by image analysis. Kidney epithelial cells with variable focal adhesion kinase expression on collagen-coated glass [122], osteoblasts on titanium [125] and NIH3T3 fibroblasts on titanium alloys [119] were studied using this system.

Table 2.

Adhesion strength of fibroblasts measured using micromanipulation techniques

Method	Cell type	ECM conc.	Incubation time	Experiment duration	Adhesion strength	Ref.
Cytodetachment	L929	80 ng/cm^2	24 h	<10 s	1070 Pa	[121]
Micropipette	Human MCL	1 µg/ml	15 min	3 min	10 nN	[130]
			1 h		70 nN	
		5 µg/ml	15 min		30 nN	
			1 h		160 nN	

Notes: Fibronectin was the ECM protein coating on the surface in all cases. The ECM coating was applied to TCPS for the experiments in [121] and poly-D-lysine-precoated glass for those in [130]. ECM surface concentrations were not reported for the experiments in [130], thus the volumetric concentrations are listed.

Yamamoto *et al.* developed a system consisting of a stainless steel cantilever with a silicon tip to detach cells [121, 126]. The bottom of the cantilever tip was positioned 0.2 µm above the substrate, and a motorized stage was used to move the cells laterally against the probe at a rate of 20 µm/s. The cantilever deflection was measured by a fiber optic sensor. Using this instrument, they performed a rigorous study of L929 murine fibroblasts adhered to glass and polystyrene surfaces with fibronectin and collagen coatings [121]. A force-displacement curve was recorded for each cell studied, which allowed the calculation of detachment energy in addition to the adhesion strength. The force required to detach the cells was divided by the cell area to calculate the average shear stress for each cell. Results from their experiments examining adhesion on fibronectin-coated TCPS are summarized in Table 2.

While the systems discussed above utilize probes that have rectangular cross sections, like conventional AFM cantilevers, probes with circular cross sections, such as fibers and micropipettes are also frequently used. For example, a 75 µm diameter, 2 cm long glass fiber, with a much smaller (7 µm diameter) carbon filament attached to it, was used to examine the detachment of chondrocytes from glass surfaces [120]. The displacement of the carbon filament was measured optically by a dual photodiode to determine the force. The fiber was oriented with its length perpendicular to the cell and its bottom positioned 2 µm above the substrate and was then moved by a motorized stage against the cell at a rate of 1 µm/s. The fiber stiffness was determined from repeated tests in which the fiber was moved against another fiber of the same material but of different length, and the force was calculated *via* equation (10). A subsequent variation of this basic system involved changes in the design and orientation of the fiber in order to better facilitate its use with an inverted microscope [127, 128] In a similar experimental design, micropipettes (O.D. 2 µm) have been used to detach individual chondrocytes from a variety of polymer surfaces [118]. In these experiments, nonlinear probe deflection occurred due the large forces induced on the probe and the relatively low stiffness of the probe compared to the cell. Optical images were captured of the micropipette deflection during the cell detachment process, and the forces were calculated using an inverse finite element analysis mechanics model. Using the same method, endothelial cells that had been exposed to hydrostatic pressure while on gelatin-coated glass substrates were also studied [129].

2.3.2. Micropipette Aspiration

Micropipette aspiration is a widely-used method for the measurement of the mechanical properties of single cells and has been employed over the past several years to measure the adhesion strength of spreading cells. The micropipettes in these studies typically possess an inner radius on the order of micrometers and are used to create a suction force on the cell membrane. The cell is detached by either increasing the aspiration pressure or by translating the pipette or substrate (Fig. 9). The force imparted on the cell by aspiration is

$$F = \pi R^2 \Delta P, \tag{11}$$

where R is the micropipette inner radius and ΔP is aspiration pressure. The size of the micropipette and the direction of its movement determine the type of loading the cell experiences, so the size and angle between the probe and the cell must be measured for proper force analysis.

Micropipette aspiration has been used in a number of experiments to measure the adhesion strength of human ligament fibroblasts adhered to a variety of ECM surface coatings in the presence of various treatments affecting cellular function [130–133]. A similar procedure was used in each of the studies. A small micropipette (I.D. 3–7 μm) was brought into contact with a cell and an aspiration pressure was applied. The pipette was then moved away from the cell at a specified rate (2–4 μm/s). If the cell did not detach, then the aspiration pressure was increased and the process repeated. Using equation (11), the adhesion strength was then defined as the minimum force necessary to detach the cell. The number of

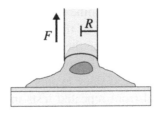

Figure 9. Schematic of an adhesion strength test using micropipette aspiration. A micropipette is brought into contact with the cell and an aspiration pressure is applied. A portion of the cell is suctioned into the pipette, which is then translated away from the surface, detaching the cell. A normal force is typically applied, but the exact loading direction depends on the translation of the micropipette relative to the cell.

cells detached for a given treatment in these studies was regularly on the order of hundreds of cells. In one study, a dependence of adhesion strength on both seeding time and fibronectin concentration was identified [130]. Results of this study are summarized in Table 2. This technique was subsequently employed to examine the adhesion strength of osteoblasts on plasma-treated titanium [134]. In another experiment, osteoblasts adhered to fibronectin-coated glass were allowed to take up titanium particles prior to the adhesion strength measurements [135]. Ligament fibroblast adhesion to collagen-coated surfaces was also studied with this method [136], as was the effect of epidermal growth factor on adhesion strength [137].

A similar method has been employed to study chondrocytic HCS 2/8 cells adhered to polyelectrolyte multilayer films [138, 139]. Micropipettes, 7 μm in diameter, were translated at approximately 3 μm/s during the tests. Cai and colleagues have also used this technique to analyze the adhesion strength of hepatocytes in comparison to that of hepatocellular carcinoma cells [140] and other hepatocytes treated with cytochalasin D, a chemical that inhibits actin polymerization [141]. Tenocytes adhered to fibronectin- and collagen-coated poly-D-lysine (PDL) on poly(DL-lactide-*co*-glycolide) (PLGA) substrates were also studied [142]. Similarly, Gao *et al.* analyzed endothelial cells adhered to silk fibroin substrates [143].

A slightly different method based on micropipette aspiration was used to detach bone marrow cells from fibronectin-coated polystyrene [144]. In this particular procedure, the inside diameter of the micropipette (14–16 μm) was similar in size to the cells being analyzed. The pipette was oriented vertically above a cell and was positioned such that part of the interior edge of the micropipette was in contact with the cell. A negative flow was initiated through the micropipette, and the cell was gradually suctioned away from the surface. The adhesion strength was defined as the suction force needed to completely detach the cell. In another variation of the standard micropipette technique, chondrocytes adhered to very thin (30 μm) glass substrates coated with silk fibroin [145] were loaded using a stationary micropipette to partially aspirate a cell and then translating the adhered glass surface away from the pipette. Adhesion strength was determined by measuring the deflection of the glass substrate immediately prior to detachment, similar to the cytodetachment studies described above.

Micropipettes have also been used to measure the adhesion strength of muscle cells, but the procedures employed are significantly different than those described above. Muscle cells seeded on patterned lines of collagen were detached in a two-step process [146]. The cells were rectangular-shaped due to the patterning, and one end was initially detached and aspirated with the micropipette (2.5–5 μm dia.), which was positioned at an angle with respect to the surface. With the pressure held constant, the micropipette was translated such that the cell was peeled off. In a modification of this technique, micropipettes with much larger diameters were used (75 μm), and after the initial detachment of the cell end, a negative flow rate was applied to the pipette such that fluid entered the pipette while flowing over the cell

[147, 148]. Simultaneously, the cell was translated so that it was peeled from the surface by the flow. This modification affects the way in which the load on the cell is controlled during detachment and allows a constant force to be applied as the cell is peeled.

3. Discussion

3.1. Cell Adhesion in Engineering, Biology and Medicine

Most of the cell adhesion strength studies discussed above fall into two general categories: those that are focused on evaluating biomaterials and those that involve analyzing the response of cells to specific biochemical treatments. Biomaterials are fundamental to medicine and biomedical technology, and understanding cellular interactions with them are crucial to their implementation. For example, cells adhered to orthopedic implants experience a variety of physical forces, and understanding how these forces along with the surface properties of the implant relate to adhesion is central in the design and functionality of such devices [149]. In addition, technologies such as implantable biosensors [11], drug delivery systems [150] and tissue engineering scaffolds [151] involve materials that must control cell adhesion in order to function properly.

Cell adhesion is intertwined with nearly all aspects of cell function, from proliferation to protein production, and ultimately, cell death. Integrins provide the dominant receptor between cells and their external surroundings, and changes in their behavior are associated with many diseases. Cancer, for example, is characterized by uncontrolled cell growth and migration, both of which are directly related to adhesion [152, 153]. Cell adhesion is also essential to regenerative processes such as wound healing, in which a number of different cells migrate to the wound area in order to prevent infection and rebuild the damaged tissue [154, 155]. Analyzing the effects of biochemical treatments on cell adhesion is an important route towards improved understanding of pathological cell behavior and developing better therapeutics.

3.2. Summary of Experimental Results

From the large body of cell adhesion strength measurements that have been completed with the different methods discussed, three general trends can be identified: (1) adhesion strength increases, to a certain extent, as incubation time increases (e.g., [21, 130]), (2) adhesion strength is enhanced when an ECM coating is applied to the substrate (e.g., [82, 121]), and (3) increasing the surface concentration of the ECM coating increases the adhesion strength to a certain extent (e.g., [37, 73]). While these are general trends that have been observed in many studies, there are many exceptions to these broad observations. Most importantly, the greatest change in adhesion strength with respect to time occurs only up to the point at which the cell becomes fully spread [21], and there is often only a particular range over which increases in the ECM concentration lead to increases in adhesion strength [37, 73]. Beyond or below these ranges, significant changes may not be seen.

Tables 1 and 2 summarize measurements of the adhesion strength of fibroblasts performed using various techniques, and the general trends identified above can be observed in these data, particularly in the cases in which multiple treatments (e.g., different ECM protein concentrations) were examined in a single study. It is difficult to compare adhesion strength values from different works, as there are many conditions such as adhesion time and seeding density which must be constant to allow for meaningful comparisons. Furthermore, comparing strength measurements from techniques which apply different types of loading is challenging. The type of stress applied (normal *versus* shear) as well as the area over which the stress is applied on the cell both can affect the magnitude of the measured strength values.

Cell adhesion is a biphasic process in which an initial binding period is followed by a period of spreading, focal adhesion development, and adhesion strengthening [20]. The short-term adhesion strength of most cells is typically weak enough such that any of the methods covered here can be used to detach them. Cells adhered for short time periods are not as geometrically heterogeneous and will not be involved in dividing or migration. Thus, they are well-suited to be studied in population assays, which cannot fully capture the heterogeneity in a cell population. In contrast, cells which have been allowed to spread for many hours possess well-developed adhesive components. As a result, spread cells in a population may possess different mechanical properties depending on whether that cell is migrating or involved in another process. Therefore, it can be useful to analyze these cells one at a time, as in a micromanipulation study, or by using a flow chamber in which individual cell properties and adhesion strength can be measured for each cell.

3.3. Comparison of Methods

All of the methods discussed above are well-established and suitable for measuring cell adhesion strength, but each has unique strengths and weaknesses. Table 3 presents a summary of the methods, their force or stress measurement range based on that found in the literature above, and a comparison of their advantages and limitations. In the centrifuge assay, the highest force that can be realistically generated using a benchtop centrifuge is on the order of 1 nN. As seen in equation (1), the force could be increased by increasing the radius of the centrifuge rotor or by increasing the rotation speed. It may be possible to extend the rotor length a small amount depending on the centrifuge model, but not enough to change the applied force meaningfully. Ultracentrifuges can achieve very high spinning rates (10 000–100 000 RPM) and thus allow higher stresses to be applied. However, they are not found in most laboratories, are expensive, and require delicate operation due to the extremely high speeds involved. In addition, the rotors used in ultracentrifuges are not designed to handle multiwell plates. Another limitation of general centrifuge assays is that since only a single force is applied per experiment, many experiments

Table 3.

Comparison of advantages and limitations of common adhesion strength assays

Method	Typical measurement range	Advantages	Limitations and challenges
Centrifugation	1–2000 pN	• Many conditions can be examined in parallel • Common lab equipment	• Low maximum force • Only a single force can be applied per experiment
Spinning disk	1–350 Pa	• Range of stresses in a single experiment • High stresses	• No direct visualization of cells • Custom-made apparatuses
Radial flow chamber	1–20 Pa	• Range of stresses in a single experiment • Direct visualization of cells	• Low maximum stress
Parallel plate flow chamber	1–30 Pa	• Straightforward construction and operation • Direct visualization of cells	• Low maximum stress
Microfluidic rectangular channels	1–400 Pa	• High maximum stress • Direct visualization of cells	• Cell culture in microenvironment
Cytodetachment	1–1000 nN	• Detailed force history data • Quick detachment of cells	• Costly, specialized equipment • Probe calibration required
Micropipette aspiration	1–1000 nN	• Common lab equipment • Precise force application	• Severe cell deformation prior to detachment • Alignment of probe and cell

must often be performed in order to generate enough data to determine the force at which 50% of the cells are detached. It should be noted that the stress exerted on a given cell depends on its volume and spread area, so all the cells may not experience the same detachment stress. Therefore, the centrifuge assay is best suited for the screening of cell adhesion to many surfaces in parallel, and it is almost exclusively limited to short-term adhesion studies.

The spinning disk method is a versatile technique that is suited for both short- or long-term adhesion studies, as shear stresses well over 100 Pa have been achieved in laminar flow conditions. In contrast to the centrifuge assay, though, spinning disk apparatuses are typically custom-built, and therefore the instruments must be constructed and validated prior to use. A key limitation of the spinning disk compared to the two flow chamber methods is that it does not allow for direct observation of the cells using microscopy during the test.

The radial flow chamber suffers primarily from the inability to generate shear stresses high enough to detach well-adhered cells while maintaining well-defined and repeatable experimental conditions. Standard parallel plate flow chambers are limited in the same respect. In addition, inconsistencies in their assembly can lead to significant variations in the channel geometry, especially for small channel heights [156]. These issues can be overcome by using microfluidic devices as illustrated in Fig. 7. However, if these devices are made from PDMS, significant deformation of the channel will occur at the flow rates required to detach spread cells. Therefore, computational modeling or experimental validation of the channel geometry during the test must be included for accurate measurements. In addition, the physics of the microfluidic environment is significantly different than that of a typical dish, multiwell plate, or macroscale chamber, which can create difficulties in culturing cells in the channels [157]. All rectangular channel assays apply a single shear stress value for a given flow rate unless the channel geometry includes a taper, which allows a variation in the stress applied.

Micromanipulation techniques allow for detailed measurements of the force required to separate single cells from a substrate. In addition, they enable greater latitude in terms of choosing which cells in a particular sample to test. While many studies only report the force at detachment, the force during the entire separation process can be recorded with many of the micromanipulation techniques. Such detailed data can be particularly useful in understanding the failure of the cell-substrate interface, but it makes the test more time-intensive and can make it impractical to analyze many cells in order to identify trends in adhesion strength with a variable such as ECM concentration.

In cases other than in which an AFM is used, the instruments used in micromanipulation studies are often custom-built. Also, because the forces are concentrated, it may be difficult for the cell loading to remain consistent when characterizing spread cells that are heterogeneous in their shape and topography. In cytodetachment experiments, the elastic properties and geometry of the probe must be well-defined, and care must be taken if image analysis is used to determine the amount of deflection. In the micropipette aspiration experiments, a part of the cell is usually suctioned directly into the pipette prior to detachment, so repeatability can be a concern in this case as well. Also, if the aspiration process needs to be repeated many times before detachment, as in [130, 138, 142], the response of the cells due to the multiple loadings may alter the results.

3.4. Analysis of Loading in Adhesion Assays

The type and method of loading in the various adhesion assays is a key characteristic that differentiates the methods and must be considered when selecting a measurement technique and interpreting data. The type of loading can be classified based on the direction of loading (normal *versus* shear), how the load is distributed over the cell (at a point, over the surface or on the body of the cell), and the time over which the loading is applied to the cell. As the measured adhesion strength can vary with the type of loading, it is important to understand how the loading varies in the different assays.

The centrifuge assay and some of the micromanipulation approaches apply normal stresses to the cell-substrate interface. The centrifuge assay applies a body force to the entire cell, and in virtually all cases this force is applied normal to the substrate. Some of the micropipette techniques also apply a normal force, but in these cases small-diameter pipettes are used so the force is applied to a specific portion of the cell membrane rather than to the entire cell. In the centrifuge tests, a constant load is applied for a fixed period of time, typically less than 10 minutes, while in the micropipette pull-off force tests the load is typically increased monotonically until failure occurs.

In the flow assays, the primary stress applied is shear; however, the cells actually experience multiple forces [158]. The flow over the cell imparts a pressure that acts downwards on the cell and a shear stress that acts tangentially along the cell surface. Therefore, a force in the direction of the flow and another force perpendicular to it are applied to the cell. A small torque is also applied as the cell has a finite height, and shear stresses are applied across the height. The finite height of the cell, usually on the order of a few micrometers, can affect the actual stress that the cell experiences and cause it to deviate from that given by equation (9) [110, 158]. This is especially important when channels with smaller heights, such as microfluidic channels, are used [159]. The duration of loading in these experiments can have an impact on the measured strengths. A fixed flow rate, and hence stress, at a particular location in the chamber can be applied for a set period of time [82]. Alternatively, computer-controlled syringe pumps allow applying shear stresses that monotonically increase with time [110, 111]. The duration and rate are important as cells have rate-dependent mechanical properties and can respond to shear stress by remodeling their structural and adhesive components [29, 160, 161].

Cytodetachment techniques can apply forces in a variety of directions, but they are commonly used to apply a shear force to the cells [121, 123, 126]. The exact positioning and shape of the probe is important in determining the force imparted on the cell. The force is localized at the point of contact, so the bonds nearest to this location will experience a different loading than those located farther away. A key challenge in micromanipulation techniques is determining how the load from the probe is transferred to the cell. The cell is not a homogeneous structure, thus the location of probe contact and the adhesion between the cell and the probe can determine if the load is applied directly to the cytoskeleton or just the cell membrane. This will affect how the applied force is transferred to and distributed at the interface. For reasons similar to those discussed above for the shear flow assay, the loading rate in micromanipulation tests can have a strong effect on the measured adhesion strength.

4. Conclusion

In this review, we have covered the most frequently used methods to measure the strength of cells adhered to surfaces, described the mechanical loading they apply to the cells, and analyzed the magnitudes of the loads that can be applied. In addition, we have provided comparisons of adhesion strength measurements performed on fibroblasts using the different methods. Each method has advantages and limitations, and these should be carefully considered when attempting to quantify adhesion strength. Furthermore, the type of data that can be obtained depends on whether population characteristics are measured or if single cells are tested one at a time. Finally, the direction and duration of the loading can have consequences on the adhesion strength measurement, as cells can respond to mechanical stimuli. Cell adhesion continues to be an important topic with the ongoing development of cell-based treatments and biomaterials, and quantitative measurements of adhesion strength will continue to be a central part of advances in these areas.

References

1. S. Huang and D. E. Ingber, *Nature Cell Biol.* **1**, E131–E138 (1999).
2. J. C. Adams and F. M. Watt, *Development* **117**, 1183–1198 (1993).
3. D. A. Lauffenburger and A. F. Horwitz, *Cell* **84**, 359–369 (1996).
4. B. M. Gumbiner, *Cell* **84**, 345–357 (1996).
5. R. P. McEver, *Thromb. Haemost.* **86**, 746–756 (2001).
6. M. C. Montoya, D. Sancho, M. Vicente-Manzanares and F. Sanchez-Madrid, *Immunol. Rev.* **186**, 68–82 (2002).
7. B. Felding-Habermann, *Clin. Exp. Metastasis* **20**, 203–213 (2003).
8. C. M. Verfaillie, J. B. Mccarthy and P. B. Mcglave, *J. Clin. Invest.* **90**, 1232–1241 (1992).
9. K. Anselme, *Biomaterials* **21**, 667–681 (2000).
10. U. Hersel, C. Dahmen and H. Kessler, *Biomaterials* **24**, 4385–4415 (2003).
11. N. Wisniewski, F. Moussy and W. M. Reichert, *Fresenius J. Anal. Chem.* **366**, 611–621 (2000).
12. G. E. Muschler, C. Nakamoto and L. G. Griffith, *J. Bone Joint. Surg. Am.* **86A**, 1541–1558 (2004).
13. C. Chai and K. W. Leong, *Molecular Therapy* **15**, 467–480 (2007).
14. V. Vogel and M. Sheetz, *Nature Rev. Mol. Cell Biol.* **7**, 265–275 (2006).
15. R. G. Flemming, C. J. Murphy, G. A. Abrams, S. L. Goodman and P. F. Nealey, *Biomaterials* **20**, 573–588 (1999).
16. D. E. Discher, P. Janmey and Y.-L. Wang, *Science* **310**, 1139–1143 (2005).
17. R. O. Hynes, *Cell* **110**, 673–687 (2002).
18. K. Burridge and M. Chrzanowska-Wodnicka, *Annu. Rev. Cell Dev. Biol.* **12**, 463–518 (1996).
19. B. Geiger, J. P. Spatz and A. D. Bershadsky, *Nature Rev. Mol. Cell Biol.* **10**, 21–33 (2009).
20. A. J. Garcia and N. D. Gallant, *Cell Biochem. Biophys.* **39**, 61–73 (2003).
21. N. D. Gallant, K. E. Michael and A. J. Garcia, *Mol. Biol. Cell* **16**, 4329–4340 (2005).
22. E. Gutierrez and A. Groisman, *Anal. Chem.* **79**, 2249–2258 (2007).
23. Z. Yin, E. Giacomello, E. Gabriele, L. Zardi, S. Aota, K. M. Yamada, B. Skerlavaji, R. Doliana, A. Colombatti and R. Perris, *Blood* **93**, 1221–1230 (1999).

24. M. Dembo and Y.-L. Wang, *Biophys. J.* **76**, 2307–2316 (1999).

25. N. Q. Balaban, U. S. Schwarz, D. Riveline, P. Goichberg, G. Tzur, I. Sabanay, D. Mahalu, S. Safran, A. Bershadsky, L. Addadi and B. Geiger, *Nature Cell Biol.* **3**, 466–472 (2001).

26. J. L. Tan, J. Tien, D. M. Pirone, D. S. Gray, K. Bhadriraju and C. S. Chen, *Proc. Natl Acad. Sci. USA* **100**, 1484–1489 (2003).

27. J. Shim, E. Hagerman, B. Wu and V. Gupta, *Acta Biomater.* **4**, 1657–1668 (2008).

28. D. Debavelaere-Callens, L. Peyre, P. Campistron and H. F. Hildebrand, *Biomol. Eng.* **24**, 521–525 (2007).

29. J. T. Butcher, A. M. Penrod, A. J. Garcia and R. M. Nerem, *Arterioscler. Thromb. Vasc. Biol.* **24**, 1429–1434 (2004).

30. G. I. Bell, *Science* **200**, 618–627 (1978).

31. C. Zhu, *J. Biomech.* **33**, 23–33 (2000).

32. Y. F. Missirlis and A. D. Spiliotis, *Biomol. Eng.* **19**, 287–294 (2002).

33. L. Marcotte and A. Tabrizian, *Ingenierie et Recherche BioMedicale (IRBM)* **29**, 77–88 (2008).

34. D. R. McClay, G. M. Wessel and R. B. Marchase, *Proc. Natl Acad. Sci. USA* **78**, 4975–4979 (1981).

35. L. Y. Koo, D. J. Irvine, A. M. Mayes, D. A. Lauffenburger and L. G. Griffith, *J. Cell Sci.* **115**, 1423–1433 (2002).

36. E. Giacomello, J. Neumayer, A. Colombatti and R. Perris, *Biotechniques* **26**, 758–766 (1999).

37. C. D. Reyes and A. J. Garcia, *J. Biomed. Mater. Res.* **67A**, 328–333 (2003).

38. L. S. Channavajjala, A. Eidsath and W. C. Saxinger, *J. Cell Sci.* **110**, 249–256 (1997).

39. O. Thoumine, A. Ott and D. Louvard, *Cell Motil. Cytoskeleton* **33**, 276–287 (1996).

40. O. Thoumine and A. Ott, *Biorheology* **34**, 309–326 (1997).

41. M. M. Lotz, C. A. Burdsal, H. P. Erickson and D. R. McClay, *J. Cell Biol.* **109**, 1795–1805 (1989).

42. J. W. Piper, R. A. Swerlick and C. Zhu, *Biophys. J.* **74**, 492–513 (1998).

43. J. R. Capadona, D. M. Collard and A. J. Garcia, *Langmuir* **19**, 1847–1852 (2003).

44. S. M. Cutler and A. J. Garcia, *Biomaterials* **24**, 1759–1770 (2003).

45. B. G. Keselowsky, D. M. Collard and A. J. Garcia, *J. Biomed. Mater. Res.* **66A**, 247–259 (2003).

46. B. G. Keselowsky, D. M. Collard and A. J. Garcia, *Biomaterials* **25**, 5947–5954 (2004).

47. C. D. Reyes and A. J. Garcia, *J. Biomed. Mater. Res.* **65A**, 511–523 (2003).

48. C. D. Reyes, T. A. Petrie and A. J. Garcia, *J. Cell. Physiol.* **217**, 450–458 (2008).

49. G. M. Harbers, L. J. Gamble, E. F. Irwin, D. G. Castner and K. E. Healy, *Langmuir* **21**, 8374–8384 (2005).

50. G. M. Harbers and K. E. Healy, *J. Biomed. Mater. Res.* **75A**, 855–869 (2005).

51. K. J. Bundy, L. G. Harris, B. A. Rahn and R. G. Richards, *Cell Biol. Int.* **25**, 289–307 (2001).

52. N. J. Hallab, K. J. Bundy, K. O'Connor, R. L. Moses and J. J. Jacobs, *Tissue Eng.* **7**, 55–71 (2001).

53. D. C. Giliberti, K. A. Anderson and K. C. Dee, *J. Biomed. Mater. Res.* **62**, 422–429 (2002).

54. P. Feugier, R. A. Black, J. A. Hunt and T. V. How, *Biomaterials* **26**, 1457–1466 (2005).

55. Y. A. Çengel and J. M. Cimbala, *Fluid Mechanics: Fundamentals and Applications*. McGraw-Hill Series in Mechanical Engineering, McGraw-Hill (2006).

56. T. A. Horbett, J. J. Waldburger, B. D. Ratner and A. S. Hoffman, *J. Biomed. Mater. Res.* **22**, 383–404 (1988).

57. A. J. Garcia, P. Ducheyne and D. Boettiger, *Biomaterials* **18**, 1091–1098 (1997).

58. N. D. Gallant, J. R. Capadona, A. B. Frazier, D. M. Collard and A. J. Garcia, *Langmuir* **18**, 5579–5584 (2002).

59. Q. Shi and D. Boettiger, *Mol. Biol. Cell* **14**, 4306–4315 (2003).

60. D. D. Deligianni, N. D. Katsala, P. G. Koutsoukos and Y. F. Missirlis, *Biomaterials* **22**, 87–96 (2001).

61. A. J. Garcia, P. Ducheyne and D. Boettiger, *Tissue Eng.* **3**, 197–206 (1997).

62. A. J. Garcia, P. Ducheyne and D. Boettiger, *J. Biomed. Mater. Res.* **40**, 48–56 (1998).

63. D. Deligianni, P. Korovessis, M. C. Porte-Derrieu and J. Amedee, *J. Spinal Disord. Tech.* **18**, 257–262 (2005).

64. M. H. Lee, C. S. Adams, D. Boettiger, W. F. DeGrado, I. M. Shapiro, R. J. Composto and P. Ducheyne, *J. Biomed. Mater. Res.* **81A**, 150–160 (2007).

65. K. E. Michael, D. W. Dumbauld, K. L. Burns, S. K. Hanks and A. J. Garcia, *Mol. Biol. Cell* **20**, 2508–2519 (2009).

66. A. Datta, Q. Shi and D. E. Boettiger, *Mol. Cell. Biol.* **21**, 7295–7306 (2001).

67. A. Datta, F. Huber and D. Boettiger, *J. Biol. Chem.* **277**, 3943–3949 (2002).

68. L. Lynch, P. I. Vodyanik, D. Boettiger and M. A. Guvakova, *Mol. Biol. Cell* **16**, 51–63 (2005).

69. T. Miller and D. Boettiger, *Langmuir* **19**, 1723–1729 (2003).

70. A. J. Garcia, J. Takagi and D. Boettiger, *J. Biol. Chem.* **273**, 34710–34715 (1998).

71. A. S. Goldstein and P. A. DiMilla, *Biotechnol. Bioeng.* **55**, 616–629 (1997).

72. A. S. Goldstein and P. A. DiMilla, *AIChE J.* **44**, 465–473 (1998).

73. A. S. Goldstein and P. A. DiMilla, *J. Biomed. Mater. Res.* **59**, 665–675 (2002).

74. A. S. Goldstein and P. A. DiMilla, *J. Biomed. Mater. Res.* **67A**, 658–666 (2003).

75. J. P. Bearinger, D. G. Castner, S. L. Golledge, A. Rezania, S. Hubchak and K. E. Healy, *Langmuir* **13**, 5175–5183 (1997).

76. A. Rezania, C. H. Thomas, A. B. Branger, C. M. Waters and K. E. Healy, *J. Biomed. Mater. Res.* **37**, 9–19 (1997).

77. A. Rezania, C. H. Thomas and K. E. Healy, *Ann. Biomed. Eng.* **25**, 190–203 (1997).

78. A. Rezania and K. E. Healy, *Biotechnol. Prog.* **15**, 19–32 (1999).

79. A. Rezania and K. E. Healy, *J. Orthop. Res.* **17**, 615–623 (1999).

80. M. R. Kreke, A. S. Badami, J. B. Brady, R. M. Akers and A. S. Goldstein, *Biomaterials* **26**, 2975–2981 (2005).

81. T. Sordel, F. Kermarec-Marcel, S. Garnier-Raveaud, N. Glade, F. Sauter-Starace, C. Pudda, M. Borella, M. Plissonnier, F. Chatelain, F. Bruckert and N. Picollet-D'hahan, *Biomaterials* **28**, 1572–1584 (2007).

82. G. A. Truskey and J. S. Pirone, *J. Biomed. Mater. Res.* **24**, 1333–1353 (1990).

83. T. G. van Kooten, J. M. Schakenraad, H. C. van der Mei and H. J. Busscher, *J. Biomed. Mater. Res.* **26**, 725–738 (1992).

84. R. L. W. Messer, C. M. Davis, J. B. Lewis, Y. Adams and J. C. Wataha, *J. Biomed. Mater. Res.* **79A**, 16–22 (2006).

85. R. M. Schinagl, M. S. Kurtis, K. D. Ellis, S. Chien and R. L. Sah, *J. Orthop. Res.* **17**, 121–129 (1999).

86. T. G. van Kooten, J. M. Schakenraad, H. C. van der Mei and H. J. Busscher, *Cells Mater.* **1**, 307–316 (1991).

87. T. G. van Kooten, J. M. Schakenraad, H. C. van der Mei and H. J. Busscher, *Biomaterials* **13**, 897–904 (1992).

88. T. G. van Kooten, J. M. Schakenraad, H. C. van der Mei, A. Dekker, C. J. Kirkpatrick and H. J. Busscher, *Med. Eng. Phys.* **16**, 506–512 (1994).

89. T. G. van Kooten, J. M. Schakenraad, H. C. van der Mei and H. J. Busscher, *J. Biomater. Sci. Polym. Ed.* **4**, 601–614 (1993).

90. C. Boura, S. Muller, D. Vautier, D. Dumas, P. Schaaf, J. C. Voegel, J. F. Stoltz and P. Menu, *Biomaterials* **26**, 4568–4575 (2005).

91. J. H. Lee, S. J. Lee, G. Khang and H. B. Lee, *J. Colloid Interface Sci.* **230**, 84–90 (2000).

92. K. M. Renshaw, D. E. Orr and K. J. L. Burg, *Biotechnol. Prog.* **21**, 538–545 (2005).

93. R. Kapur and A. S. Rudolph, *Exp. Cell Res.* **244**, 275–285 (1998).

94. G. Khang, S. J. Lee, Y. M. Lee, J. H. Lee and H. B. Lee, *Korean Polym. J.* **8**, 179–185 (2000).

95. Y. Q. Wan, J. Yang, J. L. Yang, J. Z. Bei and S. G. Wang, *Biomaterials* **24**, 3757–3764 (2003).

96. J. Yang, Y. Q. Wan, J. L. Yang, J. Z. Bei and S. G. Wang, *J. Biomed. Mater. Res.* **67A**, 1139–1147 (2003).

97. G. A. Truskey and T. L. Proulx, *Biomaterials* **14**, 243–254 (1993).

98. D. J. Iuliano, S. S. Saavedra and G. A. Truskey, *J. Biomed. Mater. Res.* **27**, 1103–1113 (1993).

99. J. S. Burmeister, J. D. Vrany, W. M. Reichert and G. A. Truskey, *J. Biomed. Mater. Res.* **30**, 13–22 (1996).

100. Y. Xiao and G. A. Truskey, *Biophys. J.* **71**, 2869–2884 (1996).

101. V. D. Bhat, G. A. Truskey and W. M. Reichert, *J. Biomed. Mater. Res.* **40**, 57–65 (1998).

102. B. P. Chan, V. D. Bhat, S. Yegnasubramanian, W. M. Reichert and G. A. Truskey, *Biomaterials* **20**, 2395–2403 (1999).

103. V. D. Bhat, G. A. Truskey and W. M. Reichert, *J. Biomed. Mater. Res.* **41**, 377–385 (1998).

104. M. A. Brown, C. S. Wallace, C. C. Anamelechi, E. Clermont, W. M. Reichert and G. A. Truskey, *Biomaterials* **28**, 3928–3935 (2007).

105. S. Usami, H. H. Chen, Y. H. Zhao, S. Chien and R. Skalak, *Ann. Biomed. Eng.* **21**, 77–83 (1993).

106. M. J. Powers, R. E. Rodriguez and L. G. Griffith, *Biotechnol. Bioeng.* **53**, 415–426 (1997).

107. S. C. Heilshorn, K. A. DiZio, E. R. Welsh and D. A. Tirrell, *Biomaterials* **24**, 4245–4252 (2003).

108. Y. Xia and G. M. Whitesides, *Angew. Chem. Int. Ed.* **37**, 551–575 (1998).

109. T. Gervais, J. El-Ali, A. Gunther and K. F. Jensen, *Lab Chip* **6**, 500–507 (2006).

110. K. V. Christ, K. B. Williamson, K. S. Masters and K. T. Turner, *Biomedical Microdevices* **3**, 443–455 (2010).

111. H. Lu, L. Y. Koo, W. M. Wang, D. A. Lauffenburger, L. G. Griffith and K. F. Jensen, *Anal. Chem.* **76**, 5257–5264 (2004).

112. E. W. K. Young, A. R. Wheeler and C. A. Simmons, *Lab Chip* **7**, 1759–1766 (2007).

113. X. Zhang, P. Jones and S. J. Haswell, *Chem. Eng. J.* **135**, S82–S88 (2008).

114. K. W. Kwon, S. S. Choi, S. H. Lee, B. Kim, S. N. Lee, M. C. Park, P. Kim, S. Y. Hwang and K. Y. Suh, *Lab Chip* **7**, 1461–1468 (2007).

115. O. Thoumine and J.-J. Meister, *Eur. Biophys. J.* **29**, 409–419 (2000).

116. O. Thoumine, P. Kocian, A. Kottelat and J.-J. Meister, *Eur. Biophys. J.* **29**, 398–408 (2000).

117. N. Walter, C. Selhuber, H. Kessler and J. P. Spatz, *Nano Lett.* **6**, 398–402 (2006).

118. Y. J. Kim, J.-W. Shin, K. D. Park, J. W. Lee, N. Yui, S.-A. Park, K. S. Jee and J. K. Kim, *J. Biomater. Sci. Polym. Ed.* **14**, 1311–1321 (2003).

119. C.-C. Wang, Y. C. Hsu, F. C. Su, S. C. Lu and T. M. Lee, *J. Biomed. Mater. Res.* **88A**, 370–383 (2009).

120. K. A. Athanasiou, B. S. Thoma, D. R. Lanctot, D. Shin, C. M. Agrawal and R. G. LeBaron, *Biomaterials* **20**, 2405–2415 (1999).

121. A. Yamamoto, S. Mishima, N. Maruyama and M. Sumita, *J. Biomed. Mater. Res.* **50**, 114–124 (2000).

122. C.-C. Wu, H.-W. Su, C.-C. Lee, M.-J. Tang and F.-C. Su, *Biochem. Biophys. Res. Commun.* **329**, 256–265 (2005).

123. G. Sagvolden, I. Giaever, E. O. Pettersen and J. Feder, *Proc. Natl. Acad. Sci. USA* **96**, 471–476 (1999).

124. C.-H. Chang, J.-D. Liao, J.-J. J. Chen, M.-S. Ju and C.-C. K. Lin, *Nanotechnology* **17**, 2449–2457 (2006).

125. C. C. Wang, Y. C. Hsu, M. C. Hsieh, S. P. Yang, F. C. Su and T. M. Lee, *Nanotechnology* **19**, (335709)1–10 (2008).

126. A. Yamamoto, S. Mishima, N. Maruyama and M. Sumita, *Biomaterials* **19**, 871–879 (1998).

127. G. Hoben, W. Huang, B. S. Thoma, R. G. LeBaron and K. A. Athanasiou, *Ann. Biomed. Eng.* **30**, 703–712 (2002).

128. W. Huang, B. Anvari, J. H. Torres, R. G. LeBaron and K. A. Athanasiou, *J. Orthop. Res.* **21**, 88–95 (2003).

129. Y. J. Kim, S. A. Park, Y. J. Lee, J. W. Shin, D.-H. Kim, S.-J. Heo, K. D. Park and J.-W. Shin, *J. Biomed. Mater. Res.* **85B**, 353–360 (2008).

130. K.-L. P. Sung, M. K. Kwan, F. Maldonado and W. H. Akeson, *J. Biomech. Eng.* **116**, 237–242 (1994).

131. K.-L. P. Sung, L. L. Steele, D. Whittermore, J. Hagan and W. H. Akeson, *J. Orthop. Res.* **13**, 166–173 (1995).

132. K.-L. P. Sung, D. E. Whittemore, L. Yang, D. Amiel and W. H. Akeson, *J. Orthop. Res.* **14**, 729–735 (1996).

133. K.-L. P. Sung, L. Yang, D. E. Whittemore, Y. Shi, G. Jin, A. H. Hsieh, W. H. Akeson and L. A. Sung, *Proc. Natl Acad. Sci. USA* **93**, 9182–9187 (1996).

134. D. J. Nugiel, D. J. Wood and K.-L. P. Sung, *Tissue Eng.* **2**, 127–140 (1996).

135. S. Y. Kwon, H. Takei, D. P. Pioletti, T. Lin, Q. J. Ma, W. H. Akeson, D. J. Wood and K.-L. P. Sung, *J. Orthop. Res.* **18**, 203–211 (2000).

136. L. Yang, C. M.-H. Tsai, A. H. Hsieh, V. S. Lin, W. H. Akeson and K.-L. P. Sung, *J. Orthop. Res.* **17**, 755–762 (1999).

137. J. M. McKean, A. H. Hsieh and K. L. P. Sung, *Biorheology* **41**, 139–152 (2004).

138. L. Richert, P. Lavalle, D. Vautier, B. Senger, J.-F. Stoltz, P. Schaaf, J.-C. Voegel and C. Picart, *Biomacromolecules* **3**, 1170–1178 (2002).

139. L. Richert, Y. Arntz, P. Schaaf, J.-C. Voegel and C. Picart, *Surface Sci.* **570**, 13–29 (2004).

140. Z.-Z. Wu, P. Li, Q.-P. Huang, J. Qin, G.-H. Xiao and S.-X. Cai, *Biorheology* **40**, 489–502 (2003).

141. K. F. Shao, Z.-Z. Wu, B.-C. Wang, M. Long and S.-X. Cai, *Colloids Surfaces B* **19**, 55–59 (2000).

142. T.-W. Qin, Z.-M. Yang, Z.-Z. Wu, H.-Q. Xie, H. Qin and S.-X. Cai, *Biomaterials* **26**, 6635–6642 (2005).

143. Z. Gao, S. Wang, H. Zhu, C. Su, G. Xu and X. Lian, *Mater. Sci. Eng. C* **28**, 1227–1235 (2008).

144. G. Athanasiou and D. Deligianni, *J. Mater. Sci. Mater. Med.* **12**, 965–970 (2001).

145. K. Yamamoto, N. Tomita, Y. Fukuda, S. Suzuki, N. Igarashi, T. Suguro and Y. Tamada, *Biomaterials* **28**, 1838–1846 (2007).

146. H. J. Ra, C. Picart, H. Feng, H. L. Sweeney and D. E. Discher, *J. Cell Sci.* **112**, 1425–1436 (1999).

147. M. A. Griffin, A. J. Engler, T. A. Barber, K. E. Healy, H. L. Sweeney and D. E. Discher, *Biophys. J.* **86**, 1209–1222 (2004).

148. M. A. Griffin, H. Feng, M. Tewari, P. Acosta, M. Kawana, H. L. Sweeney and D. E. Discher, *J. Cell Sci.* **118**, 1405–1416 (2005).
149. X. Liu, P. K. Chu and C. Ding, *Mater. Sci. Eng. R* **47**, 49–121 (2004).
150. G. Voskerician, M. S. Shive, R. S. Shawgo, H. von Recum, J. M. Anderson, M. J. Cima and R. Langer, *Biomaterials* **24**, 1959–1967 (2003).
151. M. P. Lutolf and J. A. Hubbell, *Nature Biotechnol.* **23**, 47–55 (2005).
152. U. Cavallaro and G. Christofori, *Nature Rev. Cancer* **4**, 118–132 (2004).
153. S. K. Mitra and D. D. Schlaepfer, *Curr. Opin. Cell Biol.* **18**, 516–523 (2006).
154. G. Gabbiani, *J. Pathol.* **200**, 500–503 (2003).
155. N. Laurens, P. Koolwijk and M. P. M. de Maat, *J. Thromb. Haemost.* **4**, 932–939 (2006).
156. J. A. McCann, S. D. Peterson, M. W. Plesniak, T. J. Webster and K. M. Haberstroh, *Ann. Biomed. Eng.* **33**, 328–336 (2005).
157. G. M. Walker, H. C. Zeringue and D. J. Beebe, *Lab Chip* **4**, 91–97 (2004).
158. L. A. Olivier and G. A. Truskey, *Biotechnol. Bioeng.* **42**, 963–973 (1993).
159. D. P. Gaver and S. M. Kute, *Biophys. J.* **75**, 721–733 (1998).
160. P. F. Davies, A. Robotewskyj and M. L. Griem, *J. Clin. Invest.* **93**, 2031–2038 (1994).
161. E. A. Osborn, A. Rabodzey, C. F. Dewey and J. H. Hartwig, *Am. J. Physiol. Cell Physiol.* **290**, C444–C452 (2006).

Use of the Atomic Force Microscope to Determine the Strength of Bacterial Attachment to Grooved Surface Features

J. Verran [a,*], **A. Packer** [a,b], **P. J. Kelly** [b] **and K. A. Whitehead** [a,b]

[a] School of Biology, Chemistry and Health Science, Manchester Metropolitan University, Chester Street, Manchester M1 5GD, UK
[b] Surface Engineering Group, Dalton Research Institute, Manchester Metropolitan University, Manchester M1 5GD, UK

Abstract
Wear of food contact surfaces through abrasion and cleaning alters surface roughness, and introduces topographical features which affect the retention of microorganisms. In order to identify the effect of features of specific dimensions, surfaces with defined grooved features ('smooth'; 1.02 µm wide, 0.21 µm deep grooves; and 0.59 µm wide, 0.17 µm deep grooves) that are representative of the dimensions of linear features identified on worn hygienic surfaces were fabricated. These surfaces were conformally coated with titanium using the physical vapour deposition (PVD) technique of magnetron sputtering, which provided the surface with a uniform chemistry. Atomic force microscopy (AFM) was used to investigate the strength of attachment of the Gram-negative rod-shaped bacterium *Escherichia coli* and the Gram-positive coccal-shaped bacterium *Staphylococcus sciuri* to the surfaces. The AFM tip was scanned across the surface, where an increase in applied force caused an increase in cell detachment. *E. coli* cells were easily removed from the surfaces by the addition of water alone, without the application of lateral force, but the ease of removal of coccal-shaped bacteria varied depending on the underlying topography. *S. sciuri* retention was highest on the titanium coated 1.02 µm wide grooved surfaces and least on the titanium coated 0.59 µm wide grooved surfaces, indicating that the *Staphylococcus sciuri* cells were most strongly held within the features of dimensions comparable to the cells. This work demonstrates that surface features can significantly affect the removal of attached cells, particularly when the size of the topographic features closely matches that of the bacterium.

Keywords
Atomic Force Microscopy (AFM), force, titanium, grooved structures, bacteria, surface topography, roughness

1. Introduction

The environment in food preparation plants is considered a significant source of microorganisms, where open surfaces, in particular, are the areas frequently involved

* To whom correspondence should be addressed. Tel.: +44 161 247 1206; Fax: +44 161 247 6325; e-mail: j.verran@mmu.ac.uk

Surface and Interfacial Aspects of Cell Adhesion
© Koninklijke Brill NV, Leiden, 2010

in the contamination of food products [1]. Microbial cells in this environment can stay viable, adhere to the equipment surfaces and contaminate any substance which comes into contact with them [2]. These attached bacteria may be able to colonise the surface as a biofilm, under appropriate growth conditions [3]. Microorganisms that may cause problems in this context include *Escherichia coli*, a rod-shaped (1 μm × 2 μm) Gram-negative microorganism, and *Staphylococcus sciuri*, a Gram-positive coccal-shaped (1 μm diameter) bacterium similar to the pathogen *Staphylococcus aureus*.

The adhesion of a microorganism to a surface is influenced by various factors related to the structural and physiological characteristics of the cell and the physical and chemical properties of the surface [4]. Bacterial adhesion to a surface is a multi-factor process [5]. An important contributing factor to the retention of microorganisms on surfaces is the surface topography [1, 6–8]. Surface features including scratches due to abrasion and pits due to impact damage provide niches in which microorganisms are protected from shear forces and hygiene and cleaning measures, thus allowing the entrapped microbial cells time to attach to the surface [9]. On abraded or impact damaged materials, microorganisms retained after cleaning regimes would be unlikely to be removed with cleaning [10].

Measurement of surface topography has traditionally been made using engineering values. The R_a value is used to describe the roughness of a surface, and is the average departure of the surface profile from a mean centre line in micrometers [11], although it only describes variation in topography in the vertical direction, with no account of changes in the lateral direction. Thus this two-dimensionally derived descriptor does not differentiate between linear and circular features encountered by the profile probe (i.e., scratches and pits). Further, the size of the probe scanning the surface will affect the measurements obtained, due to different probe dimensions and resolution. For hygienic surfaces, the atomic force microscope (AFM) and the white light interferometer have provided useful information on a scale more related to microorganisms than the more easily accessible solid stylus profilometers. In the engineering sector, there have been moves towards using the S_a value. This is a parameter that describes the overall topography of a specified surface area, thus providing a three-dimensional perspective, and has been used in our laboratories. However, within the biological arena, the R_a value remains the more commonly used parameter, thus facilitating comparison. In the oral environment, surfaces with R_a values of <0.2 μm are deemed to accumulate plaque less well [12]. However, although surfaces with R_a values of less than 0.8 μm are assumed to be hygienic [13], the effect of wear on a hard impervious surface is likely to be more apparent on the nanometer scale, measurable by AFM, but not by other profilometry methods [7, 8, 14, 15], since wear tends to occur on the nanometer scale, where fine scratches alter the appearance of topography and not necessarily the R_a value [6]. However, using the AFM, the dimensions of specific features can be measured, and the contribution to the profile used to calculate R_a can be deduced.

Some studies have shown that there appears to be a minimum roughness value below which microbial cells are not retained [16]. Hence characterization of the effect of surface topography on microbial retention may enable definition of a 'cut-off' roughness value for retention for a given substratum-microbial combination [17]. Much of this previous work has been carried out on surfaces that had random roughness features across the surface. To understand the fundamental mechanisms influencing bacterial retention it is desirable to produce surfaces whose features are of appropriate dimensions, and have regular and defined patterns. Studies concerned with bacterial attachment have tended to focus on flow cells where there is a known shear force applied across the test surface [18], the use of laminar flow [19] or the application of air bubbles [20] during/after which cell removal is monitored. Our system is static, with controlled, direct and known forces applied to the bacterial cells *via* the AFM stylus enabling the strength of microbial attachment to the surface to be assessed.

Atomic force microscope studies usually focus on displacement forces perpendicular to the substratum surface. By constructing a probe where a fungal or bacterial spore is used instead of an inert 'tip', AFM has been used to quantify the adhesion of *Aspergillus niger* spores (fungi) [21], *Saccharomyces cerevisiae* (yeast) [22] and *Bacillus mycoides* spores (bacteria) to glass surfaces [23]. Bacteria have also been fixed onto AFM probes and then used to probe surfaces [24, 25]. Few studies have been carried out to determine the lateral force required to remove cells [26–29], although this is more representative of the forces that may be applied during a cleaning process [30]. AFM force measurements used in a lateral/parallel fashion may be used to describe ease of cell removal in terms of forces that are proportional to the strength of cell attachment. During scanning, the perpendicular force between the tip and the surface is kept low so that the cantilever and tip move in response to changes in surface topography. However, by deliberately increasing the perpendicular tip–surface force, the AFM tip can be used to displace weakly attached cells [30]. The AFM can image down to the nanometer level, with high force resolution, hence measuring forces directly involved in cell–surface interactions.

This study has investigated the lateral force required for the removal of bacterial cells from surfaces of defined chemistry and topography with features/dimensions representative of the wear of hygienic surfaces [7, 12, 13], in order to investigate the strength of attachment of the bacteria to the surface. The aim of this work was to determine the lateral force required to remove coccal- and rod-shaped bacteria from surfaces with defined topographical features.

2. Methods

2.1. Substrata Preparation

Surfaces used had presented specific topographical features: 'smooth' surfaces (i.e., R_a value < 5 nm) and surfaces with unidirectional grooved features of regular size. These surfaces were selected since they presented regular, fabricated linear features of dimensions that had previously been observed on in-use worn surfaces. To produce the smooth surfaces, mirror finished polished silicon wafers (Montco Silicon Technologies, Spring City, PA, USA), were cut into 1 cm^2 pieces using a diamond scribing pen. Contaminant particles on these surfaces were removed using nitrogen gas before magnetron deposition of a titanium thin film.

For the production of grooved surfaces, digital versatile discs (DVDs) and compact discs (CDs) were stripped of their protective coating by soaking overnight in 30% sodium hydroxide (BDH, UK). These surfaces had previously been shown to represent grooved features with dimensions typical of minor wear observed on hygienic food contact surfaces [31]. Samples were then soaked for 30 min in sterile distilled water, and thoroughly rinsed. The stripped samples were then dried in a laminar flow hood, and cut into 1 cm^2 pieces ready for thin film deposition.

2.2. Substrata Coating

In order for the effects of topography on microbial retention to be ascertained, a uniform surface chemistry was required. To achieve this, the selected substrata were coated with titanium *via* magnetron sputtering to provide surfaces of uniform chemistry, with varying underlying topographies. Magnetron sputtering is an atomistic deposition process with concurrent bombardment of the growing film by energetic ions and neutrals. Consequently, fully dense coatings, which are conformal to the substrate surface, are produced [29, 32]. The titanium coatings were deposited onto the substrate surfaces by biased magnetron sputtering in a modified Edwards E306A coating system rig. Sputtering took place from a single 150 mm diameter × 10 mm thick, 99.5% pure titanium target. Prior to the deposition of the titanium coatings, the substrata were sputter cleaned at −1000 V DC for 10 min. During deposition, the substrates were biased at −50 V to ensure the formation of a dense, conformal coating. This conformality was confirmed by fracturing coatings and examining cross-sections using SEM, which revealed thickness, structure and substrate coverage. The coating was stable in aqueous medium.

2.3. Roughness Measurements

Roughness parameters, images and determination of feature dimensions were obtained using an AFM (Explorer, Veeco Instruments, Cambridge, UK) operated in contact mode using silicon nitride pyramidal shaped tips with a nominal spring constant of 0.05 N/m. Averages of R_a values obtained from five linear scans taken across individual AFM images were calculated, using replicate 20 μm × 20 μm samples and a scan rate of 20.03 μm/s with 300 pixel resolution.

2.4. Contact Angle Measurements

Contact angle measurements were made at room temperature using the sessile drop technique and 5 μl volumes of water on a Kruss goniometer and data analysis system. Five microliter droplets of HPLC grade water (BDH, Poole, UK) were deposited onto a horizontal sample using a syringe. Contact angles of the droplets were calculated automatically using the analytical software.

2.5. Scanning Electron Microscopy (SEM) and Energy Dispersive X-Ray (EDX)

Titanium coated surfaces were checked for conformity of film deposition using SEM/EDX characterisation. Chemical analysis of substrata was carried out to a 1 μm depth using a Link Pentafet detector (Oxford Instruments, Buckinghamshire, UK), with Inca software (Oxford Instruments, Buckinghamshire, UK). Images of substrata were obtained using a JEOL JSM 5600LV scanning electron microscope (JEOL Ltd, Herts, UK). Replicates were carried out in triplicate.

2.6. Maintenance and Preparation of Microorganisms

Escherichia coli CCL 410 (a non-pathogenic O157:H7 strain) and *Staphylococcus sciuri* CCL 101 were kindly provided by Brigitte Carpentier (AFSSA, France). *S. sciuri* was used as a representative of the *Staphylococcus* genus associated with the food industry. Stock cultures of microorganisms were stored at $-80°C$ and stored and maintained according to Cabellero *et al.* [33].

In preparation for assays, stock cultures were inoculated onto nutrient agar (NA) (Oxoid, UK) and incubated at 37°C for 24 h. Sub-cultures, maintained as stock cultures were prepared on fresh agar before use and were maintained at 4°C. Fresh stocks were prepared from the frozen stock cultures every four weeks. A single colony of *E. coli* or *S. sciuri* was inoculated from an agar plate into 100 ml of nutrient broth (Oxoid, UK) and incubated for 18 h with shaking at 37°C. Cells were harvested by centrifugation ($3600 \times g$ for 12 min) and washed three times in 10 ml sterile distilled water. The resultant cell suspension was adjusted to an optical density (OD) 1.0 at 540 nm corresponding to concentrations of *E. coli* $5.73 \pm 0.82 \times 10^8$ and *S. sciuri* $0.64 \pm 0.66 \times 10^9$ colony forming units ml^{-1} (CFU ml^{-1}).

2.7. Strength of Cell Attachment

Using an AFM, the force required to remove cells under a liquid from a surface can be quantitatively measured using a perpendicular force applied to the tip of the AFM cantilever. By extrapolation of force curve measurements, and tip dimensions, this compressive force can be translated into the lateral force of interaction. Ten microliters of bacteria were added to the test surfaces and dried for one hour in a microbiological class 2 laminar flow hood and for the additional 23 h at room temperature in a sterile container.

An Explorer AFM was used to determine the strength of attachment of cells to the surface (Veeco Instruments, Cambridge, UK). The cantilevers were pyramidal probes with front and back angles of 35° (Veeco Instruments Ltd, Cambridge, UK). Before each experiment the spring constant of the cantilever was determined by measuring the mechanical response of the cantilever to thermal noise as a function of time using the AFM software. AFM was operated in contact mode and measurements were carried out at a rate of 20.03 μm/s at a scan size of 20 μm × 20 μm. Substrata with dried cells were placed on the AFM and a dry scan of the sample was taken to ensure the presence of cells in the area of analysis. One hundred microliters of HPLC grade water (BDH, UK) were placed on the sample and the AFM laser was re-aligned. The cantilever was brought into contact with the surface and a measurement of the force applied to the cantilever was obtained from force–distance curves. To convert the cantilever deflection into a perpendicular force, the spring constant, the value of the gradient in the constant compliance region of the force curve and the zero of the force were defined [21].

The cantilever deflection is then converted into a force (F) using Hooke's law:

$$F = kd, \tag{1}$$

where k is the cantilever spring constant, and d is the cantilever deflection. The curve can be corrected by plotting F as a function of $(z - d)$, where z is the vertical displacement of the piezoelectric scanner [34]. Further calculations were carried using the methodology of Deupree and Schoenfisch [35] to determine the lateral force of interaction of the cantilever tip with the cell. The applied force normal to the plane of interaction can be calculated from the equation;

$$F_{app} = kd \sin(\theta + \phi), \tag{2}$$

where the angles θ and ϕ are parameters of probe geometry and cantilever orientation, respectively [35]. The angles θ and ϕ are specific to the probe type used. The lateral component of the applied force was determined using;

$$F_{lat} = F_{app} \cos(\theta), \tag{3}$$

giving the value of the shear lateral force that detaches cells from the surface [35]. To determine the force required for bacterial removal, scans were repeated with increasing force applied to the cantilever tip. Three replicates of the detachment experiments were carried out. After each scan the remaining bacteria were counted manually and plotted as a percentage as a function of the lateral force applied.

2.8. Statistics

Statistical tests were carried out using a two-tailed distribution t-test with two sample homoscedastic variances. The results are reported as mean ± standard deviation.

3. Results

3.1. Surface Analysis

SEM/EDX analysis across five fields demonstrated that the sputtered coatings were conformal and uniform in composition across the sample surfaces (data not shown). The surfaces were characterised using linear AFM profiles and were shown to have R_a values of 0.003 μm ± 0.0001 μm for the smooth titanium surface, 0.024 μm ± 0.001 μm for the 0.59 μm wide featured surface, and 0.026 μm ± 0.001 μm for the 1.02 μm wide featured surface (images of the featured surfaces are shown in Fig. 1). Feature dimensions of the grooves were measured peak to peak and were found to be 0.59 ± 0.03 μm wide with depth 0.17 ± 0.002 μm (DVD) and 1.02 ± 0.03 μm wide with depth 0.21 ± 0.01 μm (CD). Feature widths and depths

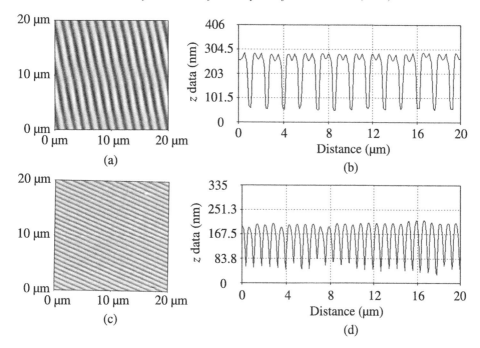

Figure 1. AFM images (20 µm × 20 µm scans) of surfaces demonstrating surface topographies and linear profiles, showing feature dimensions in the vertical 'z' scale: (a), (b) 0.59 µm wide featured titanium surface, and (c), (d) 1.02 µm wide featured titanium surface.

between the two surfaces were significantly different ($p < 0.001$ and $p < 0.05$, respectively). No significant variation ($p < 0.001$) in surface wettability was observed, with an average contact angle of $91° \pm 3.7$ being determined for the samples (data not shown).

3.2. Strength of Cell Attachment

E. coli is a rod-shaped bacterium (1 µm × 2 µm) which has flagella, whilst *S. sciuri* is a coccal-shaped bacterium (1 µm diameter) and does not have flagella. When using the AFM to determine the strength of attachment of cells to the surface, cells were manually counted from AFM images. Attached *E. coli* cells were removed from the surfaces as soon as water was added, and for all three surfaces no cells were left to scan. For *S. sciuri*, bacteria were clearly visible on all surfaces with a low lateral tip force (Figs 2(a), 3(a) and 4(a)). As the force was increased the cells began to be removed from the titanium coated silicon surface and the titanium coated 0.59 µm wide grooved surfaces (Figs 2–4). However, the number of cells attached to the titanium coated 1.02 µm wide grooved surface remained constant (Fig. 4). Significantly more ($p < 0.001$) cells were removed from the titanium coated silicon and the titanium coated 0.59 µm wide grooved surface than the titanium coated 1.02 µm wide grooved surface at an applied force of 19 nN. The titanium coated 0.59 µm wide grooved surface displayed the lowest level of reten-

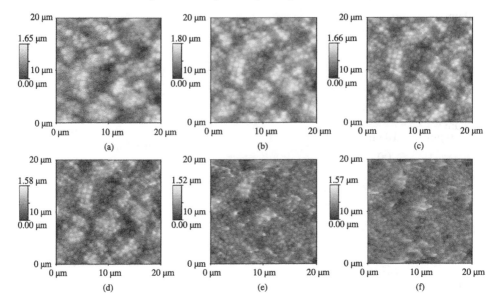

Figure 2. AFM images of *Staphylococcus sciuri* cells attached to a titanium coated smooth silicon surface with increasing lateral tip force (nN) (a) 1, (b) 5, (c) 13, (d) 18, (e) 26, (f) 28, demonstrating that even at the highest force some cells are retained on the surface.

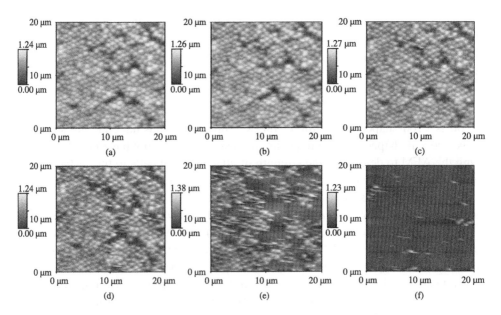

Figure 3. AFM images of *Staphylococcus sciuri* cells attached to a titanium coated 0.59 μm wide grooved surface with increasing lateral tip force (nN) (a) 3, (b) 6, (c) 8, (d) 12, (e) 16, (f) 19, demonstrating that cells are easily removed from the surface with increasing force.

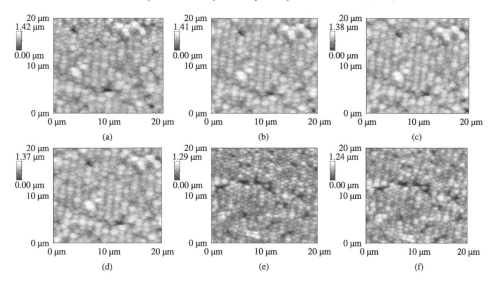

Figure 4. AFM images of *Staphylococcus sciuri* cells attached to a titanium coated 1.02 μm wide grooved surface with increasing lateral tip force (nN) (a) 3, (b) 5, (c) 8, (d) 12, (e) 17, (f) 21, demonstrating that even at highest force a high number of cells are retained on the surface and can be seen to be wedged in surface features.

tion, where a force of 19 nN was sufficient enough to remove 95% of the cells from the surface (Fig. 5(b)). Adhesion of the cells to the titanium coated silicon surface was stronger, with 65% of the bacterial cells remaining at a force of 19 nN (Fig. 5(a)). However, on the titanium coated 1.02 μm wide grooved surface, 98% of the cells remained when a force of 19 nN was applied (Fig. 5(c)).

4. Discussion

A number of studies have produced surfaces with topographies in the lower micrometer scale [36–40]. In our laboratories ion beam assisted magnetron sputtering was used to produce surfaces with defined topographies and chemistry using a template method [40, 41], thus producing surfaces with controlled chemistry, topography and wettability. No significant difference was found between the surface wettabilities for the different topographies and this is in agreement with Busscher *et al.* [42] who found no influence of surface roughening on contact angles for surfaces with roughness values < 0.1 μm. Thus a range of surfaces with controlled chemistry and wettability, and defined topography were produced.

The aim of this study was to compare the strength of adhesion of differently shaped bacterial cells (rods and cocci) to surfaces with defined grooved features of different widths *in situ via* AFM. By approximating Hooke's law and determining cantilever compression a quantitative method can be applied for direct measurement of the lateral force required for detachment of the cells. *E. coli* would not remain on the surfaces once water was applied, and therefore scanning could not be

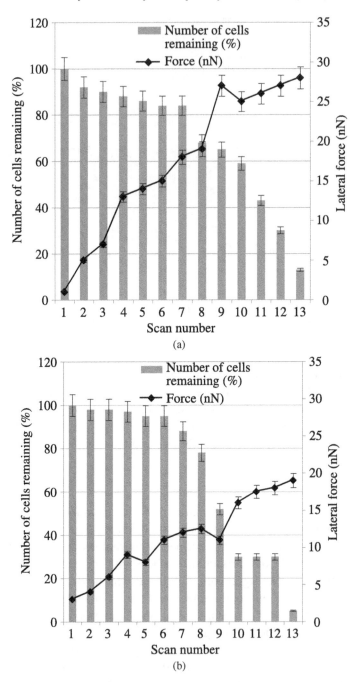

Figure 5. Removal of *Staphylococcus sciuri* cells with increasing lateral tip force on (a) titanium coated silicon surface, (b) titanium coated 0.59 μm wide grooved surface, (c) titanium coated 1.02 μm wide grooved surface demonstrating that the size of surface features affects bacterial retention: as the lateral force increases (b) cells are more easily removed from the 0.59 μm wide surface, whilst (c) a high percentage of cells are retained on the 1.02 μm wide surface.

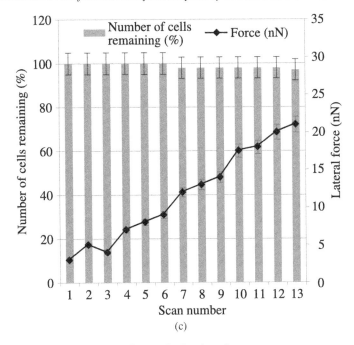

Figure 5. (Continued).

done. Conversely, it is not possible to remove the cells from the surface in a dried state since they are too well bound to the surface. Although the smoothest surface had the lowest R_a value (0.003 μm), and both the 0.59 μm wide and 1.02 μm wide surfaces had similar R_a values (0.024 μm and 0.026 μm respectively) *S. sciuri* cells were more easily removed from the 0.59 μm wide grooved surface than from the titanium coated silicon surface, with the 1.02 μm wide grooved surface exhibiting the lowest level of bacterial removal. This phenomenon may be explained by considering the surface area available for attachment of the bacterium (Fig. 6). The smooth titanium coated silicon surface provided a greater area of contact than the 0.59 μm wide grooved surface between the cell and the substratum. The 0.59 μm wide grooved surface provided only a small contact area between the edges of the groove and the bacterial cell. The greatest surface area for adhesion is provided by the 1.02 μm wide grooved surface, since the groove is nearer the same order of size as the *Staphylococcus sciuri* cell. This means that the *Staphylococcus sciuri* cell can 'sit' in the feature with around half of its circumference in contact with, or being protected by, the substrate surface. Therefore, displacement of coccal cells with a lower level of surface contact requires less force as there will be less resistance due to lower cell–surface binding energies. Thus it was determined that the size of the surface defect is important with respect to the size of the cell, and its subsequent retention.

Some workers have suggested that there is no relationship between surface roughness (in terms of R_a) and the ability of bacteria to attach [43–45]. How-

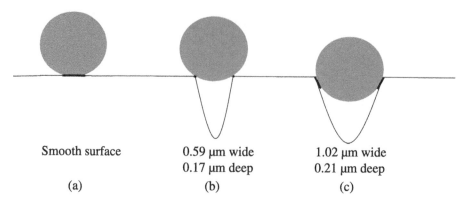

Smooth surface   0.59 µm wide   1.02 µm wide
                        0.17 µm deep   0.21 µm deep
    (a)   (b)   (c)

Figure 6. The size of the surface feature and the area of contact between the cell and the surface results in coccal cells becoming more strongly attached in surface features of dimensions similar to that of the cell. (a) The contact between a coccal cell of approximately one micrometer diameter and a smooth surface. (b) When the cell is lying on top of a surface feature of diameter smaller than the cell, then the cell–surface contact area is reduced, thus the cell would be removed more easily. (c) When the surface feature is of diameter similar to that of the cell, then the cell is retained strongly within the feature and would be difficult to remove.

ever, others have suggested that the greater the degree of surface roughness, the greater the retention of microorganisms [46–49]. According to the attachment point theory, organisms smaller than the scale of the surface micro-texture will have a greater adhesion strength because of multiple attachment points on the surface when compared to microorganisms that are of a larger scale than the surface roughness [50–53]: the converse has also been suggested [54], although it has also been shown that surface nanotopographies influence the attachment of cells to glass surfaces [55]. However, although it has also been suggested that surface features whose dimensions greatly exceed those of the microorganisms will have little effect on retention [56], it is likely that this would only be true if the profiles of these features were also extremely smooth. It is more likely that within 'macro-' features (diameter > 30 µm) there will be features of 'micro-' (<10 µm, >0.5 µm) and 'nano-' (<0.5 µm) dimensions [6] which may themselves enhance the retention of microorganisms of similar sizes [31, 57].

Thus, there have been conflicting reports on this topic although it has been speculated that a minimum R_a that would influence cell attachment could be defined. Using the AFM, Medilanski et al. [16] have demonstrated that microbial adhesion to grooved featured surfaces was minimal at $R_a = 0.16$ µm, whereas both smoother and rougher surfaces gave rise to increased cellular adhesion. Our study used surfaces with R_a values between 0.003 µm and 0.026 µm and demonstrated that there was increased cell adhesion due to surface feature size rather than R_a value. Along with R_a values, the shape of the surface features, especially those within microbial sizes, has been shown to influence microbial attachment and retention since the R_a value gives a two-dimensional profile of the surface and does not reflect the shape of surface topographies [16, 41]. Whitehead et al. [29] demonstrated that when the lateral AFM tip force was increased, coccal-shaped *Staphylococcus aureus* were removed more easily from smooth surfaces than from featured surfaces. In contrast, rod-shaped *P. aeruginosa* cells were removed more easily from 0.5 μm 'pit' featured surfaces [29], demonstrating that the shape of the cell with respect to the shape of the substratum features affected the ease of removal of the cell from the surface: on smooth surfaces the cocci had a smaller cell:surface contact area, whereas the rods had a larger cell:surface contact area. Conversely, on featured surfaces the cocci had a larger cell:surface contact area, whereas rods that lay across features had a smaller cell:surface contact area. Thus in agreement with our current work, displacement of the coccal cells from a flat surface required less force as there will be less resistance due to lower cell–surface binding energies. It has been suggested that cells will attach to a surface in such a manner as to maximise cell–surface contact [58]. Edwards and Rutenberg [58] suggested that the total adhesion energy of a bacterium with a substratum will depend on the interaction potential, surface geometry and bacterial shape, since a bacterium may deform to increase the amount of surface close to the substratum thereby increasing binding energy (since the binding strength is reduced by small grooves), but will increase when the groove exceeds a critical radius that is close to the bacterial size. Thus enhanced bacterial retention on different surface features could be due to an increase in bacterial attachment sites (for a given surface area), leading to stronger bacterial attachment and enhanced protection from shear forces such as those encountered during cleaning [10]. This was evidenced in our current work. Edwards and Rutenberg [58] further recognised that the cross-sectional shape of a groove has a large effect on binding potential, with the depth of the surface feature being less important than its cross-sectional shape. Scratches that are large enough to decrease binding, but too small for the bacterium to fit into them will reduce the contact area of the bacterium aligned parallel to them and reduce cell contact area in other orientations. Thus along with R_a values, a precise description of the surface features is essential.

All the test substrata had R_a values below the normally accepted 'hygienic' level of 0.8 μm [13]. The results show that R_a values below this value required different amounts of force to remove the bacteria. Therefore, it appears that it is the size of the feature in relation to the size of the bacterium, rather than the R_a value itself that affects the strength of adhesion of the cell to the substrate surface, as the feature size will affect the surface area in contact with the cell.

5. Conclusions

Substrata with defined linear surface topographies and uniform surface chemistry have been produced. Through use of an AFM under liquid the application of an increasing lateral force led to the removal of coccal cells. The shape of the microorganism in relation to the shape and dimension of the surface feature affected the ease of cell removal from a surface, depending on the surface area in contact with the cell. The presence of micrometer and sub-micrometer grooved features on a surface strongly influences bacterial adhesion, and potentially, surface cleanability.

Acknowledgements

The authors would like to thank Dr Brigitte Carpentier (AFSSA, France) for her kind gift of the microorganisms. The research performed has been part of the project FOOD-CT-2005-007081 (PathogenCombat) supported by the European Commission through the Sixth Framework Programme for Research and Development.

References

1. J. H. Taylor and J. T. Holah, *J. Appl. Bacteriol.* **81**, 262 (1996).
2. C. K. Bower, J. McGuire and M. A. Daeschel, *Trends Food Sci. Technol.* **7**, 152 (1996).
3. C. G. Kumar and S. K. Arand, *Intl J. Food Microbiol.* **42**, 9 (1998).
4. C. Jullien, T. Benezech, B. Carpentier, V. Lebret and C. Faille, *J. Food Eng.* **56**, 77 (2002).
5. G. M. Bruinsma, M. Rustema-Abbing, J. de Vries, B. Stegenga, H. C. van der Mei, M. L. van der Linden, J. M. M. Hooymans and H. J. Busscher, *Invest. Ophthalmol. Visual Sci.* **43**, 3646 (2002).
6. J. Verran, R. D. Boyd, K. E. Hall and R. H. West, *Biofouling* **18**, 167 (2002).
7. S. H. Flint, J. D. Brooks and P. J. Bremer, *J. Food Eng.* **43**, 235 (2000).
8. J. Verran, D. L. Rowe and R. D. Boyd, *J. Food Protection* **64**, 1183 (2001).
9. R. L. Taylor, J. Verran and G. C. Lees, *J. Mater. Sci.-Mater. Medicine* **9**, 17 (1998).
10. D. A. Timperley, R. H. Thorpe and J. T. Holah, in: *Biofilms—Science and Technology*, L. F. Melo, T. R. Bott, M. Fletcher and B. Capdeville (Eds), p. 379. Kluwer Academic Publishers, Dordrecht (1992).
11. BS1134-1 (1988).
12. C. M. L. Bollen, B. N. A. Vandekerckhove and M. Quirynen, *J. Dental Res.* **76**, 1147 (1997).
13. S. H. Flint, P. J. Bremer and J. D. Brooks, *Biofouling* **11**, 81 (1997).
14. R. D. Boyd, D. Cole, D. Rowe, J. Verran, S. J. Coultas, A. J. Paul, R. H. West and D. T. Goddard, *J. Adhesion Sci. Technol.* **14**, 1195 (2000).
15. R. D. Boyd, J. Verran, D. Rowe, D. Cole, K. E. Hall, C. Underhill, S. Hibbert and R. West, *Appl. Surface Sci.* **172**, 135 (2001).
16. E. Medilanski, K. Kaufmann, L. Y. Wick, O. Wanner and H. Harms, *Biofouling* **18**, 193 (2002).
17. J. Verran, K. Bayley, G. Sandoval and K. A. Whitehead, in: *Proc. 6th Meeting of the Biofilm Club*, D. A. A. McBain, M. Brading, A. Rickard, J. Verran and J. Walker (Eds), p. 65. Bioline, Cardiff (2003).
18. M. R. Nejadnik, H. C. van der Mei, H. J. Busscher and W. Norde, *Appl. Environm. Microbiol.* **74**, 916 (2008).
19. M. Paulsson, M. M. Kober, C. Freij-Larsson, M. Stollenwerk, B. Wesslen and A. Ljungh, *Biomaterials* **14**, 845 (1993).
20. N. P. Boks, W. Norde, H. C. van der Mei and H. J. Busscher, *Microbiol.* **154**, 3122 (2008).
21. W. R. Bowen, R. W. Lovitt and C. J. Wright, *Colloids Surfaces A* **173**, 205 (2000).
22. W. Bowen, R. Lovitt and C. Wright, *J. Mater. Sci.* **36**, 623 (2001).
23. W. R. Bowen, A. S. Fenton, R. W. Lovitt and C. J. Wright, *Biotechnol. Bioeng.* **79**, 170 (2002).
24. Y.-L. Ong, A. Razatos, G. Georgiou and M. M. Sharma, *Langmuir* **15**, 2719 (1999).
25. S. K. Lower, C. J. Tadanier and J. Hochella, *Geochim. Cosmochim. Acta* **64**, 3133 (2000).

26. A. Senechal, S. D. Carrigan and M. Tabrizian, *Langmuir* **20**, 4172 (2004).

27. A. Roosjen, N. P. Boks, H. C. van der Mei, H. J. Busscher and W. Norde, *Colloids Surfaces B* **46**, 1 (2005).

28. G. Dagvolen, I. Giaver, E. O. Pettersen and J. Feder, *Proc. Natl Acad. Sci. USA* **96**, 471 (1999).

29. K. A. Whitehead, D. Rogers, J. Colligon, C. Wright and J. Verran, *Colloids Surfaces B* **51**, 44 (2006).

30. R. D. Boyd, J. Verran, M. V. Jones and M. Bhakoo, *Langmuir* **18**, 2343 (2002).

31. J. Verran, A. Packer, P. Kelly and K. A. Whitehead, *Int. J. Food Microbiol*, in press.

32. P. J. Kelly and R. D. Arnell, *Vacuum* **56**, 159 (2000).

33. L. Caballero, K. A. Whitehead, N. S. Allen and J. Verran, *J. Photochem Photobiol A: Chemistry* **202**, 92 (2009).

34. Y. F. Dufrene, C. J. P. Boonaert, H. C. van der Mei, H. J. Busscher and P. G. Rouxhet, *Ultramicroscopy* **86**, 113 (2001).

35. S. M. Deupree and M. H. Schoenfisch, *Langmuir* **24**, 4700 (2008).

36. J. A. Schmidt and A. F. von Recum, *Biomaterials* **12**, 385 (1991).

37. J. A. Schmidt and A. F. von Recum, *Biomaterials* **13**, 1059 (1992).

38. J. Gold, B. Nilson and B. Kasemo, *J. Vac. Sci. Technol.* **A13**, 2638 (1995).

39. A. M. Green, J. A. Jansen, J. P. C. M. van der Waerden and A. F. vin Recum, *J. Biomed. Mater. Res.* **28**, 647 (1994).

40. K. A. Whitehead, J. S. Colligon and J. Verran, *Intl. Biodet. Biodeg.* **54**, 143 (2004).

41. K. A. Whitehead, J. Colligon and J. Verran, *Colloids Surfaces B* **41**, 129 (2005).

42. H. J. Busscher, A. W. J. van Pelt, P. de Boer, H. P. de Jong and J. Arends, *Colloids Surfaces B* **9**, 319 (1984).

43. L. P. M. Langeveld, A. C. Bolle and J. E. Vegter, *Netherland Milk Dairy J.* **42**, 149 (1972).

44. E. Vanhaecke, J.-P. Remon, M. Moors, D. de Rudder and A. van Peteghem, *Appl. Environm. Microbiol.* **56**, 788 (1990).

45. S. Flint and N. Hartley, *Intl Dairy J.* **6**, 223 (1996).

46. J. T. Holah and R. H. Thorpe, *J. Appl. Bacteriol.* **69**, 599 (1990).

47. C. M. L. Bollen, B. N. A. Vandekerckhove and M. Quirynen, *J. Dental Res.* **76**, 1147 (1997).

48. J. Verran and C. J. Maryan, *J. Prosthetic Dentistry* **77**, 535 (1997).

49. W. G. Characklis, G. A. McFeters and K. C. Marshall, in: *Biofilms*, W. G. Characklis and K. C. Marshall (Eds). John Wiley & Sons, New York (1990).

50. M. S. Chae, H. Schraft, L. T. Hansen and R. Mackereth, *Food Microbiol.* **23**, 250 (2006).

51. D. Howell and B. Behrends, *Biofouling* **22**, 401 (2006).

52. A. J. Scardino, E. Harvey and R. de Nys, *Biofouling* **22**, 55 (2006).

53. K. Shellenberger and B. E. Logan, *Environm. Sci. Technol.* **36**, 184 (2002).

54. W. B. Freeman, J. Middis and H. M. Muller-Steinhagen, *Chem. Eng. Process.* **27**, 1 (1990).

55. N. Mitik-Dineva, J. Wang, R. C. Mocanasu, P. R. Stoddart, R. J. Crawford and E. P. Ivanova, *Biotechnology J.* **3** 536 (2008).

56. J. Verran, D. L. Rowe and R. D. Boyd, *Intl Biodeterioration Biodegradation* **51**, 221 (2003).

57. K. A. Whitehead and J. Verran, *Trans. Inst. Chem. Engrs. C* **84**, 253 (2006).

58. K. J. Edwards and A. D. Rutenberg, *Chem. Geol.* **180**, 19 (2001).

Continuous Photobleaching to Study the Growth Modes of Focal Adhesions

Alex G. F. de Beer [a,*], **Günter Majer** [a], **Sylvie Roke** [a] **and Joachim P. Spatz** [b]

[a] Max Planck Institute for Metals Research, Heisenbergstrasse 3, D-70569 Stuttgart, Germany
[b] Department of Biophysical Chemistry, Institute for Physical Chemistry, University of Heidelberg, INF 274, D-69120 Heidelberg, Germany

Abstract

We combine total internal reflection fluorescence microscopy with continuous photobleaching to study the growth mode of focal adhesions in rat embryonic fibroblasts (REF-52). We measured GFP-labelled $\beta 3$ integrin, which exhibits a diffusion rate of 10^{-14}–10^{-13} m^2/s when it is not bound to a focal adhesion. We show that exchange of $\beta 3$ integrin between focal adhesions and the surrounding membrane occurs only at the edges of the focal adhesion. We, therefore, conclude that focal adhesions show an edge-on/edge-off growth, in contrast to the bulk-on/bulk-off growth mode that is assumed in many theoretical treatments.

Keywords

Focal adhesion, photobleaching, FRAP, TIRF, integrin

1. Introduction

Focal adhesions (FAs) form the bridge between the actin skeleton of a cell and the surrounding tissue of the cell. They are formed when precursor contacts at the edges of lamellipodia [1–4] elongate under internal stress from the actin cytoskeleton [5–8]. FAs are active mechano-sensory elements [9–13] and play a role in sensing force [8, 14–16]. FAs adapt their size based on the amount of force they are subjected to [17–21]: typically an FA grows or shrinks until the force per unit area reaches a preset level. The exact mechanism behind this force-induced growth phenomenon is currently unknown. Several models exist, which are based upon thermodynamic and mechanical assumptions [22–25]. These models either assume that growth takes place throughout the FA (bulk-on growth) [22, 25] or at the edges of the FA (edge-on growth) [23, 24]. These differences in growth modes, however, should lead to different properties, such as different FA size distributions [26].

The kinetic binding and release rates of FA proteins were determined with Fluorescence Recovery After Photobleaching (FRAP) studies [27–35]. In these studies,

---

* To whom correspondence should be addressed; Tel.: +49 (0) 711 689 3683; e-mail: debeer@mf.mpg.de

the transmembrane protein integrin showed a delayed recovery. It was suggested that the diffusion of integrin into an FA may be severely hindered by the dense structure of the FA. Since integrin forms the base of an FA, such a diffusional hindrance could be the basis of an edge-on growth mechanism.

In this paper, we present an alternative method to study the turnover dynamics of integrin within FAs. We use Continuous Photobleaching (CP) [36–40] in combination with Total Internal Reflection Fluorescence microscopy (TIRF). We use the evanescent wave of a TIRF microscope to create a limited illumination area. Only those (membrane-associated) fluorescent proteins within this area are subject to photobleaching. We show it is possible to probe the dynamics of Green Fluorescent Protein (GFP) labelled integrin $\beta 3$ within and outside of FAs and to directly visualize protein renewal within FAs. Both forming and mature FAs show an edge-on/edge-off growth behavior.

2. Theory of Continuous Photobleaching

Figure 1(top) shows a schematic of CP at a biological sample: an adherent cell is illuminated in TIRF mode. Only the part of the cell membrane that is within the order of a light wavelength from the glass surface is illuminated, and, therefore, only the fluorescent proteins within this part are subject to photobleaching. Unbound fluorescent integrin (U) can diffuse throughout the cell membrane, while bound proteins (B) remain at a fixed position until they are released. Figure 1(bottom) shows a schematic of the different processes a fluorescently labelled integrin molecule can undergo: unbound integrin (U) can diffuse in and out of the illuminated area. Whenever it resides inside the bleaching area, it bleaches at a bleaching rate λ. Bound proteins (B) are immobile and always subject to photobleaching. We denote the subpopulations of U and B that are fluorescent (bleached) with a subscript F (X). The total detected fluorescence scales with ($U_F + B_F$). Proteins bind to the structure at a rate that depends on the binding rate k_{on} and the number of available binding sites S, whereas the protein release rate is proportional to k_{off}, so the the relation between U and B is:

$$\frac{\mathrm{d}B}{\mathrm{d}t} = k_{on}US - k_{off}B. \tag{1}$$

S typically depends on the number of proteins already bound. For a fixed number of external binding sites L, for instance, the number of available sites is $S = L - B$.

There are four possible states of a molecule, and every state has a certain transition probability to a different state. All possible transitions are given by four differential equations, which are summarized by the following matrix:

$$\begin{pmatrix} \frac{\mathrm{d}U_F}{\mathrm{d}t} \\ \frac{\mathrm{d}U_X}{\mathrm{d}t} \\ \frac{\mathrm{d}B_F}{\mathrm{d}t} \\ \frac{\mathrm{d}B_X}{\mathrm{d}t} \end{pmatrix} = \begin{pmatrix} D\nabla^2 - \lambda - k_{on}S & 0 & k_{off} & 0 \\ \lambda & D\nabla^2 - k_{on}S & 0 & k_{off} \\ k_{on}S & 0 & -\lambda - k_{off} & 0 \\ 0 & k_{on}S & \lambda & -k_{off} \end{pmatrix} \begin{pmatrix} U_F \\ U_X \\ B_F \\ B_X \end{pmatrix}, \tag{2}$$

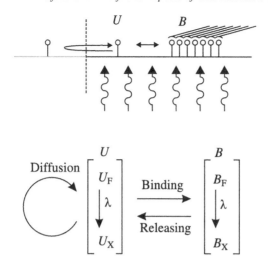

Figure 1. (Top) Schematic of continuous photobleaching. A cell membrane is partly illuminated and contains a number of proteins that can either diffuse freely (U) or are fixed in a structure (B). (Bottom) Schematic of three processes governing image intensity: Free proteins (U) may bind to a structure to become bound proteins (B). Both populations undergo bleaching at a rate λ, so that fluorescent proteins (U_F and B_F) turn into dark proteins (U_X and B_X). The free proteins can diffuse in and out of the illuminated area, while the fixed proteins cannot, which leads to a different effective bleaching rate.

where D is the diffusion coefficient and $\nabla^2 = d^2/dx^2 + d^2/dy^2$ is the two-dimensional Laplacian operator. There is no general solution to the above system of differential equations. We can, however, analyze the system for three cases that are relevant to our experiments: (1) for unbound proteins in the absence of an FA, (2) a mature FA that has reached a stable size, and (3) a newly forming FA.

2.1. Continuous Photobleaching in the Absence of Focal Adhesions

In the first case no protein structure is present ($k_{on} = 0$) and the fluorescence intensity is given by a single differential equation [40]:

$$\frac{dU_F}{dt} = D\nabla^2 U_F - \lambda U_F, \tag{3}$$

where D is the diffusion coefficient and λ varies as a function of position. An illumination edge appears, for instance, where the membrane of an adherent cell curves away from the glass surface: the part in contact with the surface is illuminated, while other (dorsal) parts are not. We can describe this with a bleaching rate that is zero for $x < 0$ and λ for $x > 0$. The solution to the above equation for such an edge is determined by the length scale $l_D = \sqrt{D/\lambda}$. l_D is a measure of the distance a fluorophore can diffuse into the illuminated area before it is bleached. A change in illumination intensity changes the bleaching rate and thereby also this length scale: at mild bleaching conditions molecules can diffuse some distance into the illuminated area before bleaching. At higher illumination intensities, all molecules will be quickly bleached after entering the bleaching volume, so that only a sharp fluorescent rim is visible at the edge. Close to the edge of the illuminated area the number of fluorescent molecules U_F is independent of the value of D [40]:

$$U_F(t)|_{x=0} = U_F(0)e^{-\lambda t/2} I_0\left(\frac{\lambda t}{2}\right), \tag{4}$$

where I_0 is the zeroth-order modified Bessel function of the first kind. This decay is much slower compared to the bleaching rate λ. Moving away from the edge into the bleached area increases the rate of fluorescence loss until at distances larger than l_D, it is close to the bleaching rate λ.

2.2. Continuous Photobleaching at Mature Focal Adhesions

In the second case, we consider a mature FA that is constant in size and is in equilibrium with its surroundings. The proteins bound to an FA are bleached at a rate λ. The actual rate of fluorescence loss is reduced by an exchange between the proteins bound at the FA and the unbound proteins surrounding it. Because these unbound proteins typically have a higher fluorescent fraction, the expected intensity loss is slightly delayed. The differential equation for fluorescent, bound proteins becomes:

$$\frac{dB_F}{dt} = -\lambda B_F - k_{off}B_F + k_{on}U_F(L - B) \tag{5}$$

$$= -\lambda B_F - k_{off}B_F + u_F\left(k_{off}B + \frac{dB}{dt}\right), \tag{6}$$

where $u_F = U_F/U$ is the fraction of unbleached free proteins. In this equation, k_{off} takes the role of an exchange rate between bound and unbound proteins, each of which has been bleached to a fraction u_F or $b_F = B_F/B$:

$$\frac{dB_F}{dt} = -\lambda B_F + k_{off}(u_F - b_F)B + u_F\frac{dB}{dt}. \tag{7}$$

In the above formula, the first term on the right-hand side denotes a decrease in fluorescence due to bleaching, the second term leads to a delay in fluorescence loss whenever u_F is larger than b_F. Finally, the last term describes an increase in intensity in a growing FA. Laplace transformation yields a general solution to equation (6):

$$B_F(t) = e^{-(k_{off}+\lambda)t}B_F(0)$$
$$+ \int_0^t e^{(k_{off}+\lambda)(\tau-t)}u_F(\tau)\left(k_{off}B(\tau) + \frac{dB(\tau)}{dt}\right)d\tau. \tag{8}$$

Whenever B remains constant, the fluorescence decay within an FA will be a decreasing function. The exchange rate $k_{off}B$ along with u_F leads to a B_F that decays slower than $B_F(0)\exp(-\lambda t)$. Figure 2(a) show decay curves (solid lines) for different values of k_{off}/λ and an u_F that corresponds to equation (4). The dashed

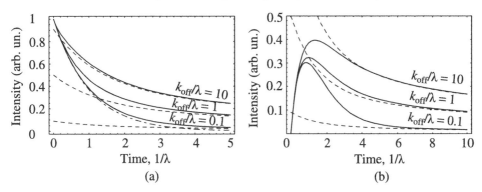

Figure 2. Plots of the simulated continuous photobleaching intensity (solid curves) for (a) an FA at equilibrium and (b) a growing FA. Different ratios of the exchange rate (governed by the ratio of k_{off} and λ) yield different curve shapes. The dashed lines show the limits for large time values t.

lines show curves that are proportional to u_F. For large time values t, B_F approaches these curves. A higher value of k_{off} leads to a faster approach to this curve, but also an overall higher value of B_F due to the elevated level at which bleached molecules are replaced by unbleached ones.

2.3. Growing Focal Adhesions

Finally, we consider an FA that forms during the experiment, so that the initial values of both B_F and B_X are zero. The fluorescence intensity is determined by three time scales: λ determines the rate of fluorescence loss, k_{off} determines the rate of exchange between bound and unbound proteins and finally a third rate, μ, determines the rate of growth of the structure. Figure 2(b) shows $B_F(t)$ for different values of k_{off}/λ (solid lines). Here, we assumed a structure that shows an exponential growth to a steady state $B(t) = B(\infty)(1 - \exp(-\mu t))$, with $\mu \approx \lambda$ as well as an u_F that corresponds to equation (4). The dashed lines show the limits for large time values, which is equal to $B_F(t) = u_F(t)B(t)k_{off}/(k_{off} + \lambda)$. Again, a higher exchange rate between free and bound proteins leads to a slower decay of fluorescence intensity.

3. Materials and Methods

REF-52 fibroblasts stably expressing GFP-β3 integrin were a kind gift from Dr. B. Geiger (Weizmann Institute of Science, Rehovot, Israel). Cells culture was performed using Dulbecco's modified Eagle's medium (DMEM) supplemented with 10% bovine fetal serum (FBS) and 2 mM L-glutamine. The cells were grown to 80%–90% confluency and passaged every 2–3 days. Prior to imaging, cells were washed with warm sterile phosphate buffered saline (PBS), detached from the vessel by a 3 min 0.25 % trypsin–EDTA treatment and centrifuged at 900 rpm for 3 min. The cell pellet was resuspended in DMEM without phenol red supplemented with 10% FBS and 50 mM Hepes buffer. All cell culture material used to passage and to perform experiments was from Invitrogen Germany. Cells were then seeded onto glass surfaces and allowed to adhere for up to one hour.

Digital images were obtained using a Zeiss axiovert 200M inverted microscope (Zeiss, Germany), in combination with a 100× oil immersion objective (N.A. 1.45, Zeiss, Germany). A Zeiss TIRF module provides illumination through the objective, using the 488 nm laser line from a laser launch module (Intelligent Imaging Innovations, Göttingen, Germany). The microscope was placed within a custom-built climate chamber that was kept at 37°C for the duration of the experiment. An acousto-optical tunable filter (NEOS Technologies, Melbourne, Florida) allowed for attenuation of illumination intensity. Images were acquired using an EMCCD camera with a 16 μm pixel size (iXon DV897, Andor Technology, Belfast, N. Ireland). The fibroblasts were plated on glass slides and were allowed to settle for one hour, after which measurements were started. Adherent cells were exposed to a continuous, constant light intensity in TIRF mode, during which images were obtained every 0.4 s.

4. Results

Figures 3(a)–(c) show CP at unbound proteins. During the experiment, the illumination power was initially set to 1 mW (corresponding to a bleaching lifetime > 60 s). Figure 3(a) shows the initial intensity distribution in the cell. The areas of highest intensity correspond to FAs, which undergo the fastest bleaching. After two minutes, the fluorescence from most FAs has diminished in intensity. Outside the FAs, the

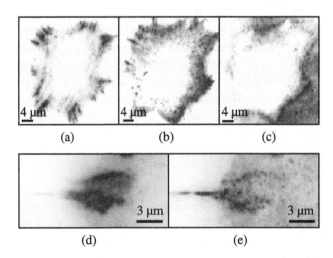

(a)   (b)   (c)

(d)   (e)

Figure 3. (a)–(c) Demonstration of continuous photobleaching. A cell showing FAs (a) is bleached under TIRF illumination at a bleaching lifetime of >60 s, which shows the appearance of a rim at the edge of the visible cell area (b). At higher intensity (bleaching lifetime < 60 s), the width of the rim decreases (c). (d)–(e) Continuous photobleaching at an FA. An FA (d) is bleached at a bleaching lifetime of 120 s. After five minutes, only a faint rim outlining the original FA remains (e). All images are shown as negatives, so that dark areas correspond to high intensity.

formation of a rim is visible (Fig. 3(b)), caused by the diffusion of unbound proteins in and out of the illuminated area. When the intensity was subsequently raised to 10 mW (bleaching lifetime < 60 s) the width of the rim reduced significantly over the course of a minute (see Fig. 3(c)). When we interpret the width of the rim in terms of the diffusion length scale $l_D = \sqrt{D/\lambda}$, we can extract a diffusion constant for β3-integrin. Though the presence of FAs and irregularities in cell shape prevent an accurate analysis, we obtain a diffusion constant in the order of 10^{-14}–10^{-13} m^2/s, which is consistent with earlier observations [41, 42].

Figures 3(d)–(e) show close-ups of a single FA in a photobleaching experiment. Figure 3(d) shows the FA before bleaching, while Fig. 3(e) shows the same FA after five minutes of photobleaching at mild bleaching lifetimes (approx. 120 s). Clearly, a difference in decay rate is observed as the center of the FA is bleached more thoroughly than the edges, which remain visible as a faint rim.

Figure 4(a) shows a first decay moment (T_1) map of a photobleaching experiment. The first decay moment T_1 is a numeric figure (in units of time) defined by

$$T_1(x, y) = \frac{\sum_t t I(x, y; t)}{\sum_t I(x, y; t)}, \tag{9}$$

where $I(x, y; t)$ is the intensity at position (x, y) and time value (frame number) t. The number can be interpreted as the moment at which the total integrated intensity from the start to this moment is exactly half of the integrated intensity for the entire measurement. Furthermore, when the first moment expressed in term of fractions of total exposure time, a constant intensity will lead to a value of 0.5, an increasing intensity will lead to a value > 0.5 and a decreasing intensity will lead to a value < 0.5. For a perfect exponential decay, T_1 approaches the exponential decay time as the total exposure length increases.

In Figure 4(a) red areas indicate a value of T_1 close to 0.5, for which the intensity remains at a constant (low) value throughout the experiment. White areas indicate a T_1 larger than 0.5, which corresponds to an increasing intensity. Finally, yellow areas correspond to a T_1 that is smaller than 0.5. The brightest of these areas align with the positions of the FAs, for which photobleaching is fastest.

Figure 4(b) shows a cut through one of the FAs present from the start of the imaging. From this image, it is clear that the center part of the FA (the area marked by 'I') shows the fastest decay, compared to the surroundings. At the edges of the FA (the areas marked by 'II'), a transition area appears, with a width of approximately 1 µm in which the decay decreases until it reaches the level of the surroundings. This difference in fluorescence decay can be attributed to a limitation in the mobility of the unbound proteins: at the edges of the FA, the unbound proteins can freely exchange with the unbound proteins in the membrane, leading to a difference between the fluorescent fractions b_F and u_F, and, therefore, to a delay in photobleaching. At the center of the FA, the mobility of the released integrin remains low, so that it is effectively trapped within the FA. The unbound proteins at the FA, therefore, bleach

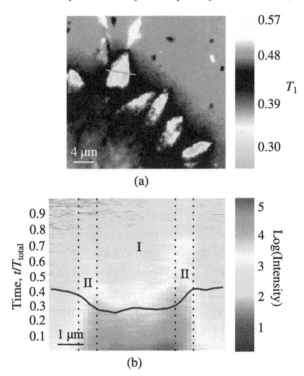

(a)

(b)

Figure 4. First decay moment representation of a photobleaching time series (a). The different colors represent T_1 values and give an indication of the bleaching lifetime at every pixel. Values close to 0.5 (red) indicate a roughly constant intensity, values > 0.5 (white) indicate an increasing intensity, while values < 0.5 (black and yellow) occur when the intensity decreases over time. A slice through an FA (b) as a function of time shows the fluorescence loss within the FA. The blue line denotes the corresponding T_1 value for every pixel in the slice. At the center of the FA (area marked 'I') the decay time is shortest. A transition occurs within 1 μm of the FA edge (the two areas marked 'II'). In these areas, the decay time gradually increases from the low value at the center of the FA to the high value outside the FA.

at the same rate as the bound proteins, so that the overall bleaching rate at this point is close to $\lambda = 0.0025$ s^{-1}.

The thickness of the edge in area 'II' is of the order of 500 nm. Although this value is close to the diffraction limit and the edge is not likely to be very sharp, we can still derive an upper limit for the effective diffusion constant of $\beta 3$ integrin of 6.3×10^{-16} m^2/s, which is one to two orders of magnitude lower than that of integrin outside the FA.

When FAs form during an experiment, an initial rise in intensity is followed by a subsequent loss due to photobleaching. Figure 5 shows the intensity curve for an FA that forms during photobleaching. An FA forms at $t = 320$ s and rises to full intensity in the next five minutes. From this point onward, fluorescence loss due to photobleaching is larger than fluorescence gain due to FA growth, so that the overall intensity decreases again. We can fit the model of Section 2.3 to this graph

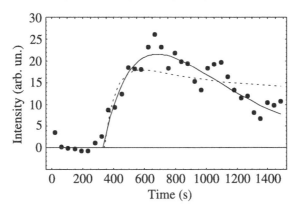

Figure 5. Data fits for a growing FA, showing the best data fits for $k_{off} = 0$ s^{-1} (solid curve) and $k_{off} = 0.012$ s^{-1} (dashed curve). Fitted values for λ (0.0026 and 0.0030 s^{-1}, respectively) were consistent with the overall bleaching rate of 0.0025 s^{-1}. Best-fit values for the FA growth rate μ were 0.0023 s^{-1} and 0.01 s^{-1}, respectively. The mismatch of the fit with the data in the $k_{off} = 0.012$ s^{-1} case suggests that the role of exchange in the growing FA is negligible, pointing to a limited effective exchange of proteins between FA and cell membrane.

by adjusting the parameters μ, λ and k_{off}. Two data fits are shown in Fig. 5: one for an exchange rate $k_{off} = 0.012$ s^{-1}, the reported off-rate for integrin [43] (dashed line), the other in absence of such an exchange ($k_{off} = 0$, solid line). The data fits corroborate the effect observed above: bound proteins that are released at the rate of k_{off} remain trapped in the dense area of an FA. Such proteins bleach at the same rate as bound proteins, so that effectively, the effect of the release rate k_{off} is zero.

5. Discussion

We have shown continuous photobleaching at FAs using the evanescent wave in a TIRF setup as a bleaching volume. The surface selectivity of TIRF illumination selectively bleaches molecules on the part of the cell membrane that is in contact with a glass substrate. The upper side of the cell is outside the range of the evanescent wave and thus remains unilluminated. We have shown that diffusion characteristics of membrane proteins can be measured in a way similar to previous CP experiments, by measuring the penetration depth into the bleached area of the molecules. We obtain values of $D = 10^{-14}$–10^{-13} m^2/s, which are comparable to previous measurements [41, 42]. TIRF CP measurements at FAs show a limited exchange rate between proteins at the center of the FA and unbound proteins outside the FA. Released proteins at the center of the FA have a limited mobility and are effectively trapped inside the FA. Proteins at the edges of an FA can exchange with the unbound proteins in the membrane outside the FA. As a result, the bleaching rate at the edge of the FA is lower than at the center. The dense structure that hinders the diffusion of unbound proteins is present immediately upon formation of the FA.

The limited exchange between FA and surroundings is observed for growing FAs as well as mature ones.

We suggest this limited exchange is the result of a lower effective diffusion rate (less than 6.3×10^{-16} m^2/s) of integrin within an FA. A dynamic equilibrium between mobile and bound integrin allows a gradual redistribution when a small section of an FA is bleached. The view of slowly diffusing integrin is supported by Tsuruta *et al.*, who showed partially bleached FAs may recover by redistribution within the FA on timescales that are insufficient for the recovery a fully bleached FA [29]. It is also consistent with earlier findings [41] that show that integrin remains (relatively) stationary within an FA. Reduced diffusion rates are not limited to integrin: Digman *et al.* [44] showed a similar reduction in diffusion rate for paxillin, a cytosolic protein closely associated with integrin.

The lowered diffusion rate and the limited exchange with surroundings leads to the conclusion that integrin is added/removed to FAs in an edge-on/edge-off fashion. Such a growth mode has already been incorporated in a number of theoretical models that treat FA growth [23, 24, 26]. An edge-on growth mode would also suggest that integrin at the core of an FA is effectively isolated from the surrounding membrane. Any mechanosensory effects, such as force-induced growth [18], should, therefore, take place on the FA edge, while integrin at the core is limited to reorganization.

6. Conclusion

In summary, we have demonstrated an application of CP using TIRF illumination. This variant allows us to follow, in parallel, relative exchange rates and protein mobilities of membrane proteins associated with cellular structures. We have developed a theoretical framework for describing the fluorescence loss in stable FAs as well as created a model for the fluorescence changes in growing FAs. Our experimental results showed that for membrane proteins information on diffusion rates can be obtained, as well as spatial information on protein exchanges. Our findings show the FA structure to be too dense to allow unhindered diffusion, as has been suggested in literature before [26, 34].

Acknowledgements

The authors would like to thank Benjamin Geiger and Ralf Richter for their scientific support and useful discussions related to this work. This work is part of the research program of the Max Planck Society and was supported by the Landesstiftung Baden-Württemberg in the frame of the program "Spitzenforschung in Baden-Württemberg" and by the National Institute of Health, though the NIH Roadmap for Medical Research (PN2 EY016586).

References

1. C. D. Nobes and A. Hall, *Cell* **81**, 53–62 (1995).
2. E. A. Clark, W. G. King, J. S. Brugge, M. Symons and R. O. Hynes, *J. Cell Biol.* **142**, 573–586 (1998).
3. K. Rottner, A. Hall and J. V. Small, *Curr. Biol.* **9**, 640–648 (1999).
4. R. Zaidel-Bar, C. Ballestrem, Z. Kam and B. Geiger, *J. Cell Sci.* **116**, 4605–4613 (2003).
5. M. Chrzanowska-Wodnicka and K. Burridge, *J. Cell Biol.* **133**, 1403–1415 (1996).
6. K. Burridge and M. Chrzanowska-Wodnicka, *Annu. Rev. Cell. Dev. Biol.* **12**, 463–518 (1996).
7. R. J. Pelham and Y. L. Wang, *Proc. Natl Acad. Sci. USA* **94**, 13661–13665 (1997).
8. D. Riveline, E. Zamir, N. Q. Balaban, U. S. Schwarz, T. Ishizaki, S. Narumiya, Z. Kam, B. Geiger and A. D. Bershadsky, *J. Cell Biol.* **153**, 1175–1185 (2001).
9. B. Geiger and A. D. Bershadsky, *Curr. Opin. Cell Biol.* **13**, 584–592 (2001).
10. A. D. Bershadsky, N. Q. Balaban and B. Geiger, *Annu. Rev. Cell Dev. Biol.* **19**, 677–695 (2003).
11. B. Wehrle-Haller and B. A. Imhof, *Trends Cell Biol.* **12**, 382–389 (2002).
12. D. E. Ingber, *Proc. Natl Acad. Sci. USA* **100**, 1472–1474 (2003).
13. V. Vogel and M. Sheetz, *Nature Rev. Mol. Cell Biol.* **7**, 265–275 (2006).
14. P. F. Davies, A. Robotewskyj and M. L. Griem, *J. Clin. Invest.* **93**, 2031–2038 (1994).
15. D. Choquet, D. P. Felsenfeld and M. P. Sheetz, *Cell* **88**, 39–48 (1997).
16. H. B. Wang, M. Dembo, S. K. Hanks and Y. L. Wang, *Proc. Natl Acad. Sci. USA* **98**, 11295–11300 (2001).
17. C. G. Galbraith and M. P. Sheetz, *Proc. Natl Acad. Sci. USA* **94**, 9114–9118 (1997).
18. N. Q. Balaban, U. S. Schwarz, D. Riveline, P. Goichberg, G. Tzur, I. Sabanay, D. Mahalu, S. A. Safran, A. D. Bershadsky, L. Addadi and B. Geiger, *Nature Cell Biol.* **3**, 466–472 (2001).
19. K. A. Beningo, M. Dembo, I. Kaverina, J. V. Small and Y. L. Wang, *J. Cell Biol.* **153**, 881–887 (2001).
20. S. Munevar, Y. L. Wang and M. Dembo, *Biophys. J.* **80**, 1744–1757 (2001).
21. J. L. Tan, J. Tien, D. M. Pirone, D. S. Gray, K. Bhadriraju and C. S. Chen, *Proc. Natl Acad. Sci. USA* **100**, 1484–1489 (2003).
22. T. Shemesh, B. Geiger, A. D. Bershadsky and M. M. Kozlov, *Proc. Natl Acad. Sci. USA* **102**, 12383–12388 (2005).
23. A. Nicolas, B. Geiger and S. A. Safran, *Proc. Natl Acad. Sci. USA* **101**, 12520–12525 (2004).
24. A. Besser and S. A. Safran, *Biophys. J.* **90**, 3469–3484 (2006).
25. N. D. Gallant and A. J. Garcia, *J. Biomech.* **40**, 1301–1309 (2007).
26. N. S. Gov, *Biophys. J.* **91**, 2844–2847 (2006).
27. C. Ballestrem, B. Hinz, B. A. Imhof and B. Wehrle-Haller, *J. Cell Biol.* **155**, 1319–1332 (2001).
28. M. Edlund, M. A. Lotano and C. A. Otey, *Cell Motil. Cytoskeleton* **48**, 190–200 (2001).
29. D. Tsuruta, M. Gonzales, S. B. Hopkinson, C. Otey, S. Khuon, R. D. Goldman and J. C. R. Jones, *FASEB J.* **16**, 866 (2002).
30. B. L. Sprague, R. L. Pego, D. A. Stavreva and J. G. McNally, *Biophys. J.* **86**, 3473–3495 (2004).
31. D. M. Cohen, B. Kutscher, H. Chen, D. B. Murphy and S. W. Craig, *J. Biol. Chem.* **281**, 16006–16015 (2006).
32. S. L. Gupton and C. M. Waterman-Storer, *Cell* **125**, 1361–1374 (2006).
33. T. P. Lele, J. Pendse, S. Kumar, M. Salanga, J. Karavitis and D. E. Ingber, *J. Cell. Physiol.* **207**, 187–194 (2006).
34. T. P. Lele, C. K. Thodeti, J. Pendse and D. E. Ingber, *Biochem. Biophys. Res. Comm.* **369**, 929–934 (2008).

35. H. Wolfenson, Y. I. Henis, B. Geiger and A. D. Bershadsky, *Cell Motil. Cytoskel.* **66**, 1017–1029 (2009).
36. R. Peters, *Cell Biol. Int. Rep.* **5**, 733–760 (1981).
37. R. Peters, A. Brunger and K. Schulten, *Proc. Natl Acad. Sci. USA — Biol. Sci.* **78**, 962–966 (1981).
38. P. Nollert, H. Kiefer and F. Jahnig, *Biophys. J.* **69**, 1447–1455 (1995).
39. C. Dietrich and R. Tampe, *Biochim. Biophys. Acta* **1238**, 183–191 (1995).
40. C. Dietrich, R. Merkel and R. Tampe, *Biophys. J.* **72**, 1701–1710 (1997).
41. J. L. Duband, G. H. Nuckolls, A. Ishihara, T. Hasegawa, K. M. Yamada, J. P. Thiery and K. Jacobson, *J. Cell Biol.* **107**, 1385–1396 (1988).
42. H. Hirata, K. Ohki and H. Miyata, *Cell Motil. Cytoskel.* **59**, 131–140 (2004).
43. F. Y. Li, S. D. Redick, H. P. Erickson and V. T. Moy, *Biophys. J.* **84**, 1252–1262 (2003).
44. M. A. Digman, C. M. Brown, A. R. Horwitz, W. W. Mantulin and E. Gratton, *Biophys. J.* **94**, 2819–2831 (2008).

Cell Adhesion to Ordered Pores: Consequences for Cellular Elasticity

Andreas Janshoff [a,*], **Bärbel Lorenz** [a], **Anna Pietuch** [a], **Tamir Fine** [a],
Marco Tarantola [a], **Claudia Steinem** [b] **and Joachim Wegener** [c]

[a] Georg-August-University, Institute of Physical Chemistry, Tammannstr. 6,
37077 Goettingen, Germany
[b] Georg-August-University, Institute of Organic and Biomolecular Chemistry, Tammannstr. 2,
37077 Goettingen, Germany
[c] University of Regensburg, Institute of Analytical Chemistry, Chemo- & Biosensors,
93053 Regensburg, Germany

Abstract

The adhesion of MDCK II cells to porous and non-porous silicon substrates has been investigated by means of fluorescence and atomic force microscopy. The MDCK II cell density and the average height of the cells were increased on porous silicon substrates with regular 1.2 µm pores as compared to flat, non-porous surfaces. In addition, we found a substantially reduced actin cytoskeleton within confluent cells cultured on the macroporous substrate compared to flat surfaces. The perturbation of the cytoskeleton relates to a significantly reduced expression of integrins on the porous area. The loss of stress fibers and cortical actin is accompanied by a dramatically reduced Young's modulus of 0.15 kPa compared to 6 kPa on flat surfaces as revealed by site-specific force–indentation experiments.

Keywords

Porous silicon, cell adhesion, force spectroscopy, indentation

1. Introduction

Most mammalian cells require a surface for attachment, growth, and proliferation. Ongoing research suggests that cells respond and develop in a surface-specific manner depending on extracellular matrix (ECM) composition, elasticity and topology. The ECM, to which cells attach, is a porous, viscoelastic composite that comprises semiflexible and stiff elastic fibers of cell-secreted proteins embedded in a polysaccharide hydrogel. In addition to the chemical cues provided by these proteins, the cell also acts on the mechanical properties of the matrix. The influence of rigidity and surface topography on cellular responses, and the investigation of

---

* To whom correspondence should be addressed. E-mail: ajansho@gwdg.de

Surface and Interfacial Aspects of Cell Adhesion
© Koninklijke Brill NV, Leiden, 2010

mechanical interactions between cells and their growth substrates, is therefore of great interest [1, 2]. Depending on the cell type, surface properties influence cell polarity, motility, morphology, proliferation rate, and attachment strength. Mechanical signals play a significant role in the signaling system regulating cell and tissue development as well as physiology. Surface stimuli encompass the viscoelasticity of the substrate, surface patterning, wettability, roughness, and porosity [1–16]. So far, model substrates with defined pore geometry and distribution have rarely been used to obtain quantitative data on cell spreading and morphological details such as cytoskeleton development and elastic properties of the cell body [11, 12]. Lee and coworkers [13, 14] reported on fibroblasts cultured on track-etched micropores (0.2–8.0 μm) and found that cell adhesion and growth rate decreased gradually with increasing pore size, while Gold and coworkers introduced a model system based on microfabricated cell culture substrates consisting of interconnected channels [11, 12]. They found that fibroblasts bridge 0.8–1.8 μm wide channels and that channel periodicity alters fibroblasts' morphology and attachment density. Highest cell density was found on flat unstructured surfaces. Even pores as large as 5 μm are spanned by fibroblasts as reported by Richter *et al.* [15]. Yamamoto and coworkers report that porcine aortic endothelial cells adhere to honeycomb patterned films with pore sizes larger than 1–2 μm by activating integrins preferably on the rim [16].

Previously, we investigated the mechanical properties of apical and basal cell membranes from Madin-Darby Canine Kidney (MDCK II) cells deposited on porous substrates. Thereby, we discovered that cells cultured on these substrates behave distinctly differently from those cultured on flat surfaces without pores [17, 18]. The question arose to what extent the elastic properties of whole living cells were altered by the presence of a regular porous matrix.

Here, we studied the mechanical properties of MDCK II cells cultured on silicon substrates that possessed a porous and non-porous, flat area on the same chip. Both parts exhibited identical surface properties and hence were suitable to visualize and quantify the impact of pores on cell morphology and cellular elasticity. We found a substantially increased cell density (20–30%) and cell height (2 μm) accompanied by a loss of stress fibers on the porous part of the surface compared to the flat, unstructured silicon substrate. Force–indentation curves taken on the cell bodies attached to the porous surface show a considerable decrease in the cells' Young's modulus compared to the cells cultured on non-porous substrates. This example shows how the substrate topology impacts the mechanical properties of confluent epithelial cells.

2. Materials and Methods

2.1. Porous Substrates

Porous silicon substrates (5 mm × 5 mm) were purchased from fluXXion B.V. (Eindhoven, the Netherlands). Each substrate contained 140 porous rectangular shaped areas separated by solid silicon (see Fig. 2). The pores were 800 nm deep, open on both sides, and 1.2 μm in diameter [17–19]. For calculation of the porosity only the porous part of the silicon chip was considered. The special geometry of the silicon sieves was used as a coordinate system to localize individual cells, which were subjected to fluorescence and atomic force microscopy. Prior to use, the substrates were coated with a thin layer of chromium (3 nm) followed by gold (60 nm) using thermal evaporation of the metals [18]. From previous studies, we know that MDCK II cells adhere and grow excellently on gold covered surfaces without coating with ECM proteins. Hence, no coating was carried out prior to seeding of the cells.

2.2. Cell Culture

The general procedure for MDCK II cell culture is detailed elsewhere [17, 18, 20, 21]. In brief, the epithelial cell line MDCK II derived from canine kidney was kept in a humidified cell culture incubator at 37°C and 5% CO_2. For cell culturing we used MEM medium (Biochrom, Berlin, Germany) supplemented with 10% fetal calf serum (PAA, Coelbe, Germany), 2 nM L-glutamin (Gibco/Invitrogen, Karlsruhe, Germany) and 100 μg/ml penicillin/streptomycin (Biochrom). The cell culture medium was renewed every third day. For the subculture of a confluent cell layer the cells were incubated for 10 min at 37°C with a 0.05% trypsin–EDTA solution followed by centrifugation and resuspension in fresh medium. For inoculation of the porous substrate a suspension of MDCK II cells was prepared from a confluent cell layer by washing the cells with phosphate-buffered saline without divalent cations (PBS devoid of Ca^{2+} or Mg^{2+} ions (PBS^{--}), Biochrom) followed by incubation with 2 ml EDTA solution (1% EDTA in PBS) for 10 min, and subsequently, trypsinization with trypsin–EDTA solution (0.05% trypsin and 0.02% EDTA in PBS for 10 min). Trypsinization was terminated by adding an excess of culture medium. After centrifugation, the cells were resuspended in fresh culture medium. Cell counting was performed using trypan blue solution (0.4%, Fluka/Aldrich, Seelze, Germany) and a Neubauer counting chamber. Prior to cell seeding, porous substrates were sterilized in an argon plasma for 15 s. MDCK II cells were disseminated on the substrates in a concentration of 29 000 cells per chip (area of the chip: 0.25 cm^2). The cells were allowed to grow for 2 days in a humidified cell culture incubator at 37°C and 5% CO_2 to form a confluent cell layer. Besides wildtype MDCK II cells, we also studied genetically modified YFP MDCK II cells, which express yellow fluorescent protein (YFP) labeled porcine ABC transporter. YFP MDCK II cells were used to avoid external labeling of the membrane in some cases. The ABC transporter is localized in the cell membrane, which allows identification of the membrane if illuminated with blue light.

2.3. Atomic Force Microscopy

AFM measurements were carried out with a MFP-3D microscope (Asylum Research, Santa Barbara, CA, USA) using either MSCT cantilevers from Veeco Instruments (Santa Barbara, CA, USA) with a nominal spring constant of 0.01 N/m

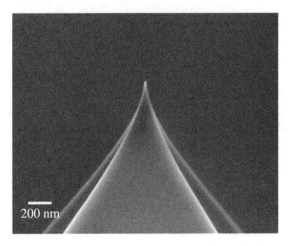

Figure 1. TEM image of the tip shape of the Olympus Biolever used for imaging and indentation experiments.

or Olympus Biolevers™ purchased from Atomic Force (Mannheim, Germany) exhibiting a spring constant of approximately 0.006 N/m (Fig. 1). The exact spring constant was determined by thermal noise calibration [22].

Generally, contact mode was used for imaging by applying minimal normal forces (200–300 pN) to the surface. Therefore, images were taken with minimal deflection set-point. All images were recorded in PBS after equilibration of the system for about 2 h. A homebuilt Teflon chamber especially designed to hold the fragile porous substrates was used.

2.4. Fluorescence Labelling of MDCK II Cells

The fluorescent markers used in these experiments for the visualization of the actin cytoskeleton were phalloidin conjugates (TRITC phalloidin, Sigma Aldrich, Seelze; Alexa Fluor 488 phalloidin, Invitrogen, Karlsruhe, Germany). The toxicity of phalloidin is due to its specific and irreversible binding to filamentous actin (F-actin) preventing its depolymerization and, hence, leading to stabilization of the actin cytoskeleton. For staining with Alexa Fluor 488 phalloidin, a 6.6 mM methanol solution was diluted to a final concentration of 0.33 µM in PBS. Each sample was incubated with 100 µl of the solution for 30 min at room temperature. TRITC phalloidin dissolved in methanol was used at a final concentration of 50 µg/ml. For labeling the cell nucleus, the fixated cells were incubated with a DAPI (4′,6′-diamindino-2-phenylinodole, Sigma Aldrich) solution (100 ng/ml in PBS) for 10 min at ambient temperature. Prior to staining with TRITC phalloidin, Alexa Fluor 488 phalloidin, or DAPI, cells were fixated with 1 ml 4% paraformaldehyde (PFA) in PBS for 45 min.

For membrane identification, samples were stained with the lipophilic dye DiI-C18 (1,1-dioctadecyl-3,3,3,3-tetramethylindocarbocyanine perchlorate) purchased from Sigma Aldrich. Above the phase transition temperature of the membrane

(26°C) DiI incorporates into lipid bilayers resulting in a red fluorescence. DiI-C18 solutions in ethanol (5.0 mg/ml diluted to 5.0 µg/ml in PBS) and dimethylsulfoxide (DMSO) (2.5 mg/ml diluted to 2.5 µg/ml in PBS) were used to stain samples for 40 min. After removal of the dye, solution samples were washed with PBS (3 × 1 ml PBS per substrate).

Labeling of the Na^+/K^+-ATPase was achieved by means of indirect immunofluorescence, using antigen–antibody complexes and fluorescently labeled secondary antibodies. After washing the fixated samples with PBS, they were permeabilized with 0.2% Triton X-100 solution (10 min, room temperature). For a reduction of non-specific binding of the first antibody, binding sites were saturated with 3% BSA (bovine serum albumin) solution in PBS by incubating the samples for 20 min at room temperature. Then, a monoclonal mouse antibody (Upstate Biotechnology, Lake Placid, NY, USA) was diluted in a 3% BSA solution in PBS to 50 µg/ml. After overnight incubation at 4°C, samples were thoroughly washed (4×) and incubated with a Alexa Fluor 546 polyclonal bovine anti-mouse antibody (2 µg/ml; Chemicon Inc., Pittsburgh, PA, USA) for 90 min at 37°C. After four subsequent washing steps samples were fixated with PFA for 4–5 min at room temperature. A final washing step provided samples ready for fluorescence microscopy inspection using an Olympus BX 51 upright fluorescence microscope (Olympus, Hamburg, Germany) equipped with water immersion objectives.

Immunostaining of integrins was carried out to visualize adhesion plaques on porous and non-porous surfaces. Fixated cells were incubated with the diluted primary antibody in PBS^{--} (Integrin αV, Purified Mouse Anti-CD51, BD Biosciences, Erembodegem, Belgium; 2 µg/ml) for 60 min at ambient temperature. After rinsing three times with PBS the secondary antibody (Alexa Fluor 488 goat anti-mouse IgG, Invitrogen, Karlsruhe, Germany) diluted in PBS (2 µg/ml) was applied for 30 min.

3. Results and Discussion

The topography of regularly ordered porous substrates with a pore diameter of 1.2 µm and a porosity of 48% as used for culturing MDCK II cells is displayed in Fig. 2 [17–19]. Notably, the substrates are only partly porous (grey area in Fig. 2(a)) and, therefore, allow to investigate the impact of surface topology on cellular attachment and morphology on a joint substrate. The thickness of the porous layer is only 800 nm and the cells are in contact with the culture medium, both with their apical and basal sides.

Optical inspection of the cellular monolayer grown on the silicon wafers using fluorescence microscopy reveals that the cells are more densely packed on the porous part of the sample as compared to the adjacent nonporous, flat surface. Counting of cell nuclei (Fig. 3(a)) revealed approximately 20–30% more cells on the porous area as compared to an adjacent flat surface. A co-staining of the nuclei (DAPI) and immuno-labeling of the Na^+/K^+-ATPase at the porous to non-porous

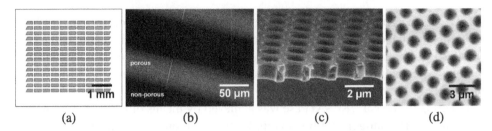

(a)   (b)   (c)   (d)

Figure 2. (a) Organization of the silicon wafer displaying porous (grey) and non-porous areas (white). (b) Optical micrograph of the silicon wafer showing both non-porous and porous surfaces. (c) SEM image of the porous silicon wafer (pore diameter 1.2 μm). (d) Atomic force microscopy image of the porous substrate.

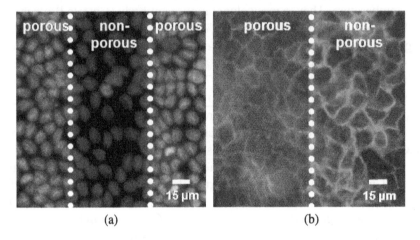

(a)   (b)

Figure 3. (a) DAPI staining of MDCK II cells cultured on a porous silicon chip displaying porous and non-porous areas. Cell density on the porous part is 20–30% higher than on the non-porous area. (b) Fluorescence micrograph showing an overlay of immuno-staining for Na^+/K^+-ATPase (cell boundaries) and DAPI staining of the nucleus. The white dotted lines indicate the boundary between the porous and non-porous part of the substrate.

border is shown in Fig. 3(b). We attribute the increased cell density to a reduced number of adhesion points on the pore rim. As a consequence, spreading of cells is kinetically hampered on porous surfaces due to the reduced surface area (52%) available for attachment leading to a higher density of cells on the surface since more and more cells arrive from the suspension covering the surface before spreading might occur. The spreading kinetics is generally related to the adhesion energy that, in turn, is a function of the available surface area, which is reduced on the porous matrix.

The height of the cell layer on the porous compared to the flat substrate was estimated by recording consecutive z-stacks with a confocal laser scanning microscope (Fig. 4). We found that cells cultured on the porous substrate were considerably higher (approx. 20%) than on the flat adjacent silicon surface (Table 1). MDCK II cells cultured on flat substrates usually display a height of (4.0 ± 0.5) μm, while

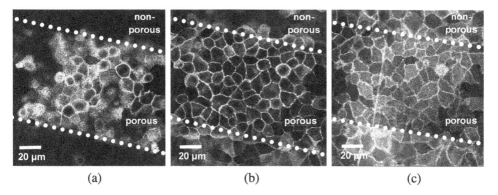

(a)   (b)   (c)

Figure 4. Consecutive confocal fluorescence images of YFP MDCK II cells on gold-coated porous silicon. As the YFP-labeled ABC-transporter is ubiquitous in the cell membrane, the bright fluorescence marks membrane domains. Images are recorded on the same sample section by changing the focal plane (z-stack) from the top of the cells at 6 μm above substrate level (a) to a focal plane 2.5 μm below (b), and to the focal plane identical with the substrate level (c).

Table 1.
MDCK II cells cultured on porous and non-porous silicon wafers. Impact of substrate topology on cell height, cell density, and Young's modulus.

	Porous substrate	Non-porous substrate
Cell height	6 ± 0.5 μm	4 ± 0.5 μm
Cell density	0.0056 ± 0.0005 cells/μm^2	0.0035 ± 0.0005 cells/μm^2
Young's modulus	0.15 ± 0.01 kPa	5.8 ± 1.6 kPa

those adhering to the porous area show a height of (6.0 ± 0.5) μm. These findings suggest that the cells' volume might indeed be conserved independent of the substrate's morphology.

Along the line, quantitative analysis of the fluorescence intensity of Na$^+$/K$^+$-ATPase immuno-staining in membranes reveals a higher intensity originating from cells grown on porous substrate compared to adjacent cells cultured on a flat substrate. We attribute this to the increased height and density of the cells on the porous matrix since the Na$^+$/K$^+$-ATPase is predominately expressed within the lateral membrane of the cell pointing towards adjacent cells in the epithelial layer. The overall amount of lateral membrane and hence the Na$^+$/K$^+$-ATPase increases accordingly. In summary, cells on porous surfaces adopt a more compact morphology due to a reduced number of adhesion sites that prevents spreading of the cells.

Most importantly, however, are our findings concerning the expression of the actin cytoskeleton as a function of substrate topography. Figure 5 illustrates that actin stress fibres are predominately visible within cells cultured on non-porous substrates and nearly disappear on the adjacent porous substrate. Evidently, either an alteration of the cell–cell contacts or cell–substrate contacts might be responsible

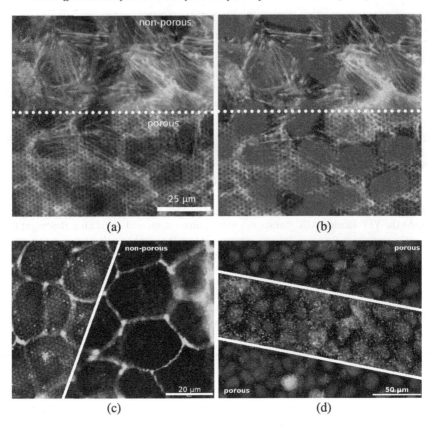

Figure 5. (a) F-actin labeling of confluent MDCK II cells with Alexa Fluor 488 phalloidin. The dotted line indicates the border between the porous and non-porous substrate. Scale bar is 25 μm. (b) DNA labeling with DAPI (dark grey cell nuclei) superimposed with (a). (c) ZO-1 staining of MDCK II cells cultured on porous silicon. (d) Immunolabeling of integrin αV (small dots) overlayed with DAPI fluorescence (cell nuclei).

for the perturbation of the actin skeleton. ZO-1 staining of confluent cell monolayers covering both porous and non-porous silicon substrates clearly reveals that cell–cell contacts are continuously established regardless of the substrate's morphology (Fig. 5(c)).

However, immunolabeling of integrins indicates that substantially less integrins (αV) are expressed on the porous surface with less clustering compared to the flat silicon substrate (Fig. 5(d)). We conclude that due to the reduced number of focal contacts the F-actin network, which is intracellularly connected to the adhesion plaques, is less developed. This has consequences for a comparison with the *in vivo* situation since the ECM displays a morphology somewhere in between a flat unstructured surface and a regularly patterned honeycomb-like porous substrate. Hence, the number of focal contacts, which in turn regulates the cytoskeleton, is controlled by substrate topology and therefore determines the mechanical properties of adherent cells to a large extent. This led us to the conclusion that alteration

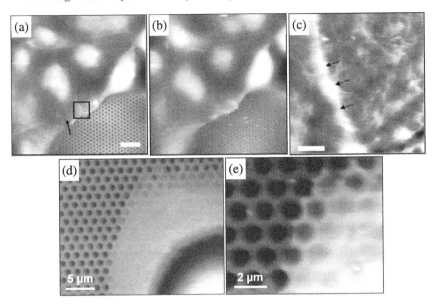

Figure 6. Atomic force microscopy images (contact mode) of a subconfluent monolayer of MDCK II cells cultured on ordered silicon pores (diameter of pores: 1.2 µm). The cells were first imaged with high force (>2 nN, scale bar is 20 µm) (a) followed by a scan with lower imaging force (<1 nN) (b). The arrow depicts a spot where the pores appear underneath the cell membrane imaged at high normal force. (c) Magnification of the area depicted by the red framed box in (a). The arrows highlight the thickened cell boundaries. Scale bar is 3 µm. (d), (e) Pseudo 3-D images showing the preferred adhesion on the pore rims and a thickening of cell protrusions on the rims.

of the cytoskeleton impacts cellular elasticity correspondingly. We expected a reduced overall Young's modulus on the porous substrate due to lack of cortical F-actin. Hence, we carried out atomic force microscopy imaging (Fig. 6) and force–indentation experiments (Fig. 7) to assess the elastic modulus of the adherent cells as a function of substrate topology. Force–indentation curves were taken on the center of the cell body. The initial 200–600 nm of indentations were subjected to nonlinear curve fitting according to the theoretical framework derived by Sneddon [22, 23]. While the Hertz model describes the elastic deformation of two elastic, nonadhesive, and homogeneous spherical surfaces in contact, the relationships derived by Sneddon also hold for different indenter geometries. Here, the load force F as a function of indentation z of a spherical indenter against a flat surface with Young's modulus E is given by:

$$F(z) = \frac{4E}{3(1 - v^2)} \sqrt{R_{\text{tip}} z^3}. \tag{1}$$

R_{tip} denotes the radius of the indenter and v the Poisson ratio. Figure 6 depicts atomic force microscopy images of cells cultured on porous substrates. The image clearly shows that cells form typical cell–cell contacts as observed on flat substrates (Fig. 6(a)–(c)) but exhibit a different spreading behavior. Figure 6(d) and (e) demon-

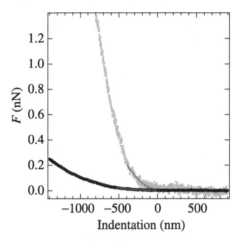

Figure 7. Force–indentation curves on confluent MDCK II cells cultured on non-porous substrates (grey) and on porous silicon wafers (black) with 1.2 μm pores and 48% porosity. The data were exclusively obtained from the cell bodies. The red lines represent the result of the nonlinear fitting according to equation (1). The average Young's modulus of cells on porous substrates is (0.15 ± 0.01) kPa, while (5.8 ± 1.6) kPa was measured on non-porous, flat surfaces.

strate that cell protrusions occur preferentially on the pore rims displaying a larger thickness as compared to the pore-spanning area. This observation and the finding that no integrins are detected on the pores suggests that adhesion solely occurs on the rim producing free-standing cell membranes across the pores and, as a consequence, there are less adhesion points than on a flat surface. In fact, cells exhibit an almost hexagonal shape on the porous substrate reproducing the hexagonal lattice of the pores. This interesting feature highlights how cells macroscopically react to changes on small length scales.

Force–indentation curves were recorded on the center of the cell body of a confluent MDCK II monolayer either cultured on the porous area or a flat substrate. Fitting the first 200–600 nm of indentation according to equation (1) results in Young's modulus of (0.15 ± 0.01) kPa for confluent MDCK II cells on porous silicon. An almost 40 times larger Young's modulus was found on non-structured flat surfaces (5.8 ± 1.6) kPa in good accord with previous data and results published by others [22, 24, 25]. We attribute this extraordinary reduction in stiffness to the lack of actin filaments as observed in the fluorescence micrographs. Considering that adhesion plaques are associated with the structure and dynamics of the actin cytoskeleton, it is not surprising that the cytoskeleton is less pronounced on porous substrates lacking 48% of attachment sites.

One might argue that the difference in cell heights between MDCK II cells cultured on the porous part and those grown on a flat sample is indirectly responsible for the stiffness difference.

It is conceivable that at larger indentation depth the mechanical response of the cell is mainly governed by the less compliant nucleus instead of the softer cell membrane and underlying cytosol. Force–indentation experiments on isolated nuclei revealed Young's modulus of about 8 kPa [26]. The argument holds particularly if the cell is probed on its center, where the nucleus is closer to the apical membrane if cultured on the non-porous substrate. As a consequence, the cells on the pores would appear softer because of their increased height and, hence, a larger distance between the apical side membrane and the nucleus. However, the indentation depth chosen for fitting the data to equation (1) (Fig. 7) is kept to only a fraction of a micrometer. At such low penetration depths the nucleus should have no impact on the measured Young's modulus. Additionally, we carried out force–distance curves on the cell's center and next to the nucleus obtaining essentially similar results at low indentation depth (Fig. 8). In fact, we found that the compliance is slightly larger if probed on the cell's center as compared to a position away from the nucleus. These findings ensure that our measurements display elastic properties rather than morphological peculiarities.

It is well known that adherent cells respond to their environment, i.e., to ECM composition and texture, ligand density, topography, and elastic properties by altering cell shape, cytoskeletal organization, and motility [1–4]. The modulation of adhesion sites is accomplished by an intricate interplay between integrins, membrane, and the cytoskeleton, mainly F-actin but also microtubuli and intermediate filaments. It has been shown that integrins belong to the primary sensors, which translate the environmental information, such as chemical composition and surface topography as well as rigidity, into a differential signal, to which the cells act in a specific way. Spatz and coworkers showed that activation of integrins and hence successful formation of focal contacts is a function of the distance between the adhesion sites [2, 4, 7]. The authors positioned RGD-functionalized gold nanoparticles in a quasi-hexagonal pattern with tunable interparticle distances. The particles

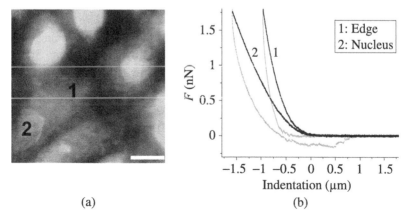

(a)   (b)

Figure 8. (a) AFM topography image of living MDCK II cells cultivated on porous chip. The cells were double labeled with DAPI and Phalloidin-Alexa488. Scale bar is 10 μm. (b) Force–indentation curves taken on positions 1 (cell nucleus) and 2 (periphery) as marked in (a).

with a size of 6 nm were small enough to allow only a single integrin to bind. Plating of fibroblasts turned out to be sensitive to the spacing of the nanoparticles. Above a maximal interparticle distance of 50–70 nm, integrin signaling is perturbed and, as a consequence, fibroblasts poorly adhere and eventually die by apoptosis. This clearly suggests that integrin clustering is essential for integrin-mediated adhesion. Since focal adhesions are basically actin–integrin links it follows that the organization of the actin cytoskeleton, such as the formation of stress fibers, is considerably hampered by perturbed integrin signaling. We attribute our findings to the fact that epithelial cells cultured on porous silicon display a higher compliance due to a poorly developed actin cytoskeleton as a response to perturbed integrin signaling. Although staining of integrins revealed their presence along the rims of the porous structure, the overall amount of integrins is substantially reduced on the porous matrix compared to the flat areas. Zinger *et al.* [28] found that cells responded to nanoscale roughness by a higher cell thickness and a delayed apparition of the focal contacts. The authors also report that condensation of actin cytoskeleton occurs on all microstructured surfaces. Interestingly, Yamamoto and coworkers [8, 9, 16] found that porcine aortic endothelial cells (PAECs) adhere better to honeycomb (HC)-patterned films from poly (ε-caprolactone) (PCL) coated on the rims with fibronectin compared to flat substrates. The authors attribute this to enhanced integrin activation on the microstructured films. The expression of focal adhesion kinase (FAK) which is autophosphorylated at the tyrosine residue (pFAK) was 3 times higher on the HC-patterned film. This implies that integrin signaling was further activated on the HC-patterned film. The authors suggest that the cellular response to HC-patterned films originates from the regularly aligned adsorption pattern of fibronectin on the regular pore structure of the film.

They are two important differences between Yamamoto's work and our study. First, their honeycomb structure exhibits a larger pore size (~6–12 μm) providing a different length scale that might allow the cells to enter the porous matrix and thus find additional contacts at the adjacent pore walls. The larger pore radii might also stimulate a distinct environmental sensing of the cells since adhesion is initiated in the perinuclear region. Second, the authors covered the pore rims and the flat reference substrate with fibronectin, which forms ring-like structures around the honeycomb openings and hence offers a very high ligand density. We did not coat our substrates with ECM since it is known that MDCK cells produce their own matrix proteins and grow nicely on gold surfaces [27]. Despite these differences, Yamamoto and coworkers also found that actin fiber staining was rather diffuse on the HC-patterned film compared to that on the flat film [16]. Fluorescence images show the absence of well-defined actin filaments. This might imply that, although integrin signaling is activated, the organization of the cytoskeleton is perturbed by the porous substrate. Hence, shape and cytoskeletal organization are strongly connected to the topological nature of the substrate. This raises the question at what spatial, temporal and chemical resolution does adhesion-mediated signaling occur and how physical features impact specific pathways. Our findings clearly indicate that a reduced area does not inhibit adhesion but considerably influences the shape, elasticity, and density of the cells. Spreading of seeded cells is kinetically slowed down leading to a smaller footprint and, under conservation of volume, a greater height. Cell–cell junctions are not affected while the elastic modulus of the cell is greatly reduced due to a less developed cortical actin cytoskeleton.

4. Conclusions

We showed that epithelial cells cultured on regularly ordered porous substrates exhibit a dramatically reduced actin cytoskeleton accompanied by a substantially reduced Young's modulus as compared to the same cells cultured on a flat substrate. The causality of these two observations is evident since most of the cell's response to mechanical probing originates from cortical actin filaments. We claim that cells on pores display many intermediate properties of cells in suspension and cells adhered to flat and hard substrates. Our findings imply that substrate dependency of cellular morphology extends from viscoelasticity of the surface to topology.

References

1. D. E. Discher, P. Janmey and Y. L. Wang, *Science* **310**, 1139 (2005).
2. B. Geiger, J. P. Spatz and A. D. Bershadsky, *Nature Rev. Mol. Cell Biol.* **10**, 21 (2009).
3. A. Curtis and M. Riehle, *Phys. Med. Biol.* **46**, R47 (2001).
4. J. P. Spatz and B. Geiger, *Methods Cell Biol.* **83**, 89 (2007).
5. V. Vogel and M. Sheetz, *Nature Rev. Mol. Cell Biol.* **7**, 265 (2006).
6. A. J. Engle, S. Sen, H. L. Sweeney and D. E. Discher, *Cell* **126**, 677 (2006).
7. M. Arnold, M. Schwieder, J. Blummel, E. A. Cavalcanti-Adam, M. Lopez-Garcia, H. Kessler, B. Geiger and J. P. Spatz, *Soft Matter* **5**, 72 (2009).
8. S. Yamamoto, M. Tanaka, H. Sunami, K. Arai, A. Takayama, S. Yamashita, Y. Morita and M. Shimomura, *Surface Sci.* **600**, 3785 (2006).
9. K. Arai, M. Tanaka, S. Yamamoto and M. Shimomura, *Colloids Surfaces A* **313–314**, 530 (2008).
10. J. Huang, S. V. Grater, F. Corbellini, S. Rinck, E. Bock, R. Kemkemer, H. Kessler, J. Ding and J. P. Spatz, *Nano Lett.* **9**, 1111 (2009).
11. S. Petris, C. Gretzer, B. Kasemo and J. Gold, *J. Biomed. Mater. Res.* **66A**, 707 (2003).
12. N. Tymchenko, J. Wallentin, S. Petronis, L. M. Bjursten, B. Kasemo and J. Gold, *Biophys. J.* **93**, 335 (2007).
13. S. J. Lee, Y. M. Lee, C. W. Han, H. B. Lee and G. Khang, *J. Appl. Polym. Sci.* **92**, 2784 (2002).
14. J. H. Lee, S. J. Lee, G. Khang and H. B. Lee, *J. Biomater. Sci. Polym. Edn* **10**, 284 (1999).
15. E. Richter, G. Fuhr, T. Mueller, S. Shirley, S. Rogaschewski, K. Reimer and C. Dell, *J. Mater. Sci. Mater. Med.* **7**, 85 (1996).
16. S. Yamamoto, M. Tanaka, H. Sunami, E. Ito, S. Yamashita, Y. Morita and M. Shimomura, *Langmuir* **23**, 8114 (2007).
17. T. Fine, I. Mey, C. Rommel, J. Wegener, C. Steinem and A. Janshoff, *Soft Matter* **5**, 3262 (2009).
18. B. Lorenz, I. Mey, S. Steltenkamp, T. Fine, C. Rommel, M. M. Mueller, A. Maiwald, J. Wegener, C. Steinem and A. Janshoff, *Small* **5**, 832 (2009).
19. I. Mey, M. Stephan, E. K. Schmitt, M. M. Mueller, M. Ben-Amar, C. Steinem and A. Janshoff, *J. Am. Chem. Soc.* **131**, 7031 (2009).

20. A. Sapper, J. Wegener and A. Janshoff, *Anal. Chem.* **78**, 5184 (2006).
21. J. Wegener, J. Seebach, A. Janshoff and H.-J. Galla, *Biophys. J.* **78**, 2821 (2000).
22. S. Steltenkamp, C. Rommel, J. Wegener and A. Janshoff, *Small* **2**, 1016 (2006).
23. I. N. Sneddon, *Int. J. Eng. Sci.* **3**, 47 (1965).
24. A. B. Mathur, A. M. Collinsworth, W. M. Reichert, W. E. Kraus and G. A. Truskey, *J. Biomech.* **34**, 1545 (2001).
25. J. H. Hoh and C.-A. Schoenenberger, *J. Cell Sci* **107**, 1105 (1994).
26. N. Caille, O. Thoumine, Y. Tardy and J.-J. Meister, *J. Biomech.* **35**, 177 (2002).
27. J. Wegener, A. Janshoff, M. Sieber and H.-J. Galla, *Eur. Biophys. J.* **28**, 26 (1998).
28. O. Zinger, K. Anselme, A. Denzer, P. Habersetzer, M. Wieland, J. Jeanfils, P. Hardouin and D. Landolt, *Biomaterials* **25**, 2695 (2004).

Protein/Material Interfaces: Investigation on Model Surfaces

Arnaud Ponche *, Lydie Ploux and Karine Anselme

Institut de Science des Matériaux de Mulhouse (IS2M), LRC 7228, Université de Haute Alsace,
Mulhouse, France

Abstract

Adhesion of mammalian cells is mediated by a protein layer adsorbed from biological fluids or the extracellular matrix. For bacteria, this 'conditioning film' also exerts an influence even if the membrane receptors are not the same. In fact, in the early stage of adhesion, whatever the surface considered, cells and bacteria are somehow in contact with proteins. These assumptions shed light on the role of protein/surface interface in the early stage of cell or bacteria adhesion. In this article, we will first review some of the techniques currently used for the analysis of such a bio-interface from a surface science point of view. Then, we will focus on the possibility to chemically model such a bio-interface. To achieve this goal and simplify the numerous interactions between a protein and cells, the biomolecule can be probed at different scales: amino acid, peptide and protein as a whole. We will examine how the surface chemistry can help to graft these fragments, what are the strategies to graft a huge molecule on a surface and what are the relevant parameters to realize a biomimetic surface, taking into account the modifications undergone by the surface during the sterilization step.

Keywords

Surface science, protein, peptide, amino-acid, grafting, cell adhesion, bacteria adhesion

List of abbreviations

APTES Aminopropyltriethoxysilane;

ATR-IR Attenuated Total Reflection/InfraRed Spectroscopy;

ATRP Atom Transfer Radical Polymerization;

BSA Bovine Serum Albumin;

DCC 1,3-Dicyclohexylcarbodiimide;

DFT Density Functional Theory;

---

* To whom correspondence should be addressed. Institut de Science des Matériaux de Mulhouse (IS2M), CNRS LRC7228, 15 rue Jean Starcky, BP 2488, 68057 Mulhouse Cedex, France. Tel.: +33 389608800; Fax: +33 389608799; e-mail: arnaud.ponche@uha.fr

Surface and Interfacial Aspects of Cell Adhesion
© Koninklijke Brill NV, Leiden, 2010

DNA DeoxyriboNucleic Acid;

EDC 1-Ethyl-3-(3-dimethylaminopropyl)carbodiimide;

ELISA Enzyme Linked Immuno-Sorbent Assay;

FT-IR Fourier Transform Infrared Spectroscopy;

HSA Human Serum Albumin;

L-DOPA L-3-(3,4-dihydroxyphenyl)alanine;

MALDI-MS Matrix-Assisted Laser Desorption/Ionisation Mass Spectrometry;

NHS N-HydroxySuccinimide;

OEG Oligo(Ethylene Glycol);

OWLS Optical Waveguide Light Spectroscopy;

PBS Phosphate Buffer Solution;

PEG Poly(Ethylene Glycol);

PLL Poly(L-Lysine);

PM-RAIRS Polarization Modulated/Reflection Absorption InfraRed Spectroscopy;

RGD Peptidic Sequence: Arginine–Glycine–Aspartic Acid;

SAMDI-MS SAMs Assisted Matrix Desorption/Ionisation Mass Spectrometry;

SELDI-MS Surface Enhanced Laser Desorption/Ionisation Mass Spectrometry;

SDS-PAGE Sodium Dodecyl Sulfate/PolyAcrylamide Gel Electrophoresis;

SPR Surface Plasmon Resonance;

STM Scanning Tunneling Microscopy;

ToF-SIMS Time of Flight Secondary Ion Mass Spectrometry;

UHV Ultra High Vacuum;

XPS X-Ray Photoelectron Spectroscopy.

1. Introduction

In 1998, B. Kasemo pointed out that biological surface science will be an important part of the future of surface science, as historic fields like semiconductors, catalysis and materials science were in the past. Ten years after, with the need for tailoring specific activities, for example in biosensors, biological surface science has become an outstanding example of multidisciplinary science, at the frontier of physics for understanding of surfaces and interactions, chemistry to assemble molecules of interest on a surface, and biology to understand cascade of chemical signals transducing in cells after adhesion [1].

Adhesion or adsorption of proteins are relevant for many biological processes and the understanding of such phenomena has become crucial, in the last years, for the development of high-throughput screening and bio-sensing devices [2]. Indeed, proteins are among the first molecules (after water) to reach the surface and adsorb on it. The entire process of mammalian cell adhesion can be divided into several major phases: cell attachment, cell spreading, actin filaments production (cytoskeletal organization) and formation of focal points. At the end of the fourth phase, the link between cell and surface is secured and cells are able to resist high stripping forces. The adhesion is mediated and regulated by membrane proteins with a main family being integrins. Integrins are heterodimers made of two extracellular chains (α and β) and two tails which cross the membrane and allow signal transducing in the intracellular area. For mammalian cells, 18 α-subunits and 8 β-subunits have been registered giving 24 combinations and so 24 different receptors [3]. An extra domain called αA or αI divides the integrin family into two groups depending on its presence or not (Fig. 1). Such complex domain is responsible for the adhesion of collagen or laminin or even cell–cell adhesion. As adhesive biomolecules, integrins present also a very high selectivity toward structure of extracellular matrix (ECM). $\alpha_1\beta_1$ integrins will bind to membrane type IV collagen, whereas $\alpha_2\beta_1$ family has higher affinity for fibrillar type I collagen [4, 5]. Up to now, subunits such as α_1 to α_6, α_v, β_1, β_3, β_5 are known to be involved in osteoblast adhesion [6]. Subunits of membrane proteins like integrins or proteoglycans will interact with a peptide sequence present in adhesive proteins adsorbed at the surface of the sample (RGD for fibronectin and vitronectin, YIGSR for laminin B1 or KRSR for heparin binding domains). Each letter of the peptidic sequence stands for one of the twenty amino acids and some sequences are summarized with their known function in Table 1. The expression of each sequence on membrane protein will further trigger cascade of biochemical events, controlling the future developments of cells [7]. As for cells, when microorganisms attach to a surface, an interface between a complex liquid medium and a solid surface is involved. At the early stage of the adhesion mechanism, membrane receptors will be sensitive to specific fragments of protein adsorbed. The substrate properties exert a strong influence on the kinetics of adsorption (polarity, hydrophilicity) and conformation will play a major role in adhesion. A conditioning film is formed within seconds of exposure in a complex medium. In liquid phase, this conditioning film has to be considered as the real surface seen by bacteria [8]. Proteins at an interface can be seen as translators, they convert an unknown surface chemistry in a layer understandable by the receptors of the cells. Understanding of this language and initiation of appropriate interactions can induce specific activities of cells [9]. The effect of the comprehension of such a chemical language can be to inhibit interactions with the D-fragment of fibrinogen known to favor inflammatory response [10, 11].

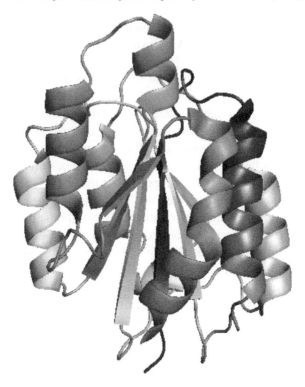

Figure 1. Conformation of adhesive domain αI of integrins showing center part with six β sheets and on each side three helical structures. (Redrawn from 1QCY.pdb files (Protein Data Bank, www.rcsb.org) with PyMol Molecular Graphics System (DeLano Scientific, San Carlos, CA USA, 2002 — www.pymol.org).)

Many strategies have been used to immobilize proteins on a solid surface. These range from adhesion by using adhesive properties of specific proteins (like adhesive proteins secreted by mussels), through development of adhesive layers known to strongly interact with proteins, to the covalent grafting using an organic divalent linker. The role of this last molecule is to initiate a covalent bond with the surface and, at the other end, to react with free chemical groups of protein such as thiols or primary amine. The last challenge to overcome for inducing a biological reaction with a biomimetic surface is to keep the reactivity of the molecule of interest after grafting or even adsorption. Many studies have shown evidence of protein denaturation upon adsorption and one can ask whether the biomolecule is still reactive when its conformation is blocked by the covalent bonds with the surface or if its mobility is reduced by the vicinity of the surface [12, 13]. In particular, structure of the hydration layer impacts the conformation of the protein in the adsorbed state and governs the activity of the protein after adsorption [14].

The purpose of this review is to summarize the major possibilities to realize model surfaces at different levels for protein interactions studies. By focusing on experimental characterization, it becomes obvious that bio-interfaces are at the fron-

Table 1.
Peptide sequences known to promote adhesion

Sequence	Amino acids	Function	References
RGD	Arginine–Glycine–Aspartic acid	Cell adhesion	[93]
YIGSR	Tyrosine–Isoleukin–Glycine–Serine–Arginine	Cell adhesion	[7]
KRSR	Lysine–Arginine–Serine–Arginine	Osteoblast adhesion	[94]
GRGDY	Glycine–Arginine–Glycine–Aspartic acid–Tyrosine	Fibroblast adhesion	[88]
GRGDSP	Glycine–Arginine–Glycine–Aspartic acid–Serine–Proline	Cell adhesion (specific interaction with fibronectin receptors)	[88]
PHSRN	Proline–Histidine–Serine–Arginine–Asparagine	Human fibrosarcoma cells adhesion (synergetic with RGD)	[48]
Cyclo-DfKRG	Cyclic–Aspartic acid–Lysine–Arginine–Glycine	Human osteoblast adhesion and differentiation	[98, 99]
FHRRIKA	Phenylalanine–Histidine–Arginine–Arginine–Isoleucine–Lysine–Alanine	Osteoblastic mineralization	[7]

tier between physical chemistry and biology. The technique able to probe quantitatively the chemistry of bio-interfaces in liquid conditions is still to be developed. For the moment, informations are gathered by a compromise between a well-controlled physical adhesion (often in ultra high vacuum conditions) and indirect informations by techniques operated in liquid phase.

This review does not intend to be exhaustive but to show the most commonly used spectroscopic techniques for biomaterials characterization as well as classical chemical routes to graft a cell-binding peptide onto a metallic surface. We will basically reduce our study to silicon and gold surfaces, the former being the most interesting for future biosensor devices and the latter a historic model surface for chemical modifications, self-assembly systems and adsorption studies.

2. Experimental Characterization of Model Surfaces

In his review on biological surface science, B. Kasemo has detailed the successive phases of adsorption when a native surface is implanted in the human body [13]. Each of the different layers relevant for cell adhesion implicates biomolecules and their effects have to be characterized in order to understand the complex phenomenon of adhesion. Biomolecules of importance can be placed on a timescale as follows: the first molecules reaching the surface are water molecules. Depending on the substrate, this water layer can create hydroxyl groups at the surface and bind strongly resulting in a hydrophilic and dense water layer. The more hydrophobic substrates will result in a heterogeneous water layer and weaker binding of water molecules. In a second stage, proteins will adsorb at the surface and the conformation will depend on the previous hydration layer (denaturation or conformational changes). Finally, cells will sense the proteinaceous layer and adhere with protein-like receptors. As we see, relevant parameters for biological adhesion are concentration, orientation and conformation of proteins at the surface.

2.1. Spectroscopic Techniques

Among all physical techniques, spectroscopic techniques such as Fourier Transform Infrared Spectroscopy (FT-IR), X-Ray Photoelectron Spectroscopy (XPS) or Time of Flight Secondary Ions Mass Spectrometry (ToF-SIMS) are the most common techniques usable to obtain chemical informations about biomolecules at a surface. In the case of proteins, many of the «biological» techniques used for characterization operate in a liquid phase or with crystalline products for X-Ray Diffraction structural determination. Due to the very low amount of protein in the vicinity or in direct contact with the surface, compared to the concentration in liquid phase, the surface signal falls in many cases below the detection limit and one has to turn towards more sensitive techniques operating in Ultra High Vacuum like XPS or ToF-SIMS. Such techniques have very low detection limit and different probe depths. XPS signal emanates from the first 8–9 nm [15] and ToF-SIMS, the most surface sensitive technique, can give information over a depth of only 1 nm depending on ion source used.

XPS has been widely used for the determination of the amount of proteins adsorbed on surfaces [16–18]. It is a surface sensitive technique where intensities of photoelectron peaks give quantitative information and binding energy is related to chemical environment. This technique is very sensitive but suffers from being operated in ultra high vacuum and is hardly sensitive to protein conformation or organization at the surface. Nevertheless, it can give information about surface coverage as detailed by Jedlicka et al. [19]. The authors decided to modify peptides based on RGD groups of fibronectin or YIGSR peptide from laminin with an alkoxysilane group. The modified peptide was then combined with a sol–gel solution and the thin film allowed to dry on a glass coverslip. XPS can give information on the surface coverage of the silane modified peptides by looking at the attenuation of the Si_{2p} signal of silica substrate. On a silica substrate, peptide amount is limited to only 10% of a monolayer (probably due to steric hindrance) and the thickness of the peptide films is in the range 10–30 Å [19]. XPS can serve as a forensic analysis technique for each step of successive chemical grafting due to its sensitivity and is particularly suitable for oligonucleotides grafted on silica surfaces [20, 21] or extracellular matrix remaining after cell detachment [22].

Techniques like Reflection Absorption Infrared Spectroscopy (RAIRS) complement very efficiently the XPS analysis in addressing molecular orientation, chemical binding and electronic structure but are more or less limited to gold surfaces [23]. Gold surfaces have been extensively studied for self-assembled monolayers, adsorption phenomena and grafting of biomolecules. A main reason can be the possibility to use thiol group to create a covalent Au–S–R bond and also the possibility to analyze it very easily using a conventional RAIRS system or Polarization Modulated-RAIRS. The gold surface is a reference for many surface science studies and Fourier Transform Infrared Spectroscopy (FT-IR) is a key technique for the surface characterization of molecules whether they are adsorbed or grafted [24]. The RAIRS characterization of gold surfaces is straightforward, but probing a monolayer on silicon is more difficult. Silicon surfaces are transparent to infrared incident light at the grazing angle giving a very low reflection signal, in contrast to a gold surface. To overcome these limitations, a multiple transmission/reflection method has been shown to give very high sensitivity for N-hydroxysuccinimide (NHS) functionalized monolayers on silicon [25]. Attenuated Total Reflection (ATR) and RAIRS are sensitive to the ordering of SAMs. The peak positions of symmetric (v_a) and asymmetric (v_{as}) CH_2 stretching modes are shifted by 5 cm^{-1} from crystalline monolayers (v_a: 2851 ± 1 cm^{-1} and v_{as}: 2818 ± 1 cm^{-1}) to disordered monolayers (v_a: 2855 ± 1 cm^{-1} and v_{as}: 2822 ± 1 cm^{-1}) [26].

Time of Flight Secondary Ions Mass Spectrometry is a surface science technique operated in Ultra High Vacuum. It analyzes the mass of secondary ions emitted from the surface under the effect of a focused primary ions gun. This technique is only quantitative with caution but for protein layer analysis, the signal recorded can give indirect information about protein conformation or orientation. It is indeed possible to follow the abundance of main ions arising from aminoacid such as $C_2NH_6^+$ ($m/z = 30$) for arginine and alanine or $C_4NH_8^+$ for proline ($m/z = 70$). These aminoacids are related to different domains of the protein molecule and the ratio of abundances of these markers can give information on the sub-domain of the fibronectin retained at the surface [27]. ToF-SIMS technique is very sensitive, the detection limit is around 10^{13} species/cm^2 (similar to XPS): for comparison, a full monolayer represents, depending on the size of the molecule, around 10^{15} species/cm^2 (i.e., ~ 1 nmole/cm^2) [28]. The coverage for peptides or proteins adsorption is often limited to a sub-monolayer. Such a low concentration limits the number of techniques usable for quantitative analysis. Statistical analysis of signals arising from amino acid composition of protein leads ToF-SIMS analysis to discriminate between mixtures of proteins adsorbed on the same surface and can lead to significant information about the competition among different biomolecules adsorbed on a metallic surface [29, 30]. The structure of the film also has an influence on the ToF-SIMS signature of the surface. If bridging occurs at the two ends of a grafted peptide, some characteristic peaks of the free peptide can disappear or vary in intensity [31]. ToF-SIMS technique has the best lateral resolution (a few tenths of a nanometer on latest generation apparatus) and thus permit to obtain chemical images that can unravel the influence of salts on adsorption of the Bovine Serum Albumin on stainless steel [32]. Spectroscopic techniques such as XPS and ToF-SIMS have the disadvantage that these operate under Ultra High Vacuum conditions, leading to a complete dehydration of protein films, which is far from being relevant for biological 'wet' conditions. In particular, the real interface or hydration layer (the first in contact with biomolecules when immersed in a complex medium) is difficult to probe. Some attempts have been made with neutron reflectivity [33] showing that a thiolated SAM with PEG end groups undergoes transition from helical structure to an amorphous state depending on solvation of PEG molecules.

Mass spectrometry (MS) and all derivative techniques are very promising for applications in biomedical field. MALDI-MS (Matrix Assisted Laser Desorption/Ionisation) and SELDI-MS (Surface Enhanced Laser Desorption/Ionisation) are directly usable on solid surfaces, even if mass spectrometry and in particular MALDI-MS has been developed for liquid samples. With increasing sensitivity and mass range, all these techniques have the main advantage to detect and identify proteins. The conventional technique used for MALDI is the so-called 'dry droplet method' where an analyte and a matrix are dissolved in a solvent and allowed to dry on a polished stainless steel plate. During evaporation of the solvent, a co-crystallization occurs, and analyte molecules remain encapsulated within matrix crystals. After laser desorption, the relative intensities of molecular ions as a function of the mass/charge ratio differentiate lysozyme (M^+: $m/z = 28\,585$), HSA (M^+: $m/z = 68\,000$), lactoferrin (M^+: $m/z = 85\,500$) and even IgG (M^+: $m/z = 152\,000$) adsorbed on a carboxymethylated dextran surface [34]. Reviews of all innovative techniques surrounding MALDI-MS have been written for biomedical applications [35] and characterization of SAMs by MALDI-MS [36, 37]. Recent developments make Laser Desorption Mass Spectroscopy a very promising technique for bio-interfaces characterization. Surface Enhanced Laser Desorption/Ionization is based on affinity technology. The principle is to characterize the analyte present in solution by retaining it on a surface derivatized with subsequent chemical probe, adding the matrix, drying and analyzing after laser desorption. The types of probes are as large as SAMs, dextran, antibody- and protein affinity probes and metal affinity or even hydrophobic interaction bio-probes [38]. When all these probes are arranged in surface arrays, it reduces considerably the time of analysis and contributes to the discovery of biomarkers of infection.

2.2. Biological Techniques

Biological characterization of materials in contact with cells or bacteria relies on fluorescence microscopy. It consists mainly in reacting a fluorescence tag with the functional group of interest. The illumination of the samples with an appropriate radiation causes the tag to fluoresce and the microscope acquires a map of localization of fluorescence tag, making the technique extremely suitable for micro-arrays [39, 40]. The main drawback of this technique arises from the limited yield of reaction between the fluorescence tag and the functional group located near the surface or very close to other groups. Fluorescence markers are usually delocalized molecules, often with aromatic rings. The steric hindrance is important and even if the yield is close to 100% in solution when the goal is to tag a protein or a peptide close to a surface, it can dramatically affect the reaction when the surface concentration is high. For this reason, a quantitative analysis of functional group of interest by fluorescence techniques is difficult to use for biological surface engineering. Fluorescence spectroscopy can follow the natural fluorescence of tryptophan residue (295 nm) and thus can be a probe for DNA hybridization, for example. The only other amino acids presenting natural fluorescence are phenylalanine and tyrosine.

A fully quantitative method used in biology is radiolabelling: a radioactive isotope is used as a tracer and integration of the nuclear signal is linked to the concentration of tagged molecules. This technique requires specific equipments and has often been used in parallel with other techniques to calibrate or verify the concentration obtained [41, 42]. More common biological characterizations of proteins involve enzyme-linked immunosorbent assay (ELISA) [43] and electrophoresis (SDS-PAGE) [44] followed by Western blot immunoassay [45]. Such techniques work in solution (as many biological techniques) but can be adapted to follow desorption of protein, or consumption of protein in the liquid phase when a surface is immersed. Unfortunately, as the amounts of proteins adsorbed at a metallic surface are fairly low compared to polymer surfaces, detachment with surfactant and direct analysis in solution fall in many cases below the detection limit.

3. Biomimetic Materials

As seen in the previous section, analyzing a biological interface is not straightforward. Due to the chemical and structural complexity of such interface, mimicking the real interface with molecules immobilized on a surface is the only way to decouple the various parameters involved in protein adhesion. The realization of such a model system is based on the different levels at which a protein can be seen: from the whole molecule to single constitutive parts of protein (amino-acid), including peptide. These model interfaces can be fabricated either by grafting or adsorption and depend strongly on the substrate used. The choice of the peptidic sequence to be grafted on a surface depends on knowledge of adhesive protein. For example, domains of fibronectin interacting with surface receptors or other proteins (heparin, fibrinogen or collagen) are summarized in Fig. 2. Eukaryotic cell binding domain (including the RGD fragment) is located at the central part of the molecule [46, 47], whereas bacteria are known to interact with NH_2 terminated heparin/fibrin binding fragment [48].

3.1. Surface Modification for Covalent Grafting

For biotechnology applications, the modified surfaces need to establish specific binding with biomolecules and reduce non-specific interactions. The establishment of a covalent bond between a metallic substrate and an organic biomolecule needs a coupling agent. This coupling agent, when dealing with bio-sensors or implants, requires to be grafted with a high density to maximize reactivity with molecules

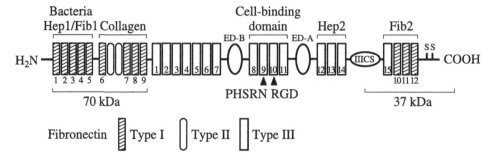

Figure 2. Structure of fibronectin showing the three repeat motifs (types I, II and III) and known binding domain. RGD and PHSRN binding sequences are located at the center of the cell binding domain.

of interest. In the case of chemical modification of a metallic surface with an organic linker, two strategies meet these requirements: self-assembled monolayers and polymer brushes [49].

3.1.1. Self-Assembled Monolayers for Biological Applications

Even if SAMs have been known for a long time, the technology employing self-assembled surfaces for biological applications has only been applied since the beginning of 1990's *via* the use of the model system involving thiols on gold. *Via* the reaction between an organic molecule and the metallic substrate, it gives a high density of alkyl groups as binding sites for further reactions [50]. Self-Assembled Monolayers (SAMs) are usually formed by immersion of the substrate into an active solution. The active molecule bears a head group which forms bond with substrate atoms (silane for silicon substrate and thiol for gold), an alkyl chain $-(CH_2)_n-$ for van der Waals interactions with neighboring chains and a surface functional group $-R$ (usually NH_2 or COOH for bio-surfaces) [26]. The molecular assembly of thiol molecules on a gold surface provides the opportunity to realize hybrid interfaces on various metallic surfaces. Each type of metallic surface has found a reacting chemical group to bind covalently with the surface. Preponderant reactive groups, found in the literature, are thiols, silanes and phosphonates [51] for various metallic surfaces. For biological applications, SAMs can be considered as a synthetic model for membrane and thus can provide information on interaction between protein and cells. It is worth noting that the property of thiols to form strong covalent bonds with a gold surface has been used to graft peptides on gold without the use of a linker. Indeed, cysteine is the only natural amino acid to bear a thiol terminated side chain. As a consequence, a peptide with cysteine residue can be directly grafted on a gold surface *via* cysteine side chain [23, 52].

The choice of the functional group $-R$ of the molecule governs further properties of the surface. For long, oligo-ethyleneglycol (OEG) terminated SAMs have proven to be protein repellent in certain conditions. Love *et al.* have expressed the following three characteristics of SAMs to be a model biological surface [53]:

(1) Limit non-specific interactions;

(2) Tune density of biomolecules or composition of ligand;

(3) React with the ligands while maintaining their biological activities.

The first characteristic can be easily achieved by tailoring the end groups of the SAMs to be oligo-ethylene glycol [54] or poly(ethylene glycol) (PEG) [55]. This aspect will be further developed in the next section about 'polymer brushes'. For oligo-ethylene glycol ended SAMs, a minimum of two units of ethylene glycol are necessary to obtain repellent property but longer size ($n = 6$) shows robust repellent properties towards fibrinogen and lysozyme, irrespective of the solvent [54]. Ethylene glycol is not the only molecule giving repellent surface. In particular, N,N-dimethyl amide functional group presents reduced affinity for protein. Alkanethiols terminated with permethylated sorbitol and acetylpiperazine were even more resistant to fibrinogen adsorption than $(EG)_n$ group [56]. A detailed, recent XPS characterization of an ethylene glycol repellent layer can be found in [57]. It shows that the layer is no more stable after 24 h incubation in PBS buffer at 37°C and loses its repellent properties. The instability is due to the hydrolysis of the siloxane bond which is only observed at a temperature of 37°C.

The second characteristic can be realized with mixed SAMs. For a gold surface, when two thiol molecules with different end groups are co-reacted from solution, the ratio of molecules immobilized onto the surface is related to the molar composition of the solution. It is then possible to tune the density of active ligands and repellent end groups to limit non-specific adsorption in the mean time. Finally, the use of mixed SAMs to realize oligonucleotide microarrays with biotinylated- and oligoethylene-alkanethiols on a gold surface has been shown to immobilize high amount of isolated streptavidin. The immobilization protocol keeps the biological activity intact and allows to realize high selectivity sensors [24].

Self-assembled monolayers can as well be modified to bear a phosphorylated analog of amino acids such as serine, threonine and tyrosine leading, after assembly, to biomimetic surfaces [58]. Nevertheless, an important point is the general tendency to idealize the structure of a so-called self-assembled monolayer. The lateral dimension of the end groups can interfere with the auto-assembly processes [53] resulting in a less ordered adlayer. For quantitative applications like biosensors, manufacturers have to ensure that the accessibility of the reacting groups is not affected by the process. The experimental conditions necessary to obtain a true monolayer are very drastic and some silanes (usually short silanes) like amino-propyltriethoxysilane (APTES or γ-APS) are known to give easily a mulitlayer structure [59]. APTES is often used as a coupling agent for biomaterials applications as it yields a high density multilayer when adsorbed from liquid phase. A better control of the structure can be obtained in vapor phase and, in this case, molecularly thick films are observed [60].

3.1.2. Polymer Brushes

Polymer brushes architecture can be used as SAMs for maximizing and controlling density of grafting. One of the main applications, for biomaterials, is the biofouling aspect of OEG or PEG. The difference between the two is related to the size of the molecule. To adsorb on a hydrophilic surface, a protein needs to remove water molecules from the immobilized hydration layer. Through conformational changes, the protein gains enough energy to overcome the energy barrier of the hydration layer. As removed water molecules turn to a liquid phase, entropy increases and this phenomenon is thermodynamically favorable. PEG macromolecules have strong interactions with water molecules, resulting in a thermodynamically stable layer. The energy barrier is too high and conformational changes of proteins are not sufficient to displace the hydration layer [61]. The biomolecule adsorption is turned unfavorable and such surfaces are called 'protein repellent' or 'protein resistant'. Coating a surface by PEG or OEG layer is then a very efficient way to remove all non-specific adhesion of proteins on a surface. Tuning the repellent properties of OEG by mixing with hydrophobic SAMs can also tune the conformation of proteins from helical to amorphous layers [62].

Experimentally, such polymer surfaces can be realized either by chemisorption on the basis of reactive end-groups of SAMs in the most conventional strategy or by physisorption involving block copolymers. Physisorption currently involves copolymers such as poly(L-lysine)-graft-poly(ethylene glycol). PLL-g-PEGs interact electrostatically (*via* the PLL positively charged block) with various negatively charged oxide layers including TiO_2, SiO_2 [63] and Nb_2O_5 [64]. On such surfaces, the resistance of HSA (Human Serum Albumin) and fibrinogen has proven to be effective over a time span of 30 h. In optimal conditions, a minimum of only 5 ng/cm^2 of HSA and fibrinogen can be attained depending on the size of the PEG and the graft ratio (typically 1–5 lysine residues for 1 PEG chain). The quantity on an uncoated oxide surface was around 600 ng/cm^2 (measured by Optical Waveguide Lightmode Spectroscopy, OWLS) [63].

Nature-inspired coatings can also meet all the requirements for immobilization of biomolecules. The composition of adhesive proteins secreted by mussels is particularly rich in catechol L-DOPA (3,4-dihydroxy-L-phenylalanine) and lysine. Researchers have anticipated the binding affinities of L-DOPA to realize PEG modified analog and generate, by simple 'dip and rinse' protocols, protein-repellent surfaces [65]. Lee *et al.* [66] realized an adherent polydopamine film by dip coating from a dopamine solution onto a wide range of organic and inorganic surfaces. Polymerization of dopamine in controlled basic conditions (Tris Buffer, pH 8.5) leads to the growth of an adhesive film on various substrates (including metals, oxides, polymers and ceramics), the structure of which is closely related to melanin. As for melanin, the exact chemical structure is not known but such a film presents the opportunity to react with nucleophiles like primary amine groups of proteins to realize a covalent immobilization [66]. A more recent article has shown that thickness of dopamin films is proportional to the time of immersion if fresh dopamine solution is regularly provided [67]. The polydopamine thin film is a good candidate for further conjugation of biomolecules (peptide *via* NH$_2$ groups, sugars *via* SH groups) or for surface engineering like photolithography or electroless deposition of metals [68]. It should be mentioned, as well, the possibility to immobilize an ATRP (Atom Transfer Radical Polymerization) precursor (usually brominated molecules) on the surface and, on the basis of the precursor, initiate an atom transfer radical polymerization. This could lead to protein-repellent polymer brushes of poly(carboxybetaine) [69] or poly(sulfobetaine) [70]. Adsorption of human fibrinogen, lysozyme and Human Chorionic Gonadotropin is reduced to lower than 0.3 ng/cm^2 (Surface Plasmon Resonance determination) with carboxybetaine. With sulfobetaine polymer brushes, the low fouling behavior is evidenced by a very low amount of fibrinogen of 0.3 ng/cm^2 and amounts of lysozyme and bovine serum albumin in the range 5–10 ng/cm^2.

Polymer brushes or SAMs are usually the initial step of a biomolecule grafting strategy on silicon or more generally metallic surfaces. As such surfaces will be in contact with cells and bacteria, it is important to estimate contamination and evaluate the effect of sterilization step on SAMs or polymer layers.

3.1.3. Chemical Modification Induced by Sterilization or Cleaning Steps

XPS and other surface-specific spectroscopic techniques can give a strong insight into carbonaceous species present at a surface. This layer is usually called organic contamination and is formed in less than a second when a high surface energy surface (metals) is in contact with the atmosphere. It is very difficult to remove all contaminants from a surface even with strong detergents, solvents or even using UV and plasma-based techniques [71]. Cleaning processes usually modify the layer by incorporating oxygenaceous species and tend to homogenize the contamination rather than really removing it [72]. The sterilization step is particularly critical when organic layer or polymer surfaces are used, leading to strong modification of the outermost surface [73, 74]. When a contaminated surface will be in contact with a complex medium like serum, the equilibrium between initially adsorbed contaminants and surface will be displaced, leading preferentially to adsorption of biomolecules. The same competition occurs with pre-adsorbed proteins like HSA or fibrinogen, when placed in contact with a solution of another protein. Displacement of equilibrium is modulated by the affinity of each protein for the surface. HSA or fibrinogen layers are easily displaced on glass whereas they are more resistant when first adsorbed on polystyrene.

3.2. Amino Acids

Amino acids are the building blocks or monomeric units of proteins. On their own, they can be energy metabolites, essential nutriments in animals, or play a biochemical role as neurotransmitters for glycine [75]. For lightest amino acids and depending on sublimation behavior, the study of adsorption in Ultra High Vacuum (UHV) leads to highly controlled deposition rate and conformation of isolated molecules.

3.2.1. Physisorption

Smallest amino acids are easily evaporated in vacuum conditions using a Knudsen cell operated at low temperature. This possibility induces a high control on the adsorption of glycine or alanine on a polycrystalline surface. Such analysis provides a further understanding of self-assembly processes and orientation of adsorbed molecules using Scanning Tunneling Microscopy (STM) and DFT calculations [76]. Working in Ultra High Vacuum provides a good control on the adsorption (flow rate, pressure and even orientation) and is an extremely clean process (absence of solvents, salts or other biomolecules). But when the evaporation of the molecule is not possible, the adsorption step must be realized in 'wet' conditions. In a way, it is a step closer to biological conditions but it also suffers from possible contamination of the surface for spectroscopic characterization and possible changes in conformation during dehydration when exposed to UHV conditions.

Among all amino acids, glycine is the smallest and has been the most studied in UHV over the last fifteen year. It has been deposited on a variety of well-controlled crystallographic surfaces like Cu(1 0 0) [77] or Pt(1 1 1) [78]. The adsorption behavior of glycine in UHV conditions is controlled by the temperature. RAIRS study shows that glycine mainly adsorbs in anionic form at room temperature on Cu(1 1 0) and the plane of carboxylate ion is oriented perpendicular to the surface. S-Alanine shows the same adsorption trends in anionic forms at room temperature but in contrast to glycine, the surface is not fully covered (only a third of surface atoms are covered, i.e., 0.33 monolayer). A high coverage of alanine can be obtained at a temperature of 420 K, with a more complex crystal phase and a saturated monolayer is present in a (3 × 2) phase at 470 K as revealed by STM [79].

Physical adsorption and desorption behaviors of cysteine or cystine on Au(1 1 1) have also been analyzed in light of the possibility of thiol groups to interact with a gold surface. In this case, voltammetry experiments can give a strong insight into the surface structure [80]. Cysteine interacts with divalent cations like Cu^{2+} and can be used to realize multilayers by electrostatic interactions [81]. Organization of the cysteine molecules on the surface can only be unraveled using synchrotron radiation for XPS measurements because of the low deposition of L-cysteine in UHV [82]. When adsorption occurs from the liquid phase, the structure of the final film can be compared to alkanethiol SAMs structure [83].

3.2.2. Chemisorption

Reaction of thiols with gold atoms can be used to realize a direct interface between cysteine (or all cysteine containing peptides) and gold surface. These L-cysteine layers present interesting electronic properties [84] and are used to modify gold electrode to induce specific complexation of Cu^{2+} ions [85]. Such chiral films, grown with amino acid chiral enantiomers can induce a discriminative crystallization of enantiomers from racemic solution [86].

3.3. Peptides

A sequence of amino acids is called a peptide and grafting a peptide (in place of the whole protein) can induce specific interaction with membrane receptors of cells and bacteria. RGD sequence, recognized by eukaryotic cell surface receptors, has been found in numerous adhesive proteins of extracellular matrices (fibronectin, vitronectin, osteopontin, collagen, fibrinogen) [87]. Biomimicking surfaces with RGD peptides is less effective in cell attachment assay than fibronectin itself. But, use of short peptides has the advantage to increase selectivity. Thus, GRGDSP short peptide becomes more specific towards fibronectin receptor [88].

3.3.1. Adsorption

As seen earlier, interfaces between metals, insulator or semiconductors and bio-molecules are of importance for further development of biosensors. A systematic study [89] on nine surfaces, three solvents and homopeptides based on the twenty amino acids has shown that adsorption was maximized for polar and charged ho-mopeptides on Si_3N_4 or SiO_2 surfaces. This adsorption behavior is strongly de-pendent on the pH of the peptide solution and roughly, maximum peptide adhesion density is obtained for pH $<$ pK_a of the free amino acid. Concentration of the pep-tide plays also a major role and adsorption is very limited when the concentration is lower than 0.01 mM [89].

In the biosensor field, semiconductor surfaces are of high interest and the adsorp-tion of biomolecules on such surfaces is a very recent field of research. Some studies have analyzed the adsorption behavior of peptides and have compared to the highly ordered SAMs obtained with alkanethiol. It is found that the amounts adsorbed are fairly low and cationic peptides interact more strongly with the surface of InP(1 0 0) [90].

We have seen that peptides can be considered as part of a protein. In order to find active adhesive site of a protein toward a surface, this protein can be digested with a trypsin into peptides fragments. The adhesion behavior of isolated fragments can then be probed toward the surface. By this method, it is shown that regions of the β-lactoglobulin rich in glutamine and aspartic acid residues are involved in the adhesion process. By varying the pH, the authors show that electrostatic interactions alone cannot explain the adhesive behavior of the peptides considered. As a matter of fact, adhesion of the peptides of interest is irreversible at pH 3.3 where carboxylic groups of glutamine and aspartic acid are non-ionized [91].

3.3.2. Top-Down Grafting

Top-down grafting is an approach where isolated biomolecules are synthesized and derivatized with suitable head groups which can further react with a surface. These can be a thiol group of a cysteine residue to bind to a gold surface or a peptide modified by a silane to bind to a silicon surface [92]. The complete grafting protocol is as follows: oxidation/cleaning of the surface, APTES immobilization, peptide coupling to amine groups of the APTES layer with carbodiimide chemistry (EDC or DCC coupling agents activate carboxylic groups of peptide). *Via* this method,

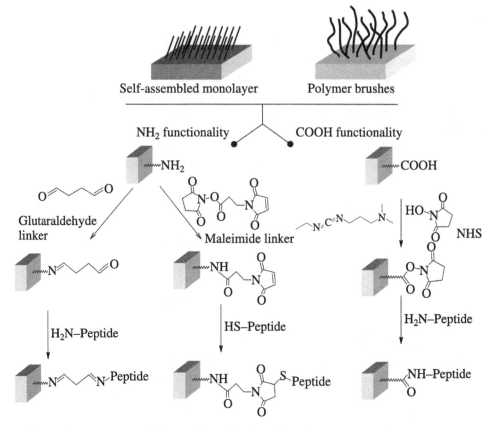

Figure 3. Principal chemical routes for peptide immobilization on SAMs and polymer brushes.

one can immobilize approximately 80 pmol of short peptide RGDS or KRSR per square centimeter on glass surfaces [93, 94]. The usual chemical routes to graft a peptide on a self-assembled layer are summarized in Fig. 3 with examples on NH$_2$- and COOH-terminated SAMs and use of linkers like glutaraldehyde and N-succinimidyl-3-maleimidylpropionate.

Even if all chemical steps are well controlled, one needs to take care of the side reactions which can take place in a complex structure. For example, if we consider the use of a bifunctional linker with a maleimide at one end and succinimidyl ester at the other (Fig. 3), the latter group reacts with primary amine of the SAMs and the maleimide group is then free to react with a thiolated peptide. This leads to covalent grafting of a peptide usually by a cysteine residue (containing a –SH group). In an ideal view and without taking side reactions into account, all succinimidyl groups will react with amine groups of the substrate and maleimide groups are oriented towards the air. But if we consider that maleimide groups can react with amine with a limited yield, this can lead to succinimidyl oriented away from the surface. Free amine of the peptide grafted by the first way reacts easily with succinimidyl groups and create a bridging of the peptide along the surface as evidenced by RAIRS [31]. This kind of immobilization of peptide along the surface will annihilate all bioactivity and block recognition sites. This emphasizes the importance of a careful control of the chemistry as well as the structure for biological applications [95]. A beautiful example of monitoring of all these reaction steps by XPS and evaluation of yield of surface reaction can be found in [96]. The results obtained by XPS and radiolabelling converge to give a value of maleimide modified APTES surface of 2.2 ± 0.5 groups/nm^2 whereas it falls to 0.2–0.4 group/nm^2 when the maleimide is reacted with peptides. Density of peptides in top-down grafting is limited by the steric hindrance of the peptide. As an example, the yield of reaction (estimated by XPS) of cysteine on maleimide surface is around 75–80% depending on the type of maleimide reactant. For larger peptide, it is constrained within the range of 15–20%.

Some specifically designed peptides can be used for supramolecular architecture. In the precursor AMBBA-Cys-Glu-Cys-Glu, the two cysteine residues are expected to attach and orient the peptide horizontally on the gold surface. Such implementation is important to obtain a specific organization and orient the glutamine so that it can give strong interaction with BSA or Chymotrypsin [97]. Up to now, most of the short peptides investigated had a linear geometry. Cyclopeptides, based on RGD motifs, have also been immobilized on hydroxyapatite and titanium surfaces. Cyclo-DfKRG tends to favor short time attachment compared to RGD linear sequence on hydroxyapatite surfaces but also induces modulated cell differentiation according to the anchor used (thiol or phosphonate) [98, 99].

3.3.3. Bottom-up Grafting or Solid Phase Synthesis

As mentioned earlier, when the peptide to be grafted has a large cross section, it limits intrachain interactions between molecules constituting the SAMs and induces a disorder. To achieve a better coverage and a good ordering, there exists another way of grafting called 'Bottom-up grafting' or *in situ* biomolecule synthesis. Usually, peptides are synthesized following the Merrifield reaction which combines each amino acid step by step by creating a peptidic bond between the carboxylic acid of the first amino acid and the amino group of the second. Protected amino acids are used to avoid side reactions. The solid support is generally made of polymer beads or hydrogels. The first step is to introduce a cleavable linker grafted on the polymer. The peptide is assembled on this linker by a succession of coupling reactions and de-protections according to the primary structure. Finally, by a cleaving step, the peptide can be released in the liquid phase. Even if recent progress in synthesis has succeeded to reach an yield of 98% for each coupling step, introduction of 20 amino acids in the structure leads to an overall yield of 0.98^{20}, i.e., 67%. The synthesis of a complete protein with thousands of amino acids is, *via* this method, hardly conceivable.

Up to now, this synthesis has been limited to solid surfaces and small peptides. The use of peptide synthesizer can, therefore, increase the productivity. The protocol maintains the same steps except that there is no need for insertion of a cleavable group in the peptide chain as we want the peptide to be firmly grafted onto the surface. A recent paper has reported the use of microwave assisted peptide synthesizer for the synthesis of a 10-unit peptide on a silicon surface [95].

3.4. Proteins

Proteins are the most complex biomolecules due to their size (up to 3000 kDa) and also due to the multiple interactions a protein can establish to adsorb onto a surface. In liquid, a protein exists in the native state but can adopt multiple conformations when adsorbed on a surface.

3.4.1. Adsorption and Denaturation

From an entropical point of view, hydrophobic interactions between a surface and a protein induce a disorganization of the water layer that is thermodynamically favorable. On the contrary, the displacement of water molecules from a hydrophilic surface needs more energy and this explains the lower amounts of proteins adsorbed on hydrophilic surfaces compared to hydrophobic surfaces. This is not to say that adsorption does not occur on a hydrophilic surface, in fact it implies other mechanisms like charge interactions or protein conformation changes providing the required amount of energy to displace water molecules [12]. Structural changes or loss of secondary structure called denaturation will result in a stronger attachment of protein and will depend on the time of residence. The energy gap between native and denaturated states has been measured by calorimetry, and it is evaluated at 10–60 kJ/mol [100], such an amount of energy corresponds to only a few hydrogen bonds. This destabilized state can then be easily reached upon adsorption. Proteins with a high internal stability, non-deformable (wide gap between the two states), and 'hard' proteins (lysozyme or ribonuclease) will be unable to gain sufficient energy to displace water layer and thus will adsorb in very low amounts on hydrophilic surfaces. For this kind of protein, higher amount adsorbed will imply electrostatic interactions with a pH dependence. On the contrary, 'soft' proteins (BSA, HSA, lactoalbumin, casein or hemoglobin) will adsorb on most surfaces by gaining energy through conformational changes. This phenomenon can be probed by studying adsorption–desorption cycle and will be considerably different between a flexible protein like BSA and egg lysozyme hard protein. As the denaturation process is driven, somehow, by the binding sites of the surface accessible to the adsorbed protein, the competition with protein in the liquid phase will limit conformational changes and thus, prevent a loss of biological activity [12]. As a specific class of proteins, enzymes tend to be more sensitive to loss of biological activity after adsorption [101]. In their review of protein adsorption onto solid surface, Nakanishi *et al.* [91] listed all the techniques applicable to measure amounts of adsorbed proteins, and the most used are the depletion method (decrease in solute concentration), Quartz Crystal Microbalance (QCM), ellipsometry, or radio-isotope labelling. These techniques are sufficiently sensitive to measure protein concentration in the range of 0.15 mg/m^2 (lactoalbumin on silica gel, depletion) to 6.3 mg/m^2 (Ferritin on gold, QCM). Detection of conformational changes upon adsorption is generally unraveled by FT-IR spectroscopy, Atomic Force Microscopy, or fluorescence of tryptophan residues [91]. Mixed SAMs are also good candidates to investigate the influence of hydrophobicity on the affinity of proteins for the surface. For example, a mixture of thiol terminated by hydroxyl or methyl groups will give a surface concentration of each thiol comparable to the solution ratio. By adjusting the ratio of the two thiols in solution, one can balance the hydrophilicity of the surface from 0% OH to 100% OH. When adsorbed from single-protein solution, fibronectin and vitronectin follow the usual trend for proteins, i.e., the more the surface is hydrophilic, the less proteins are adsorbed. When adsorbed from serum, the amount of fibronectin is lower, revealing the competition for adsorption with other proteins. On the contrary, even if vitronectin undergoes competition with other proteins from the serum, the hydrophilicity of the surface no more impacts the amount of protein and the higher amount is observed for 100% OH samples. The antigen activity evaluated on different surfaces indicates that hydrophilic surfaces preserve better cell binding activity [102]. On mono-functional SAMs, adsorption of enzyme (human glucocerebrosidase) appears to be more strongly dependent on the pH. Maximum adsorption is obtained in very acidic conditions (pH 4.5), and surface binding is maximized by electrostatic interaction. But, on the contrary, specific activity of the enzyme is enhanced near the isoelectric point of the enzyme (pH 7–8). This illustrates the influence of denaturation or conformational changes on bioactivity: proteins adsorbed at pH near the isoelectric point are less sensitive to structural rearrangements, keeping the enzyme after adsorption in a more efficient tertiary structure for bioactivity [102]. Conformation controls biological activity of the protein but, unfortunately, such information is very difficult to obtain. Some attempts were made by ToF-SIMS but only partial information could be obtained [27]. Monte Carlo simulation can help to understand denaturation or stable state of protein adsorbed [103].

3.4.2. Grafting of Proteins

Mixed SAMs have been used on gold to investigate adsorption behavior of recombinant Protein A (PrA) and its biological activity towards rabbit IgG. Mixed SAMs on gold are made by immersion in a mixture of two thiols. The surface concentration of each thiol appears to be correlated to the composition of the solution. XPS and FT-IR studies show that classical activation by NHS is more efficient when the surface carboxylic group is more accessible, i.e., when the two thiols have different sizes. Unfortunately, when the lengths of the thiols are different, phase segregation occurs. For PrA grafting, the best surface ratio was around 0.25. The authors have also shown that mixed SAMs with equal lengths lead to a much higher amount of PrA by an adsorption process and not by binding. This is confirmed by measuring binding capacity towards rabbit IgG. This affinity is low on surface where physisorption occurs even if the concentration is higher, whereas the affinity is concentration-dependent for PrA grafted on mixed SAMs [104]. We have already seen that cysteine amino acid was the only one to bear a thiol group and this property makes it a good candidate for SAMs on gold surface. The side thiol group can also be used to realize covalent grafting when it is part of the macromolecular chain of a protein. If sulfur atoms are not involved in a disulfide bond giving a superstructure to the protein, they can be exploited to bind azurin [105]. Variable fragments of antigen molecules can be grafted on gold *via* SAMs/maleimide strategy to realize biosensor for virus detection. In that case, the target is a virus, known to interact with the fusion protein grafted on the gold surface [106]. The protein is grafted on the maleimide linker *via* a cysteine tag as shown in Fig. 3.

4. Summary and Conclusions

Interfaces play a major role in most of biological reactions. When dealing with a surface reaction (like protein impact on osseointegration of implants or bacterial colonization), techniques able to analyze the interface in 'wet' conditions are scarce. For the moment, most of chemical information is obtained from spectroscopic techniques operated *ex situ* in Ultra High Vacuum. The relevance of water–surface interactions for biological adhesion was pointed out by Kasemo in 1998 [1]. We actually know theoretically the impact of water molecules or hydration layer on the conformation or ability of proteins to bind to a surface but experimental techniques able to probe a liquid–solid interface and provide physical–chemical information at an atomic scale are still to be developed.

Due to the difficulty in probing a real 'wet' interface, studies have focussed on realization of model surfaces with a simplified interface, made of peptide or amino acids. On the basis of these simplified models, one expects to understand more efficiently the behavior of proteins in the vicinity of surfaces and its influence on cell adhesion and bacterial colonization. Some studies have already shown that adsorbed proteins 'translate' the surface chemistry into interactions understandable by cell membrane proteins. Understanding of protein conformation, and interactions with cells and bacteria, will open the way to chemically engineered implants and specific cell response. Nevertheless, *in vivo* mechanisms are a step forward in complexity due to the multiple partners involved and the three-dimensional characteristics of extracellular matrices. We actually know that the rate at which fibroblasts adhere on cell-derived three-dimensional matrix is considerably higher than onto a two-dimensional substrate [107–109]. Such studies aim to control the chemistry and also the topography of the materials to be analyzed. This requires a strong expertise in surface science, chemistry, cell mechanics and biology: the need for multidisciplinary approaches in biological surface science has never been so crucial to address the requirements of bio-engineered materials with controlled biological response. Designing model peptidic surfaces with a high control over the chemistry can help to determine interactions between cells or bacteria and surfaces. For additional information about influence of chemistry and topography on cells and bacteria, readers can refer to the article of K. Anselme *et al.* or the more specific article focused on interfaces with bacteria written by L. Ploux *et al.* in this same *Journal of Adhesion Science and Technology* issue.

References

1. B. Kasemo, *Current Opinion Solid State Mater. Sci.* **3**, 451–459 (1998).
2. P. J. Walla, *Modern Biophysical Chemistry*. Wiley VCH, Weinheim (2009).
3. M. A. Arnaout, S. Goodman and J.-P. Xiong, *Current Opinion Cell Biology* **19**, 495–507 (2007).
4. M. Tulla, M. Lahti, J. S. Puranen, A.-M. Brandt, J. Käpylä, A. Domogatskaya, T. A. Salminen, K. Tryggvason, M. S. Johnson and J. Heino, *Expl. Cell Research* **314**, 1734–1743 (2008).
5. Y. Nymalm, J. S. Puranen, T. K. M. Nyholm, J. Käpylä, H. Kidron, O. T. Pentikäinen, T. T. Airenne, J. Heino, J. P. Slotte, M. S. Johnson and T. A. Salminen, *J. Biol. Chem.* **279**, 7962–7970 (2004).
6. B. Stevens, Y. Yang, A. Mohandas, B. Stucker and K. T. Nguyen, *J. Biomedical Mater. Res. B* **85**, 573–582 (2007).
7. H. Shin, S. Jo and A. G. Mikos, *Biomaterials* **24**, 4353–4364 (2003).
8. R. M. Donlan, *Emerging Infectious Diseases* **8**, 881–890 (2002).
9. G. L. Bowlin and G. Wnek, *Encyclopedia of Biomaterials and Biomedical Engineering*. Taylor & Francis (2005).
10. W.-J. Hu, J. W. Eaton and L. Tang, *Blood* **98**, 1231–1238 (2001).
11. B. G. Keselowsky, A. W. Bridges, K. L. Burns, C. C. Tate, J. E. Babensee, M. C. LaPlaca and A. J. Garcia, *Biomaterials* **28**, 3626–3631 (2007).
12. C. E. Wilson, R. E. Clegg, D. I. Leavesley and M. J. Pearcy, *Tissue Engineering* **11**, 1–17 (2005).
13. B. Kasemo, *Surface Sci.* **500**, 656–677 (2002).
14. E. A. Vogler, *Adv. Colloid Interface Sci.* **74**, 69–117 (1998).
15. A.-S. Duwez, *J. Electron Spectr. Rel. Phenom.* **134**, 97–138 (2004).
16. M. Collaud Coen, R. Lehmann, P. Groning, M. Bielmann, C. Galli and L. Schlapbach, *J. Colloid Interface Sci.* **233**, 180–189 (2001).
17. Y. F. Dufrene, T. G. Marchal and P. G. Rouxhet, *Appl. Surface Sci.* **144–145**, 638–643 (1999).
18. G. M. Harbers, L. J. Gamble, E. F. Irwin, D. G. Castner and K. E. Healy, *Langmuir* **21**, 8374–8384 (2005).
19. S. S. Jedlicka, J. L. Rickus and D. Y. Zemlyanov, *J. Phys. Chem. B* **111**, 11850–11857 (2007).
20. N. Cottenye, F. Teixeira, A. Ponche, G. Reiter, K. Anselme, W. Meier, L. Ploux and C. Vebert-Nardin, *Macromolecular Biosci.* **8**, 1161–1172 (2008).
21. A. V. Saprigin, C. W. Thomas, C. S. Dulcey, C. H. Patterson and M. S. Spector, *Surface Interface Anal.* **36**, 24–32 (2004).
22. H. E. Canavan, X. Cheng, D. J. Graham, B. D. Ratner and D. G. Castner, *Langmuir* **21**, 1949–1955 (2005).
23. R. M. Petoral and K. Uvdal, *Colloids Surfaces B* **25**, 335–346 (2002).
24. M. Riepl, K. Enander, B. Liedberg, M. Schaferling, M. Kruschina and F. Ortigao, *Langmuir* **18**, 7016–7023 (2002).
25. H-B. Liu, N. V. Venkataraman, T. E. Bauert, M. Textor and S-J. Xiao, *J. Phys. Chem. A* **112**, 12373–12377 (2008).
26. D. K. Aswal, S. Lenfant, D. Guerin, J. V. Yakhmi and D. Vuillaume, *Analytica Chimica Acta* **568**, 84–108 (2006).
27. J. B. Lhoest, E. Detrait, P. van den Bosch de Aguilar and P. Bertrand, *J. Biomed. Mater. Res.* **41**, 95–103 (1998).
28. Y. Xing, N. Dementev and E. Borguet, *Current Opinion Solid State Mater. Sci.* **11**, 86–91 (2007).
29. C. Brüning, S. Hellweg, S. Dambach, D. Lipinsky and H. F. Arlinghaus, *Surface Interface Anal.* **38**, 191–193 (2006).

30. R. E. Rawsterne, J. E. Gough, F. J. M. Rutten, N. T. Pham, W. C. K. Poon, S. L. Flitsch, B. Malt-man, M. R. Alexander and R. V. Ulijn, *Surface Interface Anal.* **38**, 1505–1511 (2006).
31. S.-J. Xiao, S. Brunner and M. Wiedland, *J. Phys. Chem. B* **108**, 16508–16517 (2004).
32. C. Poleunis, C. Rubio, C. Compere and P. Bertrand, *Appl. Surface Sci.* **203–204**, 693–697 (2003).
33. J. Fick, R. Steitz, V. Leiner, S. Tokumitsu, M. Himmelhaus and M. Grunze, *Langmuir* **20**, 3848–3853 (2004).
34. H. J. Griesser, P. Kingshott, S. L. McArthur, K. M. McLean, G. R. Kinsel and R. B. Timmons, *Biomaterials* **25**, 4861–4875 (2004).
35. B. J. Houseman and M. Mrksich, *Trends Biotechnol.* **20**, 279–281 (2002).
36. M. Mrksich, *ACS Nano* **2**, 7–18 (2008).
37. J. Su and M. Mrksich, *Langmuir* **19**, 4867–4870 (2003).
38. N. Tang, P. Tornatore and S. R. Weinberger, *Mass Spectrometry Reviews* **23**, 34–44 (2004).
39. S. Nagl, M. Schaeferling and O. S. Wolfbeis, *Microchimica Acta* **151**, 1–21 (2005).
40. P.-H. Chua, K.-G. Neoh, E.-T. Kang and W. Wang, *Biomaterials* **29**, 1412–1421 (2008).
41. C. C. Barrias, C. L. Martins, C. S. Miranda and M. A. Barbosa, *Biomaterials* **26**, 2695–2704 (2005).
42. I. C. Goncalves, M. C. L. Martins, M. A. Barbosa and B. D. Ratner, *Biomaterials* **26**, 3891–3899 (2005).
43. M.-C. Clochard, N. Betz, M. Goncalves, C. Bittencourt, J.-J. Pireaux, K. Gionnet, G. Deleris and A. Le Moel, *Nucl. Instrum. Methods Phys. Res. B* **236**, 208–215 (2005).
44. J. G. Archambault and J. L. Brash, *Colloids Surfaces B* **39**, 9–16 (2004).
45. I. Dalle-Donne, R. Rossi, D. Giustarini, A. Milzani and R. Colombo, *Clinica Chemica Acta* **329**, 23–38 (2003).
46. K. Ichihara-Tanaka, K. Tikani and K. Sekigushi, *J. Cell Sci.* **108**, 907–915 (1995).
47. W. Kang, S. Park and J.-H. Jang, *Biotechnology Letters* **30**, 55–59 (2008).
48. R.-I. Manabe, N. Oh-e, T. Maeda, T. Fukuda and K. Sekiguchi, *J. Cell Biology* **139**, 295–307 (1997).
49. W. Senaratne, L. Andruzzi and C. K. Ober, *Biomacromolecules* **6**, 2427–2448 (2005).
50. M. Mrksich and G. M. Whitesides, *Trends Biotechnol.* **13**, 228–235 (1995).
51. J. Amalric, P. H. Mutin, G. Guerrero, A. Ponche, A. Sotto and J.-P. Lavigne, *J. Mater. Chem.* **19**, 141–149 (2009).
52. E. Mateo-Marti, C. Briones, E. Roman, E. Briand, C. M. Pradier and J. A. Martin-Gago, *Langmuir* **21**, 9510–9517 (2005).
53. J. C. Love, L. A. Estroff, J. K. Kriebel, R. G. Nuzzo and G. M. Whitesides, *Chemical Reviews* **105**, 1103–1169 (2005).
54. L. Li, S. Chen, J. Zheng, B. D. Ratner and S. Jiang, *J. Phys. Chem. B* **109**, 2934–2941 (2005).
55. Z. Yang, J. A. Galloway and H. Yu, *Langmuir* **15**, 8405–8411 (1999).
56. E. Ostuni, R. G. Chapman, R. E. Holmlin, S. Takayama and G. M. Whitesides, *Langmuir* **17**, 5605–5620 (2001).
57. C. M. Dekeyser, C. C. Buron, K. Mc. Evoy, C. C. Dupont-Gillain, J. Marchand-Brynaert and A. M. Jonas, *J. Colloid Interface Sci.* **324**, 118–126 (2008).
58. A. Borgh, J. Ekeroth, R. M. Petoral Jr, K. Uvdal, P. Konradsson and B. Liedberg, *J. Colloid Interface Sci.* **295**, 41–49 (2006).
59. H. G. Hong, M. Jiang, S. G. Sligar and P. W. Bohn, *Langmuir* **10**, 153–158 (1994).
60. D. G. Kurth and T. Bein, *Langmuir* **11**, 3061–3067 (1995).
61. P. Harder, M. Grunze, R. Dahint, G. M. Whitesides and P. E. Laibinis, *J. Phys. Chem. B* **102**, 426–436 (1998).

62. C. Hoffmann and G. E. M. Tovar, *J. Colloid Interface Sci.* **295**, 427–435 (2006).

63. G. L. Kenausis, J. Voros, D. L. Elbert, N. Huang, R. Hofer, L. Ruiz-taylor, M. Textor, J. A. Hubbell and N. D. Spencer, *J. Phys. Chem. B* **104**, 3298–3309 (2000).

64. S. Pasche, S. M. De Paul, J. Voros, N. D. Spencer and M. Textor, *Langmuir* **19**, 9216–9225 (2003).

65. J.-Y. Wach, B. Malisova, S. Bonazzi, S. Tosatti, M. Textor, S. Zurcher and K. Gademann, *Chem. Eur. J.* **14**, 10579–10584 (2008).

66. H. Lee, J. Rho and P. B. Messersmith, *Adv. Mater.* **20**, 1–4 (2008).

67. F. Bernsmann, A. Ponche, C. Ringwald, J. Hemmerle, J. Raya, B. Bechinger, J.-C. Voegel, P. Schaaf and V. Ball, *J. Phys. Chem. C* **113**, 8234–8242 (2009).

68. H. Lee, S. M. Dellatore, W. M. Miller and P. B. Messersmith, *Science* **318**, 426–430 (2007).

69. Z. Zhang, S. Chen and S. Jiang, *Biomacromolecules* **7**, 3311–3315 (2006).

70. Z. Zhang, S. Chen, Y. Chung and S. Jiang, *J. Phys. Chem. B* **110**, 10799–10804 (2006).

71. F. Bretagnol, H. Rauscher, M. Hasiwa, O. Kylian, G. Ceccone, L. Hazell, A. J. Paul, O. Lefranc and F. Rossi, *Acta Biomaterialia* **4**, 1745–1751 (2008).

72. S. Caillou, P. A. Gerin, C. J. Nonckreman, S. Fleith, C. C. Dupont-Gillain, J. Landoulsi, S. M. Pancera, M. J. Genet and P. J. Rouxhet, *Electrochimica Acta* **54**, 116–122 (2008).

73. S. Fleith, A. Ponche, R. Bareille, J. Amedee and M. Nardin, *Colloids Surfaces B* **44**, 15–24 (2005).

74. L. Ploux, K. Anselme, A. Dirani, A. Ponche, O. Soppera and V. Roucoules, *Langmuir* **25**, 8161–8169 (2009).

75. D. Voet and J. G. Voet, *Biochemistry*, 3rd edn. Wiley International Edition (2004).

76. F. Höök, B. Kasemo, M. Grünze and S. Zauscher, *ACS Nano* **2**, 2428–2436 (2008).

77. S. M. Barlow, K. J. Kitching, S. Haq and N. V. Richardson, *Surface Sci.* **401**, 322–335 (1998).

78. P. Löfgren, A. Krozer, J. Lausmaa and B. Kasemo, *Surface Sci.* **370**, 277–292 (1997).

79. S. M. Barlow and R. Raval, *Surface Sci. Reports* **50**, 201–341 (2003).

80. G. Hager and A. G. Brolo, *J. Electroanal. Chem.* **550–551**, 291–301 (2003).

81. W.-W. Zhang, C.-S. Lu, Y. Zou, J.-L. Xie, X.-M. Ren, H.-Z. Zhu and Q.-J. Meng, *J. Colloid Interface Sci.* **249**, 301–306 (2002).

82. G. Gonella, S. Terreni, D. Cvetko, A. Cossaro, L. Mattera, O. Cavalleri, R. Rolandi, A. Morgante, L. Floreano and M. Canepa, *J. Phys. Chem. B* **109**, 18003–18009 (2005).

83. O. Cavalleri, L. Oliveri, A. Dacca, R. Parodi and R. Rolandi, *Appl. Surface Sci.* **175–176**, 357–362 (2001).

84. M. M. Beerbom, R. Gargaliano and R. Schlaf, *Langmuir* **21**, 3551–3558 (2005).

85. W. Yang, J. J. Gooding and D. B. Hibbert, *J. Electroanal. Chem.* **516**, 10–16 (2001).

86. D. H. Dressler and Y. Mastai, *Chirality* **19**, 358–365 (2007).

87. E. Ruoslahti and M. Pierschbacher, *Science* **238**, 491–497 (1987).

88. M. Tirrell, E. Kokkoli and M. Biesalski, *Surface Sci.* **500**, 61–83 (2002).

89. R. L. Willet, K. W. Baldwin, K. W. West and L. N. Pfeifer, *Proc. Natl Acad. Sci. USA* **22**, 7817–7822 (2005).

90. H. H. Park and A. Ivanisevic, *J. Phys. Chem. C* **111**, 3710–3718 (2007).

91. K. Nakanishi, T. Sakiyama and K. Imamura, *J. Bioscience Bioeng.* **91**, 233–244 (2001).

92. J. P. Cloarec, Y. Chevolot, E. Laurenceau, M. Phaner-Goutorbe and E. Souteyrand, *ITBM-RBM* **29**, 105–127 (2008).

93. K. C. Dee, T. T. Andersen and R. Bizios, *J. Biomed. Mater. Res.* **40**, 371–377 (1998).

94. K. C. Dee, T. T. Andersen and R. Bizios, *Biomaterials* **20**, 221–227 (1999).

95. W. K. J. Mosse, M. L. Koppens, T. R. Gengenbach, D. B. Scanlon, S. L. Gras and W. A. Ducker, *Langmuir* **25**, 1488–1494 (2009).

96. S.-J. Xiao, M. Textor and N. D. Spencer, *Langmuir* **14**, 5507–5516 (1998).

97. T. Baas, L. Gamble, K. D. Hauch, D. G. Castner and T. Sasaki, *Langmuir* **18**, 4898–4902 (2002).

98. S. Pallu, C. Bourgeta, R. Bareille, C. Labrugere, M. Dard, A. Sewing, A. Jonczyk, M. Vernizeau, M.-C. Durrieu and J. Amédée-Vilamitjan, *Biomaterials* **26**, 6932–6940 (2005).

99. M. C. Durrieu, S. Pallu, F. Guillemot, R. Bareille, J. Amedee, Ch. Baquey, C. Labrugere and M. Dard, *J. Mater. Sci. — Materials in Medicine* **15**, 779–786 (2004).

100. L. Razumovsky and S. Damodaran, *Langmuir* **15**, 1392–1399 (1999).

101. J. H. Wei, T. Kacar, C. Tamerler, M. Sarikaya and D. S. Ginger, *Small* **5**, 689–693 (2009).

102. C. C. Barrias, M. C. L. Martins, G. Almeida-Porada, M. A. Barbosa and P. L. Granja, *Biomaterials* **30**, 307–316 (2009).

103. H.-J. Hsu, S.-Y. Sheu and R.-Y. Tsay, *Colloids Surfaces B*, **67**, 183–191 (2008).

104. E. Briand, M. Salmain, C. Compere and C.-M. Pradier, *Colloids Surfaces B*, **53**, 215–224 (2006).

105. Q. Chi, J. Zhang, J. U. Nielsen, E. P. Friis, I. Chorkendorff, G. W. Canters, J. E. T. Andersen and J. Ulstrup, *J. Amer. Chem. Soc.* **122**, 4047–4055 (2000).

106. L. Torrance, A. Ziegler, H. Pittman, M. Paterson, R. Tothb and I. Eggleston, *J. Virological Methods* **134**, 164–170 (2006).

107. B. Geiger, *Science* **294**, 1601–1602 (2001).

108. E. Cukierman, R. Pankov, D. R. Stevens and K. M. Yamada, *Science* **294**, 1708–1712 (2001).

109. M. R. Dusseiller, M. L. Smith, V. Vogel and M. Textor, *Biointerphases* **1**, 1–4 (2006).

Addressable Cell Microarrays *via* Switchable Superhydrophobic Surfaces

Jau-Ye Shiu [a], **Chiung Wen Kuo** [b], **Wha-Tzong Whang** [a] **and Peilin Chen** [b,*]

[a] Department of Material Science and Engineering, National Chiao Tung University, Hsin Chu 300, Taiwan
[b] Center for Applied Sciences, Academia Sinica, 128, Section 2, Academia Road, Nankang, Taipei 115, Taiwan

Abstract
Here we describe an approach to fabricate addressable cell microarrays, which are based on the patterned switchable superhydrophobic surfaces. The switchable superhydrophobic surfaces were prepared by roughening the surfaces of fluoropolymers on the electrodes. Upon the application of 150 V to the underneath electrodes, the water contact angle on the roughened fluoropolymer surfaces could be decreased from 163° to less than 10° allowing the deposition of fibronectin, which could guide the growth of the cells. Our result indicated that it was possible to control the spatial distribution of two different cells on the cell microarrays.

Keywords
Superhydrophobic surface, protein array, cell array, nanostructure, cell adhesion

1. Introduction

In the areas of genomics and proteomics, there are increasing demands for the development of novel patterning techniques to create arrays of functional biomolecules or cells on the miniaturized devices, which could be used in various large-scale biomedical applications such as biosensing, proteomics, immunoassays or drug screening [1, 2]. Several processes have been demonstrated which are capable of patterning biomolecules with very high degree of spatial control including dip-pen lithography, inkjet printing, photolithography, nanoimprinting, etc. [3–10]. While the serial writing techniques provide individual addressability, the parallel printing processes offer an easy and fast protein patterning. However, very few of the above-mentioned techniques are capable of patterning cells. The cell microarrays, which provide the native environments for various biochemical reactions, are often used to investigate the expression of genes and the function of proteins [11]. In the past few years, many schemes have been proposed to fabricate cells microarrays

---

* To whom correspondence should be addressed. E-mail: peilin@gate.sinica.edu.tw

Surface and Interfacial Aspects of Cell Adhesion
© Koninklijke Brill NV, Leiden, 2010

[12]. One of the most popular approaches is to print biomolecules on a chip where the desired types of cells are cultured. However, in such type of cell microarray, the cells are not confined. The separation of different colonies sometime becomes problematic. Another approach is to employ micro-contact printing where the extracellular matrix (ECM) molecules such as fibronectin, vitronecin and collagens are first patterned on the surfaces [13]. Then the growth of cells on the surfaces is guided through binding to these ECM molecules. However, in these two cases, only one type of cells can be used on a chip. Here we report the use of switchable superhydrophobic surfaces to create cell microarrays where two or more types of cells can simultaneously be cultured on different areas of the same chip.

Superhydrophobic surfaces, whose water contact angles are larger than 150°, have been one of the most popular research topics for material scientists recently [14]. The studies on superhydrophobic surfaces allow investigation of the influence of surface nanostructures on the water-repellent behavior, similar to that observed in many living organs [15, 16]. The understanding of the origin of the water-repellent behavior may help in developing new industrial applications such as self-cleaning, anti-adhesion and oxidation resistant coatings [17, 18]. To prepare superhydrophobic surfaces, there are two general approaches: roughening the surfaces of hydrophobic materials, or coating the surface with layers of hydrophobic nanostructured materials [19–25]. In these processes, the surface hydrophobicity can be controlled *via* proper surface engineering. However, the surface wettability cannot be varied using these approaches once the materials are fabricated. A switchable surface is always desirable because of its great potential in many applications including fluidic manipulation, actuation and the study of cell adhesion [26–31]. In a previous publication [32], we have demonstrated a novel class of nanostructured materials, switchable superhydrophobic surfaces, for the fabrication of functional multi-component protein arrays where the electrowetting effect was employed to convert a superhydrophobic state into a complete wetted state, allowing fast but addressable protein deposition on the otherwise protein-resistant superhydrophobic surfaces. In such switchable superhydrophobic surfaces, the contact between protein solution and surface is minimized. Therefore, the protein deposition takes place only on the arrays, which are activated by applying voltage. As the protein solution stays on the top of device only for a few seconds, it is very unlikely that proteins would accidentally deposit on an area already patterned with other proteins. To pattern different types of cells on such device, we propose to prepare addressable cell microarrays by patterning the extracellular matrix (ECM) molecules, such as fibronectin, sequentially on the pre-determined areas, and then the microarray is cultured with the desired cell type. By repeating this process, two different types of cells can be cultured on to the same chip with spatial control.

2. Experimental Section

The detailed fabrication process for addressable superhydrophobic microarrays can be found in a previous publication [32]. In short, to fabricate addressable cell mi-

Figure 1. Optical image of an addressable chip containing 4 × 4 switchable superhydrophobic microarrays.

croarrays, the patterned switchable superhydrophobic surfaces were prepared on the ITO glass. A layer of 5 µm thick fluoropolymer poly [tetrafluoroethylene-co-2,2-bis(trifluoromethyl)-4,5-difluoro-1,3-dioxole] (Teflon AF, DuPont) was first coated on the ITO glass with pre-patterned electrodes, which were covered with a layer of silicon oxide (∼300 nm thick) for insulation purpose. Then a layer of photoresist (S1813, Shipley) was spun on top of the fluoropolymer and a photolithographic process was used to define the superhydrophobic area on the photoresist. The superhydrophobic microarray was fabricated using an oxygen plasma treatment (Oxford Plasmalab 80 Plus, 80 W) with O_2 as the gas (2 sccm) at a total pressure of 25 mTorr. After plasma treatment, the photoresist was removed by washing the surface with acetone. Only the areas exposed to the oxygen plasma exhibited superhydrophobic behavior, whose surface water contact angle was measured to be 163° and the surface roughness was 65 nm. The switchable superhydrophobic chip is shown in Fig. 1.

Shown in Scheme 1 is the patterning process for ECM molecules and cells. To guide the growth of the cells, ECM molecules such as fibronectins were patterned on the superhydrophobic microarray (Scheme 1(a)). A drop (∼15 µl) of fibronectin solution was pipetted onto the top of the microarray, which covered the whole superhydrophobic microarray. A platinum wire (0.1 mm in diameter) was inserted into the droplet, which served as the counter electrode. A 150 V voltage was applied to the selected ITO electrode for a few seconds to switch the surface wettability of individual superhydrophobic microarrays (Scheme 1(b)). After washing the chip with phosphate buffered saline (PBS) solution, the fibronectin patterned microarrays on the desired area could be obtained (Scheme 1(c)). The chip was then used to culture the first type of cells for a short time. Cells would attach to the area patterned with

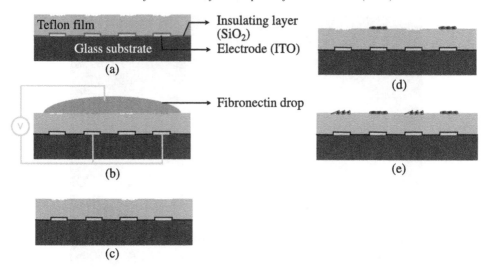

Scheme 1. (a) The switchable superhydrophobic surface is fabricated by roughening a layer of fluoropolymer on the pre-patterned ITO electrodes. (b) A drop of fibronectin solution is added to the surface and a 150 V is applied to the desired electrodes. (c) Fibronectin molecules are deposited on the array with underneath electrode activated. (d) The microarray is then used for cell culture. The cells will only attach to the area coated with fibronectin. (e) The procedure is repeated to culture the second type of cells.

fibronectin (Scheme 1(d)). The process was repeated once to culture the second type of cells on other patterned areas (Scheme 1(e)).

To create cell microarrays, a 4 × 4 switchable superhydrophobic microarray was used. The dimensions for each array were 200 μm × 200 μm. Two cell lines, NIH 3T3 and HeLa, were seeded on the patterned superhydrophobic surfaces and placed in a confocal microscope (Fluoview 1000, Olympus) equipped with an incubator (MIU-IBC-IF, Olympus) at 37°C and 5% CO_2 for 6 h. The density of the cells was about 10^5 cell/ml. Before measurement, the suspension cells were removed and the Differential Interference Contrast (DIC) or fluorescence image was taken.

3. Results and Discussion

Shown in Fig. 1 is an addressable chip containing 4 × 4 switchable superhydrophobic microarrays. In a previous experiment [32], we had demonstrated that the water contact angle on the switchable superhydrophobic surface could be decreased from 163° to less than 10° by applying 150 V to the underneath electrodes and five different proteins could be selectively deposited into individual microarrays. To produce cell microarrays with different types of cells, ECM molecules were deposited into the desired area and followed by culturing the first type of cells. After the cells were attached to the desired area, the ECM molecules could be deposited into another area and followed by culturing the second type of cells.

Before using the switchable superhydrophobic microarray for cell patterning, the chip was tested by depositing two different protein solutions. To deposit proteins on

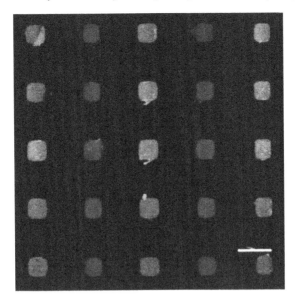

Figure 2. Fluorescence image of the patterned FITC conjugated anti-chicken IgG (green) and cy3 conjugated anti-rabbit IgG (red). Bar: 400 μm.

the switchable superhydrophobic microarray, a drop (15 μl) of phosphate buffered saline (PBS) solution containing green-fluorescent FITC conjugated anti-chicken IgG (5 μg/ml, Sigma-Aldrich) was first placed on the superhydrophobic microarray for 1 s with 150 V applied voltage, and then washed with PBS solution. A second drop of protein solution containing cy3 conjugated anti-rabbit IgG (10 μg/ml, red, Sigma-Aldrich), was then added on the chip and the procedure was repeated. The result is depicted in Fig. 2. It can be clearly seen that the areas with deposited anti-chicken IgG (green) and anti-rabbit IgG (red) were well separated and there was very little cross contamination (<2%).

Knowing that the protein could be selectively deposited on the switchable superhydrophobic microarray, the protein solution containing fibronectins (50 μg/ml) was then deposited into a 4 × 4 microarray. The chip was then placed in the cell culture dish and seeded with HeLa cells at a concentration of 10^5 cell/ml. After 6 h of incubation at 5% of CO_2 and 37°C, HeLa cells were found to attach to all the arrays patterned with fibronectins as shown in Fig. 3. Since the HeLa cells can grow even in the suspension, some HeLa cells were found to grow on the flat area (no fibronectin deposition). During the cell culture, the HeLa cells were found to migrate from the flat area to the patterned area. The HeLa cells attached to the flat area tended to aggregate. The situation was slightly different for the adherent cell line. When the fibroblast cells were seeded on the alternatively patterned microarrays, it was found that the fibroblast cells were attached exclusively on the arrays patterned with fibronectins as shown in Fig. 4. No fibroblast cell was found in the roughened region without the fibronectin deposition.

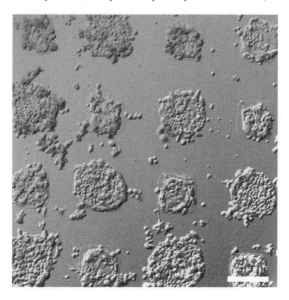

Figure 3. HeLa cells patterned on the switchable superhydrophobic microarrays. Bar: 200 μm.

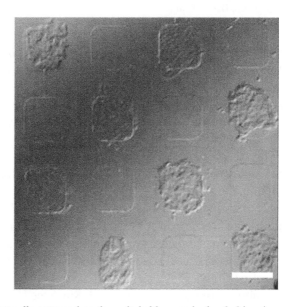

Figure 4. Fibroblast cells patterned on the switchable superhydrophobic microarrays. Bar: 200 μm.

To culture different cells on the same chip, the fibronectin solution was deposited on alternative arrays similar to those shown in Fig. 4 and then seeded with fibroblast cells. After 30 min of incubation, the fibronectin was deposited on the rest of the microarrays and the HeLa cells were added to the culture dish. To distinguish two different types of cells, the fibroblast cells were stained with a red cell tracker dye and the HeLa cells were stained with a green cell tracker dye. Shown in Fig. 5

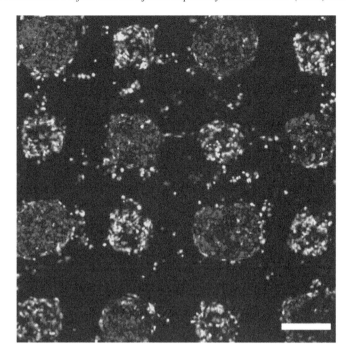

Figure 5. Fibroblast cells (red) were first patterned on switchable superhydrophobic microarray then followed by the HeLa cells (green). Bar: 200 µm.

is the fluorescence image of the cells on the switchable superhydrophobic array. It can be clearly seen that two different cells can be grown in the desired region in an addressable fashion. Therefore, we conclude that our approach can be used to co-culture two different types of cells on the same chip with spatial control. In principle, this approach can be extended to pattern more than two types of cells.

4. Conclusion

In conclusion, we have demonstrated a novel cell patterning technique using switchable superhydrophobic surfaces. It has been shown that each element on a switchable superhydrophobic microarray can be addressed individually and different types of functional biomolecules can be selectively deposited on the microarray. It has also been demonstrated that two different types of cells can be cultured on the same chip in any desired area.

Acknowledgements

This research was supported, in part, by the National Science Council, Taiwan under contract 97-2628-M-001-010-MY3 and Academia Sinica Research Project on Nano Science and Technology.

References

1. G. MacBeath and S. L. Schreiber, *Science* **289**, 1760 (2000).
2. D. S. Wilson and S. Nock, *Angew. Chem. Int. Ed.* **42**, 494 (2003).
3. L. M. Demers, D. S. Ginger, S. J. Park, Z. Li, S. W. Chung and C. A. Mirkin, *Science* **296**, 1836 (2002).
4. A. Bruckbauer, D. Zhou, L. Ying, Y. E. Korchev, C. Abell and D. Klenerman, *J. Am. Chem. Soc.* **125**, 9834 (2003).
5. T. Okamoto, T. Suzuki and N. Yamamoto, *Nature Biotechnol.* **18**, 438 (2000).
6. S. A. Brook, N. Dontha, C. B. Davis, J. K. Stuart, G. O'Neill and W. G. Kuhr, *Anal. Chem.* **72**, 3253 (2002).
7. L. M. Lee, R. L. Heimark, R. Guzman, J. C. Baygents and Y. Zohar, *Lab. Chip.* **6**, 1080 (2006).
8. J. Doh and D. J. Irvine, *J. Am. Chem. Soc.* **126**, 9170 (2004).
9. G. M. Whitesides, E. Ostuni, S. Takayama, X. Jiang and D. E. Ingber, *Annu. Rev. Biomed. Eng.* **3**, 335 (2001).
10. J. P. Renault, A. Bernard, D. Juncker, B. Michel, H. R. Bosshard and E. Delamarche, *Angew. Chem. Int. Ed.* **41**, 2320 (2002).
11. J. Ziauddin and D. M. Sabatini, *Nature* **411**, 107 (2001).
12. W. Franks, S. Tosatti, F. Heer, P. Seif, M. Textor and A. Hierlemann, *Biosens. Bioelectron.* **22**, 1426 (2007).
13. C. S. Chen, X. Jiang and G. M. Whitesides, *MRS Bulletin* **30**, 194 (2005).
14. A. Carre and K. L. Mittal (Eds), *Superhydrophobic Surfaces*. VSP/Brill, Leiden, The Netherlands (2009).
15. C. Neinhuis and W. Barthlott, *Annals Botany* **79**, 667 (1997).
16. W. Barthlott and C. Neinhuis, *Planta* **202**, 1 (1997).
17. A. Nakajima, A. Fujishima, K. Hashimoto and T. Watanabe, *Adv. Mater.* **11**, 1365 (1999).
18. N. A. Patankar, *Langmuir* **19**, 1249 (2003).
19. Z.-Z. Gu, H. Uetsuka, K. Takahashi, R. Nakajima, H. Onishi, A. Fujishima and O. Sato, *Angew. Chem. Int. Ed.* **42**, 894 (2003).
20. J. Y. Shiu, C. W. Kuo, P. Chen and C. Y. Mou, *Chem. Mater.* **16**, 561 (2004).
21. K. Tsujii, T. Onda, T. Yamamoto and S. Shibuichi, *Angew. Chem. Int. Ed.* **36**, 1011 (1997).
22. H. Y. Erbil, A. L. Demirel, Y. Avci and O. Mert, *Science* **299**, 1377 (2003).
23. J. Y. Shiu, W. T. Whang and P. Chen, *J. Adhesion Sci. Technol.* **22**, 1883 (2008).
24. H. Li, X. Wang, Y. Song, Y. Liu, Q. Li, L. Jiang and D. Zhu, *Angew. Chem. Int. Ed.* **40**, 1743 (2002).
25. L. Feng, S. Li, H. Li, J. Zhai, Y. Song, L. Jiang and D. Zhu, *Angew. Chem. Int. Ed.* **41**, 1221 (2002).
26. J. Lahann, S. Mitragotri, T.-N. Tran, H. Kaido, J. Sundaram, I. S. Choi, S. Hoffer, G. A. Somorjai and R. Langer, *Science* **299**, 371 (2003).
27. B. Gallardo, V. K. Guta, F. D. Eagerton, L. I. Jong, V. S. Craig, R. R. Shah and N. L. Abbott, *Science* **283**, 57 (1999).
28. R. A. Hayes and B. J. Feenstra, *Nature* **425**, 383 (2003).
29. K. Ichimura, S.-K. Oh and M. Nakagawa, *Science* **288**, 1624 (2000).
30. S. Minko, M. Muller, M. Motornov, M. Nitschke, K. Grundke and M. Stamm, *J. Am. Chem. Soc.* **125**, 3896 (2003).
31. C. S. Chen, M. Mrksich, S. Huang, G. M. Whitesides and D. E. Ingber, *Science* **276**, 1425 (1997).
32. J. Y. Shiu and P. Chen, *Adv. Func. Mater.* **17**, 2680 (2007).

Part 3

Surface Treatments to Control Cell Adhesion and Behavior

A Versatile Gradient of Biomolecules for Regulating Cell Behaviour

J. Racine [a], E. Luong-Van [a], Y. Sadikin [a], R. K. C. Kang [a], Y. S. Chu [b], V. Racine [b], J. P. Thiery [b] and W. R. Birch [a,*]

[a] Institute of Materials Research and Engineering, A*STAR (Agency for Science, Technology and Research), 3 Research Link, Singapore 117602
[b] Institute of Molecular and Cell Biology, A*STAR (Agency for Science, Technology and Research), 61 Biopolis Drive, Singapore 138673

Abstract

Interactions between cells and surface-immobilized gradients of biomolecules provide a tool for discerning key parameters that direct cell behaviour. The implementation of a tuneable, grafted polymer scaffold on polystyrene and poly(ethylene terephthalate) is described. This is developed by UV-ozone activation of the surface, followed by *in situ* 'grafting from' of acrylic acid (AA). Wide ranges of poly(acrylic acid) (pAA) graft densities and lengths are explored by quantifying the surface density of carboxylic acid (–COOH) groups. The reactivity of (–COOH) moieties is used to immobilize streptavidin (SAV), either *via* covalently bound biotin or by carbodiimide-mediated reaction with (–NH$_2$) moieties. Biotinylated biomolecules, immobilized on SAV, can thus be presented to cultured cells. Immobilized fluorescent SAV and biotin indicate that controlled variations in pAA surface density are translated into tuneable surface density gradients of the immobilized biomolecules. Cyclo-RGD peptide sequences immobilized on the pAA scaffold promote the attachment of cultured murine sarcoma S180 cells, giving rise to higher spreading and motility with higher RGD surface density. This proof-of-concept illustrates how engineered surfaces can provide a simple tool for presenting gradients of biotinylated molecules, which regulate and enable the study of cell-extracellular matrix interactions.

Keywords

Poly(ethylene terephthalate) (PET), polystyrene (PS), poly(acrylic acid), grafted polymer scaffold, gradient of biomolecules, biotin, streptavidin, RGD peptide, S180 murine sarcoma cells

1. Introduction

Concentration gradients of peptides are ubiquitous in early developmental biology, where they provide a key influence in regulating cell behaviour. Signals emitted from specific locations in an embryo spread across adjacent groups of cells, determining their differentiation and the location of tissues [1–4]. Recent experiments

---

* To whom correspondence should be addressed. Tel.: +65-6847 4033; e-mail: w-birch@imre.a-star.edu.sg

Surface and Interfacial Aspects of Cell Adhesion
© Koninklijke Brill NV, Leiden, 2010

have found gradients of proteins *in vivo* during the early development of the embryo [5]. However, the underlying molecular mechanisms leading to cell differentiation and tissue development still need to be elucidated. An example of note would be the gradient of fibroblast growth factor 10 (Fgf10) forming the somite of a mouse embryo [6]. With the certainty that *in vivo* gradients are a key mechanism in regulating cell fate, specific experiments applying surface gradients can be designed to probe the early stages of embryonic development.

Artificial tools have been designed to shed light on biological gradient interactions. To this end, several approaches have implemented gradients of immobilized molecules. Examples include microfluidics [7, 8] and microcontact printing [9] to study the early development of the nervous system or the UV-activated fixing of a gradient to a hydrogel scaffold to follow cell migration [10].

Lee *et al.* used an alternative approach, based on a corona discharge, to generate wettability gradients on a polymer surface [11]. This surface activation was later used to graft poly(acrylic acid) (pAA) chains, which were subsequently modified to expose simple moieties for cell culture applications [12]. A wide variety of polymer materials have since been functionalized with pAA scaffolds. These include poly(ethylene terephthalate) (PET) [13], polyethylene [14], poly(dimethyl siloxane) elastomer (PDMS) [15], and poly(tetrafluoroethylene) (PTFE) [16]. Surface activation [17] can be achieved by corona discharge, plasma treatment [13], or ozone-generating ultraviolet radiation (UV) [18]. Ozone-generating UV can be used to generate two-dimensional patterns by exposure through a photolithography mask. A plasma discharge can similarly activate three-dimensional structures, as found in tissue culture scaffolds. The subsequent AA polymerization reaction, which develops the grafted pAA scaffold, can either be thermally activated [12] or UV-induced [18].

The implementation of surface grafted pAA scaffolds in cell culture studies generally requires immobilization of biomolecules. While a surface gradient has been used to guide neurite growth [19], bare pAA is generally toxic to cultured cells, necessitating coating the scaffold with collagen, either by charge-absorption [20] or by covalent reaction using a carbodiimide link [21]. For the latter, flexibility of biomolecule immobilization was further enhanced by using biotin and streptavidin (SAV) [22]. Biotin hydrazide was covalently bound to the –COOH moieties on pAA by reaction with carbodiimide. SAV is a well-characterized tetrameric protein (60 kDa) with a high binding affinity for biotin (dissociation constant, $K_a = 2.5 \times 10^{13}$ M^{-1}) and SAV contains four biotin-binding sites [23]. Multiple binding sites allow the biotin-immobilized SAV molecule to immobilize biotinylated ligands. The biotin-SAV (B-SAV) bond, which is rapidly formed, remains stable over a wide pH and temperature range [24], and also resists drying.

In the present study, we have developed an alternative method to generate a controlled gradient of immobilized biomolecules by tuning the polymer scaffold architecture, through variations in both the length and graft density of the pAA chains. A poly(ethylene terephthalate), PET, film was used for cell culture experiments. PET was selected for its biocompatibility and its non-biodegradability [18, 19, 25, 26]. Its surface was activated with ozone-generating UV radiation and AA polymerization was induced by UV radiation beyond 310 nm. The combination of UV activation, which influences the graft density, and UV polymerization, which controls chain length, determines the architecture of the pAA scaffold.

Vermette *et al.* [24] characterized neutravidin attached to pAA polymer chains using X-ray photoelectron spectroscopy (XPS), atomic force microscopy (AFM) and enzyme-linked immunosorbent assays (ELISA). Neutravidin is a deglycosylated form of avidin, sharing the same biotin-binding properties as SAV. XPS experiments were used to characterize each step of the multilayer fabrication and they confirmed the presence of bound neutravidin. AFM measurements with a biotinylated AFM tip were used to confirm specific binding of biotin to the surface-immobilized neutravidin. The activity of immobilized neutravidin, either covalently bound to the –COOH of pAA or *via* biotin anchored to pAA, was probed by ELISA.

Although Vermette *et al.* examined the availability of immobilized molecules, thus providing information potentially pertinent to their bioactivity, their study did not probe the influence of the surface properties on cultured cells. Moreover, Vermette *et al.* made no attempt to vary the graft density or chain length of the pAA scaffold. The present study explores the relationship between the architecture of the polymer scaffold and its ability to generate gradients of biomolecules. SAV is immobilized, either by biotin covalently-bound to the pAA scaffold (henceforth abbreviated as B-SAV) or by direct covalent binding (henceforth abbreviated as DC), and it is used to immobilize biotinylated ligands. Fluorescence, measured using the same scaffold grafted onto PS, probes the surface density of immobilized SAV and biotin, demonstrating that gradients can be generated over wide-ranging UV activation and UV polymerization times. To verify the biocompatibility and bioactivity of these surfaces, immobilized cyclo-RGD peptide sequences are presented with pAA scaffolds grafted onto PET. These surfaces are suitable for cell attachment, influencing the spreading and motility of cultured Murine S180 cells. This cell line was selected for exhibiting a similar behaviour to neural crest cells, which may be considered a model system for probing mechanisms regulating the *in vitro* and *in vivo* migration of cells [27].

2. Experimental

2.1. Materials

Biaxially oriented PS and PET films, 125 and 100 μm thick, respectively, were purchased from Goodfellow (Cambridge, UK). Acrylic acid (AA) solution, acetic acid and iron(II) sulfate heptahydrate (AISH) were purchased from Merck (Singapore). Toluidine Blue O (TBO), (+) — biotin hydrazide, (+) — biotin, SAV from *Streptomyces avidinii*, N-(3-dimethylaminopropyl)-N′-ethylcarbodiimide hydrochloride (EDC), and phosphate buffer saline (PBS) were purchased from Sigma-Aldrich (Singapore). Peroxidase-conjugated IgG fraction of monoclonal mouse anti-biotin and anti-biotin conjugated to horseradish peroxidase (anti-biotin-HRP) were sourced from Jackson ImmunoResearch Laboratories (USA). Cyclo [Arg-Gly-Asp-D-Phe-Lys(Biotin)] (cyclo-RGD) was purchased from Peptides International (USA). N-hydroxysuccinimide (NHS) and 3,3',5,5'-tetramethylbenzidine (TMB) was purchased from Pierce (USA). Hydrochloric acid rinsing solution (PBS) was prepared by diluting 37% stock (HCl, reagent grade, Sigma-Aldrich, Singapore) to 20 mM with deionised water, produced by a Milli-Q Gradient A10 (Millipore, USA) and used throughout the experiments. Phosphate buffered saline (PBS) was diluted to 1X from 10X stock solution (Invitrogen, Singapore). Aqueous sodium acetate (NaOAc) was prepared from reagent grade (Sigma-Aldrich, Singapore) to 10 mM concentration. AlexaFluor® 488-conjugated SAV (Alexa-SAV) and biotin-4-fluorescein (biotin-FITC) were purchased from Molecular Probes® (Invitrogen, Singapore).

2.2. Polymer Surface Activation and Grafting of Acrylic Acid

The surface of the polymer film was activated by ozone-generating UV radiation, produced by a SEN Lights Corporation (Japan) UVL20PS-6 lamp, emitting 183 and 254 nm wavelengths [28]. To generate a step-gradient, film was vacuum-mounted on a nanopositioning stage with a step resolution of 100 nm (Bayside, USA) translated by a controller (Galil Motion Control, USA). A 6 mm-wide slit was translated across the sample, with each 6 mm-wide band receiving a different UV exposure time of 1, 2, 5, 10, 15 and 30 min. For fluorescence measurements, the polymer surface was activated through a mask with a grid pattern. Following UV activation, the surface was placed in contact with aqueous 10 wt% AA with 0.01 M AISH. AA polymerization was induced by soft UV radiation ($\lambda > 310$ nm), generated by an ultraviolet lamp (model 2000C-EC, Dymax, USA) 18 cm from the sample surface. UV polymerization times of 1, 2, 5, 7, 10 and 14 min were used to develop the pAA scaffold. The polymer film was then rinsed overnight in 45°C water to remove non-grafted pAA homopolymer and AA monomers from the surface.

2.3. Quantifying the Carboxylic Acid Groups by TBO Staining

Aqueous 0.5 mM TBO, adjusted to pH 10, was used to stain the carboxylic acid (–COOH) groups on the grafted pAA scaffold by incubating the surface for 2 h. After a brief rinse in water, TBO was desorbed in a 50% v/v aqueous acetic acid solution. The dissolved TBO was quantified by comparing optical transmission at 620 nm with known concentrations of aqueous TBO. Assuming a 1:1 interaction between TBO and –COOH, quantification of the desorbed TBO stain was used to calculate the surface density of grafted AA units.

2.4. SAV Immobilization

For B-SAV immobilization, biotin-hydrazide was covalently bound to the –COOH moieties of the pAA scaffold *via* EDC coupling. A freshly-prepared 5M solution of EDC with 2 mg/ml biotin hydrazide in PBS was placed in contact with pAA grafted to the polymer film surface for one hour at room temperature. The sample was then rinsed three times in PBS and the covalently-bound biotin exposed to 0.1 mg/ml SAV in PBS for one hour at room temperature. The surface was then rinsed in HCl and twice in PBS.

For DC immobilization, covalent binding of SAV to the –COOH groups was achieved by first incubating the pAA scaffold for 10 min in a room temperature solution of 14 mg/ml NHS and 20 mg/ml EDC in NaOAc at pH 5. After washing five times in NaOAc, the surface was incubated for one hour with a 0.1 mg/ml SAV solution in NaOAc at pH 5. The surface was then rinsed once in HCl and twice in PBS.

2.5. Fluorescence Experiments

Fluorescent labelling provided a semi-quantitative evaluation of the immobilized biotin and SAV. These experiments were carried out with PS film, due to its lower background fluorescence. The pAA scaffolds grafted onto PS and PET films were found to be essentially identical, as measured by TBO staining (data not shown). The pAA scaffold was prepared and biotinylated as described above. It was then incubated for one hour in a room temperature 0.1 mg/ml Alexa-SAV solution, followed by rinsing in HCl and PBS. After blow-drying the sample was imaged under a fluorescence microscope (model DMI 6000, from Leica, Germany) with an X10 objective (model HC PL APO 10X/0.4 PH1, from Leica, Germany) and equipped with a cooled CCD camera (model CoolSNAP HQ^2, from Photometrics, USA), which was interfaced with MetaMorph® version 7.5.3 software (Molecular Devices, USA). To probe available biotin-binding sites on the biotin-bound SAV, the surface was incubated with 0.02 mM biotin-FITC for one hour at room temperature. The sample was rinsed once in HCl and thrice in PBS before blow-drying and fluorescence measurement.

Analysis of the fluorescence intensity was performed using ImageJ software (version 1.37, rsb.info.nih.gov/ij/). The fluorescence background was determined from the non-pAA-grafted region, and was subtracted from the signal emitted by the pAA-functionalized surface. Samples were made in triplicate and three fluorescence measurements were made on each sample.

2.6. ELISA

Peroxidase-conjugated IgG fraction monoclonal mouse anti-biotin was used to detect immobilized biotin. ELISA experiments were performed in 96-well tissue culture polystyrene plates. Circular discs of PET film, 8 mm in diameter, were functionalized with a grafted pAA scaffold, generated using 10 min of UV activation and 14 min of UV polymerization. SAV was immobilized using B-SAV or DC, as described above. The discs were placed in individual wells and incubated for 1 h at room temperature with 500 μl of 0.16 μg/ml anti-biotin-HRP. They were then rinsed three times in PBS buffer and 200 μl of TMB substrate was added over 20 min. One hundred microliters of the transformed substrate solution from each well was rapidly transferred to another plate and its absorption at 405 nm was read using a microplate reader (Bio-Rad, Singapore).

2.7. Cell Culture and Time-Lapse Video Microscopy

A step-gradient on PET film was generated using five UV activation times: 0, 1, 2, 5 and 10 min, with a single UV polymerization time of 14 min. Biotinylated cyclo-RGD peptide sequences were affinity-bound to B-SAV, as described above. S180 murine sarcoma cells, prepared following the protocol by Chu *et al.* [29], were plated onto this functional surface and cultured in serum-free Dulbecco's Modified Eagle's Medium. The cell cultures were observed with an inverted microscope, equipped with an incubator (model DMI 6000 B, from Leica, Germany) and a motorized stage (model MAC 5000 XY, from Ludl Electronics Products, USA). The atmosphere in the incubator was humidified and set to 37°C with 6% CO_2. Images were taken through an X20 phase objective at five minute intervals using a cooled CCD camera (CoolSNAP HQ^2, sourced from Roper Scientific, Germany), run with MetaMorph® version 7 software (Universal Imaging Corporation, USA). The sample was initially positioned to encompass a field of view containing at least 20 cells.

2.8. Cell Motility Analysis

Cell trajectories were analysed using the Track Objects module of MetaMorph software. Following an initial manual scoring, cell position was updated by maximizing the image correlation across cell neighbourhoods, taken from two successive frames. Cell outlines were defined by analysing phase contrast images. The software's graphical interface allows a manual override of the cell position. Custom-designed scripts for Matlab software generated cell displacement and its associated speed. For cell displacement, a mean filter applied over 5 frames reduced localisation errors, which are primarily generated by intracellular displacements that remodel the image of the cell interior and lead to computational errors in image correlation. Removing these errors is critical, since they give rise to instantaneous cell displacements that mask the true cell speed.

The trajectories of individual cells are plotted for 40 consecutive frames, starting from the origin of the graph. For each RGD surface density, cell trajectories are used to calculate absolute cell speeds, which are divided into two categories: within 3 h and beyond 3 h from cell seeding. Since S180 cells require about three hours to complete their spreading and begin their migration [30], this cut-off estimates that 90% of the cells have transitioned from spreading to migrating across the surface. The average over all cell speed values for a single RGD surface density is computed within each category. Error bars reflect the standard error, defined as the standard deviation of average speed for individual cells divided by the square root of the number of cells.

3. Results and Discussion

3.1. Characterizing the pAA Scaffold

Peroxide reactive sites are induced by ozone-generating UV radiation and provide grafting sites for pAA chains. The use of a chromium mask on a quartz support allows patterning of the UV exposure and variation of the UV exposure time across the polymer film surface. Earlier XPS measurements confirmed the presence of peroxides [18], which are generated by a radical-based photo-oxidation mechanism [31]. These reactive groups have a two-day half-life [32], allowing ample time for pAA grafting or wettability measurements. Figure 1(a) reports sessile water drop contact angle on the activated polymer surface. A decreasing contact angle with UV activation time results from the formation of hydrogen-bonding and ionisable groups, whose surface density may be estimated from water contact angles measured under octane [33]. Assuming one hydrogen bond is formed with each ionisable surface group, Fig. 1(b) shows the result of this calculation: a rapidly rising surface density of activating sites, reaching saturation at approximately 0.8 sites per square nanometre. The surface-presented peroxide moieties provided initiators for the *in situ* AA polymerization (also known as 'grafting-from' polymerization), which developed the grafted pAA scaffold. Activation energy for the AA polymerization was supplied by soft UV radiation. A previous study confirmed the presence of grafted pAA by attenuated total reflection-Fourier transform infrared spectroscopy [19].

TBO was used to quantify the surface density of immobilized COOH. This process was optimized with respect to pH by using the criteria of darkest staining and maximal contrast across step-gradient bands (data not shown). Its implementation at pH 10 ensured that the pAA brush was fully charged [34].

To probe the influence of UV activation and polymerization times on the architecture of the pAA scaffold, we quantified the –COOH surface density, which

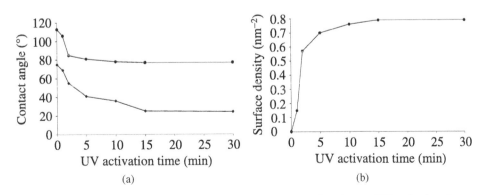

Figure 1. Wettability (contact angle) and surface density of reactive sites *vs* UV activation time on a bare PET film. (a) wettability of sessile water drops, measured in air (filled diamonds) and under octane (filled circles); (b) surface density of charged or hydrogen-bonding sites, calculated from sessile water drop contact angles measured under octane.

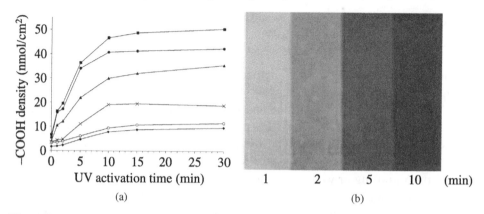

Figure 2. (a) –COOH surface density *vs* UV activation time from TBO titration; each curve corresponds to a constant polymerization time: 14 min (filled squares), 10 min (filled circles), 7 min (filled triangles), 5 min (crosses), 2 min (open circles), 1 min (filled diamonds); (b) image of TBO-stained pAA scaffold grafted onto a PET film. The bands are 6mm wide and correspond to UV activation times of 1, 2, 5 and 10 min, with 14 minutes' UV polymerization time applied to the entire sample.

corresponds to the number of AA units per unit area and represents the product of the chain graft density and chain length. Figure 2(a) plots these data as –COOH surface density *versus* UV activation time for fixed values of the UV polymerization time. The surface density saturates for UV activation > 10 min, indicating a maximum chain graft density. For constant UV activation, a quasi-linear increase in surface density with UV polymerization time is indicative of a steadily increasing chain length. The surface density range, from approximately 5 to 50 nmol/cm^2, converts to 30 to 300 AA repeat units per square nanometre.

The maximum density of surface grafted sites, 0.8 nm^{-2} from wettability measurements, compares favourably with the pAA graft densities reported by Steffens *et al.* [35] and Wu *et al.* [36], of 1 nm^{-2} and 0.85 nm^{-2}, respectively. In comparison, studies of non-polyelectrolyte polymer brushes have found lower surface-graft densities: 0.3 nm^{-2} for PS [37] and 0.7 nm^{-2} for poly(methyl methacrylate) [38]. Assuming 0.8 pAA chains per square nanometre, the surface density of 300 AA monomers per square nanometre yields an average chain length of 375 AA repeat units, which converts to a chain length of 112.5 nm (using 0.3 nm for the AA monomer length [35]). However, the distribution of grafted chain lengths remains undetermined in this study, leaving open the possibility of short pAA chains interspersed with longer mushroom-like chains.

Figure 2(b) shows an image of a step-gradient with 6 mm bands, whose width allows verification of the AA surface density by TBO staining. The presence of –COOH at zero UV activation time is attributed to non-specific adsorption of pAA chains, which are not rinsed off, as confirmed by the absence of TBO staining on a bare PET film.

These data demonstrate that by adjusting the UV activation and polymerization times one can generate a wide range of pAA scaffold architectures and –COOH surface densities.

3.2. Fluorescence of SAV Immobilized on the pAA Scaffold

Fluorescence measurements probed the translation of a pAA surface gradient into an immobilized SAV gradient. For these measurements, PET films were UV activated by exposure through a photomask, generating a grid pattern. The contrast provided by this pattern allows direct comparison between pAA-immobilized Alexa-SAV and its non-specific adsorption on the surface of the polymer film.

The images in Fig. 3 demonstrate successful immobilization of Alexa-SAV using either B-SAV or DC. HCl and PBS rinsing reduces the level of non-specific SAV adsorption, contributing to sharp pattern edges. Fluorescence intensities plotted in Fig. 4 show increasing Alexa-SAV immobilization with –COOH surface density, for both B-SAV and DC immobilization.

The density of binding sites present on the pAA scaffold ranges from 30 to 300 per square nanometre, as measured by TBO titration. This is in excess of what is necessary to bind a single layer of SAV molecules, which have dimensions of $4.2 \times 4.2 \times 5.8$ nm^3. Establishing an SAV gradient suggests that SAV binds to the extended pAA chains, penetrating below the exposed topmost region of the chains.

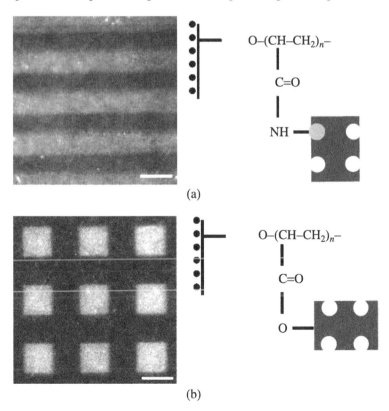

(a)

(b)

Figure 3. Fluorescence images (left) of Alexa-SAV, immobilized using (a) B-SAV and (b) DC, onto a pAA scaffold generated using UV activation and polymerization times of 30 and 10 min, respectively. The scale bar is 200 μm. Schematic diagrams on the right show B-SAV and DC immobilization.

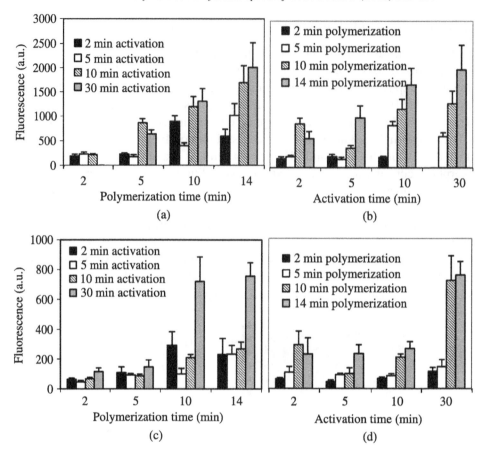

Figure 4. Fluorescence intensity of Alexa-SAV. (a) and (b) B-SAV-immobilization, (c) and (d) DC-immobilization. Plots (a) and (c) show bars of constant UV activation *vs* UV polymerization time. Plots (b) and (d) show bars of constant UV polymerization *vs* UV activation time.

Accessibility of the pAA scaffold has been implied by Ying *et al.*, who report its efficiency for immobilizing galactose ligands [39]. For SAV grafted onto the outer surface of the pAA scaffold, a single monolayer would yield a surface density of 7 pMol/nm^2 [40]. Thus, it is likely that SAV immobilized on the pAA scaffold significantly exceeds this surface density.

Figure 4(a) shows a monotonic increase in the surface density of biotin-bound SAV with UV polymerization time when using UV activation times of 10 and 30 min. For shorter UV activation times of 2 and 5 min, the SAV surface density increases non-monotonically with UV polymerization time. In Fig. 4(b), SAV surface density increases non-monotonically with UV activation for all constant UV polymerization times below 14 min. The minimum in SAV surface density for 5 min of UV activation and 10 min of UV polymerization might reflect a specific scaffold architecture, associated with a different pAA chain configuration.

DC-immobilized SAV shows a step-like increase in surface density with UV polymerization time, as shown in Fig. 4(c). This is in contrast to B-SAV immobilization data (Fig. 4(a)), demonstrating that B-SAV and DC immobilizations control how the pAA scaffold architecture translates into an SAV gradient. For UV polymerization times of 2 and 5 min, the SAV surface density remains low, irrespective of the UV activation time, suggesting a minimal pAA chain length requirement for higher SAV surface densities. This may be due to shorter chains immobilizing SAV to the outer surface of the grafted pAA brush. Figure 4(d) shows that UV activation times below 30 min, at fixed UV polymerization time, do not increase the SAV surface density, implying that DC immobilization is not suitable for generating an SAV gradient by tuning the UV activation exposure time.

3.3. A pAA Scaffold That Repels Non-specific SAV Adsorption

An interesting and unexpected data point arises from 30 min of UV activation and 2 min of UV polymerization. This corresponds to a high density of graft sites and short pAA chain lengths, leading to the presumption of a high-density short polymer brush. The surface of this pAA scaffold appears to be repellent, yielding a lower fluorescence signal than non-specific SAV adsorption on the adjacent, non-pAA-functionalized background. The 300 ± 100 a.u. signal generated by non-specific SAV adsorption lies well above the detection limit, indicating a repellent, or non-binding surface. This pAA scaffold architecture may be suitable for biomedical applications, providing an alternative to such processes as Pluronic® surfactant physisorption onto a surface with controlled wettability [16].

3.4. Using Immobilized SAV to Bind Fluorescent Biotin

The purpose of surface-immobilized SAV is to bind biotinylated ligands. A surface functionalized with a pAA scaffold enables patterning and control over the surface density of immobilized biomolecules. This was verified by immobilizing biotin-FITC with surface-presented SAV, bound onto pAA scaffolds generated by different combinations of UV activation and UV polymerization times by using DC or B-SAV. Fluorescein isothiocyanate (FITC), while being a smaller molecule than a protein or a polysaccharide, may provide a suitable model for the immobilization of a biotinylated biomolecule. The increase in SAV surface density with UV polymerization time (Fig. 4(a)) for 30 min of UV activation time is matched by a monotonic increase in bound FITC (Fig. 5(a)). This shows a successful translation of the B-SAV gradient into a biotin-FITC gradient. However, for 10 min of UV activation time, the surface density of immobilized biotin-FITC increases non-monotonically. It increases sharply when the UV polymerization time is increased from 5 to 10 min and a subsequent decrease for 30 min of UV polymerization time (Fig. 5(b)). This trend in immobilized FITC contrasts with the monotonic, quasi-linear increase in SAV surface density, seen in Fig. 4(b). Further differences between surface density gradients of SAV and immobilized biotin-FITC are evidenced from a comparison of Figs 4 and 5. These differences confirm the need to directly probe the surface

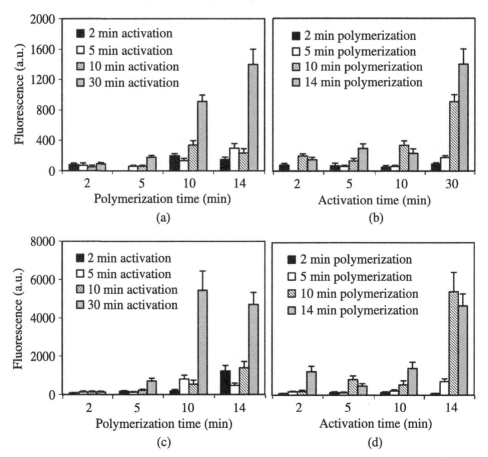

Figure 5. Fluorescence intensity of biotin-FITC, bound to SAV, immobilized using: (a) and (b) B-SAV-immobilization, (c) and (d) DC-immobilization. Plots (a) and (c) show bars of constant UV activation *vs* UV polymerization time, while plots (b) and (d) show bars of constant UV polymerization *vs* UV activation time.

density of biomolecules used in an application, without relying solely on the –COOH surface density to estimate the gradient of immobilized biomolecules. For generating a biomolecule gradient by varying UV polymerization time, these results indicate that a UV activation time of 30 min, when used with B-SAV immobilization, yields the most faithful translation of pAA surface density into a gradient of immobilized biotinylated ligands. Conversely, a constant UV polymerization time of 10 min used with B-SAV immobilization is recommended for translating of the pAA gradient into a biotin-FITC gradient by varying the UV activation time.

Diverse biological applications may require a variety of surface density gradients. Figure 5 indicates that the pAA scaffold architecture can present a wide range of bio-molecular surface densities. The range of surface densities may be inferred from Fig. 5(b), where constant UV activation times of 5 and 30 min generate a four-fold or a fifteen-fold increase, respectively, in the fluorescence of surface-

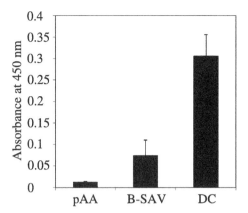

Figure 6. Colorimetric detection of HRP activity. Biotin-HRP was bound to a pAA scaffold, pre-pared using UV activation and polymerization times of 10 and 14 min, respectively. pAA refers to non-specifically adsorbed biotin-HRP. B-SAV and DC refer to biotin-HRP bound to SAV, which was coupled to the pAA scaffold using B-SAV and DC, respectively.

immobilized FITC with increasing polymerization time. This fluorescence range may be considered as representing the surface density immobilized biotinylated biomolecules.

3.5. Activity of Biotin-HRP Immobilized via B-SAV- or DC-Immobilized-SAV

Surface-immobilized SAV allows the co-immobilization of biotinylated peptides. This has been shown to be useful in biological applications [41]. Given the observed differences between translating B-SAV- and DC-immobilized SAV gradients into an FITC surface density (Section 3.4), the availability of SAV-immobilized bio-molecules may also be influenced by B-SAV or DC immobilization. The ELISA assay measures the oxidative activity of the SAV-immobilized biotin-HRP enzyme by colorimetric change of the TMB substrate solution. This model system may be used to probe the availability of immobilized biomolecules. Figure 6 shows a higher activity for DC-immobilized SAV, as compared to B-SAV immobilization. The ratio of HRP activities is comparable to the biotin-FITC fluorescence ratio on the same pAA scaffold (generated by 10 min of UV activation and 14 min of UV polymerization, Fig. 5(b) and 5(d)). The enhanced immobilization of biotiny-lated ligands is attributed to a higher number of unoccupied biotin-binding sites on DC-immobilized SAV. The scaffold-bound biotin present in B-SAV immobilization occupies biotin-binding sites on the SAV, thus reducing the number of available binding pockets for immobilizing biotinylated ligands. This result is in agreement with Vermette *et al.* [24].

3.6. Influence of Cyclo-RGD Surface Density on Cell Morphology and Motility

The culture of mouse sarcoma S180 cells was used to assess the biocompatibil-ity and bioactivity of the pAA scaffold presenting cyclo-RGD, bound to B-SAV-immobilized SAV. The influence of different RGD surface densities was probed with three cell behaviour parameters, which are regulated by the cell's ability to attach to the cell culture surface: cell morphology, cell trajectory, and quantifying cell speed. Images of adsorbed cells taken 30 min after the start of the experiment (Fig. 7(a)) show a higher surface density of cells with increasing cyclo-RGD surface density. These cells also exhibit more flattened and spread morphologies with increasing RGD surface density. Images of the cells taken after 16 h confirm this trend. Figure 7(b) shows longer cell trajectories on the RGD-functionalized surface than on bare PET. This increase in cell motility is reflected in the cell speed data (Fig. 7(c)). Cells in contact with the surface for less than three hours show higher speeds than on bare PET, with the exception of the pAA scaffold generated with 2 min of UV activation. For times longer than three hours, cell speeds remain lower on bare PET *versus* the functionalized surface. The reduced difference in cell speeds is attributed to the cells migrating on a previously deposited extra-cellular matrix, whose structure and composition reflect prior cellular interactions with the underlying surface.

This cell culture assay confirms that the pAA scaffold functionalized with biotin-bound SAV and cyclo-RGD is non-toxic and shows enhanced cell adhesion over a bare PET surface. Moreover, the pAA scaffold allows the SAV-immobilized cyclo-RGD to retain its biological activity. The step-gradient used in this experiment allowed cells to be cultured in contact with different surface properties on a continuous polymer film.

4. Summary and Conclusions

The tuneable immobilization of biomolecules on polymeric materials is demonstrated by implementing grafted pAA scaffold, grafted onto PS and PET substrates. Control over the graft density and length of pAA chains was verified by titration of the –COOH surface density and the translation of pAA gradients into surface gradients of SAV and immobilized biotinylated ligands was explored by fluorescence measurements. This immobilization may be implemented on a variety of polymeric materials, ranging from PDMS to PTFE, both on planar surfaces as well as three-dimensional structures. An advantage of this technique is that photolithography patterning allows the variation of surface properties from length scales comparable to cell dimensions to macroscopic length scales. Fluorescence gradients indicate the importance of quantifying the surface density of surface-presented biomole-

---

Figure 7. Spreading and motility of S180 cells, cultured on bare PET and biotinylated RGD, imobilized *via* SAV, bound to the pAA scaffold using B-SAV. The scaffold was prepared using UV activation times of 1, 2, 5 and 10, with 14 min of UV polymerization time. Numbers above (a) and (b) refer to the UV activation time. Zero refers to cells cultured on a bare PET film. (a) images of seeded cells, showing their morphology after 30 min (top row) and 16 h (bottom row). Scale bar is 5 μm. (b) plot of 40 cell trajectories for each cell culture surface; (c) calculated cell speeds, for times less and more than 3 h after starting the experiment.

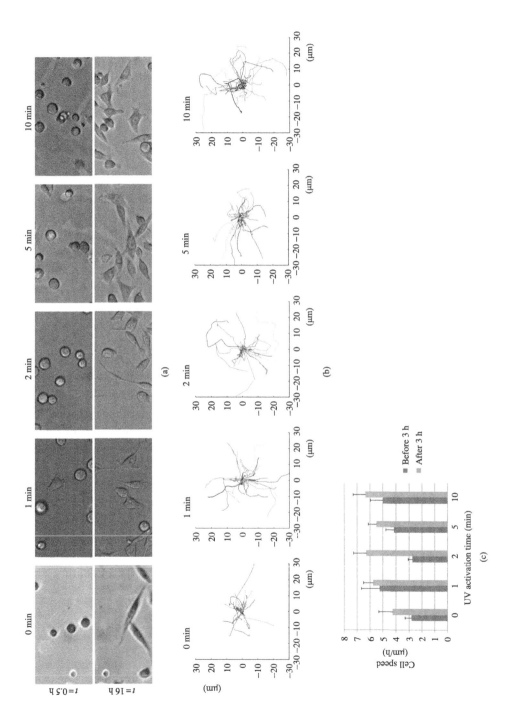

cules, while using the AA surface density titration to verify reproducibility of the polymer scaffold.

Biotin and SAV were selected for their widespread application and the flexibility they offer to immobilize (biotinylated) biomolecules. Two routes for SAV immobilization gave differing results, with B-SAV generating a more monotonic increase in the surface density, while DC favours availability of the immobilized biomolecules. The latter is in agreement with Vermette *et al.* [24], who probed B-SAV and DC immobilizations without tuning the pAA scaffold architecture. Low non-specific surface adsorption was found for a pAA scaffold generated by 30 min of UV activation and 2 min of UV polymerization time. This indicates a non-binding surface, which was not expected for a surface-grafted polyelectrolyte.

Observing the spreading and quantifying the motility of S180 murine sarcoma cells tested a PET film with SAV bound to pAA scaffolds. The cells were cultured on different surface densities of cyclo-RGD peptide sequences, which provide cell attachment points. The RGD-functionalized surface increased the surface density of attached cells, their initial spreading and their motility during migration. This enhancement with respect to bare PET confirmed the biocompatibility and bioactivity of the pAA-functionalized cell culture surface.

Presenting biomolecules on a tuneable cell culture surface facilitates studies of cell adhesion. The use of surface-bound SAV allows the binding of one or more biotinylated biomolecules. Immobilising surface gradients of morphogens with an engineered scaffold is expected to reveal their role in cell signalling. An example is surface-presented activin. This would determine how cells react to their position in a concentration gradient, thus exploring whether cells respond individually to a local activin concentration or if they need to interact with their neighbours to generate a concentration-related response [42]. The formation of tissues in a developing embryo relies on cell differentiation in response to protein gradients, which are mediated by extra-cellular glycans. This regulation of cell evolution may be viewed as a model for tissue engineering from pluripotent cell lines, such as embryonic stem cells, induced pluripotent stem cells, or adult mesenchymal stem cells. The secretion of Sonic hedgehog (Shh) morphogen by a notochord induces distinct ventral cell identities in the adjacent neural tube *via* a concentration-dependent response [43]. Thus, a gradient of surface-presented Shh may determine its role in this phenomenon [44].

Acknowledgements

The authors thank Professors D. Fernig and M. Textor for fruitful discussions and Ms. V. Tjong for contributing her knowledge of surface fabrication. The authors acknowledge funding support from the Agency for Science, Technology and Research (A*STAR), Singapore.

References

1. M. Kerszberg and L. Wolpert, *Cell* **130**, 205 (2007).
2. J. B. A. Green, H. V. New and J. C. Smith, *Cell* **71**, 731 (1992).
3. J. B. Gurdon, A. Mitchell and D. Mahony, *Nature* **376**, 520 (1995).
4. C. Tickle, *Seminars in Cell and Developmental Biology* **10**, 345 (1999).
5. A. W. Powell, T. Sassa, Y. Wu, M. Tessier-Lavigne and F. Polleux, *PLoS Biology* **6**, 1047 (2008).
6. J. M. Veltmaat, F. Relaix, L. T. Le, K. Kratochwil, F. G. Sala, W. van Veelen, R. Rice, B. Spencer-Dene, A. A. Mailleux, D. P. Rice, J. P. Thiery and S. Bellusci, *Development* **133**, 2325 (2006).
7. C. J. Wang, X. Li, B. Lin, S. Shim, G. L. Ming and A. Levchenko, *Lab on a Chip* **8**, 227 (2008).
8. S. Lang, A. C. von Philipsborn, A. Bernard, F. Bonhoeffer and M. Bastmeyer, *Anal. Bioanal. Chem.* **390**, 809 (2008).
9. A. C. von Philipsborn, S. Lang, J. Loeschinger, A. Bernard, A. David, D. Lehnert, F. Bonhoeffer and M. Bastmeyer, *Development* **133**, 2487 (2006).
10. M. C. Delfini, J. Dubrulle, P. Malapert, J. Chal and O. Pourquié, *Proc. Natl. Acad. Sci. USA* **102**, 11343 (2005).
11. J. H. Lee, H. G. Kim, G. S. Khang, H. B. Lee and M. S. Jhon, *J. Colloid Interface Sci.* **151**, 563 (1992).
12. J. H. Lee, H. W. Jung, I. K. Kang and H. B. Lee, *Biomaterials* **15**, 705 (1994).
13. B. Gupta, J. G. Hilborn, I. Bisson and P. Frey, *J. Appl. Polym. Sci.* **81**, 2993 (2001).
14. S. Sano, K. Kato and Y. Ikada, *Biomaterials* **14**, 817 (1993).
15. S. D. Lee, G. H. Hsiue and C. Y. Kao, *J. Polym. Sci.: Part A Polym. Chem.* **34**, 141 (1996).
16. E. T. Kang, K. L. Tan, K. Kato, Y. Uyama and Y. Ikada, *Macromolecules* **29**, 6872 (1996).
17. K. L. Mittal (Ed.), *Polymer Surface Modification: Relevance to Adhesion*, Vol. 5. VSP/Brill, Leiden (2009).
18. B. Li, Y. Ma, S. Wang and P. M. Moran, *Biomaterials* **26**, 1487 (2005).
19. B. Li, Y. Ma, S. Wang and P. M. Moran, *Biomaterials* **26**, 4956 (2005).
20. I. Bisson, M. Kosinski, S. Ruault, B. Gupta, J. Hilborn, F. Wurm and P. Frey, *Biomaterials* **23**, 3149 (2002).
21. S. D. Lee, G. H. Hsiue, P. C. T. Chang and C. Y. Kao, *Biomaterials* **17**, 1599 (1996).
22. P. Sadurní, A. Alagón, R. Aliev, G. Burillo and A. S. Hoffman, *J. Biomater. Sci., Polym. Ed.* **16**, 181 (2005).
23. P. C. Weber, D. H. Ohlendorf, J. J. Wendoloski and F. R. Salemme, *Science* **243**, 85 (1989).
24. P. Vermette, T. Gengenbach, U. Divisekera, P. A. Kambouris, H. J. Griesser and L. Meagher, *J. Colloid Interface Sci.* **259**, 13 (2003).
25. M. Buttiglione, F. Vitiello, E. Sardella, L. Petrone, M. Nardulli, P. Favia, R. d'Agostino and R. Gristina, *Biomaterials* **28**, 2932 (2007).
26. X. S. Jiang, C. Chai, Y. Zhang, R. X. Zhuo, H. Q. Mao and K. W. Leong, *Biomaterials* **27**, 2723 (2006).
27. A. Beauvais, C. A. Erickson, T. Goins, S. E. Craig, M. J. Humphries, J. P. Thiery and S. Dufour, *J. Cell. Biol.* **128**, 699 (1995).
28. J. R. Vig, *J. Vac. Sci. Technol. A* **3**, 1027 (1985).
29. Y. S. Chu, W. A. Thomas, O. Eder, F. Pincet, E. Perez, J. P. Thiery and S. Dufour, *J. Cell Biol.* **167**, 1183 (2004).
30. S. Dufour, A. Beauvais-Jouneau, A. Delouvée and J. P. Thiery, *J. Cell Biol.* **146**, 501 (1999).
31. P. Gijsman, G. Meijers and G. Vitarelli, *Polym. Degrad. Stabil.* **65**, 433 (1999).
32. A. Welle and E. Gottwald, *Biomedical Microdevices* **4**, 33 (2002).

33. A. Carré, V. Lacarrière and W. Birch, *J. Colloid Interface Sci.* **260**, 49 (2003).
34. P. Gong, T. Wu, J. Genzer and I. Szleifer, *Macromolecules* **40**, 8765 (2007).
35. G. C. M. Steffens, L. Nothdurft, G. Buse, H. Thissen, H. Höcker and D. Klee, *Biomaterials* **23**, 3523 (2002).
36. T. Wu, P. Gong, I. Szleifer, P. Vlčzek, V. Šubr and J. Genzer, *Macromolecules* **40**, 8756 (2007).
37. R. Ivkov, P. D. Butler, S. K. Satija and L. J. Fetters, *Langmuir* **17**, 2999 (2001).
38. Y. Tsujii, K. Ohno, S. Yamamoto, A. Goto and T. Fukuda, *Adv. Polym. Sci.* **197**, 1 (2006).
39. L. Ying, C. Yin, R. X. Zhuo, K. W. Leong, H. Q. Mao, E. T. Kang and K. G. Neoh, *Biomacromolecules* **4**, 157 (2003).
40. N. P. Huang, J. Vörös, S. M. De Paul, M. Textor and N. D. Spencer, *Langmuir* **18**, 220 (2002).
41. H. Hatakeyama, A. Kikuchia, M. Yamato and T. Okano, *Biomaterials* **27**, 5069 (2006).
42. J. B. Gurdon, H. Stanley, S. Dyson, K. Butler, T. Langon, K. Ryan, F. Stennard, K. Shimizu and A. Zorn, *Development* **23**, 5309 (1999).
43. C. E. Chamberlain, J. Jeong, C. Guo, B. L. Allen and A. P. McMahon, *Development* **135**, 1097 (2008).
44. E. Dessaud, L. L. Yang, K. Hill, B. Cox, F. Ulloa, A. Ribeiro, A. Mynett, B. G. Novitch and J. Briscoe, *Nature* **450**, 717 (2007).

Evaluation of Cell Behaviour on Atmospheric Plasma Deposited Siloxane and Fluorosiloxane Coatings

Malika Ardhaoui [a], Mariam Naciri [b], Tracy Mullen [c], Cathal Brugha [a],
Alan K. Keenan [c], Mohamed Al-Rubeai [b] and Denis P. Dowling [a,*]

[a] Surface Engineering Research Group, School of Electrical, Electronic and Mechanical Engineering,
University College Dublin, Belfield, Dublin 4, Ireland
[b] School of Chemical and Bioprocess Engineering, University College Dublin, Belfield,
Dublin 4, Ireland
[c] Conway Institute of Biomolecular and Biomedical Research, University College Dublin, Belfield,
Dublin 4, Ireland

Abstract
For developing functional biomaterials, an understanding of the biological response at material surfaces is of key importance. In particular, surface chemistry, roughness and cell type influence this response. Many previous reports in the literature have involved the study of single cell types and their adhesion to surfaces with a limited range of water contact angles. The objective of this study was to investigate the adhesion of five cell lines on surfaces with contact angles in the range of 20° to 115°. This range of water contact angles was obtained using siloxane and fluorosiloxane coatings deposited using atmospheric plasma deposition. These nm thick coatings were deposited by nebulizing liquid precursors consisting of poly(dimethylsiloxane) (PDMS) and a mixture of perfluorodecyl acrylate/tetraethylorthosilicate (PPFDA/TEOS) into the atmospheric plasmas. Cell adhesion studies were carried out with the following cell types: Osteoblast, Human Embryonic Kidney (HEK), Chinese hamster ovary (CHO), Hepatocytes (HepZ) and THP1 leukemic cells. The study demonstrated that cell adhesion was significantly influenced by the type of cell line, water contact angle and coating chemistry. For example, the sensitivity of cell lines to changes in contact angle was found to decrease in the following order: Osteoblasts > Hepatocytes > CHO. The HEK and THP-1 inflammatory cells in contrast were not found to be sensitive to changes in water contact angle.

Keywords
Atmospheric plasma, siloxane/fluorosiloxane coatings, wettability, cell adhesion, cell line

1. Introduction

Cell adhesion is involved in various natural phenomena such as embryogenesis, maintenance of tissue structure, wound healing, immune response, metastasis as

---

* To whom correspondence should be addressed. Tel.: +353 (0)1 716 1747; Fax: +353 (0)1 269 6035; e-mail: denis.dowling@ucd.ie

Surface and Interfacial Aspects of Cell Adhesion
© Koninklijke Brill NV, Leiden, 2010

well as tissue integration of biomaterials. It has previously been established that the surface properties of biomaterials strongly influence cellular behaviour. In particular, parameters such as surface energy, roughness and chemical composition significantly influence these interactions [1–3].

Many research groups have studied the effect of water contact angle on the interactions of biological species with surfaces. A large number of studies have concluded that cells tend to attach onto hydrophilic, rather than onto hydrophobic surfaces [4–9]. In contrast, other reports demonstrated that fibroblast cells adhered and proliferated at the highest rate when cultured on hydrophobic surfaces [10]. Others observed that cells adhere optimally on moderately wettable surfaces and cell adhesion and proliferation rates were lower on more hydrophilic or more hydrophobic surfaces [11–13]. These conflicting results may be explained by the fact that the tests were performed on different substrates (metals, ceramics, polymers, etc.) and surface topographies [10, 12, 14–18]. Moreover, many of these studies were carried out on surfaces with a relatively narrow range of water contact angle. It has also been demonstrated that different cell lines exhibit different levels of sensitivity to material surfaces. For example, Jansen *et al.* [19] observed that human fibroblasts were more sensitive to surface wettability than epithelial cells. Johann *et al.* [20] also showed that HEK cells were not sensitive to changes in the surface wettability of PDMS.

The aim of this work was to investigate the effect of cell type and water contact angle on the cell-surface interactions. In order to obtain a wide contact angle range, siloxane and fluorosiloxane coatings were deposited using the atmospheric plasma processing technique. This involved nebulising precursors of the liquid monomers into a helium or helium/oxygen plasmas. The chemical functionality of the precursor is retained in the deposited coating [21]. The level of coating oxidation is controlled by modifying the exposure of the precursor to the plasma. Siloxane coatings with water contact angles in the range 20° to 97° were obtained. Similarly, fluorosiloxane coatings with water contact angles ranging from 63° to 115° were also deposited. In order to investigate the effect of this wide range of water contact angles on cell adhesion, the following cell types were chosen: Osteoblast, Human Embryonic Kidney (HEK), Chinese Hamster Ovary (CHO), Hepatocyte (HepZ) and Human Monocytic Leukemia (THP-1) cells. For the THP-1 inflammatory cells, monocytes/macrophages adhesion and CD14 expression were also evaluated.

2. Materials and Methods

2.1. Surface Preparation and Characterization

The siloxane and fluorosiloxane coatings were deposited on polystyrene (PS) cut from Petri dishes. Uncoated PS and tissue culture grade polystyrene (TCPS) were used as references. PS is an example of a hydrophobic surface with a low surface energy and a correspondingly relatively high contact angle of approximately 90°

[22, 23]. The surface of TCPS has been modified to impart an oxygen-containing surface chemistry that has a water contact angle in the range of 60°–70° [22–24].

2.1.1. Atmospheric Plasma Deposition of Siloxane Coatings

Siloxane coatings were deposited on the PS substrates using the Labline™ (Dow Corning Plasma Solutions, Midleton, Co. Cork, Ireland) atmospheric pressure plasma system [25]. This reel-to-reel web system combines the liquid delivery of precursors with an atmospheric pressure dielectric barrier discharge plasma. It comprises two vertical plasma chambers made up of 30×32 cm^2 electrodes consisting of a conductive liquid housed in a dielectric perimeter. The PS samples were mounted onto the poly(ethylene terephthalate) (PET) web with a double sided tape and passed through the plasma chamber at speeds of approximately 1 m/min. A helium (He) flow rate of 40 l/min and oxygen (O$_2$) flow rate of 0.25 l/min were used. The input power to the electrodes was maintained at 1000 W. The liquid PDMS precursor was nebulized into the helium (He) plasma. The precursor flow rate and the number of depositions (passes through the deposition chamber) were varied in order to build up layers of coating using the same total flow rate (Table 1). This effectively increased the exposure to the plasma during deposition, while maintaining the same volume of precursor.

2.1.2. Atmospheric Plasma Deposition of Fluorosiloxane Coatings

The deposition of fluorosiloxane coatings onto PS substrates was carried out using an atmospheric plasma jet system known as PlasmaStream™ (Dow Corning Plasma Solutions, Midleton, Co. Cork, Ireland) [26]. The atmospheric pressure discharge is formed from a modified PTI 100 W rf power supply between two metallic electrodes. The plasma operates at a frequency of approximately 15–25 kHz, with maximum output voltages of between 11.8 and 14.9 kV. A Teflon tube is mounted at the orifice of the jet and under the flow of He gas, the plasma extends out from the base of this tube. This Teflon tube is 75 mm long and has a diameter of 15 mm. The substrate to nozzle distance was maintained at 2 mm in this study. The jet is moved over the substrates using a computer numerical control (CNC) system with

Table 1.
Influence of the number of passes through the LabLine™ deposition chamber and addition of O$_2$ into the plasma on siloxane (PDMS) coating water contact angles

PDMS Flow rate (µl/min)	O$_2$	No. of passes	Water contact angle (°)
50	No	1	97
50	Yes	1	73
25	Yes	2	65
17	Yes	3	40
10	Yes	5	20

Table 2.
Influence of the number of plasma jet passes over the surface using the PlasmaStream™ system on the fluorosiloxane coatings water contact angles

Computer numerical control (CNC) speed (mm/s)	No. of passes	Water contact angle (°)
150	3	63
10	3	76
10	6	78
50	9	103
10	15	115

speed of 25 mm/s. A manual valve rotameter is used to control the He gas flow (10 μl/min) and a syringe pump supplies reactive precursor liquids to an atomiser positioned between the electrodes. The fluorosiloxane coatings were deposited from a mixture of two precursors 1H,1H,2H,2H-perfluorodecyl acrylate (PPFDA) and tetraethylorthosilicate (TEOS) mixed in equal volumes. The flow rate of this precursor mixture into the nebulizer was 5 μl/min. The TEOS was added to PPFDA because this alkoxysilane acts as a crosslinking agent and improves the adhesion and crosslinking of the coating [27]. As detailed in Table 2 the number of passes and the CNC line speed were varied in order to deposit fluorosiloxane coatings with a range of water contact angles. In a previous study we demonstrated that broadly similar siloxane coating chemistries are deposited in both the PlasmaStream and Labline systems [28].

2.1.3. Surface Characterisation

Contact angle measurements were carried out using a Dataphysics OCA 20 Video-Based Contact Angle Meter. A 1 μl drop of de-ionised water was allowed to sit on the surface for approximately 10 s before the water contact angle was measured. Measurements were made at three different locations on the coating and averaged. Three samples of each coating type were studied. All contact angle measurements were carried out 10 days after the coating was deposited. This was done to minimize the influence of any activation effect that the plasma has on the polymer substrate [29]. It has been reported that significant hydrophobic recovery of siloxane materials can occur. This hydrophobic recovery is due to a reorientation of hydrophilic groups from the surface into the bulk by torsion about sigma bonds, thereby replacing hydroxyl groups by methyl groups in the surface region.

Coating thickness was determined by a Woolam M2000 (J. A. Woolam Co., Inc., USA) variable wavelength ellipsometer on silicon wafer samples passed through the plasma along with the PS samples. Measurements were made at three different locations on three silicon wafer samples. Siloxane coating thickness was maintained at approximately 5 nm, while the thickness of the plasma jet deposited fluorosiloxane coatings varied between 40 and 150 nm.

The surface roughness of the coated surfaces was measured using a Wyko NT1100 optical profilometer (Veeco, USA). Three measurements were obtained on each of the PS surfaces, each over an area of 256 μm by 290 μm. Measurements were carried out on three randomly selected fields on each sample and the arithmetic mean surface roughness (R_a) averaged. Three coatings of each type were examined.

The FT-IR analysis was carried out using a Bruker Vertex-70 (Bruker Optik GmbH, Germany) system with a liquid nitrogen cooled MCT detector and a KBr beam splitter. Spectra were collected in the range of 4400–400 cm^{-1} using a spectral resolution of 4 cm^{-1} and an overlay of 64 scans per sample cycle.

2.2. Cells Culture Assay

The cell adhesion studies were carried out approximately 2 weeks after coating deposition. This ensured that any no influence of plasma activation on the chemistry of the coated samples was minimized as detailed earlier.

2.2.1. Mammalian Cells Adhesion

The adhesion of four different mammalian cells (Osteoblasts, HEK, CHO and HepZ) was evaluated on the siloxane surfaces. The MG63 cells were supported in Minimum Essential Medium (Gibco) supplemented with 10% fetal bovine serum (FBS; Gibco), penicillin:streptomycin (100 U: 100 μg/ml) (Gibco), L-glutamine (1% v/v) (Gibco) and non-essential amino acids (1% v/v) (Sigma). The HEK and HepZ cells were cultured in Dulbecco's Modified Eagle's Medium (D5671, Sigma) supplemented with 5% fetal bovine serum (FBS; Lonza) and L-glutamine (4 mM) (Sigma). For CHO cells non-essential amino acids (NEAA, M7145, Sigma) were added. All cells were maintained at 37°C, in 5% CO_2, at 100% humidity in incubators. The cells were seeded onto prepared siloxane surfaces at a density of 2.5×10^5 cells/well. Osteoblast cell adhesion was evaluated after 4 hour's incubation; however, for HEK, CHO and HepZ cells, adhesion was evaluated after 2 h and 30 min. After this time had elapsed, non-adherent cells were removed by washing the surface gently with warm phosphate buffered saline (PBS). Cells were then detached using trypsin for Osteoblast cells and Accutase™ for the other mammalian cells and counted to determine the extent of adhesion.

2.2.2. THP-1 Human Cells Adhesion and Inflammatory Response

The adhesion and the inflammatory response of Human acute monocytic leukemia cell line (THP-1) and macrophage-like cells were also evaluated. The monocytic cells were a kind gift from Dr Paola Maderna of the Mater Misericordiae Hospital, Dublin. The haematopoietic cell line was maintained in suspension culture in RPMI complete media (Promocell, UK) with 10% (v/v) heat inactivated fetal bovine serum (FBS) (Gibco, UK), 2 mM L-glutamine, 100 U penicillin and 100 μg/ml streptomycin. The cells were grown in a humidified atmosphere with 5% CO_2. THP-1 cells were differentiated into macrophage-like phenotype by incubation with the phorbol ester, phorbal-12-myristate-13-acetate (PMA). The THP-1 cells were seeded at a density of 7.5×10^5 cells/well per coated disc and incubated for 24 and 48 h. At the end of the incubation period the cells were rinsed with PBS to remove non-adherent cells and treated with glucosaminidase solution [7.5 mM p-nitro-phenyl-N-acetyl-β-D-glucosaminide, 0.1 M sodium citrate and 5% (v/v) Triton X-100 (pH5)] for two hours at 37°C in a humidified atmosphere. The reaction was stopped by the addition of 80 mM glycine containing 5 mM EDTA (pH 10.4). An aliquot of supernatant was transferred to a 96-well plate and the optical density was measured at 405 nm in a multi-well plate reader. The number of cells adhered to the polymeric surface (cells/cm^2) were calculated by standard curve of known cell amounts and the adhesion level was then determined.

Enzyme-linked immunosorbent assays (ELISAs) were used to evaluate the inflammatory response of the THP-1 human acute monocytic cell line on the tested surfaces. A Human s-CD14 Immunoassay (Quatikine® Cat No. DC140) was used for the determination of soluble CD14 (sCD14) concentrations in cell culture serum. All samples were assayed for the presence of sCD14 by ELISA, according to the manufacturer's instructions (R&D Systems Europe Ltd., Abingdon, UK).

The cell adhesion values are presented as the mean ± standard error. Statistical significance was evaluated using the Student's t test for paired comparison; $p < 0.05$ was considered significant.

3. Results and Discussions

3.1. Influence of Deposition Conditions on Siloxane and Fluorosiloxane Water Contact Angle

The volume of the precursor used to deposit the PDMS coating on the PS and silicon wafer substrates in the Labline system was kept constant at 50 μl. By varying the flow rate and the duration of exposure to the He/O$_2$ plasma, the chemistry of the deposited coatings could be systematically altered (Table 1). With the increased exposure, a higher level of siloxane coating oxidation occurs [30]. This change in chemistry facilitated the systematic change in water contact angle given in Table 1 and Fig. 1. In the absence of oxygen, a hydrophobic siloxane coating is deposited; in contrast, with increased exposure of the precursor to the He/O$_2$ plasma, the coating becomes more hydrophilic.

As shown in Table 2, fluorosiloxane coatings deposited using the PlasmaStream system from the PPFDA/TEOS precursor mixture exhibited an increase in water contact angle from 63 to 115°, with higher levels of plasma exposure (increased number of passes). This indicates higher levels of fluorine at the film surface for coatings deposited with higher plasma exposure. The coating chemistry was assessed by FT-IR and compared to the TEOS coating. The resulting spectrum (Fig. 2) included peaks at 1180 cm^{-1} and 1135 cm^{-1} corresponding to –CF$_3$ and –CF$_2$ bonds.

Examination of the PS surfaces before and after the application of both the siloxane and fluorosiloxane coatings by optical profilometry demonstrated that there was

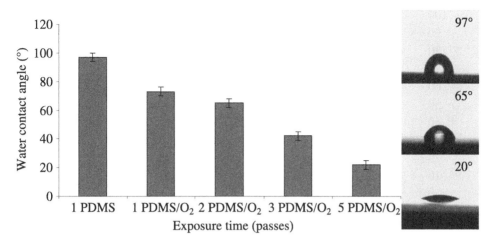

Figure 1. The effect of increasing PDMS precursor plasma exposure time on the coating water contact angle. The number of passes through the chamber is shown, i.e., 1–5.

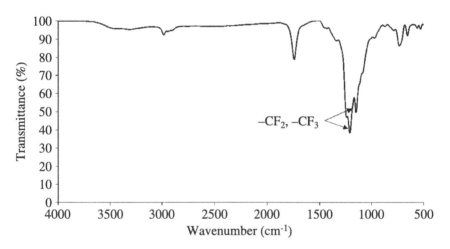

Figure 2. FT-IR spectrum of PPFDA/TEOS coating.

no significant change in the surface roughness (R_a). The R_a value of the siloxane coated surfaces remains at approximately 8 ± 2 nm, while that of the fluorosiloxane coated surfaces roughness varies between 10 and 20 ± 2 nm. Thus, any changes in cell adhesion are associated with only changes in surface chemistry and not topography.

3.2. Mammalian Cell Adhesion to Siloxane Coatings

3.2.1. Osteoblast Cell Adhesion
The effect of water contact angle on Osteoblast cell adhesion to PS, TCPS and Labline deposited PDMS coatings is given in Fig. 3. While cell adhesion was observed on all the siloxane surfaces, optimal adhesion of Osteoblast cells was ob-

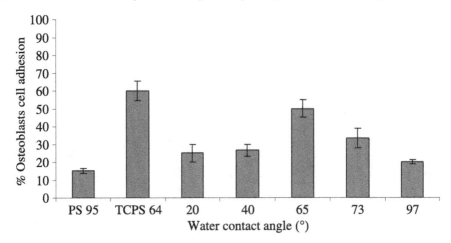

Figure 3. Influence of siloxane water contact angle on Osteoblast cell adhesion.

served for the siloxane coating with water contact angle of approximately 65°. Cell adhesion was found to progressively decrease on more hydrophobic or hydrophilic surfaces. As demonstrated in Fig. 3 also, significantly lower Osteoblast cell adhesion is observed on PS compared with the TCPS surfaces. This as outlined earlier is associated with the enhanced oxygen-containing surface chemistry of the TCPS.

3.2.2. HEK Cell Adhesion

The adhesion of HEK cells on the different siloxane coatings is summarized in Fig. 4. Similar to the other cell types, HEK cells adhere well on all the siloxane coatings. This is in agreement with the observation in the literature that oxidized PDMS and other hydroxyl containing surfaces are very favorable for cell attachment [31–33]. In contrast, Johann et al. [20] report that no HEK cell adhesion was observed on PDMS surfaces, although relatively good adhesion was observed on the hydrophobic PDMS coating ($\theta = 97°$) examined in this study.

The surface wettability and chemistry appear to have a minor influence on the ability of these cells to adhere. This is in agreement with the observation in the literature by Johann et al. [20] who observed that HEK cells were not sensitive to variations in the water contact angle of functionalised PDMS surfaces.

3.2.3. CHO Cell Adhesion

As shown in Fig. 5, CHO cells exhibit better adhesion on the more hydrophilic, compared with the more hydrophobic siloxane surfaces. This behaviour is similar to results previously reported for these type of cells [34, 35]. An interesting observation with respect to Fig. 5 is that comparing the adhesion of the cells on TCPS and a siloxane coating with similar water contact angle ($\theta = 65°$), the adhesion was higher on the TCPS. This would indicate that the type of oxygenated chemistry of the TCPS surfaces is more favourable to the adhesion for these cells. It is interesting to note that in a previous study by Lee et al. [36] the adhesion of CHO cells was found to be higher on H_2O plasma treated PS ($\theta = 61°$), compared with that of the

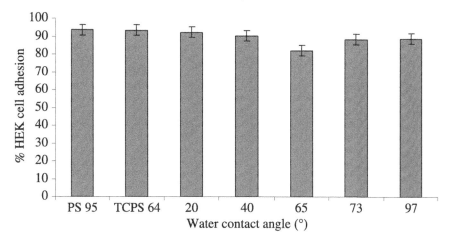

Figure 4. Influence of siloxane water contact angle on HEK cell adhesion.

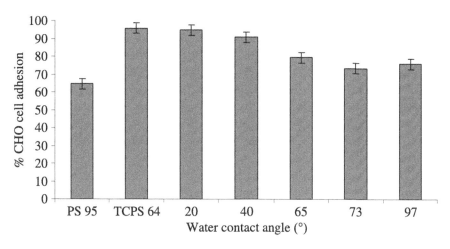

Figure 5. Influence of siloxane water contact angle on CHO cell adhesion.

O_2 plasma treated polymer ($\theta = 53°$). The authors indicated that hydroxyl groups positively influence the adhesion and spreading of the CHO cells.

3.2.4. Hepatocyte Cell Adhesion

Figure 6 demonstrates that the Hepatocyte cells are more adherent on hydrophilic siloxane surfaces. For the siloxane surface with water contact angle of 20°, for example, a relative adhesion of approximately 80% is obtained. In contrast, only 20% adhesion is observed on hydrophobic siloxane surface with water contact angle of 97°. Similar observations were reported recently by Nakazawa et al. [37]. They demonstrated that the Hepatocytes strongly adhered to the hydrophilic siloxane surface (contact angle $28 \pm 3°$); however, cell adhesion to the hydrophobic siloxane surface (contact angle $120 \pm 3°$) was strikingly inhibited. Optimal Hepatocytes adhesion was observed on TCPS (96%).

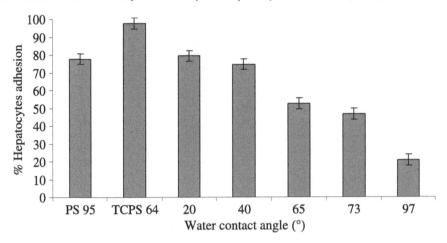

Figure 6. Influence of siloxane water contact angle on Hepatocytes (HepZ) cell adhesion.

Comparing the adhesion of the four mammalian cell types investigated, adhesion is, in general, better on the more hydrophilic siloxane surfaces. This behaviour can be attributed to the presence of hydroxyl/carboxylic groups generated during the plasma polymerization of PDMS in the presence of oxygen. This is in agreement with previous reports that oxidized PDMS and other hydroxyl containing surfaces are favorable for cell attachment [20, 31–33]. This may be explained as due to the variation in the integrin–fibronectin bonding affinities on these surfaces, which decrease in the order –OH > –CH$_3$ [38]. Fibronectin is able to adsorb onto surfaces with a wide range of physicochemical properties and plays a dominant role in the adhesion of most cultured cell lines [32, 38, 39].

Comparing the level of cell adhesion on the different siloxane surfaces, the following results are highlighted. In general, the level of adhesion decreases in the following order: HEK > CHO > Hepatocytes > Osteoblasts. This demonstrates that the adhesion of cells depends not only on the biomaterial surface properties but also on the cell line examined. Osteoblast and Hepatocytes cells are more sensitive to surface wettability than CHO cells. On the siloxane coatings tested, the percentage of the adhering Osteoblast cells varies from 13% to 50%. However, it varies only between 73% and 95% in the case of CHO cells. HEK cells are, however, not sensitive to the water contact angle, adhering similarly on all the siloxane coatings. Lee *et al.* [40] have previously demonstrated higher level of adhesion of CHO cells on corona treated polyethylene surfaces, compared to endothelial cells and fibroblasts. These differences in cell adhesion can be explained as due to the differences in type, quantity, conformation and activity of the adhesive proteins synthesized by each cell on the substrates during adhesion and proliferation. Moreover, some cells secrete growth factors and some other molecules that promote cell adhesion [41].

3.2.5. THP-1 Cell Adhesion and Inflammatory Response
Monocyte and macrophage cells adhesion and foreign body giant cell (FBGC) formation are vital processes in the inflammatory and wound-healing responses to

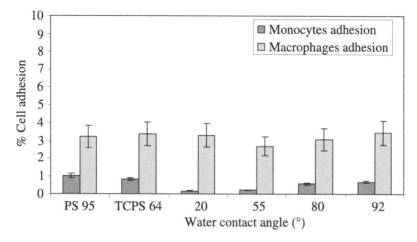

Figure 7. Influence of siloxane coating water contact angle on the adhesion of THP-1 monocytic and macrophages cells (24 h incubation).

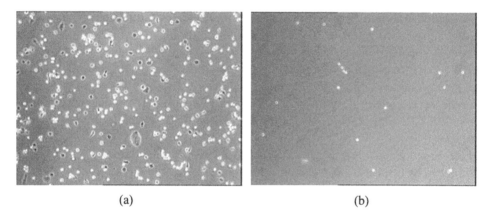

(a)   (b)

Figure 8. Representative data comparing the adhesion of (a) PMA-differentiated THP-1 macrophage cells and (b) THP-1 monocytic cells on PDMS coated tissue culture polystyrene ($\theta = 92°$) after 4 h incubation. The images were captured on Nikon TMS microscope (area of 1.289 mm^2).

implanted biomaterials [42]. To investigate the influence of the surface wettability on the adhesion of THP-1 monocyte and macrophage, the adhesion and inflammatory response of these cells were tested on siloxane and fluorosiloxane coatings. As detailed in Fig. 7, THP-1 monocytic cells adhesion is not significantly influenced by the siloxane wettability variation. The PMA-differentiated THP-1 macrophage cells exhibited enhanced adhesion compared with the THP-1 monocytic cells. Figure 8, showing optical microscopy images of THP-1 monocytic cells and PMA-differentiated THP-1 macrophage cells grown on siloxane coated PS, confirm these results. No effect of wettability on macrophage cells adhesion was observed.

THP1 monocytic cell adhesion studies were also carried out on fluorosiloxane surfaces (Fig. 9). The level of adhesion was found to be three times higher on

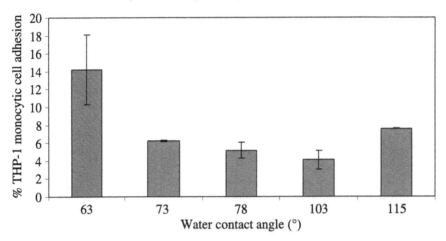

Figure 9. Influence of water contact angle on THP-1 monocytic cells adhesion on fluorosiloxane coatings.

moderately hydrophilic fluorosiloxane coating ($\theta = 63°$) than that on the more hydrophobic coating surface ($\theta = 103°$). The slightly higher adhesion on more hydrophobic fluorosiloxane surface ($\theta = 115°$) could be due to the relatively rougher surface of the coating ($R_a = 60$ nm ± 3) compared with that of the other fluorosiloxane coating surfaces ($R_a = 10$–20 nm). In a previous study, Eloy et al. [43] reported that the development of inflammatory cells was prevented when the substrate was treated in CF_4 plasma. This was attributed to the hydrophobic property of these surfaces.

The immune system provides the first response to an infection by initiating an inflammatory response and the monocyte surface molecule CD14 is a key element in this response system. In this work, the effect of the surface chemistry on the inflammatory response of THP-1 cells was investigated by evaluating CD14 secretion on TCPS, PS, siloxane (55°) and fluorosiloxane (103°) surfaces (Fig. 10). Higher CD14 secretion was observed on fluorosiloxane and untreated PS surfaces. Low CD14 secretion was observed on the more moderately hydrophilic siloxane surface. These observations confirm the major effect of the surface chemistry on the inflammatory response of the THP-1 cells [44, 45].

4. Conclusions

In this study, the effect of siloxane and fluorosiloxane coating water contact angles on the adhesion of five cell lines was investigated. The cells studied were: Osteoblasts, Human Embryonic Kidney (HEK), Chinese hamster ovary (CHO), Hepatocytes (HepZ) and THP1 inflammatory cells. The study has demonstrated the following:

1. All the cells tested adhered to the siloxane surfaces; however, the sensitivity of the cells to contact angle variation depends on the cell line. For three cell lines

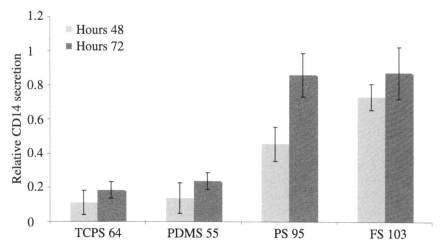

Figure 10. Effect of the surface chemistry on the CD14 secretion by THP-1 cells (48 and 72 h incubation). Note PDMS 55 and FS103 refer to the siloxane and fluorosiloxane coatings along with their contact angles.

the sensitivity to changes in contact angle decreased in the following order: Osteoblasts > Hepatocytes > CHO. The percentage of the adhering Osteoblast cells on the siloxane surfaces varies from 13 to 50% but only from 73 to 95% in the case of CHO cells. The HEK and THP-1 inflammatory cells, in contrast, were found not be sensitive to changes in water contact angle.

2. The level of adhesion is cell type, contact angle and chemistry dependent. Osteoblast cells adhere better on moderately hydrophilic surfaces ($\theta = 64°$). However, CHO and HepZ cells were found more adhesive on hydrophilic siloxane surfaces ($\theta = 20°$). HEK and THP-1 monocytic cell adhesion was found to be relatively insensitive to the water contact angle of the siloxane coatings tested.

3. The adhesion of the THP-1 monocytic cells was found to be sensitive to surface chemistry. For example, higher levels of adhesion were observed on fluorosiloxane compared with siloxane surfaces with similar water contact angles.

4. The adhesion of PMA-differentiated THP-1 macrophage cells on PS, TCPS and siloxane surfaces is higher than that observed for monocyte cells.

5. The inflammatory response of THP-1 cells is surface chemistry dependent. The CD14 expression is four times higher on fluorosiloxane and untreated PS surfaces than on siloxane and TCPS surfaces.

These results demonstrate the importance of the biomaterial surface wettability and chemistry for cell attachment and confirm the major and complex role of the cell-surface interaction in determining cell adhesion. The results are significant for the assessment and design of new surfaces for implant device applications.

Acknowledgement

This work is partly supported by the Science Foundation Ireland under grant No. 08/SRC/11411.

References

1. Y. X. Wang, J. L. Robertson, W. B. Spillman Jr. and R. O. Claus, *Pharm. Res.* **21**, 1362 (2004).
2. K. C. Dee, D. A. Puleo and R. Bizios, *An Introduction to Tissue-Biomaterial Interaction*, Vol. 3, pp. 37–52. Wiley, New York, NY (2002).
3. J. A. Hunt and M. Shoichet, *Curr. Opin. Solid State Mater. Sci.* **5**, 161 (2001).
4. G. Altankov, F. Grinnell and T. Groth, *J. Biomed. Mater. Res.* **30**, 385 (1996).
5. G. Altankov and T. Groth, *J. Mater. Sci. Mater. Med.* **5**, 732 (1994).
6. Y. Arima and H. Iwata, *Biomaterials* **28**, 3074 (2007).
7. J. Wei, M. Yoshinari, S. Takemoto, M. Hattori, E. Kawada, B. Liu and Y. Oda, *J. Biomed. Mater. Res.* **81B**, 66 (2007).
8. Y. Ozdemir, N. Hasirci and K. Serbetci, *J. Mater. Sci: Mater. Med.* **13**, 1147 (2002).
9. M. Chen, P. O. Zamora, P. Som, L. A. Pena and S. Osaki, *J. Biomater. Sci. Polym. Ed.* **14**, 917 (2003).
10. Y. W. Wang, Q. Wu and G. Q. Chen, *Biomaterials* **24**, 4621 (2003).
11. J. H. Lee, G. Khang, J. W. Lee and H. B. Lee, *J. Colloid Interface Sci.* **205**, 323 (1998).
12. Y. Tamada and Y. Ikada, *J. Biomed. Mater. Res.* **28**, 783 (1994).
13. K. E. Geckeler, R. Wacker and W. K. Aicher, *Naturwissenschaften* **87**, 351 (2000).
14. K. M. Swart, J. C. Keller, J. P. Wightman, R. A. Draughn, C. M. Stanford and C. M. Michaels, *J. Oral. Implantol.* **18**, 130 (1992).
15. J. D. Bumgardner, R. Wiser, S. H. Elder, R. Jouett, Y. Yang and J. L. Ong, *J. Biomater. Sci. Polym. Ed.* **14**, 1401 (2003).
16. T. A. Horbett, M. B. Schway and B. D. Ratner, *J. Colloid Interface Sci.* **104**, 28 (1985).
17. P. B. Van Wachem, A. H. Hogt, T. Beugeling, J. Feijen, A. Bantjes, J. P. Detmers and W. G. Van Aken, *Biomaterials* **8**, 323 (1987).
18. J. M. Schakenraad, J. Arends, H. J. Busscher, F. Dijk, P. B. van Wachem and C. R. Wildevuur, *Biomaterials* **10**, 43 (1989).
19. J. A. Jansen, J. P. C. M. Van Der Waerden and K. de Groot, *Biomaterials* **12**, 25 (1991).
20. R. M. Johann, Ch. Baiotto and Ph. Renaud, *Biomed. Microdevices* **9**, 475 (2007).
21. L. A. O'Hare, L. O'Neill and A. J. Goodwin, *Surface Interface Anal.* **38**, 1519 (2006).
22. P. Van der Valk, A. W. J. Van Pelt, H. J. Busscher, H. P. de Jong, C. R. H. Wildevuur and J. Arends, *J. Biomed. Mater. Res.* **17**, 807 (1983).
23. J. G. Steele, G. Johnson, C. McFarland, B. A. Dalton, T. R. Gengenbach, R. C. Chatelier, P. A. Underwood and H. J. Griesser, *J. Biomater. Sci. Polym. Ed.* **6**, 511 (1994).
24. S. I. Ertel, A. Chilkoti, T. A. Horbett and B. D. Ratner, *J. Biomater. Sci. Polym. Edn.* **3**, 163 (1991).
25. L. A. O'Hare, A. Hynes and M. R. Alexander, *Surf. Interf. Anal.* **39**, 926 (2007).
26. L. O'Neill, N. Shephard, S. R. Leadley and L. A. O'Hare, *J. Adhesion* **84**, 562 (2008).
27. P. R. Underhill and D. L. DuQuesnay, in: *Silanes and Other Coupling Agents*, K. L. Mittal (Ed.), Vol. 2, pp. 149–158. VSP, Utrecht (2000).
28. D. P. Dowling, A. R. Ramamoorthy, M. Rahman, D. A. Mooney and J. M. D. MacElroy, *Plasma Proc. Polym.* **6**, S483 (2009).
29. H. Hillborg and U. W. Gedde, *Polymer* **39**, 1991 (1998).

30. B. Twomey, D. P. Dowling, G. Byrne, L. O'Neill and L. O'Hare, *Plasma Proc. Polym.* **4**, 450 (2007).
31. K. Faid, R. Voicu, M. Bani-Yaghoub, R. Tremblay, G. Mealing, C. Py and R. Barjovanu, *Biomed. Microdevices* **7**, 179 (2005).
32. G. K. Toworfe, R. J. Composto, C. S. Adams, I. M. Shapiro and P. Ducheyne, *J. Biomed. Mater. Res. Part A* **71**, 449 (2004).
33. A. S. G. Curtis and H. McMurray, *J. Cell Sci.* **86**, 25 (1986).
34. A. Prokop, Z. Prokop, D. Schaffer, E. Kozlov, J. Wikzwo, D. Cliffel and F. Baudenbacher, *Biomed. Microdevices* **6**, 325 (2004).
35. J. H. Lee and H. B. Lee, *Polymer (Korea)* **16**, 680 (1992).
36. J. H. Lee, H. G. Kim, G. S. Khang, H. B. Lee and M. S. Jhon, *J. Colloid Interface Sci.* **151**, 563 (1992).
37. K. Nakazawa, Y. Izumi and R. Mori, *Acta Biomaterialica* **5**, 613 (2009).
38. M. H. Lee, P. Ducheyne, L. Lynch, D. Boettiger and R. J. Composto, *Biomaterials* **27**, 1907 (2006).
39. F. Crea, D. Sarti, F. Falciani and M. Al-Rubeai, *J. Biotechnol.* **121**, 109 (2006).
40. J. H. Lee, G. Khang, J. W. Lee and H. B. Lee, *J. Colloid Interface Sci.* **205**, 323 (1998).
41. H. Morita, H. Yasumitsu, Y. Watanabe, K. Miyazaki and M. Umeda, *Cell Struct. Funct.* **18**, 61 (1993).
42. M. Anderson and K. M. Miller, *Biomaterials* **5**, 5 (1984).
43. R. Eloy, D. Parrat, T. M. Duc, G. Legeay and A. Bechetoille, *J. Cataract. Refract. Surg.* **19**, 364 (1993).
44. L. Chou, J. D. Firth, D. Nathanson, V. J. Uitto and D. M. Brunette, *J. Biomed. Mater. Res.* **31**, 209 (1996).
45. Y. Z. Zhang, L. M. Bjursten, C. Freij-Larsson, M. Kober and B. Wesslen, *Biomaterials* **17**, 2265 (1996).

Similarities between Plasma Amino Functionalized PEEK and Titanium Surfaces Concerning Enhancement of Osteoblast Cell Adhesion

K. Schröder [a,*], B. Finke [a], H. Jesswein [b], F. Lüthen [b], A. Diener [b], R. Ihrke [a], A. Ohl [a], K.-D. Weltmann [a], J. Rychly [b] and J. B. Nebe [b]

[a] Leibniz Institute for Plasma Science and Technology (INP), F.-Hausdorff Straße 2, 17489 Greifswald, Germany
[b] University of Rostock, Biomedical Research Center, Department of Cell Biology, Schillingallee 69, 18057 Rostock, Germany

Abstract

The application of gas discharge plasmas for different functionalization and coating strategies is discussed with respect to cell adhesion to polymeric and metallic surfaces. Poly(ether ether ketone) (PEEK) and titanium (Ti) were selected as typical bone cell-contacting biomaterials. The surfaces were equipped with nitrogen-based chemical functionality, mainly amino groups. The behavior of human MG-63 osteoblasts was investigated with respect to cell adhesion and growth on plasma-treated PEEK and Ti surfaces. MG-63 cells adhere faster and occupy a wider cell area on the plasma-treated compared to untreated surfaces, which is not integrin receptor mediated. Although different plasma treatments were applied to functionalize polymeric and metallic surfaces with amino groups, similar cell adhesion properties were achieved.

Keywords

Plasma polymers, amino functionalization, gas discharge plasma, cell adhesion, human osteoblasts

1. Introduction

The improvement of bone cell (osteoblast) adhesion to artificial materials surfaces is of fundamental importance for the improvement of hip, knee, dental, and other bone-contacting implants. Today, metallic implants, especially made from titanium (Ti) and its alloys, are the state of the art [1]. Besides, implants made from poly(ether ether ketone) (PEEK) are considered as a possible alternative for certain purposes [2–5] because of their mechanical characteristics like stiffness and modulus which match better the characteristics of bone [2]. A further improvement

* To whom correspondence should be addressed. Tel.: +49 3834 554 428; Fax: +49 3834 554 301; e-mail: schroeder@inp-greifswald.de

Surface and Interfacial Aspects of Cell Adhesion
© Koninklijke Brill NV, Leiden, 2010

of bone cell attachment both to Ti and PEEK implants is still a challenging task although a number of successful surface modifications have been developed. In particular, improved control of surface chemical properties for acceleration of implant in-growth is of current interest. It has been shown that success of implantation procedure can be improved this way, i.e., by effectively enhancing bone apposition [6] and soft tissue integration [7]. In this special case, chemical modification of titanium with sulfuric and hydrochloric acids has been used to increase wettability of titanium dental implants [8]. Numerous other surface modification strategies have been developed including different types of surface activation and hydrophilization [9], coatings with calcium phosphates like hydroxyapatite (HA), and immobilization of cell adhesion molecules (CAMs) by organic chemical groups (for instance, hydroxyl or carboxyl groups).

For good reasons, almost every solution approach is largely determined by the chemical properties of the implant materials. However, this must not ignore the fact that the desired surface properties are determined basically by requirements imposed by the contacting tissue. Remember that implants with complex functions, e.g., joints, often consist of several parts which are made from different materials. Thus, different materials should be modified in a similar manner.

Indeed, for chemically very different materials such as Ti and PEEK it is not impossible to develop such strategies. Both materials differ considerably in their surface properties. Ti is covered by an oxide layer which is partially terminated by OH functional groups. PEEK exhibits chemical inertness. While titanium is well known for its good protein adhesion properties, PEEK is rather non-adhesive. Osteoblast cell adhesion is acceptable but not yet optimal on titanium surfaces [10] and it is impaired on PEEK [11]. Hence, simply coating PEEK (or other inert materials) with Ti should be generally useful. In fact, this procedure resulted in a significantly higher bone apposition [12]. In addition, this approach would allow the transfer of some modification strategies developed for titanium to other materials. However, it remains questionable whether coating with titanium is in general a promising approach since additional treatment is needed in any case.

Other surface modification strategies with general usefulness are coatings with calcium hydroxide [13] or with calcium phosphates [13, 14] by means of electrochemical processes or by plasma spraying of hydroxyapatite (HA) [15]. These are really bio-inspired coatings since bone contains these components. Electrochemically deposited calcium phosphate coatings are resorbable and exist only during the first period of osseointegration as a matrix for the early immobilization of osteoblast-like cells and for development of vascularized bone tissue on the implant [16]. Besides, deposited brushite phase with less HA may act as a precursor for newly precipitated calcium phosphates in the cell culture [16]. Therefore, these surfaces seem to be clinically advantageous concerning bone remodeling properties. Plasma sprayed calcium phosphate coatings are known to promote earlier and stronger fixation of the implants. Actually, such coatings are not limited to metallic substrates. Also on PEEK surfaces, precipitation of HA resulting in a 50 μm thick layer significantly increased the osteogenic MC3T3E1 cell viability measured by the 3-(4,5-dimethylthiazol-2-yl)-2,5-diphenyl-tetrazoliumbromide (MTT) assay [17]. Vacuum plasma spray coatings with HA (VPS–HA) were also investigated on PEEK [18, 19].

A drawback of phosphate coatings is resorption. This hinders direct control of tissue–implant interaction on the molecular level. For this reason, immobilization of cell adhesion molecules (CAMs) by organic chemical groups is another surface modification strategy of general interest [20, 21]. The ligands are provided for specific cell adhesion receptors, the integrins [10]. It is promising because it can also trigger a specific cellular behavior. The general applicability of such methods depends on the required extent of specificity. Important questions are: is it necessary to have specific densities of selected CAMs like fibronectin, laminin or vitronectin, or CAM groups like YIGSR (for laminin) or RGD (for collagen)? Is it already sufficient just to have certain surface densities of available functional chemical groups like hydroxyl, carboxyl, thiol or amino groups? Can a specific distribution of functional groups influence the performance?

Here, we add some specific information to the discussion of these questions especially for the amino functionalization. Amino functionalized surfaces are of special interest for the control of CAM attachment due to their outstanding basic character and positive charging in aqueous environment at physiological pH. There are fundamental differences between Ti and PEEK surfaces concerning chemical reactivity, which suggests use of different chemical surface functionalization procedures for practical reasons. Ti cannot be equipped with amino groups directly. However, coating by amino group containing allylamine plasma polymer is possible and has shown advantageous effects on osteoblast adhesion on Ti [10]. For PEEK, direct amino functionalization is much easier using ammonia plasma [22]. Modification with OH- and NH_2-functional groups for improved adsorption of the CAM fibronectin was shown [23]. If the presence of amino groups solely would have the determining influence on cell adhesion, this should become clear by comparing bone cell adhesion to amino functionalized Ti and PEEK surfaces. In addition, concomitant oxidation processes could provide further valuable information. Ti already has a specifically oxidized, i.e., hydroxylated, surface, while specific oxidation of PEEK is difficult.

With these questions in mind, here we compare the growth of MG-63 osteoblastic cells on amino modified Ti and PEEK surfaces. The human MG-63 bone cell line is a widely used model for the study of interactions of bone cells with implant materials [24, 25]. After brief descriptions of surface modification, surface characterization, cell culture, and cell characterization methods we present results on surface chemical properties and cell adhesion in more detail. We demonstrate that indeed the presence of amino groups solely seems to have great influence on cell adhesion. Compared to this effect, the differences in the modification methods and substrates seem to have only minor effect.

2. Materials and Methods

2.1. Surface Functionalization of PEEK

Poly(ether ether ketone) foils with a thickness of 250 μm were purchased from Reichelt (number 48928, Reichelt Chemietechnik, Heidelberg, Germany). The untreated foils were used as reference and named PEEK–control.

Plasma functionalization of PEEK was performed in a low pressure microwave (2.54 GHz) reactor V55G (Plasma-Finish, Schwedt, Germany). The details of the reactor are described in [26]. Briefly, the microwave plasma source is located on top of the chamber and the PEEK foils were placed in the downstream region, 5 cm below the quartz glass window over which the microwave energy is coupled into the plasma. Pure ammonia (40 sccm) was used to produce functionalization rich in amino groups [26]. Plasma processes were run with applied microwave power of 500 W at a pressure of 0.2 mbar. The treatment duration was varied up to 600 s. Samples treated this way are abbreviated PEEK–5NH$_2$, PEEK–30NH$_2$, or PEEK–180NH$_2$, where the number denotes the treatment duration in seconds.

2.2. Surface Functionalization of Titanium

Circular disks of titanium (99+% purity) were received (DOT GmbH, Rostock, Germany) in diameters of 30 mm and 11 mm. The surfaces were polished down to a roughness R_a of about 0.19 μm. These samples were not plasma-treated and were called Ti–control.

The Ti–control surfaces were treated in downstream oxygen plasma (continuous wave, 500 W, 50 Pa, 100 sccm O$_2$ and 25 sccm Ar) in a commercial V55G low pressure microwave plasma reactor (Plasma Finish, Schwedt, Germany). These samples were named Ti–oxygen.

Ti–oxygen samples were further coated with a plasma polymer (about 60 nm thick) without breaking the vacuum according to [10]. Briefly, allylamine (H$_2$C=CH–CH$_2$–NH$_2$) was plasma polymerized in a pulsed, microwave excited, low pressure gas discharge plasma (2.45 GHz, 500 W, 50 Pa, pulse parameters $t_{on} = 0.3$ s and $t_{off} = 1.7$ s) for 144 s (effective treatment time). Ti samples coated this way were marked as Ti–NH$_2$.

2.3. Physicochemical Surface Characterization: XPS Analysis and Determination of Surface Energy

The elemental surface chemical composition and chemical binding properties were determined by X-ray photoelectron spectroscopy (XPS), using an AXIS ULTRA spectrometer from Kratos, Manchester, UK. The measurement conditions were described in detail previously [10]. Briefly, monochromatic Al K$_\alpha$ line at 1486 eV (150 W), implementation of charge neutralization and pass energy of 80 eV were used for estimating the chemical elemental composition. Each surface composition value represents an average of three XPS measurements.

C$_{1s}$ peaks were measured with a pass energy of 10 eV (high energy resolution) to investigate chemical functional groups in more detail. The peak positions were pre-

R–OH + (F_3C–CO–O–CO–CF_3) (g) → (R–O–CO–CF_3) + CF_3COOH

Figure 1. Chemical derivatization reaction for the determination of hydroxyl group density on Ti–control and Ti–oxygen.

R–NH_2 + HC(=O)–⟨◯⟩–CF_3 (g) → R–N=CH–⟨◯⟩–CF_3 + H_2O

Figure 2. Chemical derivatization for the determination of primary amino group density.

assigned according to [27]: The C–C/C–H component of the C_{1s} peak was adjusted to 285.0 eV; C–NH to 285.7 ± 0.1 eV; C≡N, C=N, and C–O to 286.6 ± 0.2 eV; C=O to 287.2 ± 0.3 eV; and N–C=O to 288.3 ± 0.3 eV.

Hydroxyl group density cannot be measured with XPS directly. Samples were reacted in the gas phase with trifluoroacetic anhydride (TFAA) in a vacuum chamber for 15 min according to [28, 29], see Fig. 1.

This way every hydroxyl group is labelled by three fluorine atoms. The number of hydroxyl groups is equal to one third of F content measured by XPS.

Chemical derivatization with 4-trifluoromethylbenzaldehyde (TFBA) in a saturated gas phase (40°C for 2 h) was used for labelling of primary amino groups [26]. Three fluorine atoms label one primary amino group according to Fig. 2.

The polar and dispersion components of surface energy were calculated from measurements of contact angles with different liquids. Water, ethylene glycol, and methylene iodide contact angles were determined using the OCA 30 contact angle measuring system (Data Physics Instruments GmbH, Filderstadt, Germany) with the sessile drop method (using software SCA20). The surface energy was calculated using the methods of Owens, Wendt, Rabel and Kaelble [30, 31]. These measurements were always performed within 30 min after sample preparation.

For the determination of film thickness, an uncoated stripe was prepared on a silicon wafer by masking with cellulose acetate, plasma coating and removing the mask after the end of the process. The height of this trench was measured with a surface profiler Dektak3ST' (Veeco, Santa Barbara, CA, USA).

Surface roughness was determined with a scanning probe microscope diCP2 (Veeco, Santa Barbara, CA, USA) in the non-contact mode before and after plasma treatment. The average roughness R_q was calculated as the geometric average of measurements on a 4 µm × 4 µm wide area.

2.4. Cell Culture and Preparation of Osteoblastic MG-63 Cells

The osteoblast cultivation was already described [10, 24, 32]. Briefly, osteoblastic cells of the osteosarcoma cell line MG-63 (ATCC, LGC Promochem, Wesel, Germany) were seeded with a density of 3×10^5 cells/cm^2 onto the Ti plates or PEEK foils. The samples were not sterilized before use, but they were decontaminated by the plasma processes. Cells were allowed to settle at room temperature (RT) for 10 min before Dulbecco's modified Eagle medium (DMEM, Gibco Invitrogen, Carlsbad, CA, USA) with 1% gentamicin was added. Then cells were cultured at 37°C in a 5% CO_2 atmosphere.

As a further control for our cell investigations, collagen I (rat tail, 20 µg/cm², TEBU, Frankfurt/Main, Germany) coated cover glass or Petri dishes were used and named COL–control. Collagen is a ligand for integrin adhesion receptors [33] and, therefore, represents a positive control in our investigations.

The actin cytoskeleton staining was done as described [10, 24]. Briefly, after 1 h of cell culture on the specimens, cells were fixed with 4% paraformaldehyde (PFA) for 10 min at RT, followed by permeabilization of the cell membrane with 0.1% Triton X-100 in phosphate buffer solution (PBS) (Merck KGaA, Darmstadt, Germany), for 10 min (RT). Then, actin was stained with BODIPY® FL phallacidin (Invitrogen Molecular Probes®, Carlsbad, CA, USA) (dilution in PBS 1:40, 30 min, RT). To microscopically observe the cells, specimens were fixed onto microscope slides and before covering with cover slips cells were embedded in mounting medium (30 g glycerine, 12 g poly(vinyl alcohol), 30 ml distilled water, 0.5 g phenol, completed with 60 ml 0.1 M Tris-buffer at pH 8.5). For microscopic analysis we used a 63× oil immersion objective of the confocal laser scanning microscope (LSM 410, Carl Zeiss, Jena, Germany; excitation wavelength 488 nm, emission bandpass 510–525 nm).

The integrin expression after 48 h of cell culture was measured by flow cytometry (BD FACSCalibur, argon-ion laser, BD Biosciences, Heidelberg, Germany) as described [24]. Therefore, cells were labelled by anti-integrin antibodies specific for the β- and α-subchains: β_1 (CD29), β_3 (CD61), α_2 (CD49b), α_3 (CD49c) and α_5 (CD49e) (all from Coulter Immunotech, Marseille, France; dilution of the antibody in PBS 1:6) or for control with mouse IgG1 (BD Biosciences, Heidelberg, Germany) at 4°C for 30 min. After washing, cells were incubated with FITC-conjugated anti-mouse IgG (Fab2) fragment (Sigma-Aldrich Chemie, St. Louis, MO, USA, dilution of the antibody in PBS 1:16).

The adhesion of cells on the Ti-specimens and calculation in percent after acquisition of non-adherent, suspended cells in the flow cytometer was described elsewhere [25]. The adhesion of MG-63 cells on the PEEK foils was determined as follows: PEEK foils (40 mm × 40 mm) with differently modified surfaces were placed in 60 mm Petri dishes. The dishes were partitioned with a flexiPERM slide into 4 chambers (In Vitro Systems & Services GmbH, Osterode, Germany). The cells were seeded into the chambers (200 µl/chamber) and cultivated at 37°C and 5% CO_2. After 5, 10, 15 and 30 min non-adherent cells were aspirated, the chambers were rinsed with 200 µl FACS-PBS (BD Biosciences, Heidelberg, Germany) and the samples were counted by flow cytometry (FACScan, BD Biosciences, Heidelberg, Germany) equipped with an argon-ion laser) in comparison to a blank (cell suspension). For data acquisition and analysis, the software CellQuest Pro 4.0.1 (BD Biosciences, Heidelberg, Germany) was used and adherent cells were then calculated in (%). Statistical analysis was performed with SPSS 10.0 (SPSS Inc., Chicago, IL, USA) with $p < 0.05$ considered as significant.

3. Results and Discussion

3.1. Functionalization of PEEK

PEEK is a polymer material which can be easily functionalized with standard plasma processes such as argon or oxygen plasma treatments. Such treatments make the material wettable. For example, the contact angle could be reduced from $75.2 \pm 3.9°$ for untreated material to less than $10°$ by functionalization of PEEK foil in microwave Ar/O_2 plasma for 30 s. Similar results were obtained in other cases, e.g., using RF plasma excitation [17, 34]. In contrast, the ammonia plasma treatment is not as effective in hydrophilization of PEEK surfaces. The contact angle of ammonia plasma-treated PEEK samples strongly depends on treatment duration (Fig. 3). After plasma treatments, the contact angles increase, but do not exceed $70°$. The reduction of contact angles is not accompanied by changes in the roughness (R_q) of the PEEK foil. PEEK–60NH$_2$ exhibits a roughness of 4 nm, which does not differ from the untreated foil (5.5 nm).

Of course, ammonia plasma treatment creates different functional groups and wettability depends on their nature and density. Plasma chemical dissociation of ammonia generates hydrogen radicals [35] which can remove oxygen in PEEK by OH radical formation and, therefore, reduce the O/C ratio. Furthermore, plasma processes generate carbon-based surface radicals, which can react with oxygen or

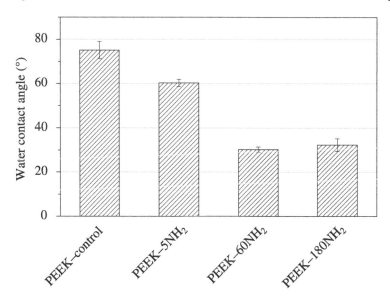

Figure 3. Contact angle measurements. Comparison of differently treated PEEK surfaces: untreated (PEEK–control) and ammonia plasma-treated PEEK for 5 s (PEEK–5NH$_2$), for 60 s (PEEK–60NH$_2$), and for 180 s (PEEK–180NH$_2$).

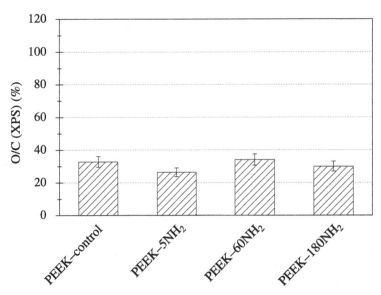

Figure 4. Results of XPS O/C measurements on ammonia plasma-treated PEEK surfaces after 0 s (PEEK–control), 5 s (PEEK–5NH$_2$), 60 s (PEEK–60NH$_2$) and 180 s (PEEK–180NH$_2$) treatment times.

water from air and, therefore, increase the O/C ratio. These processes compensate each other, so that oxygen content is not much changed by the ammonia plasma treatment (Fig. 4).

In contrast, nitrogen content changes considerably. Nitrogen should neither be found on untreated PEEK nor on the oxygen plasma-treated surfaces. But it was detected in small concentrations as a contamination from foil manufacturing. A high N/C ratio was detected already after only a few seconds of ammonia plasma treatment and a decrease of nitrogen content after some minutes of treatment (Fig. 5). Maximum nitrogen was found for treatment times of 5–30 s. Of course, only part of this nitrogen belongs to amino groups.

A maximum amino group density of 2.8% –NH$_2$/C was detected for 5 s plasma treatment, which is not much reduced with longer plasma treatment durations (Fig. 6). Correspondingly, cell adhesion is enhanced and is comparable to COL–control for all treatment conditions (see Fig. 7) during the first 30 min of cell culture. A quantitative differentiation regarding cell adhesion between the conditions is not possible due to the generally high uncertainty in the results.

Interestingly, integrin expression did not show any differences among the samples (Fig. 8), and the two tested integrin subunits β_1 and β_3, which normally bind to the extracellular matrix proteins collagen and bone sialo protein, respectively. This might be interpreted as a hint for a basically unspecific action mechanism of the amino groups independent of integrin receptor binding, which could be a charge mediated attachment of the whole cell by additional adhesion mechanisms, e.g., the hyaluronan coat [25].

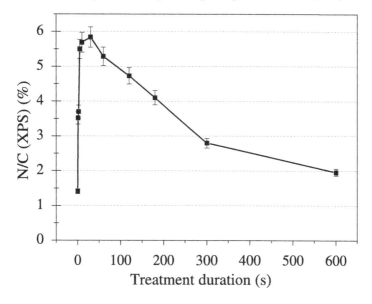

Figure 5. Results of XPS N/C measurements on PEEK surfaces after different durations of treatment with ammonia plasma.

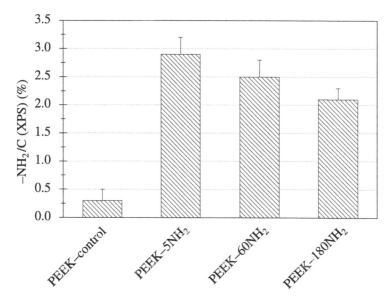

Figure 6. Amino group density $-NH_2/C$ on plasma-treated PEEK surfaces after 0 s (PEEK–control), 5 s (PEEK–5NH$_2$), 60 s (PEEK–60NH$_2$) and 180 s (PEEK–180NH$_2$) treatment times.

3.2. Functionalization of Titanium

The procedure applied in the present work to coat Ti with an amino-functional layer starts with cleaning in oxygen plasma. This treatment produces surfaces with en-hanced surface energy for improved adhesion of the plasma polymer layer. Such

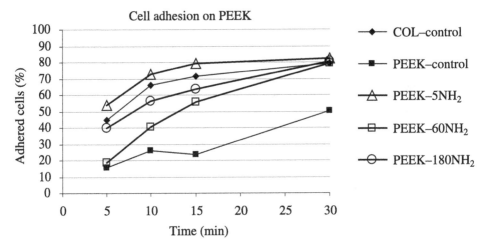

Figure 7. Adhesion of osteoblasts on functionalized PEEK. Note that NH_2-groups on the PEEK surface result in similar cell adhesion as on collagen-coating and is significantly enhanced for all PEEK modifications (and COL–control) at 15 and 30 min *vs* the control.

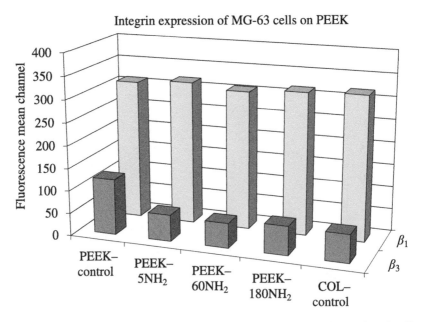

Figure 8. Integrin expression of osteoblasts on PEEK surfaces ammonia plasma functionalized for 60 and 180 s is compared with untreated PEEK–control and COL–control. Note that no significant changes in the β-integrin subunits β_1 and β_3 could be observed although cell adhesion in the initial phase was increased.

'oxidized' surfaces carry a negative surface charge, which is interesting for answering the question posed above. Therefore, they were included in the investigations.

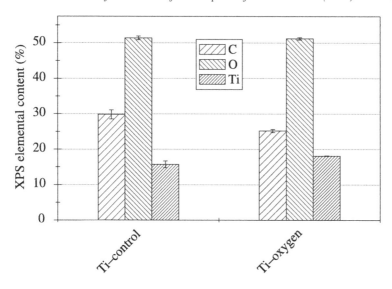

Figure 9. Chemical composition of Ti–oxygen compared to Ti–control determined by XPS measurements.

The initial Ti–control is characterized by a native oxide layer of about 4–5 nm thickness, which is covered by noticeable amounts (typically 30% C) of different hydrocarbons [36] (Fig. 9).

Oxygen plasma treatment yields a 'clean' Ti surface. This surface is characterized not only by reduced carbon content (Figs 9 and 11) but also by an enhanced hydroxyl group density. Here, on Ti–oxygen about 8.5% –OH/Ti was found compared to 3.8% –OH/Ti for the Ti–control. Ti–control surfaces are hydrophobic with typical water contact angles of $78 \pm 3°$ and a surface energy about 35 mJ/m^2 (Fig. 10). Ti–oxygen surfaces had a decreased water contact angle of $12 \pm 3°$ and an increased surface energy of 67 mJ/m^2, which is typical for clean TiO$_2$ surfaces [36].

The coating with the plasma polymer also creates hydrophilic surfaces. Ultrathin coated (50 ± 10 nm thickness) Ti–NH$_2$ surfaces reveal a water contact angle of $48 \pm 3°$ and a surface energy of about 53 mJ/m^2 (Fig. 10). This is of the same order as PEEK–NH$_2$. In contrast to plasma functionalized PEEK, the water contact angle decreases with storage duration reaching 40° after 100 days. Ti–NH$_2$ deposited on Ti–oxygen smoothed the surface, the roughness R_q was reduced from 4.1 to 2.3 nm. Note that these films cover the surface completely. No Ti could be detected by XPS (Fig. 11).

The N/C ratio of the films was $27.5 \pm 2.5\%$, which is near the (theoretical) N/C value of poly(allylamine) (33.3%). Compared to the functionalization of PEEK, oxygen was detected in the coating. Immediately after plasma deposition, the O/C ratio was 4%. High resolution XPS C$_{1s}$ (Fig. 12) spectrum of Ti–NH$_2$ verifies the existence of different C–C, C–H, and C–N bonds and also oxygen-containing bonds.

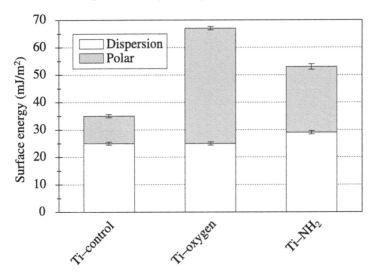

Figure 10. Surface energy and its components of the investigated surfaces Ti–control, Ti–oxygen and Ti–NH$_2$.

Amino group density cannot be calculated directly from such peak fits. It was calculated from derivatization reaction with TFBA to be $2.5 \pm 0.5\%$ –NH$_2$/C. The stability of these coatings was proved *in vitro* both in air for a storage time of one year and also in aqueous solution in an ultrasonic water bath for 10–15 min. XPS measurements, however, verify changes in the surface composition after *in vivo* [37] investigations 56 days after implantation.

As already mentioned, natural Ti surfaces exhibit already acceptable but not yet optimal biocompatibility. Basically, this fact is a result of the OH termination of the oxide layer. Hence, increasing OH should be advantageous. Indeed, our Ti–oxygen sample with twofold OH content exhibits a small but significant increase of cell adhesion (Fig. 13). Interestingly, this increased adhesion still can be outperformed by Ti–NH$_2$. This is a good demonstration that the chemical nature of surface functionalization rather than surface energy (see Fig. 10) alone determines cell adhesion. This conclusion is further bolstered by results of cytoskeleton visualization (Fig. 14).

The comparison of the actin cytoskeleton of MG-63 osteoblasts on oxygen- and NH$_2$-functionalized titanium revealed that oxygen plasma already induces cell's actin formation comparable to collagen-coated controls (Fig. 14). However, effects are much stronger on Ti–NH$_2$. In addition, as in the case of PEEK, integrin expression did not show any significant differences between Ti–NH$_2$ and the controls (Fig. 15). This underlines the assumption of basically unspecific, charge mediated attachment of the cells by additional adhesion mechanisms [25].

3.3. Improved Cell Growth on Amino Functionalized PEEK and Ti Surfaces

With respect to our initial question concerning a possibly determining influence of amino groups on cell adhesion it is interesting to directly compare results on PEEK

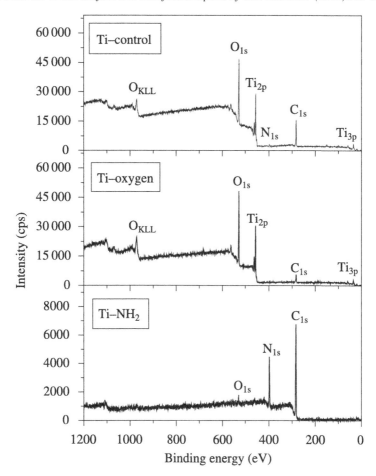

Figure 11. XPS wide scan spectra of Ti–oxygen and Ti–NH$_2$ compared to Ti–control.

and Ti. Figure 16 demonstrates the actin cytoskeleton organization on PEEK–5NH$_2$, Ti–NH$_2$ and the respective control substrates. Clearly, amino functionalization results in an acceleration of actin polymerization and organization after 60 min of cell culture. Similar trends are obvious on both materials. To a certain extent, the effect obscures the influence of the different substrate materials. These results are a clear demonstration of the determining role of amino functionalization.

For cell adhesion we also observe similar trends on Ti and PEEK (Fig. 17). The controls exhibit less adhesion while amino functionalization of the surfaces results in higher and comparable adhesion. Both amino functional surfaces also outperform the collagen coatings, although collagen is the gold standard concerning receptor mediated adhesion [32, 33].

An interesting detail of the actin cytoskeleton architecture is the high density of long actin filaments which are organized throughout the cell, especially on Ti–NH$_2$. A similar effect could be seen for the comparison of Ti–oxygen *vs* Ti–NH$_2$ (Fig. 14). It seems that Ti–NH$_2$ outperforms all other modifications to a certain

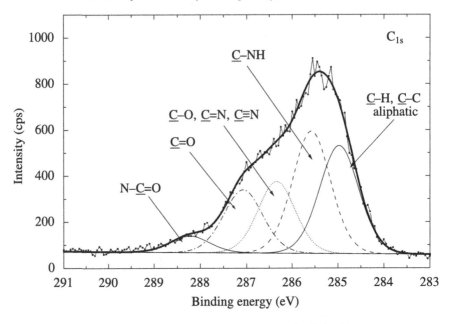

Figure 12. Highly resolved XPS C_{1s} peak fit of Ti–NH_2.

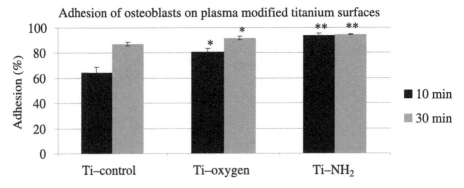

Figure 13. Adhesion of osteoblasts on functionalized Ti after 10 and 30 min. Note that not only amino groups on the surface induce increased attachment of the cells but also oxygen plasma (*$p < 0.01$, **$p < 0.005$ *vs* Ti–control).

degree. This is inasmuch noticeable since the amino group surface densities on PEEK–NH_2 and on Ti–NH_2 are comparable. However, the chemical structures of the modifications are different. While for PEEK–NH_2, the amino groups are located two-dimensionally on flat surfaces, the amino groups on Ti–NH_2 are distributed three-dimensionally in a 50 nm thin film which in addition may be susceptible to swelling. In effect, the polyelectrolyte layer on Ti–NH_2 may present a higher capacity for electrostatic interactions with proteins and other biomolecules. No changes could be observed for the beta-integrin subunits of the osteoblasts (Figs 8 and 15) as well as for the alpha-integrin subunits ($\alpha_2, \alpha_3, \alpha_5$: not shown) which would indicate specific cell binding reactions to the surface.

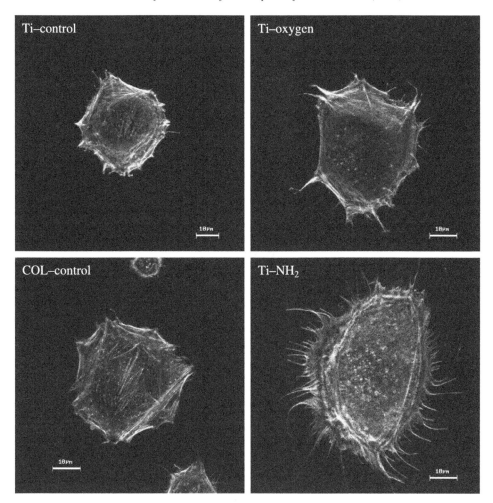

Figure 14. Comparison of the actin cytoskeleton organization of MG-63 osteoblasts on Ti–oxygen and Ti–NH_2 after 60 min. Note that oxygen plasma already induces cell's actin formation and the increase of the cell area compared to collagen-coated controls but not to the same extent as amino groups on the surface. Confocal microscopy (LSM 410, Carl Zeiss, Jena, Germany).

The results presented here therefore confirm and expand screening of polymer surface modifications with different functional groups, which demonstrated the advantage of amino groups [38, 39]. Amino groups also played a major role in our earlier investigations on Ti regarding the hyaluronan-dependent adhesion in the first cell's attachment phase [10, 25, 32]. Hyaluronan is negatively charged and the amino functionalized Ti surface acts as a positively charged counterpart. Probably, this is a very strong counterpart since the morphology of osteoblasts demonstrates an extremely flattened phenotype on NH_2-modified surfaces [40], resulting convincingly in improved time-dependent adhesion strength [41].

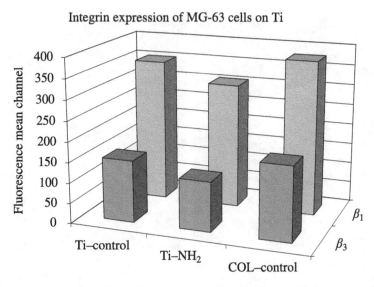

Figure 15. Integrin expression of osteoblasts on amino functionalized Ti surfaces. Note that no significant changes in the β-integrin subunits β_1 and β_3 could be observed although cell adhesion in the initial phase was increased.

4. Summary

Low pressure ammonia plasma treatment of PEEK, and allylamine plasma polymer deposition on Ti were used to evaluate plasma-assisted amino functionalization as a potential surface modification strategy with general usefulness for improvement of bone implant materials biocompatibility. In particular, acceleration of early bone cell adhesion was the focus of interest. PEEK and Ti were chosen for these investigations since they are widely used bone replacement materials with very different cell adhesion characteristics and very different surface chemistries. Samples of PEEK with different surface amino group densities were prepared using different treatment times. They were compared with untreated PEEK and collagen I (positive control). Samples of Ti with allylamine plasma polymer coating were compared with pristine oxide covered Ti, oxygen plasma-treated Ti, collagen I coated Ti, and treated and untreated PEEK. The chemical properties of all surfaces were characterized by surface energy determination from contact angle measurements and by XPS. MG-63 human osteoblastic cells were used as representatives of bone cells to obtain informations on cell adhesion, actin cytoskeleton organization and integrin expression during the early phase of cell culture. On amino functionalized PEEK, cell adhesion is enhanced during the first phase of cell culture and is comparable to adhesion on collagen control samples for all treatment conditions. A differentiation between the conditions is not possible as integrin expression did not show any differences. Oxygen plasma-treated Ti with twofold surface OH content compared to pristine Ti exhibited a significant increase of cell adhesion. This increased adhesion still could be outperformed by allylamine plasma polymer coated Ti. Integrin ex-

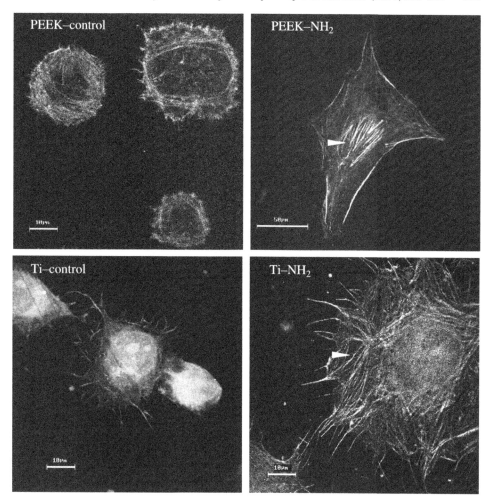

Figure 16. Organization of the actin cytoskeleton of MG-63 osteoblasts after 60 min on amino functionalized PEEK (above) and titanium (below). Note that amino groups on the surface enlarge not only the cell size with time but also the actin filament formation (arrowheads). Confocal microscopy (LSM 410, Carl Zeiss, Jena, Germany).

pression did not show any differences between all samples. The direct comparison of cell culture on different PEEK and Ti samples showed that amino functionalization resulted in a similar acceleration of actin polymerization and organization on both materials. The same was true for cell adhesion. Indeed, the presence of amino groups solely has great influence on cell adhesion. This seems to be a general effect, since differences in modification methods and substrates were of minor influence.

Acknowledgements

The authors thank Urte Kellner, Uwe Lindemann and Gerd Friedrichs (INP) as well as Annelie Peters (University of Rostock) for excellent assistance. This study was

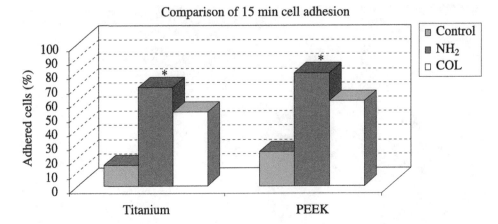

Figure 17. Comparison of 15 min MG-63 cell adhesion on NH_2 functionalized PEEK and NH_2-functionalized Ti implant surfaces. Note that the amino groups (NH_2) on PEEK as well as on Ti surfaces result in the same significant increase ($^* p < 0.05, n = 3$–4) of attached cells *vs* the untreated surfaces (control), respectively.

supported by the Federal State of Mecklenburg-Vorpommern, the Helmholtz Association of German Research Centres (UR 0402210), and by the Federal Ministry of Education and Research (grant no. 13N9779, Campus PlasmaMed).

References

1. M. Long and H. J. Rack, *Biomaterials* **19**, 1621 (1998).
2. S. M. Kurtz and J. N. Devine, *Biomaterials* **28**, 4845 (2007).
3. N. Gomathi, A. Sureshkumar and S. Neogi, *Current Sci.* **94**, 1478 (2008).
4. K. B. Kwarteng and C. Stark, *SAMPE Quart.* **22**, 10 (1990).
5. K. A. Jockish, S. A. Brown, T. W. Brauer and K. Merritt, *J. Biomed. Mater. Res.* **26**, 133 (1992).
6. D. Buser, N. Broggini, M. Wieland, R. K. Schenk, A. J. Denzer, D. L. Cochran, B. Hoffmann, A. Lussi and S. G. Steinemann, *J. Dental Res.* **83**, 529 (2004).
7. F. Schwarz, M. Herten, M. Sager, M. Wieland and J. Becker, *J. Clin. Periodontol.* **34**, 78 (2007).
8. F. Rupp, L. Scheideler, N. Olshanska, M. de Wild, M. Wieland and J. Geis-Gersdorfer, *J. Biomed. Mater. Res. A* **76**, 323 (2006).
9. C. von Wilmowsky, L. Müller, R. Lutz, U. Lohbauer, F. Rupp, F. W. Neukam, E. Nkenke, K. A. Schlegel and F. A. Müller, *Adv. Eng. Mater.* **10**, B61 (2008).
10. B. Finke, F. Lüthen, K. Schröder, P. D. Müller, C. Bergemann, M. Frant, A. Ohl and J. B. Nebe, *Biomaterials* **28**, 4521 (2007).
11. O. Noiset, C. Henneuse, Y.-J. Schneider and J. Marchand-Brynaert, *Macromolecules* **30**, 540 (1997).
12. S. D. Cook and A. M. Rust-Dawicki, *J. Oral Implantol.* **21**, 176 (1995).
13. C. Moseke, W. Braun and A. Ewald, *Adv. Eng. Mater.* **11**, B1 (2009).
14. P. Becker, P. Zeggel, F. Lüthen, B. Nebe, J. Rychly and H. G. Neumann, *Key Eng. Mater.* **218–220**, 653 (2002).
15. L. M. Sun, C. C. Berndt, K. A. Gross and A. Kucuk, *J. Biomed. Mater. Res.* **58**, 570 (2001).

16. P. Becker, B. Nebe, F. Lüthen, J. Rychly and H. G. Neumann, *J. Mater. Sci. Mater. Med.* **15**, 437 (2004).

17. S. W. Ha, M. Kirch, F. Birchler, K.-L. Eckert, J. Mayer, E. Wintermantel, C. Sittig, I. Pfund-Klingenfuss, M. Textor, N. D. Spencer, M. Guecheva and H. Vonmont, *J. Mater. Sci. Mater. Med.* **8**, 683 (1997).

18. S.-W. Ha, J. Mayer, B. Koch and E. Wintermantel, *J. Mater. Sci. Mater. Med.* **5**, 481 (1994).

19. S.-W. Ha, K. L. Eckert, F. Birchler, A. Gisep, H. Gruner and E. Wintermantel, in: *Bioceramics*, R. Z. LeGeros and J. P. LeGeros (Eds), Vol. 11, pp. 193–196. World Scientific Publishing Co., Singapore (1998).

20. M. Morra, C. Cassinelli, A. Capri, R. Giardino and M. Fini, *Biomed. Pharmacother.* **60**, 365 (2006).

21. M. Morra, C. Cassinelli, G. Cascardo, L. Mazzucco, P. Borzini, M. Fini, G. Giavaresi and R. Giardino, *J. Biomed. Mater. Res. A* **78**, 449 (2006).

22. D. Briem, S. Strametz, K. Schroeder, N. M. Meenen, W. Lehmann, W. Linhart, A. Ohl and J. M. Rueger, *J. Mater. Sci. Mater. Med.* **16**, 671 (2005).

23. O. Noiset, Y.-J. Schneider and J. Marchand-Brynaert, *J. Biomater. Sci. Polym. Edn* **10**, 657 (1999).

24. F. Lüthen, R. Lange, P. Becker, J. Rychly, U. Beck and J. B. Nebe, *Biomaterials* **26**, 2423 (2005).

25. B. Nebe, F. Lüthen, B. Finke, C. Bergemann, K. Schröder, J. Rychly, K. Liefeith and A. Ohl, *Biomol. Eng.* **24**, 447 (2007).

26. K. Schröder, A. Meyer-Plath, D. Keller, W. Besch, G. Babucke and A. Ohl, *Contrib. Plasma Phys.* **41**, 562 (2001).

27. G. Beamson and D. Briggs, *High Resolution XPS of Organic Polymers: the Scienta ESCA 300 Database*. Wiley, Chichester (1992).

28. L. J. Gerenser, J. F. Elman, M. G. Mason and J. M. Pochan, *Polymer* **26**, 1162 (1985).

29. J. F. Friedrich, W. E. S. Unger, A. Lippitz, I. Koprinarov, G. Kühn, St. Weidner and L. Vogel, *Surf. Coat. Technol.* **116–119**, 772 (1999).

30. D. K. Owens and R. C. Wendt, *J. Appl. Polym. Sci.* **13**, 1741 (1969).

31. W. Rabel, *Farbe und Lacke* **77**, 997 (1971).

32. J. B. Nebe and F. Lüthen, in: *Metallic Biomaterial Interfaces*, J. Breme, C. J. Kirkpatrick and R. Thull (Eds), pp. 179–182. Wiley-VCH, Weinheim (2008).

33. R. O. Hynes, *Cell* **69**, 11 (1992).

34. S. W. Ha, R. Hauert, K. H. Ernst and E. Wintermantel, *Surf. Coat. Technol.* **96**, 293 (1997).

35. A. A. Meyer-Plath, K. Schröder, B. Finke and A. Ohl, *Vacuum* **71**, 391 (2003).

36. M. Textor, C. Sittig, V. Franchinger, S. Tosatti and D. Brunette, in: *Titanium in Medicine*, D. M. Brunette, P. Tengvall, M. Textor and P. Thompson (Eds), pp. 171–229. Springer, Berlin (2001).

37. A. Hoene, U. Walschus, B. Finke, S. Lucke, B. Nebe, K. Schröder, A. Ohl and M. Schlosser, *Acta Biomaterialia* **6**, 676 (2010).

38. J. H. Lee, H. W. Jung, I. K. Kang and H. B. Lee, *Biomaterials* **15**, 705 (1994).

39. S. Kamath, D. Bhattacharyya, C. Padukudru, R. B. Timmons and L. P. Tang, *J. Biomed. Mater. Res. A* **86**, 617 (2008).

40. H. Jesswein, A. Weidmann, B. Finke, R. Lange, U. Beck, S. Stählke, K. Schroeder and J. B. Nebe, *Eur. J. Cell Biol.* **88**, Suppl. 59, 48 (2009) (ISSN 0171-9335).

41. H. Jesswein, B. Finke, R. Lange, S. Stählke, U. Beck, K. Schröder and J. B. Nebe, *BIOmaterialien* **S1**, 74 (2009).

Helium Plasma Treatment to Improve Biocompatibility and Blood Compatibility of Polycarbonate

Nageswaran Gomathi [a], **Debasish Mishra** [b], **Tapas Kumar Maiti** [b] and **Sudarsan Neogi** [a,*]

[a] Department of Chemical Engineering, India Institute of Technology Kharagpur, Kharagpur-721302, India
[b] Department of Biotechnology, Indian Institute of Technology Kharagpur, Kharagpur-712302, India

Abstract

Radiofrequency discharge of helium gas at low pressure was employed to modify surface properties of polycarbonate. The effects of process parameters on wettability and plasma etching were determined by monitoring surface energy and weight loss, respectively. Quadratic equations for surface energy and weight loss, in terms of process variables, namely power, pressure, flowrate and treatment time were developed. Multiple response optimization was performed using central composite design (CCD) of response surface methodology (RSM) to maximize the surface energy and minimize the weight loss. Helium plasma treated polycarbonate resulted in increased hydrophilicity. From optical emission spectroscopic studies helium was identified as excited and metastable atom and ions which caused surface chemistry and morphology changes. Enhanced biocompatibility in terms of increased cell adhesion and proliferation was observed for all plasma treatment conditions. Confluent cell growth was observed with helium plasma treated polycarbonate. Both reduced platelet adhesion and increased partial thromboplastin time (increased to 204 s from 128 s corresponding to untreated polycarbonate) confirm the improved blood compatibility of plasma treated polycarbonate.

Keywords

Helium plasma treatment, response surface methodology, polycarbonate, biocompatibility, blood compatibility

1. Introduction

Polymers, as a result of their desirable physical properties, are of great interest for various applications such as food packaging, automobiles, biomedical applications, etc. [1]. They are widely used as biomaterials and find specific applications

---

* To whom correspondence should be addressed. Tel.: +91-03222 28 3936; Fax: +91-03222 28 2250; e-mail: sneogi@che.iitkgp.ernet.in

Surface and Interfacial Aspects of Cell Adhesion
© Koninklijke Brill NV, Leiden, 2010

in orthopedic implant, cardiac implant, dental implant, soft tissue implant, biosensors and biomedical devices [2]. Nevertheless, problems with respect to suitable surface properties of these materials limit their biomedical applications. Indeed, the hydrophobic nature of many polymeric surfaces limits their adhesion property. Surface properties such as surface energy, surface chemistry, surface topography, surface charges, crystallinity and water content in the surface layer are the factors that greatly influence the interaction between the biological environment and artificial materials [3–7]. Among the above-mentioned surface properties, the surface energy of the biomaterials plays an important role in protein adsorption, adhesion and spreading of cells and platelets. Due to the complex nature of adsorbed proteins, even small changes in surface energy produce remarkable changes in the biological response toward the surface [8]. Biocompatibility of polymers is influenced by various physical and chemical properties. When a synthetic material interacts with biological environment, proteins are adsorbed on its surface *via* hydrophobic and electrostatic interactions. Depending on the surface chemistry, non-adhesive proteins like albumin or adhesive proteins like fibronectin are adsorbed on the surface. Besides, surface chemistry influences the properties associated with an adsorbed protein such as its bond strength, conformation, exchange and restructuring.

The increasing demand for the desirable polymers surface properties has led to the development of various surface modification methods without altering their bulk properties. The treatments commonly used to improve the surface properties of the polymers are physical (physical adsorption and Langmuir–Blodgett film method), chemical (oxidation by acid treatment, ozone treatment and chemisorption) and ionized gas surface treatment methods [9–13]. The common ionized gas treatment methods are corona, flame, ion beam, and plasma treatment methods. Biocompatibility and blood compatibility of polymers are improved by a number of methods including surface modification by plasma or ion beam, polymer grafting and biological methods of immobilizing anticoagulants [14–18]. Surface modification of polymers by helium plasma treatment involves formation of reactive sites on the surface through formation of free radicals by ion bombardment. It has proved to be useful for enhancing adhesion property, biocompatibility, printability and dyeability. Helium plasma activates the polymer surface by leaving free radicals on its surface which then interact with oxygen from the air. Thus oxygen containing functional groups are incorporated onto the polymer surface during post-plasma reactions. The aim of this paper is to show how helium plasma treatment alters the wettability, surface chemistry, surface morphology, biocompatibility and blood compatibility of polycarbonate. Various process parameters are optimized using response surface methodology. Active species causing changes in the surface properties of polycarbonate are also investigated.

2. Materials and Methods

2.1. Plasma Treatment

Polycarbonate samples, with a thickness of 1 mm, were cut into 5 cm × 2.5 cm sheets, cleaned with isopropyl alcohol in an ultrasonic cleaner for 15 min and then dried in a hot air oven for 30 min at a temperature of 50°C. The cleaned polymer samples were treated in a plasma reactor, M-PECVD-1A [S], procured from M/s. Milman Thin Film Systems, Pune, India. The two electrodes, powered top electrode with perforations and grounded bottom electrode, are capacitively coupled. Substrates were placed on the grounded electrode and the helium gas was introduced at the desired flowrate into the reactor through the top perforated powered electrode. RF power was supplied to the reactor at the frequency of 13.56 MHz to generate plasma and the samples were treated for the required duration. A detailed description of the plasma treatment of polymers using this approach is presented elsewhere [19]. Plasma was characterized by the emission spectra in the visible and UV wavelength ranges using an optical emission spectrophotometer (Model: HR4000-CG-UV-NIR, Ocean Optics Inc., USA). Emissions from excited species were collected at the viewport of the PECVD reactor.

2.2. Characterization

Contact angle measurements were done on polymer surfaces before and after plasma treatment, to study the wettability changes, by the sessile drop method using Rame-Hart 500-F1 advanced goniometer (Rame-Hart Instrument Co., Netcong, NJ, USA) at ambient humidity and temperature. In order to determine the surface energy of the polymer, static contact angles of three liquids, namely deionized water, formamide and diiodomethane, with different surface tension values covering a wide range of polar and dispersion components were measured on the material surface. Each contact angle reported in this work is the average of the values obtained from at least five different positions on the polymer sample surface. To study the effect of etching, the samples were weighed before and after plasma treatment with a precision electronic balance (A&D Co. Ltd., Model: HR-60, Tokyo, Japan). The percentage weight loss was calculated as follows:

Percentage weight loss = (Initial weight − Final weight) × 100/Initial weight.

The effects of various plasma variables on the surface chemistry of polymers were investigated from FTIR spectra of untreated and modified polymers obtained in attenuated total reflectance (ATR) mode using a Thermo Nicolet Nexus 870 FTIR (Thermo Nicolet Corporation North America, Madison, WI) equipped with a ZnSe crystal. The samples were recorded in the range 4000–600 cm^{-1} with a resolution of 4 cm^{-1}. The surface morphology of untreated and various plasma treated polymers at the central runs (S. no.: 17–24 in Table 2) was examined with a JEOL JSM-5800 scanning electron microscope (JEOL, Tokyo, Japan). SEM examination of polymers was conducted to provide a qualitative assessment of surface

Table 1.
Coded levels and ranges of variables

Variables	Range and coded levels				
	Axial $(-\alpha)$	Low (-1)	Centre (0)	High $(+1)$	Axial $(+\alpha)$
Power (W)	20.0	65.0	110.0	155.0	200.0
Pressure (Pa)	13.3	16.7	20.0	23.3	26.7
Flowrate (sccm)	5.0	10.0	15.0	20.0	25.0
Treatment time (min)	2.0	4.0	6.0	8.0	10.0

roughness. Samples were coated with a thin conductive layer of gold in vacuum conditions prior to analysis.

2.3. Experimental Design

The effects of the process variables, namely power, pressure, helium gas flowrate and treatment time, on the surface energy and weight loss were studied *via* central composite design. The selection of variables and their levels based on the limitations of the PECVD reactor used and preliminary experimental studies are presented in Table 1. Six replicates were employed at center point to represent process variations. The experiments were run in random order to make the observations as independently distributed random variables which is not affected by any system drift [20].

The second-order model as given in equation (1) was proposed to predict the responses surface energy and weight loss.

$$Y = \beta_0 \sum_{i=1}^{4} \beta_i X_i + \sum_{i=1}^{4} \beta_{ii} X_i^2 + \sum_{i=1}^{4} \sum_{j=i+1}^{4} \beta_{ij} X_i X_j + e, \tag{1}$$

where Y is the response, β_0 the constant coefficient, X_i ($i = 1$–4) are non-coded variables, and β_i, β_{ii} and β_{ij} (i and $j = 1$–4) are linear, quadratic, and second-order interaction coefficients, respectively, and e is the experimental error.

2.4. Statistical Analysis

The statistical analysis of the experimental design results was performed using statistical software Design-Expert 7.1 trial version. The quality of fitness of the model was checked by the coefficient of regression (R^2). When R^2 approaches unity, the empirical model represents an excellent fit to the actual data. Statistical significances of the model and the coefficients were judged by the F-test and the t-test, respectively. A value of probability less than 0.05 was taken as the level of significance.

2.5. Optimization of Process Conditions

Plasma process conditions were optimized for two independent parameters while fixing the other parameters at coded zero level. This was performed by the multiple response method using desirability function, which reflects the desirable range for each response (d_i). The desirable ranges vary between 0 and 1 corresponding to least and most desirable conditions, respectively. A geometric mean of transformed responses giving the simultaneous objective function (D) is given below:

$$D = (d_1 \times d_2 \times \cdots \times d_n)^{1/n},$$

where n is the number of responses in the measure and responses lying outside the desirability range make the overall function zero.

2.6. Cell Adhesion

L929 fibroblast cells preserved by freezing were thawed to room temperature and suspended in an appropriate amount of Dulbecco's Modified Eagle Medium (DMEM) containing 10% fetal bovine serum. The supernatant was removed by centrifuging at 300g for 10 min. After washing with DMEM, the cells were seeded in a T-25 tissue culture flask and incubated in humidified atmosphere of 5% CO_2 in air at 37°C. Confluent cells were trypsinized and seeded onto the untreated and helium plasma treated polycarbonate surfaces. Approximately 2.5×10^4 cells were seeded on each polymer sample and incubated for 72 h. The adhered cells were washed with phosphate buffered saline (PBS) and fixed with 2.5% glutaraldehyde. Series of graded alcohols were used to dehydrate the cells and dried to critical point to avoid the problem of surface tension of evaporating water from the naturally hydrated specimens. A gold sputter coating was deposited on both untreated and treated samples in vacuum before examination by SEM using secondary electrons in a JEOL JSM-5800 microscope.

2.7. MTT Assay

An MTT (3-(4,5-dimethylthiazol-2-yl)-2,5-diphenyltetrazolium bromide) assay was performed to quantitatively assess the number of L929 viable cells attached and grown on polymer film surfaces after 72 h incubation. The colored formazan product, formed by reduction of MTT during incubation, and which corresponds to the viable cells, was assayed spectrophotometrically at 595 nm using an ELISA plate reader.

2.8. In-vitro Blood Compatibility Test

2.8.1. Exposure of Materials to Platelet-Rich Plasma

Blood from human volunteers was collected into the anticoagulant: CPD-A (Citrate phosphate dextrose anticoagulant). Blood was centrifuged at 2500 rpm for 5 min to prepare platelet-rich plasma (PRP). Platelet-poor plasma was prepared by centrifuging blood at 4000 rpm for 15 min. Platelet count in PRP was adjusted between 2.0×10^8 per ml and 2.5×10^8 per ml. The test materials were placed in polystyrene culture plates and immersed in phosphate buffered saline before they were exposed to PRP. To each plate 2 ml of PRP was added and from this 0.5 ml was collected immediately for partial thromboplastin time (PTT) analysis. The remaining 1.5 ml was used to expose the materials for 30 min under agitation at 75 ± 5 rpm using a shaker incubator thermostatted at $35 \pm 2°C$. Three replicates were tested for each sample.

2.8.2. Plasma Coagulation (Partial Thromboplastin Time Assay)
The PRP samples (initial and after 30 min) were centrifuged at 4000 rpm for 15 min and platelet-poor plasma (PPP) was aspirated. The partial thromboplastin time in each sample was determined using a reagent kit obtained from Diagnostica Stago on a Start 4 coagulation analyzer (France).

2.8.3. Platelet Adhesion Test
After agitating in PRP, the samples were washed three times with PBS and each sample was fixed with 2% glutaraldehyde overnight at 4°C and then dehydrated. Samples in 100% alcohol were critical point dried, gold sputter coated and viewed using a scanning electron microscope (Hitachi S2400).

3. Results and Discussion

3.1. Wettability and Weight Loss

Untreated polycarbonate exhibited surface energy (γ_s) of 33.1 mJ/m^2, with the polar and dispersion components of $\gamma_s^p = 1.3$ mJ/m^2 and $\gamma_s^d = 31.8$ mJ/m^2, respectively. Table 2 depicts the increase in the surface energy of helium plasma treated polycarbonate for all the conditions. The surface energy for helium plasma treated polycarbonate ranged from 31.9 mJ/m^2 to 41.7 mJ/m^2. The lowest surface energy was observed at the lowest pressure and highest surface energy was observed at the lowest power. With increasing power, the surface energy initially decreased up to 110 W and then increased as shown in Fig. 1, the main effects plot showing the effect of power on surface energy. However, weight loss of polycarbonate after helium plasma treatment increased gradually with increasing power. Increased surface energy with increasing pressure attributed to increasing ion density at higher pressure was observed. Changes in the levels of flowrate of helium gas were found not to influence the surface energy. Prolonged treatment time exhibited negative effect on the surface energy. This may be due to the higher rate of etching with respect to the addition of polar groups. This same trend was also reported with argon–oxygen plasma treated polycarbonate [21]. It was reported that in the first few seconds of plasma treatment, oxidation and removal of surface contaminants occurred. Further treatment caused functional group attachment. But the prolonged treatment time caused excessive chain scission leading to a layer of low-molecular-weight fragments on the surface.

Table 2.
Surface energy and its components (mJ/m^2) and weight loss for helium plasma treated polycarbonate samples

S. no.	Power (W)	Pressure (Pa)	Flowrate (sccm)	Treatment time (min)	γ_s^p	γ_s^d	$P = \gamma_s^p/(\gamma_s^p + \gamma_s^d)$	γ_s	Wt loss (%)
1	65	16.7	10	4	2.2	34.4	0.06	36.6	0.09
2	155	16.7	10	4	6.4	29.5	0.18	35.9	0.14
3	65	23.3	10	4	9.2	30.0	0.24	39.3	0.09
4	155	23.3	10	4	3.3	35.6	0.08	38.9	0.10
5	65	16.7	20	4	5.8	33.0	0.15	38.7	0.09
6	155	16.7	20	4	4.9	30.0	0.14	34.9	0.12
7	65	23.3	20	4	6.7	31.4	0.17	38.1	0.06
8	155	23.3	20	4	7.0	28.3	0.20	35.3	0.11
9	65	16.7	10	8	2.6	32.3	0.07	34.8	0.10
10	155	16.7	10	8	3.9	30.5	0.11	34.4	0.17
11	65	23.3	10	8	4.3	32.1	0.12	36.4	0.12
12	155	23.3	10	8	6.5	30.3	0.18	36.8	0.14
13	65	16.7	20	8	3.1	33.5	0.08	36.6	0.14
14	155	16.7	20	8	5.7	30.3	0.16	36.0	0.20
15	65	23.3	20	8	3.8	33.4	0.10	37.2	0.12
16	155	23.3	20	8	4.5	32.2	0.12	36.8	0.17
17	20	20.0	15	6	11.9	29.9	0.28	41.7	0.10
18	200	20.0	15	6	3.9	33.9	0.10	37.8	0.21
19	110	13.3	15	6	4.1	27.9	0.13	31.9	0.11
20	110	26.7	15	6	6.0	30.2	0.16	36.2	0.10
21	110	20.0	5	6	7.0	28.6	0.20	35.6	0.11
22	110	20.0	25	6	4.4	32.8	0.12	37.1	0.13
23	110	20.0	15	2	3.3	33.6	0.09	36.9	0.07
24	110	20.0	15	10	7.0	28.2	0.20	35.2	0.19
25	110	20.0	15	6	4.0	28.8	0.12	32.8	0.10
26	110	20.0	15	6	4.3	28.7	0.13	33.0	0.10
27	110	20.0	15	6	4.0	30.2	0.12	34.2	0.09
28	110	20.0	15	6	4.9	29.0	0.14	33.9	0.10
29	110	20.0	15	6	4.8	30.2	0.14	35.0	0.09
30	110	20.0	15	6	4.5	30.3	0.13	34.9	0.11

3.2. Statistical Analysis

The results of thirty experiments run according to the central composite design presented in Table 2 were analyzed to study the effect of process variables: power, pressure, flowrate and treatment time on surface energy and weight loss. The following polynomial equations for surface energy and weight loss, respectively, were obtained:

$$Y_1 = 49.75 - 0.20X_1 + 0.45X_2 - 0.17X_3 - 3.11X_4$$
$$+ (1.01\text{E}{-}03)X_1X_2 - (1.81\text{E}{-}03)X_1X_3 + (4.58\text{E}{-}03)X_1X_4$$

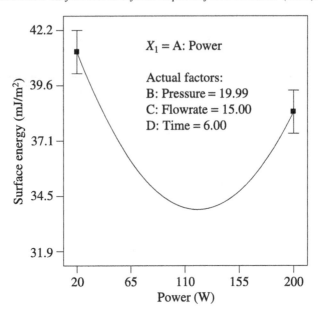

Figure 1. Main effect plot showing the effect of power on surface energy.

$$- 0.03X_2X_3 - (3.75\text{E}-04X_2)X_4 + 0.05X_3X_4 + (7.25\text{E}-04)X_1^2$$
$$+ (4.03\text{E}-03)X_2^2 + 0.02X_3^2 + 0.14X_4^2, \tag{2}$$

$$Y_2 = 0.24 - (6.80\text{E}-04)X_1 - (2.06\text{E}-04)X_2 - (8.14\text{E}-03)X_3$$
$$- 0.03X_4 - (3.31\text{E}-05)X_1X_2 + (1.11\text{E}-05)X_1X_3$$
$$+ (4.17\text{E}-05)X_1X_4 - (1.50\text{E}-04)X_2X_3 + (1.87\text{E}-04)X_2X_4$$
$$+ (8.75\text{E}-04)X_3X_4 + (6.58\text{E}-06)X_1^2 + (7.49\text{E}-05)X_2^2$$
$$+ (1.83\text{E}-04)X_3^2 + (1.77\text{E}-03)X_4^2. \tag{3}$$

Table 3 shows the significance and fitness of the models developed for surface energy and weight loss. Significance of the above models is indicated by the 'Prob > F'-value which was <0.0001 in both cases. Significance of the models was also indicated from the lack of fit values (0.8064 and 0.1175). Models fitness was evaluated by their coefficients of determination, which were found to be 0.9280 and 0.9495, respectively.

The significance of the coefficients is indicated by their p-values presented in Table 4. Except the flowrate, all other process variables were found to be significant for both surface energy and weight loss. Interaction effects among the process variables, namely power–treatment time, pressure–flowrate, and flowrate–treatment time significantly affected surface energy. The interaction effect of flowrate–treatment time affected weight loss significantly. Except for pressure, quadratic terms of all the process variables were found to be significant in affecting surface energy. Quadratic terms of power and treatment time were significant for weight

Table 3.
ANOVA results for the quadratic equations developed for the responses, surface energy and weight loss

Response	S.D.	R^2	Adj. R^2	Pred. R^2	F-value		p-value	
					Model	Lack of fit	Model	Lack of fit
Surface energy of PC	0.77	0.9280	0.8609	0.7342	13.82	0.54	<0.0001	0.8064
% weight loss of PC	0.012	0.9495	0.9024	0.7404	20.16	3.01	<0.0001	0.1175

Table 4.
Probability values of coefficient for helium plasma treated polycarbonate

Model terms	p-value for surface energy	p-value for weight loss
Model	<0.0001	<0.0001
X_1 (Power)	0.0005	<0.0001
X_2 (Pressure)	0.0001	0.0126
X_3 (Flowrate)	0.3598	0.0968
X_4 (Treatment time)	0.0065	<0.0001
$X_1 X_2$	0.4464	0.1031
$X_1 X_3$	0.0524	0.3992
$X_1 X_4$	0.0499	0.2126
$X_2 X_3$	0.0140	0.3992
$X_2 X_4$	0.9899	0.6705
$X_3 X_4$	0.0226	0.0083
X_1^2	<0.0001	<0.0001
X_2^2	0.7659	0.7102
X_3^2	0.0007	0.0548
X_4^2	0.0023	0.0057

loss. Optimized process conditions for helium plasma treated polycarbonate were obtained by solving the quadratic model with the constraints as given in Table 5. Optimum conditions obtained were found to be 65 W, 23.3 Pa, 20 sccm and 4 min with the resultant surface energy and weight loss of 38.4 mJ/m^2 and 0.067%, respectively. The corresponding desirability value was found to be 0.794. As such the optimum process conditions were close to the conditions of experimental run number 7 with surface energy and weight loss of 38.1 mJ/m^2 and 0.06%, respectively. Models developed were validated from the normal probability of residual plot as shown in Fig. 2 in which all residuals were found to lie along a straight line. Correlation of the predicted and actual surface energy and weight loss values as shown in Fig. 3 also confirmed the validity of the quadratic models.

Table 5.
Constraints for optimization for helium plasma treated polycarbonate

Constraints	Goal	Lower limit	Upper limit	Lower weight	Upper weight	Importance
Power (W)	In range	65.0	155.0	–	–	–
Pressure (Pa)	In range	16.7	23.3	–	–	–
Flowrate (sccm)	In range	10.0	20.0	–	–	–
Treatment time (min)	In range	4.0	8.0	–	–	–
Surface energy (mJ/m^2)	Maximize	31.9	41.7	1	1	3
Weight loss (%)	Minimize	0.06	0.21	1	1	3

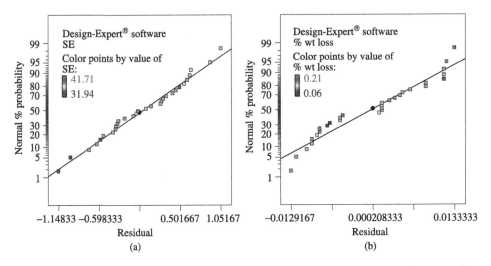

Figure 2. Normal probability plots of residuals for (a) surface energy and (b) weight loss of helium plasma treated polycarbonate (SE: surface energy).

3.3. Plasma Characterization

Various active species in helium plasma were identified from the emission spectrum of helium shown in Fig. 4. From the emission spectrum it can be seen that helium is available in the form of neutral as well as ions but with low relative intensity. This may be due to higher ionization potential of helium as compared to argon. The first ionization potential of helium is 24.59 eV whereas for argon it is just 15.76 eV with the second ionization potential of helium and argon being 54.42 eV and 27.63 eV, respectively. Helium in the metastable state was observed in the range of 389–471 nm as shown in Fig. 4(a). Helium as excited atom was also observed at 492–505 nm. Helium ion was detected from the emission spectrum shown in Fig. 4(b) at 656 nm.

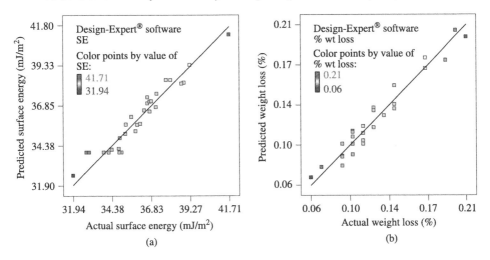

Figure 3. Predicted *versus* actual plots for (a) surface energy and (b) weight loss of helium plasma treated polycarbonate.

3.4. Surface Chemistry

Helium plasma causes surface chemistry changes mainly by physical sputtering due to the low energy ions which may not contribute directly in the surface modification [22, 23]. Physical sputtering involves collision of particles with other energetic particles during which these gain sufficient kinetic energy to overcome the surface binding energy. It also involves reactions with electron and ion bombardment and VUV and UV photons. These species break C–C or C–H bonds and form carbon radicals as reported in argon and argon–oxygen plasma treatments [21, 24]. These radicals react with near-surface polymer chain to form a stable cross-linked structure as observed in argon plasma treatment [24, 25].

It is reported that if oxygen traces are present in the discharge, atomic oxygen creates free radicals and forms oxygen containing groups. This oxygen inclusion inhibits cross-linking [26]. When the polycarbonate was treated with helium plasma, the surface chemistry changes observed were similar to the effect of argon plasma treatment [24]. Table 6 depicts the various chemical changes observed from FTIR spectra of untreated and helium plasma treated polycarbonate. The effect of hydrogen abstraction at alkyl and aromatic groups is indicated by the reduced absorbance values at 765, 1365, 1408, and 2969 cm^{-1} corresponding to alkyl C–H and at 706 and 1008 cm^{-1} corresponding to aromatic C–H. Chain scissions were also observed at carbonate sites, ether group, and aromatic C–C as presented in Table 6.

The effects of hydrogen abstraction and chain scission can also be seen from the reduced peak intensities corresponding to the respective peaks as shown in Fig. 5. Double bond formation at 1666 cm^{-1} as a result of hydrogen abstraction and hydroxyl group formation at 3675 cm^{-1} as an effect of post-treatment reaction between the free radicals and moisture from the atmosphere were also observed.

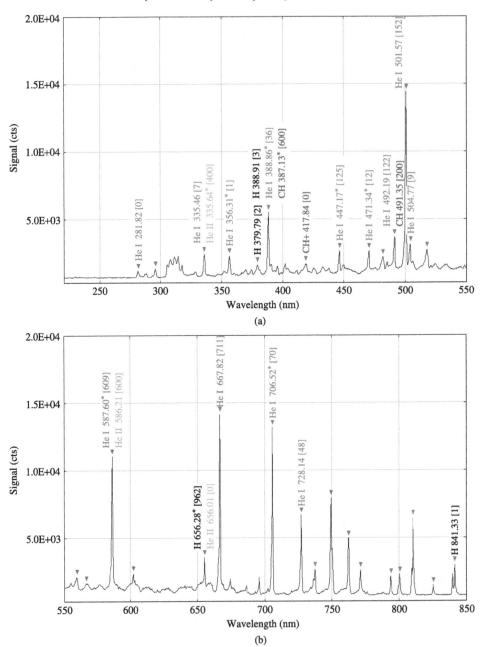

Figure 4. Emission spectrum of helium plasma.

3.5. Surface Morphology

Surface morphologies of the untreated and helium plasma treated polycarbonate samples are shown in Figs 6–10. As observed in the other cases helium plasma also

Table 6.

Surface chemistry changes in helium plasma treated polycarbonate

Groups identified	Wavenumber (cm^{-1})	Comment
Out-of-plane bending mode of hydrogen attached to phenyl ring	706	Hydrogen abstraction
Out-of-plane skeletal vibration of C–H deformation	765	Hydrogen abstraction
C–O–C vibrational mode in ether	1220	Ether bond breakage
	1154	
	1186	
Aromatic in-plane C–C stretching vibration	1502	Dearomatization
Characteristic bond of polycarbonate corresponding to C=O stretching of the carbonyl	1769	Carbonate cleavage
Aromatic CH deformation	1008	Hydrogen abstraction
Asymmetrical CH stretching vibration of CH$_3$	2968	Cleavage of methyl group
C–H bending vibration of CH$_3$	1365	Hydrogen abstraction
	1408	
Stretching vibration of C=C–OH	1666	Double bond formation
	3675	Hydroxyl formation

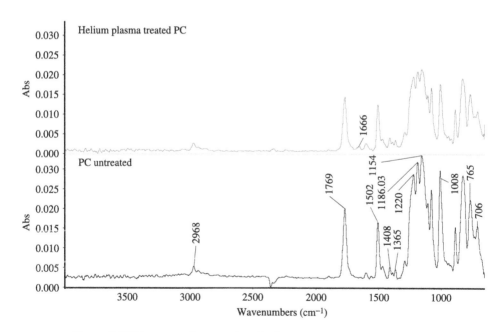

Figure 5. FTIR spectra of untreated and helium plasma treated polycarbonate.

resulted in enhanced etching effect at higher level of power (Fig. 7) [21, 23]. Due to this, a significant level of solid surface fragmentation was observed. Degradation by surface fragmentation can be seen in the SEM micrographs of the helium plasma

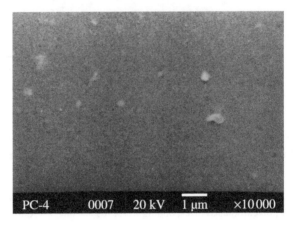

Figure 6. SEM micrograph of untreated polycarbonate.

(a)   (b)

Figure 7. SEM micrographs of helium plasma treated polycarbonate at 13.33 Pa, 15 sccm, 6 min and (a) 20 W and (b) 200 W.

(a)   (b)

Figure 8. SEM micrographs of helium plasma treated polycarbonate at 65 W, 15 sccm, 6 min and (a) 13.3 Pa and (b) 26.7 Pa.

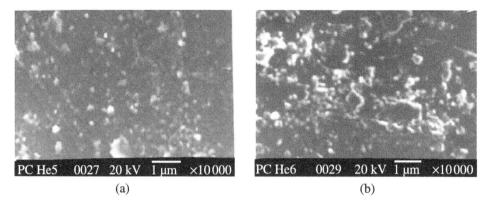

(a)   (b)

Figure 9. SEM micrographs of helium plasma treated polycarbonate at 65 W, 13.33 Pa, 6 min and (a) 5 sccm and (b) 25 sccm.

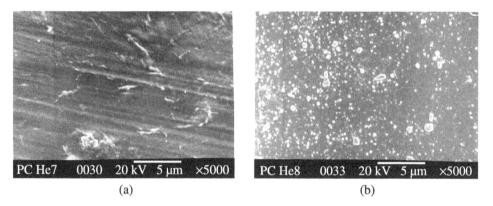

(a)   (b)

Figure 10. SEM micrographs of helium plasma treated polycarbonate at 65 W, 13.33 Pa, 15 sccm and (a) 2 min and (b) 10 min.

treated polycarbonate as small solid particles. During fragmentation low molecular weight carbon centered radicals are formed and leave as volatile products. Highly energetic VUV radiation resulted in rapid etching which causes surface fragmentation. Different levels of pressure (Fig. 8) and flowrate (Fig. 9) did not show much differences in morphology changes. However, enhanced surface roughness can be observed in helium plasma treated polycarbonate for both levels of all the process variables. Longer treatment time caused higher etching (Fig. 10) as observed in argon and argon–oxygen plasma treatments [24, 25].

3.6. Biocompatibility

Cell adhesion and viability were studied for untreated polycarbonate and helium plasma treated polycarbonate at the optimum conditions (65 W, 23.3 Pa, 20 sccm and 4 min). SEM observations revealed initial attachment, growth and spreading of L929 mouse fibroblast cells on untreated and helium plasma treated polycarbonate. On untreated surface, less filopodia extension was observed from Fig. 11(a). In

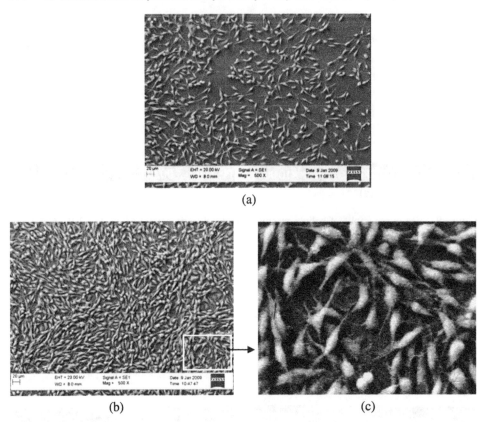

Figure 11. Cell adhesion on (a) untreated and (b) helium plasma treated polycarbonate and (c) filopodia extension in helium plasma treated polycarbonate under optimized conditions.

contrast, more dense growth with a greater number of cell contacts was observed on helium plasma treated polycarbonate (Fig. 11(b)). Cells connected to each other through the extra-cellular matrix were found to adhere and spread on top of each other. After 3 days of culture, a confluent layer of flattened cells covering the entire area of helium plasma treated polycarbonate was noticed. Cells adhered onto helium plasma treated surface provides more sites through cell–cell interaction for further fibroblast cell adhesion. Polar group incorporation and increased surface roughness provide anchorage sites on polycarbonate surface for filopodia extensions of the cells as shown in Fig. 11(c). Significant increase in optical density was also noticed as shown in Fig. 12 compared to that of untreated polycarbonate.

3.7. Blood Compatibility

Platelets adhesion contributing to surface-induced thrombosis is caused by denaturation of protein adsorbed on the surface. The extent of platelet adhesion and activation is indicative of thrombogenicity of the material. It is reported that hydrophilic surfaces are blood compatible since they exert steric repulsion to proteins

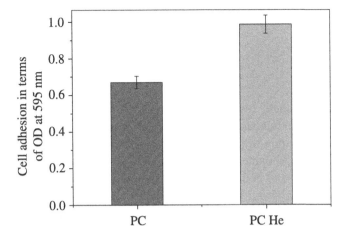

Figure 12. Cell adhesion at 72 h on helium plasma treated polycarbonate under optimized conditions.

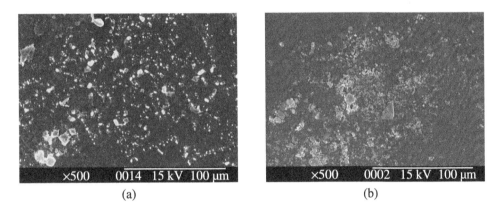

Figure 13. Platelet adhesion on (a) untreated and (b) helium plasma treated polycarbonate under optimized conditions.

that reach the surface. Partial thromboplastin time (PTT), which detects the intrinsic coagulation, is used to evaluate thrombogenicity of the material. It was found that the partial thromboplastin time of helium plasma treated polycarbonate at the optimum conditions of 65 W, 23.3 Pa, 20 sccm and 4 min was significantly prolonged to 204 s from 128 s observed for untreated polycarbonate. Increased PTT after plasma treatment indicates that activation of the intrinsic blood coagulation was suppressed by increased polarity of the surface. The helium plasma treatment resulted in increased hydrophilicity through increased polarity due to incorporation of hydroxyl group. From Fig. 13 it can be seen that platelet adhesion on helium plasma treated polycarbonate was reduced with slightly enhanced platelet aggregation. Adhesion of platelets to the surface of a biomaterial is influenced by the ratio of albumin to fibrinogen at the surface which should be high for a blood compatible material. The adsorption of specific proteins depends on the surface properties and fibrinogen adsorption on the surface was reduced with increasing hydrophilicity of the surface [27, 28].

4. Conclusion

Helium plasma treatment resulted in enhanced surface energy compared to untreated polycarbonate. Increased polarity through incorporation of oxygen functionalities and roughness due to plasma treatment increased surface energy of helium plasma treated polycarbonate. Quadratic models were developed for both surface energy and weight loss of helium plasma treated polycarbonate. Fitness and significance of the models, and significance of coefficients were evaluated and the process conditions were optimized. The models and optimum conditions were validated and were found to satisfactorily fit the data. From the helium emission spectrum, helium in the form of neutral and ions was identified which caused the surface chemistry and morphology changes. Double bond formation due to hydrogen abstraction and hydroxyl group incorporation were observed in helium plasma treated polycarbonate. Increased hydrophilicity due to increased polarity by hydroxyl group incorporation and roughness helped in increasing the adsorption of proteins on the surface which subsequently influenced cell attachment. Enhanced biocompatibility of helium plasma treated polycarbonate was observed from cell viability and proliferation test performed *via* MTT reduction by fibroblast cells adhering to the plasma treated polymers. Increased partial thromboplastin time and reduced platelet adhesion confirmed enhanced blood compatibility.

References

1. J. M. Goddard and J. H. Hotchkiss, *Prog. Polymer Sci.* **32**, 698 (2007).
2. S. V. Bhat, *Biomaterials*. Narosa Publishing House, New Delhi, India (2002).
3. L. D. Bartolo, S. Morelli, A. Bader and E. Drioli, *Biomaterials* **23**, 2485 (2002).
4. B. D. Boyan, T. W. Hummert, D. D. Dean and Z. Schwartz, *Biomaterials* **17**, 137 (1996).
5. Y. Ikada, M. Suzuki, M. Taniguchi, H. Iwata, W. Taki, H. Miyake, Y. Yonekawa and H. Handa, *Radiation Phys. Chem.* **18**, 1207 (1981).
6. M. S. Kim, G. Khang and H. B. Lee, *Prog. Polym. Sci.* **33**, 138 (2008).
7. J. C. Lin and S. L. Cooper, *Biomaterials* **16**, 1017 (1995).
8. R. L. Williams, D. J. Wilson and N. P. Rhodes, *Biomaterials* **25**, 4659 (2004).
9. S. M. Mirabedinia, H. Rahimi, Sh. Hamedifar and M. Mohseni, *Intl J. Adhesion Adhesives* **24**, 163 (2004).
10. D. L. Cho, K. H. Shin, W. J. Lee and D. H. Kim, *J. Adhesion Sci. Technol.* **15**, 653 (2001).
11. J. Behnisch, A. Hollander and H. Zimmermann, *J. Appl. Polym. Sci.* **49**, 117 (1993).
12. K. L. Mittal (Ed.), *Polymer Surface Modification: Relevance to Adhesion*, Vol. 5. VSP/Brill, Leiden (2009).
13. K. L. Mittal (Ed.), *Polymer Surface Modification: Relevance to Adhesion*, Vol. 4. VSP/Brill, Leiden (2007).
14. R. Barbucci, S. Lamponi and A. Magnani, *Biomacromolecules* **4**, 1506 (2003).

15. T. Hasebe, T. Ishimaru, A. Kamijo, Y. Yoshimoto, T. Yoshimura, S. Yohena, H. Kodama, A. Hotta, K. Takahashi and T. Suzuki, *Diamond Rel. Mater.* **16**, 1343 (2007).
16. Y. Ikada, *Biomaterials* **15**, 725 (1994).
17. G. Jin, Q. Yao, S. Zhang and L. Zhang, *Mater. Sci. Eng C* **28**, 1480 (2008).
18. H. Tan, J. Liu, J. H. Li, X. Jiang, X. Y. Xie, Y. P. Zhong and Q. Fu, *Biomacromolecules* **7**, 2591 (2006).
19. N. Gomathi and S. Neogi, *Appl. Surface Sci.* **255**, 7590 (2009).
20. R. H. Myers and D. C. Montgomery, *Response Surface Methodology*. John Wiley and Sons, New York (2002).
21. N. Gomathi and S. Neogi, *J. Adhesion Sci. Technol.* **23**, 1811 (2009).
22. S. Yang and M. C. Gupta, *Surface Coatings Technol.* **187**, 172 (2004).
23. M. J. Shenton and G. C. Stevens, *J. Phys. D: Appl. Phys.* **34**, 2761 (2001).
24. N. Gomathi, C. Eswaraiah and S. Neogi, *J. Appl. Polym. Sci.* **114**, 1557 (2009).
25. Y. Wang, Y. Jin, Z. Zhu, C. Liu, Y. Sun, Z. Wang, M. Hou, X. Chen, C. Zhang, J. Liu and B. Li, *Nucl. Instrum. Methods Phys. Res. B* **164**, 420 (2000).
26. M. Gheorghiu, F. Arefi, J. Amouroux, G. Placinta, G. Popa and M. Tatoulian, *Plasma Sources Sci. Technol.* **6**, 8 (1997).
27. I. Dion, C. Baquey, B. Candelon and J. R. Monties, *Intl J. Artificial Organs* **15**, 617 (1992).
28. C. H. Lin, W. C. Jao, Y. H. Yeh, W. C. Lin and M. C. Yang, *Colloids Surfaces B* **70**, 132 (2009).

Surface Engineering and Cell Adhesion

Gilbert Legeay [a,*], **Arnaud Coudreuse** [a], **Fabienne Poncin-Epaillard** [b],
Jean Marie Herry [c] **and Marie Noëlle Bellon-Fontaine** [c]

[a] Centre de Transfert de Technologie du Mans (CTTM), rue Thalès de Milet, 72000 Le Mans, France
[b] PCI-CNRS, UMR 6120, Université du Maine, avenue Olivier Messiaen, 72000 Le Mans, France
[c] INRA- AgroParisTech, UMR 1319 Institut MICALIS, équipe BHM, 25 avenue de la République, 91300 Massy, France

Abstract
Cell adhesion is a multi-process phenomenon involving physical, physico-chemical and biological mechanisms. The complexity of interfaces is the reason why progress in the theory of cell adhesion has been slow. Greater understanding of interaction mechanisms has been enhanced by complete knowledge of supports and of biological components, in particular the extracellular matrix, membrane walls, cell multiplication processes and apoptosis. The construction of novel surfaces with strongly hydrophilic or ultrahydrophobic properties has allowed new theoretical advances, while at the same time offering numerous and varied technological applications. These include:

- Bioadhesion with mechanical anchoring using ubiquitous surface roughness and deformability of certain micro-organisms.
- Physico-chemical bioadhesion or repellence resulting mainly from the energy characteristics of support surfaces.
- Processes of sorting and guidance by biomolecules present at the support–biofilm interface, generating biochemical responses that can induce cell multiplication or degeneration (as in cancer), or cell death.

Keywords
Bioadhesion, cell adhesion, surface engineering, biofilm, macromolecule adsorption

1. Introduction

Most materials in contact with a biological medium can be colonised by various types of cells: prokaryotes (viruses, phages or bacteria) and eukaryotes (protozoa, algae and moulds, plant and animal cells). The interactions between materials and cells that arise from this contact can cause cell adhesion, also called bioadhesion. This process might need to be either promoted or counteracted to avoid pathological hazards and/or economical consequences in various application domains. In medical fields, for example, the adhesion of animal and/or human cells to the surface of

---

* To whom correspondence should be addressed. E-mail: glegeay@cttm-lemans.com

Surface and Interfacial Aspects of Cell Adhesion
© Koninklijke Brill NV, Leiden, 2010

biomaterials will be actively sought for cell recolonisation, in regenerative medicine (bone, cornea and cardiovascular applications) and tools involved in diagnosis and dosage of target cells [1–4]. Conversely, the adhesion of cells and pathogenic bacteria to the surfaces of plastics, composites, glasses, cements, ceramics, or foodstuffs needs to be combated to avert nosocomial diseases, prevent premature explantation of surgical implants, preserve the functionality of medical instruments such as catheters, especially central venous catheters (CVC's) and urocatheters [5] and, in general, to stop infections, which cause morbidity and mortality, lengthening hospital stays and thus increasing medical costs [6].

In the agrofood area, bacterial adhesion is undesirable, in particular, when it involves pathogenic germs and/or causes spoilage. It can raise production costs (reduced equipment performance, increased volume of washing products and treatment times, etc.), losses associated with the premature deterioration of finished products, plus the costs incurred by recalling products found to contain pathogens and the consequences of food poisoning outbreaks. On the other hand, in certain biotechnology processes, microbial adhesion will be positively sought, e.g., in industrial fixed cell fermentation or in cheese and cured meat maturation [7].

Whatever its ultimate purpose and intended domain of application, control of cell adhesion is a challenge that can only be met through the development of new approaches. Among these, novel surfaces with controlled geometry and functional properties play a key role.

In this concise review we present a basic knowledge of structure of cell surfaces required for the understanding of cell–surface interactions, then briefly present mechanisms involved in cell adhesion to solid surfaces. We focus on some definitions and descriptions of surfaces, taking into account the specific features of the surrounding biological medium. Lastly we present those avenues of research in surface engineering that are currently being explored.

2. Basic Knowledge of the Cell Surface Structure

As mentioned previously there are two basic types of cells:

- Prokaryotes which do not have a real nucleus and whose DNA is free in the cytoplasm. Their size generally ranges from 1 to 10 µm. Bacteria are in this category.
- Eukaryotes that have a real nucleus which contains the DNA separated from the cytoplasm by a membrane. Their size varies from 10 to 100 µm and they include, in particular, animal cells.

All cells (Fig. 1) have a plasma membrane that isolates the interior of the cell from its environment and acts as filter. In the case of eukaryotes, this membrane forms the cell surface, while for the bacteria it is covered with a layer composed of peptidoglycan which gives the cell stress-bearing properties and a shape-determining structure.

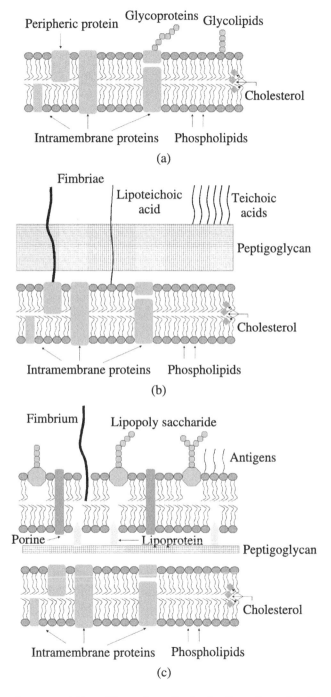

Figure 1. Schematic representation of the surface of animal cell membrane (a), Gram⁻ bacteria (b) and Gram⁺ bacteria (c).

The peptidoglycan of the Gram$^-$ bacterium is surrounded by a structure called outer membrane, whereas peptidoglycan of the Gram$^+$ cells is generally exposed to the environment. External structures like fimbriae, flagella, etc., can be found on the surface of bacteria. Such structures play a key role in the process of cell adhesion. As mentioned by Ofek and Doyle [8], in Gram$^-$ cells, adhesins are most often found to be on fimbriae, although fibrils, capsule, slimectin, outer membrane and flagella have been reported to possess adhesin characteristics. Fimbriae may be flexible, rigid or curled (curli). The fimbriae adhesins may be located at the very tip of the structure or may be found all along the filamentous appendage. In Gram$^+$ cells, the most prominent adhesins appear to be fibrils, fimbriae and a part of membrane wall [8].

3. Cell Adhesion to Solid Surfaces: a Complex Process Involving Physical, Physico-Chemical and Biological Mechanisms

Whether cells are eukaryotic or prokaryotic, we now know that their adhesion to a solid support follows a dynamic sequence of events, including (i) transport, (ii) molecular or macromolecular adsorption, (iii) physico-chemical and/or stereospecific interactions and (iv) anchoring process depending, in particular, on the topography scale and substrate nature.

These adhesion step may be followed by a colonisation of the substrate which is specific of the cells involved (migration, proliferation and differentiation for animal cells, formation of biofilms for micro-organisms such as bacteria).

3.1. Transport

Adhesion requires proximity between the adhering partners. Bioadhesion is not an exception: to adhere to a receptor support, cells must be located within a few nanometers of the host surface. Besides these physical processes (diffusion, sedimentation and Brownian motion), which depend largely on the characteristics of the suspending fluid (viscosity, temperature, pH) and the local hydrodynamics of surrounding fluid (turbulent or laminar flow or stagnant fluid), it has been observed in the case of bacteria that transport mechanisms specific to the micro-organisms themselves or linked to their physiology are important. This is the case, for example, of certain bacteria that are imparted mobility by extracellular flagella-like appendices [9], or bacteria that move under the influence of chemo-attractants such as amino acids, sugars or oligopeptides: a process called chemotaxis or haptotaxis [10].

These transport processes concern not only cells, but also molecules and macromolecules present in biological medium. Being smaller than cells, these entities can reach the receptor material faster and form a surface film of varying density and evenness called a conditioning film [11] or primary film.

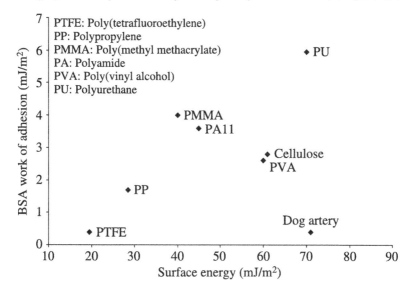

Figure 2. Work of adhesion of bovine serum albumin (BSA) to various polymer surfaces at 37°C in phosphate buffer as a function of surface energy of materials. The initial concentration of BSA is 3 g/l.

3.2. Molecular Adsorption of the Conditioning Film in Cell Adhesion

A biological medium is necessarily very complex. It is made up of water, mineral salts and naturally-occurring molecules or macromolecules (proteins, polysaccharides, biosurfactants, neutral or polyelectrolyte polymers, etc.) some of which are secreted by the cells themselves [12]. These macromolecules are transported to the receptor support very rapidly and can be adsorbed in a few seconds, modifying both its physico-chemical and topographic characteristics [13]. This adsorption depends not only on the energy and electrical characteristics of the support materials, but also on the physico-chemical properties of the surrounding fluid (pH, ionic strength, temperature, etc.). Some authors measured the work of adhesion of BSA (Bovine Serum Albumin) on various surfaces [14]. The authors indicate that adsorption was lower with (Fig. 2) (i) surfaces of lower energy (fluorinated polymers and silicones), and (ii) surfaces with the highest energy (i.e., the most hydrophilic).

Neither the composition nor the spatial structure of this adsorbed layer is set. Both can evolve, progressively and time-dependently, through movement of species by the Vroman effect (the most rapid and concentrated species can be expected to be partially replaced by more adhesive ones) or, in the case of proteins, through conformational changes [15].

3.3. Physico-Chemical and Stereospecific Interactions in Cell Adhesion

Once cells come close enough, they can interact with solid surfaces *via* physico-chemical or non-specific interactions (van der Waals, Lewis acid–base, electrostatic and Brownian interactions) and in some cases through stereospecific interactions [16]. However, as indicated by van Oss [12], these interactions can be considered as resulting from non-covalent interactions between specific entities such as antigens, antibodies, lectins, carbohydrates or enzymes and certain chemical sites on the substrates.

Concerning bacteria, for example, many studies point out the fundamental importance of the surface properties of materials for their ability or inability to favour bioadhesion and so allow cell colonisation. Flint *et al.* [17] have shown that *Streptococci* adhere to stainless steel and zinc substrates in preference to other metal or glass substrates. They noted the importance of the stainless steel grade: more bacteria adhered to 316L grade than to 304L grade. Hamadi *et al.* [18] found a close correlation between various physico-chemical properties (hydrophilicity or electron donor/acceptor characteristics) of *Staphylococcus aureus* ATCC 25923 and the number of cells adhering to glass substrate. Other works [19, 20] on adhesion of *Saccharomyces cerevisiae* to glass and polymers indicate dependence on the balance between the Lifshitz–van der Waals and Lewis acid–base interactions. Although the surface free energies of polystyrene and stainless steel are nearly the same, metal surface promoted entirely distinct behaviour, characterised by strong, though highly variable, adhesion [19]. Briandet *et al.* [20] showed that the degree of biocontamination of 304L stainless steel grade could be modified by varying its surface charge.

Bioadhesion of animal cells has also been described in some works. For example, Faucheux *et al.* [21] found influence of both surface properties of the substrate and the chemical composition of the surrounding medium on aggregation of Swiss mouse 3T3 cells.

3.4. Cellular 'Anchoring' and Adhesion According to Topography and Roughness of Receptor Surfaces

Materials in common use seldom have perfectly smooth surfaces. Grooves, scratches and other irregularities are present. It has been amply demonstrated that such surface 'defects' favour the mechanical bonding of bacteria and other unicellular entities [16, 22]. However, this 'cause (rugosity)-and-effect (mechanical anchoring)' rule is too simple: in addition to their impact on mechanical interactions, cavities, protrusions and overall unevenness also have effects on the local hydrodynamics of surrounding fluid, and thus on the transport mechanisms described above. It has been observed that deep, narrow cavities are conducive to turbulence, favouring adhesion and anchoring of cells to the bottom of cavities. When the distances between cavities and ridges are more wave-like, the probability of turbulence occurring and bioadhesion decreases [23].

After adhesion some cells can change shape to fit the contact surface. Animal cells, for example, can produce variously shaped extensions (e.g., filopods and lamellipods) that enable them to consolidate their position on the receptor site and initiate their colonisation phase [24, 25]. For certain bacteria, this anchoring can occur *via* the production of exo-polymeric substances that enable the bacteria to enter into an irreversible adhesion phase [26].

3.5. Colonisation of Solid Supports and Extracellular Matrix

When the conditions are favourable, cell adhesion is followed by cell spreading and migration in the case of eukaryotic cells and by proliferation leading to three-dimensional structures called biofilms in the case of prokaryotes such as bacteria [22]. As described above, this cell growth is influenced by surface topography. However, in both cell types, colonisation can be accompanied by the production of extracellular compounds to form an exocellular matrix of varying density and viscosity.

In animal cells, this matrix is made up of glycoproteins (collagen, fibronectin and laminin), pure proteins (elastin), glycosaminoglycans and proteoglycans [27]. The matrix forms a hydrated gel containing the cells. The proteins are produced by the cells (chondrocytes, osteocytes, etc.). The three-dimensional structure of the matrix is a network of collagen fibres held together by elastin filaments in the case of skin and blood vessels. On this network are also attached adhesion glycoproteins (mainly fibronectin) and globular collagen. The constituents of the extracellular matrix allow cells to organize into tissues [28]. The matrix plays a key role in the structural support, adhesion, mobility and regulation of the cells. Thus integrins (trans-membrane proteins) ensure the communication between the extracellular medium (matrix) and the intracellular medium through cytoskeletal filaments (actin fibers, microtubules, intermediate filaments). The spatial composition of the extracellular matrix can be modified by the presence of chemical groups on the outermost surface layer of an exogenous support. Chemical or mechanical variations in the extracellular matrix generate signals transmitted *via* integrins to the cytoskeleton: the transduction of these signals can trigger a sequence of chemical reactions throughout the adjacent cells and induce biochemical and biological responses, either favourable or unfavourable.

In the case of bacteria, the extracellular matrix, which has a structure with variable density is viscous but largely hydrated, acts as a nutrient source for the cells of the biofilm, and protects them against external aggression, e.g., from antibiotics and disinfectants... [29].

However, in certain cases this colonisation step can be stopped and certain anoïkis and apoptosis processes can be observed. Gap junctions formed by proteins between adjacent cells have been studied for many years [30]. Each gap junction channel is made up of two hemi-channels or connections formed by the oligomerization of connexin (Cx) protein subunits in the Golgi apparatus of cells [31, 32]. This protein is transported to the cell plasma membrane where it interacts with adjacent cells [33, 34]. The result is a chemical stimulus of cell membrane receptors (integrins) by chemokines, in relation with the surface properties of the material: chemical, topographic and nano-mechanical [35, 36]. The integrin–cell adhesion is essential for cell survival. When it is prevented, there can be no more bioadhesion [37] and apoptosis occurs [38–40].

4. Knowledge of the Surface Support

4.1. Hydrophobic Surface

It is commonly admitted that surfaces with a water contact angle above 90° are hydrophobic. These surfaces have a low energy with a nearly-negligible polar component (lower than 5 mJ/m^2). They are composed of chemical groups such as –C–H, C–C, –C–F, Si–O–Si. With such surfaces only van der Waals forces establish physico-chemical interactions with molecules bearing hydrophobic sequences in the surrounding medium. Hydrocarbon materials (polyolefins), fluorinated polymers and silicones belong to this category. Surfaces coated with polycations are also qualified as moderately hydrophobic [41].

4.2. Hydrophilic Surface

This is a surface with a high energy and a high polar component, due to the presence of polar groups such as –OH (alcohol), –NH$_2$ (amine), –COOH (acid), etc. It can establish polar, ionic, acid–base bonds with the surrounding medium. Materials with such properties include glass, silicon wafer, poly(vinyl alcohol), cellulose derivatives, poly(ethylene glycol), polyamides, some polyurethanes, poly(vinylpyrrolidone), etc.

4.3. Topography/Roughness

Topography (according to the Abbott–Firestone description [42]) is described by the roughness parameters R_{rms} (mean roughness) and R_Z (maximum roughness) and other parameters [43]. The roughness of supports was for a long time measured at the micrometer scale [42]. Today it is determined at the nanometer scale using techniques like atomic force microscopy (AFM). This progress in topographical knowledge has revealed the importance of nanometre–scale interactions in cell adhesion [44] and has allowed the use of chemically or topographically nanostructured surfaces to analyse such interactions.

Biological effects on cell adhesion due to roughness might be different according to the low or high hydrophobicity of the surface:

– For classical materials with surfaces of varying roughness and moderate or high surface energy (hydrophilic), the deformability of cells can facilitate and increase both adhesion and mechanical anchoring, as the cells can better adapt to the exact local topography of the support.

– For hydrophobic materials, high roughness combined with hydrophobicity makes displacement of entrapped air energetically unfavourable. The formation of micro-bubbles between the biological deposit and the surface prevents or limits the process of bioadhesion or spreading. Authors indicated that presence of bubbles creates an oxidative stress with *Escherichia coli* [45, 46]. A proteinic denaturation may happen as described with BSA in [47].

4.4. Electrical Components

When a liquid comes in contact with a surface, it can become electrically charged due to the polarisability of the molecules. This polarisability depends on the species present in the solution, the pH, etc., and is associated with diffuse electrically charged layers. Gouy and Chapman developed the theory of the diffuse double layer, taking into account thermal movement. Between these two layers there is an electrostatic potential (Nernst potential), which varies according to the distance from the surface. The value of the potential at the surface of the Nernst layer, called the zeta potential, characterises the distribution of the electrical charges on the surface. As polymers and colloids are negatively charged, this potential is globally negative (Fig. 3) [48]. This model has been tested for negatively charged Langmuir monolayers with different surface charge densities [49].

Consequently, surface charges influence the presence, the orientation and the spatial distribution of the various biological entities which possess their own electrical charge.

Some works described the effect of zeta potential and surface energy on bacterial adhesion to uncoated and saliva-coated human enamel and dentin [50]. The results showed a significant decrease in adhesion when the ionic strength of the

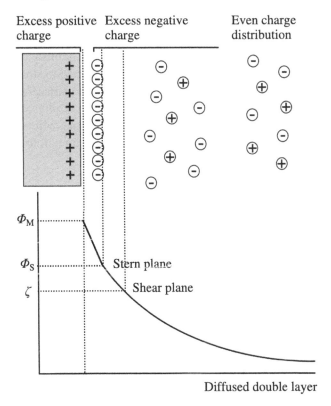

Figure 3. Double layer model according to Gouy–Chapman–Stern.

medium was lowered, due to increased electrostatic repulsion. However, the adhesion of some bacteria was independent of the ionic strength of the medium. Near the surface, spatial distribution and orientation of biomolecules was in accordance with the electrical model.

4.5. Stability of the Outermost Surface Layer

In a biological medium, the outermost surface layer of supports is not always stable and lasting. In massive supports this instability can be due to mobility of the outermost surface layer in response to the surrounding medium. In the case of supports coated with a layer of different chemical composition with varying thickness and adhesiveness (oxides, paint, film, coating, etc.), the instability can arise from molecular rearrangements. This can be due to migration of chemical groups at the outermost layer towards the bulk of the material (overturn process) [51]. For example, X-ray photoelectron spectroscopy (XPS) and more especially static secondary ion mass spectrometry (SSIMS) have revealed the presence of abnormally large amounts of polar groups, such as ester, alcohol, acid groups, on the surface of cellulosic derivatives coated on a polycarbonate support compared to the cellulose normal stoichiometry [52]. Ageing in a biological medium suggests that the overturn and/or orientation and migration of the polar groups towards the water phase occurs in such a way as to minimise the difference in surface free energy between the polymer surface and water.

In the case of materials with modified surfaces, this instability depends, in large part, on the adhesion and cohesion of the deposit. If there is no strong adhesion between the deposit and the surface, or if cohesion between adjacent layers is weak, then simple cleaning or a long exposure to a biological or physiological medium will eliminate the deposit formed, and the deposit properties will be lost [53]. Consequently, after several hours or days the biological medium will find a support with a new surface composition.

5. Surface Engineering and Bioadhesion

The modification of surfaces by surface engineering can be achieved by coating with micrometer and nanometre–scale thicknesses of a biologically active material to impart it 'anti-adhesive' properties, based on physico-chemical effects, or a bactericidal/virucidal effect. Some of these new surfaces are already under technological development in view of further applications.

5.1. Stability of the Outermost Surface Layer of New Supports and Principles of Construction

When a deposit of nano- or micrometric thickness is deposited on a massive support, the interfacial adhesion can be strong with surfaces having higher surface energy than the coating. Theses surfaces might present a strong ionic character, or present a strong acid–base character. A covalent bond can also be established with the coating in the presence of reactive functions. Multiple hydrogen bonds also lead to strong adhesion. This surface character could be weak, if only a few hydrogen bonds or weak acid–base bonds are formed (physisorption), or negligible if the support has a surface energy markedly lower than that of the deposit.

The methods by which new surfaces are constructed can be classified into two categories:

(a) One-step approach: direct deposition of a coating layer from an appropriate liquid or gaseous medium if a sufficient adhesion is spontaneously created.
(b) Two-step approach:

  – In a first step, a surface activation, using a clean process (without solvents, heavy metals, . . .) for objects with shapes of varying complexity. Usually electromagnetic discharge and irradiation are well-suited to this purpose.
  – In a second step, coating of the previously activated material by a layer of a new material that has useful physico-chemical and biological properties [54].

For example, grafting of poly(acrylamide) on polyethylene activated either by irradiation or by plasma has been the subject of many works [55–58].

5.2. New Surfaces with Physico-Chemical Effects

The construction of new surfaces with an 'antiadhesive' effect on cells is based on a modification of the physico-chemical surface characteristics of materials (in particular their hydrophilic/hydrophobic balance). Their prime utility is that they obviate the use of biocidal substances. Physiological media are by nature aqueous: they are made up of water, dissolved mineral salts and often biomolecules that display some degree of hydrophilicity. Several situations arise according to the composition of the outermost surface layer concerned:

– For supports with very hydrophilic surfaces, a film of water, by nature antiadhesive, forms immediately on contact between the support and the physiological medium. A layer with weak cohesion (H_2O) thus prevents or limits the formation of a protein deposit, and cell bioadhesion will be very low (Fig. 4).
– For hydrophobic and ultrahydrophobic surfaces, as wetting is poor, the extracellular matrix is poorly distributed, and thus will not adhere well to the support. Here again, adhesion is weak.
– For supports that are moderately hydrophobic or hydrophilic due to the presence of local ionic entities, in particular at the surface (salts, ionised biomolecules), the existing ionic layers (according to the Gouy–Chapman–Stern model) guide the selection and spatial arrangement of the biomolecules in the extracellular matrix in the thin interfacial zone [48]. An orientation of the elements that make up the extracellular matrix is observed. The local composition of the surface is difficult to analyse precisely because it is in relation with the surrounding biological medium which is complex, and is in a dynamic state. Furthermore, the interface is very thin (nanometric layer).

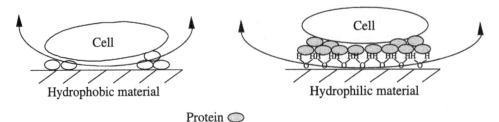

Figure 4. Non-adhesion mechanism of cells and biofilms in biological medium (1) on hydrophobic surface (on the left) preventing the deposition of the extracellular matrix proteins proteins, and (2) on very hydrophilic surface (on the right) forming a thin layer of water which suppresses the protein adhesion.

The construction of either very hydrophilic or very hydrophobic surfaces thus offers the only means to obtain surfaces with useful 'anti-adhesive' properties.

5.2.1. Novel Hydrophobic Surfaces

In some cases such hydrophobic surfaces are advantageous when bacterial or cells colonisation is unwanted. For example, in biomedical applications, cells do not adhere to PTFE used for the manufacturing of medical devices (tubing, plug, etc.) or on silicone prosthesis (for example, mammary).

In ophthalmological applications, bacteria and cells should be prevented from adhering to contact lenses or intraocular implants in order to maintain optical transmission. However, most of the polymers used (acrylic derivatives or silicones) are weakly hydrophobic and become coated with bacterial deposits within a few days. For long-term use, lenses made of poly(methyl methacrylate) (PMMA) [59] are treated with a CF_4 plasma to fluorinate its surface, creating a hydrophobic layer comparable to that of poly(tetrafluoroethylene) (PTFE) ($-(CF_2)_n-$). The surface energy of the material is less than 20 mJ/m^2. Tests conducted *in vivo* in rabbits and humans showed very low protein adhesion, inflammatory cell growth and cell debris formation. In addition, no activation of endothelial cells and adhesion of granulocytes were observed.

The activation of the complement system has also been examined in work on the cuprophan membranes used in renal dialysis [60]. It was found that the degree of complement activation was directly related to the proportion of hydrophobic sites, such as fluorocarbon groups (C–F, CF$_2$, etc.).

Recently non-leaching, permanently micro-biocidic polymeric coatings have been developed. These coatings, which contain hydrophobic polycations and which are either covalently attached or deposited onto surfaces, avidly kill human pathogenic bacteria, fungi, and viruses on contact by disrupting their lipid membranes [61, 62].

5.2.2. Novel Ultrahydrophobic Surface

A material is considered ultrahydrophobic when its wetting angle with water is higher than 150° [63]. An example is the polyethylene fluorinated by plasma method [64]. Its chemical composition is described as 'Teflon-like', the polymer

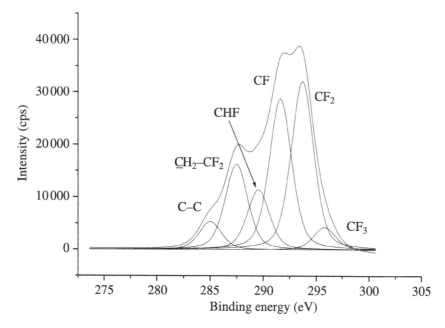

Figure 5. High-resolution XPS spectrum of the C_{1s} peak of CF_4 plasma treated polyethylene.

surface being made up of CF_2 chains and some CF, CF_3, $C-CF_x$ groups (Fig. 5). In addition, and as indicated above, roughness (at the nanometre scale) enhances water repellence. The topography favours the formation, in aqueous medium, of a layer of air or micro-bubbles, thereby prevents spreading of physiological medium and consequently limits bioadhesion of cells. This property is a result of the highly polar character of the surface and its roughness (a few nanometres to some hundreds of nanometres) (Fig. 6) [64]. Whereas these surfaces are not yet used in the bioadhesion field, they should have a strong potential for application in this field.

Such surfaces can be engineered, firstly by modifying the roughness of the surface (etching of the surface, for example, by plasma process) followed by plasma fluorination.

5.2.3. Novel Hydrophilic Surfaces

One of our first results in the surface engineering of biomaterials was the corona treatment for oxidation of cuprophan membranes for a pancreatic prosthesis [65]. Surface support oxidation corresponds to an increase in surface energy. We did not describe the physico-chemical variables, but *in vivo* implantation gave interesting results for the bioadhesion of endothelial cells in rat peritoneal cavity.

The strongly hydrophilic character of certain polymers (high surface energy and polar component) is useful, and many such commercial polymers are of pharmaceutical grade, a requirement for many biological applications (Table 1). PVP (poly(vinylpyrrolidone)) presents the highest surface energy and its polar component is elevated (respectively 63.4 mJ/m^2 and 42.0 mJ/m^2). This polymer provides

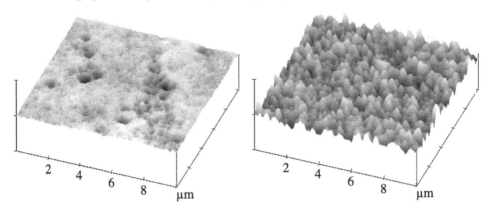

Figure 6. AFM images of hydrophobic and ultrahydrophobic PEBD surfaces treated with both O_2 and CF_4 plasmas (left: untreated surface, $R_{rms} = 8.6$, water contact angle $= 127°$; right: treated surface, $R_{rms} = 43.3$, water contact angle $= 168°$) (R_{rms}: quadratic mean of the various measured heights in μm).

Table 1.
Contact angle and surface energy of hydrophobic and hydrophilic polymers

Polymers	Contact angle (°)			Surface energy (mJ/m^2)	Polar component (mJ/m^2)
	H_2O	Formamid	CH_2I_2		
PTFE	118	107	95	9.3	0.5
Silicone	90	95	50	24.0	2.8
HPMC (E4M)	60	55	49	42.1	18.5
CMC-Na (7LF)	44	19	48	55.9	29.5
HEC	28	34	34	62.5	36.9
PVP (Kollidon K 90)	26	34	45	63.4	42
PVA:					
– grade 75	49	34	36	52.9	22.8
– grade 95	76	40	46	41.6	6.5
Poly(HEMA)	42	16	35	58.2	27.0
Biosurfactants:					
Lactobacillus helveticus				38.0	
Bacillus subtilis				27.0	

a very hydrophilic surface, and some of its cellulose derivatives (hydroxyethylcellulose (HEC), hydroxypropylmethylcellulose (HPMC) and carboxymethylcellulose-sodium (CMC-Na)) display a range of acid–base characteristics.

Some of these polymers are used to create novel surfaces of specific biological interest. Comparative biological validation has been carried out with polystyrene (PS) supports that have undergone various treatments: commercial tissue culture polystyrene (tPS), supports coated with poly(2-hydroxyethyl methacrylate) (poly-HEMA), and supports coated with a bi-layer of HPMC and CMC-Na in comparison

Table 2.
Biocontamination of glass surfaces with a PVP coating by the pathogenic bacteria *Staphylococcus aureus*. Contamination level is measured by the number of living and adherent cells after 24 h of test

Bacterial strains	*Staphylococcus aureus* (C1)	*Staphylococcus aureus* (C2)
No treatment	5.7 E05 CFU[1] cm^{-2}	2.9 E07 CFU cm^{-2}
PVP treated surface	3.0 E05 CFU cm^{-2}	2.8 E06 CFU cm^{-2}
Reduction in bacterial adhesion	46.7%	90.3%

[1] CFU = Colony Forming Unit; C1, C2 = concentrations of suspensions of *Staphylococcus aureus* used: respectively 10^6 and 10^8 bacteria per ml.

with raw polystyrene. The relative proportion of polar contribution has an influence on fibronectin adsorption and such influence can be estimated by measuring the quantity of adsorbed fibronectin and the quantity of fluorescently adsorbed labelled fibronectin. PolyHEMA and HPMC/CMC-Na were found to be hydrophilic, tPS moderately hydrophilic, and PS highly hydrophobic. After fibronectin adsorption on PS and tPS, a significant increase in the surface polar interaction was observed [66]. On polyHEMA and HPMC/CMC-Na, the polar interactions remained unchanged on fibronectin adsorption [67]. Cultures of cancer cells (melanoma and leukaemia) were grown on the same supports. The best results were obtained with a coating of HPMC overlaid with CMC-Na. The CMC-Na layer is more hydrophilic than HPMC one (surface energy $r_s = 55.9$ against 42 mJ/m^2), more basic, polar and ionic (1% solutions: pH = 6.5 against 5.1) [68]. Fibroblasts and 3T3 mouse melanoma probably fail to develop because the surfaces do not adsorb proteins (fibronectin), a physico-chemical property, which results in reduced cell proliferation, a biological property. It has been observed that the cells aggregated on these modified substrates quickly adhere. Cell differentiation then occurs and as horizontal growth is impossible, pyramidal aggregates form after 24 h (Fig. 7). They correspond to the piling up of concentric layers of melanoma cell. Cell death by apoptosis then occurs, and no colonisation of the surface is observed [69]. These new supports, for example biofunctional cellulosic membranes, have confirmed the importance of support surface chemistry [70, 71].

Other examples show the utility and efficacy of novel hydrophilic surfaces for the prevention of biocontamination by physico-chemical processes.

– *Hygiene.* It has been shown that coating a glass surface with PVP causes a sharp drop in the numbers of adherent, viable, cultivable *Staphylococcus aureus* bacteria compared with a control (Table 2). On other supports (metals, ceramics and textiles) the same effect was found. Biocontamination can thus be averted. This process is now used in hospitals and in the agrofood sector [72].

– *Biocontamination of biomedical implants.* Hydrophilic coatings have been deposited on polycarbonate membranes used to construct a bio-artificial pancreas, to prevent the adhesion of endothelial cells present in the peritoneal cavity of mam-

Figure 7. Pyramidal aggregates of melanoma cells (Swiss 3T3) after 24 h on polystyrene-coated cellulose substrate.

Figure 8. Polycarbonate membrane uncoated (right image) and coated (left image) with cellulosic derivatives, after one month in the peritoneal cavity of rats.

mals [73, 74]. The membranes are plasma-treated and then coated with a layer of cellulose (HPMC). After implanting for one month in the peritoneal cavity of rats, the treated product did not become coated with a protein film, and so was not colonised by endothelial cells, unlike the untreated material (Fig. 8). The overall functional performance of the implant was maintained for at least several months.

These results for polycarbonate were confirmed recently [75] with a correlation between wettability values and adhesion of bacteria such as *Staphylococcus epidermis*.

– *Biocontamination of ancillary equipment.* Catheters (e.g., urocatheters) are prone to abundant bacterial proliferation leading to inflammatory reactions. The

Figure 9. Adhesion test results of *Staphylococcus epidermidis* on PS substrate (left image) and on PS coated with HPMC (right image) realized under dynamic conditions for 3 days.

evaluation of biocontamination by *Escherichia coli* on urinary catheters shows reduced colonisation when the catheter was coated with a surface layer of PVP. In addition, friction coefficient is reduced as the hydrophilic deposit swells, creating a water film that lubricates the catheter. Insertion and withdrawal of the instrument is thus facilitated [76].

In the same way, it was shown that adhesion of the bacterium *Staphylococcus epidermidis* was limited on PS substrate coated with HPMC compared to pristine PS (Fig. 9).

– *Microfluidic systems and biosensors.* New applications have been proposed in the engineering of biosensors for bioelectronics. Two aspects have to be considered:

– In a complex medium, a biosensor has to recognize one specific molecule for its detection or titration. A problem is the spontaneous and non-specific bioadhesion by cells and/or biomolecules from the complex medium, which can saturate the active surface and 'blind' it: this effect can be rapid, and the sensor is soon disabled [77]. Surface treatments can be proposed to avoid bioadhesion of the various microorganisms leading to this malfunction.

– The microchannels, carrying aqueous physiological media, have to be wettable in order to improve the flow of the complex biological liquids, thus avoid the formation of air bubbles [78]. Very hydrophilic surfaces have been proposed to suit this purpose.

5.3. Novel Surfaces with Biosurfactants

Another approach under development for few years has been to limit or delay the biocontamination of food contact materials caused by coating inert surfaces *via* the adsorption of biosurfactants produced by various microorganisms (*Pseudomonas, Bacillus, Acinetobacter, Lactobacillus*, etc.) [79]. Recent studies have shown that these biomolecules, already used in various applications (oils, bioremediation, detergency, cosmetics, emulsions, etc.), can also play a major role in interfacial processes such as bioadhesion to surfaces. It was thus underlined that modifications of the properties of the surface of substratum (stainless steel, glass, polystyrene, etc.) [80, 81] induced by the adsorption of these biosurfactants could considerably reduce the microbial adhesion of bacterial pathogens such as *Listeria monocytogenes*, *Escherichia coli* and *Staphylococcus aureus*. These biosurfactants also have the advantage of being biodegradable and not very toxic, while remaining active over wide temperature and pH ranges.

5.4. Surfaces with Antimicrobial Effects (Bactericidal, Virucidal, Algaecidal)

Novel surfaces can be created by incorporating into the substrate or coating it with a layer of a bactericidal and/or virucidal and/or algaecidal substance, which is gradually released into the surrounding biological medium. The biocide has to be authorised by the regulating authorities. For example, a recent publication argues on the bactericidal effect of various poly(ethyleneimine)-based paints towards virus and the airborne human pathogenic bacteria *Escherichia coli* and *Staphylococcus aureus* [82]. With these surfaces, bioadhesion is not the most important factor. Antimicrobial agents can also be grafted. This grafting can be achieved by chemical binding to the surface and then specific antimicrobial sites react with cells. Numerous studies have been undertaken to create antimicrobial coatings by fixing various antibiotics onto several different surfaces. Maleic anhydride or methacrylic acid are fixed onto the surface after treatment (by plasma or by e-beam) and a chemical group can be used to fix antimicrobial entities [83]. More recently a combination of two antimicrobial surfaces has been proposed and the interaction of the new surface with *Staphylococcus aureus* and *Pseudomonas putida* is described [84].

In this last approach the antimicrobial mechanism is probably only chemical or biochemical. No bioadhesion mechanism is described by the authors. Oligonucleotide model surfaces allowing independent variation of topography and chemical composition have been designed to study the adhesion and biofilm growth of *Escherichia coli*. Surfaces are produced by covalent binding of oligonucleotides and immobilisation of nucleotide-based vesicles. This study produced convincing evidence that oligonucleotides could modify surfaces, independently of the topographical feature used in this study, and enhance bacteria expression without an increase in the number of adherent bacteria [85].

Similarly, the construction of functional layers has been proposed with polymers whose monomer units have intrinsic biocidal and fungicidal properties through the presence of sites bearing quaternary ammonium salts known to induce bactericidal reactions [86]. Validations with *Pseudomonas aeruginosa*, *Staphylococcus aureus*, *Candida albicans* and *Aspergillus niger* have shown the utility of this approach.

5.5. Further Novel Surfaces and Multilayers

Other model surfaces have been studied with $-NH_2$ and $-CH_3$ terminal self-assembled monolayers on silicon wafers (low roughness). A strain of *Escherichia coli K12* was able to produce a biofilm after 4–336 h on both types of substrates.

Results indicate an influence of the surface chemistry on the extent of bacterial bioadhesion and a clear impact on the dynamics of biofilm growth and structure [87]. Theses new surfaces might potentially be used to prevent the formation of biofilms on medical implants.

Another process is the building of multilayers of polyelectrolyte molecules, obtained by self-assembling of polyanions and polycations. These multilayers have been used for tissue engineering, for instance, for cartilage regrowth.

Alternative biofunctionalised multilayers and also cell-containing layers have been shown to be an essential step towards the fabrication of stratified architectures, and tuning the cellular activity is possible by controlling the position of active molecules [88].

6. Conclusions and Perspectives

Much work has been done and various interpretations published concerning the interfacial interactions and processes of adhesion and repellence between microorganisms or cells and the outermost layers of support surfaces.

The proposed working hypotheses are focused on the understanding of processes that always have an experimental, descriptive basis with the use of various types of functionalised supports. In the mechanisms underlying the micro-organism/support surface interface it is possible to discern several phases:

- A very fast physico-chemical response: wetting, hydrophilicity–hydrophobicity, attraction and sometimes construction of layers with electrical effects.
- A very fast biological response linked to the topography and roughness of the support.
- A biochemical response with emission of chemical signals that trigger cell response. In a second phase the physico-chemical variables (electrical, pH, zeta potential, hydrophilic/hydrophobic character) established in the first moments of contact between the different entities (cell, extracellular matrix and exogenous surface) will sort and spatially guide the biomolecules present. The result is the sending of various signals to the cell receptors that will initiate a cell reaction, for example, processes of repellence, multiplication by lysis, or apoptosis/anoïkis.

References

1. B. E. Rittmann, D. Stilwell and A. Ohashi, *Structure and Function of Biofilms*, pp. 49–58. Wiley, New York (2002).
2. E. Engel, O. Castaño, E. Salvagni, M. P. Ginebra and J. A. Planell, in: *Biomaterials for Tissue Engineering of Hard Tissues, Strategies in Regenerative Medicine, Integrating Biology with Materials Design*, pp. 1–42. Springer, New York (2009).
3. B. Brulez, J. P. Laboureau, V. Migonney, M. Ciobanu, G. Pavon-Djavid and A. Siove, French Patent no. PCT/FR2004/000103 (2004).
4. D. Machy, J. Jozefonvicz and D. Letourneur, Patent no. WO/2001/076652, PCT/FR2001/000964 (2001).

5. D. J. Stickler, J. N. S. Morris, R. J. C. MacLean and C. Fuqua, *Appl. Environ. Microbiol.* **64**, 3486 (1998).
6. G. Donelli and I. Francolini, *J. Chemotherapy* **13**, 595 (2002).
7. O. Habimana, M. Meyrand, T. Meylheuc, S. Kulakauskas and R. Briandet, *Appl. Environmental Microbiology* **75**, 7814–7821 (2009).
8. I. Ofek and R. J. Doyle, *Bacterial Adhesion to Cells and Tissues.* Chapman and Hall, New York (1994).
9. C. Prigent-Combaret, G. Prensier, T. T. Le Thi, O. Vidal, P. Lejeune and C. Dorel, *Environmental Microbiology* **2**, 450–464 (2000).
10. M. Katsikogianni and Y. F. Missirlis, *European Cells Mater.* **8**, 37 (2004).
11. C. Rubio, D. Costa, M.-N. Bellon-Fontaine, P. Relkin, C. M. Pradier and P. Marcus, *Colloids Surfaces B: Biointerfaces* **24**, 193 (2002).
12. C. J. van Oss, *Forces Interfaciales en Milieu Aqueux.* Masson, Paris (2004).
13. F. Fritz-Feugeas, A. Cornet and B. Tribollet, *Biodétérioration des matériaux: action des micro-organismes, de l'échelle nanométrique à l'échelle macroscopique.* Technosup Ellipses, Paris (2008).
14. Y. Ikada, M. Suzuki and Y. Tamada, *Adv. Polym. Sci.* **57**, 135–147 (1984).
15. J. Vitte, M. Benoliel, A. Pierres and P. Bongrand, *European Cells Mater.* **7**, 52 (2004).
16. R. Bos, H. C. van der Mei and H. J. Busscher, *FEMS Microbiology Reviews* **23**, 179 (1999).
17. S. H. Flint, J. D. Brooks and P. J. Bremer, *J. Food Eng.* **43**, 235 (2000).
18. F. Hamadi, H. Latrache, M. Mabrouki, A. Elghmari, A. Outzourhit, M. Ellouali and A. Chtaini, *J. Adhesion Sci. Technol.* **19**, 73 (2005).
19. G. Guillemot, G. Vaca-Medina, H. Martin-Yken, A. Vernhet, P. Schmitz and M. Mercier-Bonin, *Colloids Surfaces B: Biointerfaces* **49**, 126 (2006).
20. R. Briandet, J.-M. Herry and M.-N. Bellon-Fontaine, *Colloids Surfaces B: Biointerfaces* **21**, 299 (2001).
21. N. Faucheux, R. Warocquier-Clérout, J. L. Duval, B. Haye and M. D. Nagel, *Biomaterials* **20**, 159 (1999).
22. A. Allion, J.-P. Baron and L. Boulange-Petermann, *Biofouling* **22**, 269 (2006).
23. R. Schmidt, *Comportement des matériaux dans les milieux biologiques*, Chapter 4. Presses Polytechniques et Universitaires Romandes, Lausanne (1999).
24. S. R. Farmer, K. M. Wan, A. Ben-Ze'ev and S. Penman, *Mol. Cell Biol.* **3**, 182 (1983).
25. R. McBeath, D. M. Pirone, C. M. Nelson, K. Bhadriraju and C. S. Chen, *Cell* **6**, 483 (2004).
26. J. D. Brooks and S. H. Flint, *Int. J. Food Sci. Technol.* **43**, 2163 (2008).
27. J. M. Tarbell and E. E. Ebong, *Sci. Signal.* **1**, 8 (2008).
28. A. Stevens and J. Lowe, *Human Histology*, 2nd edn. Times Mirror International Publishers Ltd, London (1997).
29. J. O. Kamgang, R. Briandet, J. M. Herry, J. L. Brisset and M. Naïtali, *J. Applied Microbiology* **103**, 621 (2007).
30. W. R. Loewenstein, *Biochim. Biophysica Acta* **560**, 1 (1979).
31. C. D. Moorby and E. Gherardi, *Exp. Cell Res.* **249**, 367 (1999).
32. R. Bruzzone, T. W. White and D. L. Paul, *Eur. J. Biochem.* **238**, 1 (1996).
33. W. H. Evans, S. Ahmad, J. Diez, G. H. George, J. M. Kendall and P. E. Martin, *Proc. Novartis Found Symp.* **219**, 44 (1999).
34. M. Yeager, V. M. Unger and M. M. Falk, *Curr. Opin. Struct. Biol.* **8**, 517 (1998).
35. A. Folch and M. Toner, *Annu. Rev. Biomed. Eng.* **1**, 227 (2000).
36. J. Schmitz and K. E. Gottschalk, *J. Biol. Chem.* **4**, 1373 (2008).

37. J. Folkman and A. Moscona, *Nature* **273**, 345 (1978).
38. L. K. Hansen, D. J. Mooney, J. P. Vacanti and D. E. Ingber, *Mol. Biol. Cell* **9**, 967 (1994).
39. S. M. Fritsch and H. Francis, *J. Cell. Biol.* **124**, 619 (1994).
40. M. Zhan, H. Zhao and Z. Han, *Histology Histopathology* **19**, 973 (2004).
41. K. Lewis and A. M. Klibanov, *Trends Biotechnol.* **23**, 343 (2005).
42. E. J. Abbott and F. A. Firestone, *Mechanical Engineering* **55**, 569 (1933).
43. E. S. Gadelmawla, M. M. Koura, T. M. A. Maksoud, I. M. Elewa and H. H. Soliman, *J. Mater. Process. Technol.* **123**, 133 (2002).
44. M. Nardin, K. Anselme, V. Arnold, N. Cottenye, S. Fleith, C. Herrier, L. Ploux, V. Roucoules, O. Soppera and L. Vonna, *Matériaux & Techniques* **95**, 357 (2007).
45. S. Belkin, D. R. Smulski, A. C. Vollmer, T. K. Van Dyk and R. A. LaRossa, *Appl. Environ. Microbiol.* **62**, 2252 (1996).
46. A. C. Vollmer, S. Kwakye, M. Halpern and E. C. Everbach, *Appl. Environ. Microbiol.* **64**, 3927 (1998).
47. L. Lensun, T. A. Smith and M. L. Gee, *Langmuir* **18**, 9924 (2002).
48. M. Yi, H. Nymeyer and H. X. Zhou, *Phys. Rev. Letters* **101**, 038103 (2008).
49. G. Brezesinski and V. L. Shapovalov, *J. Phys. Chem. B* **110**, 10822 (2006).
50. A. H. Weerkamp, H. M. Uyen and H. J. Busscher, *J. Dental Res.* **67**, 1483 (1988).
51. M. Tsuchida and Z. Osawa, *Colloid Polym. Sci.* **272**, 1435 (1994).
52. M. Henry, C. Dupont-Gillain and P. Bertrand, *Langmuir* **19**, 6271 (2003).
53. G. Legeay and F. Poncin-Epaillard, in: *Adhesion — Current Research and Applications*, W. Possart (Ed.), pp. 175–188. Wiley, Weinheim (2005).
54. G. Legeay, F. Poncin-Epaillard and C. Arciola, *J. Artificial Organs* **29**, 453 (2006).
55. B. Gupta, J. Hilborn, I. Bisson and P. Frey, *J. Appl. Polym. Sci.* **81**, 2993 (2001).
56. B. Gupta and N. Anjum, *J. Appl. Polym. Sci.* **86**, 1118 (2002).
57. H. P. Brack, M. Wyler, G. Peter and G. G. Scherer, *J. Membrane Sci.* **214**, 1 (2003).
58. N. Anjum, B. Gupta and A. M. Riquet, *J. Appl. Polym. Sci.* **101**, 772 (2006).
59. R. Eloy, D. Parrat, T. M. Duc, G. Legeay and A. Bechetoille, *J. Cataract. Refract. Surg.* **9**, 364 (1993).
60. N. K. Man, G. Legeay, G. Jehenne, D. Tiberghien and D. de la Faye, *Artif. Organs* **14**, 44 (1991).
61. A. M. Klibanov, *J. Mater. Chem.* **17**, 2479 (2007).
62. J. Haldar, J. Chen, T. M. Tumpey, L. V. Gubareva and A. M. Klibanov, *Biotechnol. Lett.* **30**, 475 (2008).
63. A. Carré and K. L. Mittal (Eds), *Superhydrophobic Surfaces*. VSP/Brill, Leiden (2009).
64. J. Fresnais, L. Benyahia, J. P. Chapel and F. Poncin-Epaillard, *Eur. Phys. J. Appl. Phys.* **26**, 209 (2004).
65. L. Kessler, G. Legeay, C. Jesser, C. Damgé and M. Pinget, *Biomaterials* **16**, 185 (1995).
66. F. Grinnell and M. K. Feld, *J. Biological Chem.* **257**, 4888 (1982).
67. E. Velzenberger, K. El-Kirat, G. Legeay, M. D. Nagel and I. Pezron, *Colloids Surfaces B: Biointerfaces* **68**, 238 (2009).
68. E. Velzenberger, M. Vayssade, G. Legeay and M. D. Nagel, *Cellulose* **15**, 347 (2008).
69. M. Hindié, G. Legeay, M. Vayssade, R. Warocquier-Clérout and M. D. Nagel, *Biomolecular Eng.* **22**, 205 (2005).
70. N. Faucheux, J. M. Zahm, N. Bonnet, G. Legeay and M. D. Nagel, *Biomaterials* **25**, 2501 (2004).
71. R. Barrientos, S. Baltrusch, S. Sigrist, G. Legeay, A. Belcourt and S. Lenzen, *J. Hormone Metabolic Res.* **41**, 5 (2009).

72. G. Legeay, T. Honoré, M.-N. Bellon Fontaine and J.-M. Herry, PCT/FR 2005/050142 du 02/03/2005, demande internationale WO 2005/084719 A1 du 15/09/2005.
73. G. Legeay, P. Bertrand, A. Belcourt and L. Kessler, US Patent no. 7,056,726 B2 (2006).
74. L. Kessler, G. Legeay, A. Coudreuse, P. Bertrand, C. Poleunis, X. Vandeneynde, K. Mandes, P. Marchetti, M. Pinget and A. Belcourt, *J. Biomater. Sci. Polym. Edn* **14**, 1135 (2003).
75. C. Sousa, P. Teixeira, S. Bordeira, J. Fonseca and R. Oleiveira, *J. Adhesion Sci. Technol.* **22**, 675 (2008).
76. T. Le-Thi, P. Lejeune and G. Legeay, European Patent no. EP 1 070 508 A1 (2001).
77. J. Rag, G. Herzog, M. Manning, C. Volcke, B. D. MacCraith, S. Ballantyne, M. Thompson and D. W. M. Arrigan, *Biosensors Bioelectronics* **24**, 2654 (2009).
78. L. Griscom, G. Legeay and B. Le Pioufle, in: *Microtechnology in Medicine and Biology, 3rd IEEE/EMBS Proc.*, Hawaii, 194 (2005).
79. T. Meylheuc, C. J. van Oss and M.-N. Bellon-Fontaine, *J. Appl. Microbiol.* **91**, 822 (2001).
80. T. Meylheuc, J.-M. Herry and M.-N. Bellon-Fontaine, *Sciences des Aliments* **21**, 591 (2001).
81. C. Dagbert, T. Meylheuc and M.-N. Bellon-Fontaine, *Electrochimica Acta* **51**, 5221 (2006).
82. J. Haldar, D. An, L. Á. de Cienfuegos, J. Chen and A. M. Klibanov, *Proc. Natl Acad. Sci. USA* **103**, 17667 (2006).
83. R. C. Advincula, W. J. Brittain, K. C. Caster and J. Rühe, *Polymer Brushes*. Wiley (VCH), Weinheim (2004).
84. N. Aumsuwan, M. S. McConnell and M. W. Urban, *Biomacromolecules* **10**, 623 (2009).
85. N. Cottenye, F. Teixeira, A. Ponche, G. Reiter, K. Anselme, W. Meier, L. Ploux and C. Vebert-Nardin, *Macromolecular Bioscience* **8**, 1161 (2008).
86. L. Caillier, E. Taffin de Givenchy, R. Levy, Y. Vandenbergh, S. Géribaldi and F. Guittard, *European J. Medicinal Chem.* **44**, 3201 (2009).
87. L. Ploux, S. Beckendorff, M. Nardin and S. Neunlist, *Colloids Surfaces B: Biointerfaces* **57**, 174 (2007).
88. L. Grossin, D. Cortial, B. Saulnier, O. Félix, A. Chassepot, G. Decher, P. Netter, P. Schaaf, P. Gillet, D. Mainard, J. C. Voegel and N. Benkirane-Jessel, *Adv. Mater.* **21**, 650 (2008).

Part 4

Cell Adhesion in Medicine and Therapy

Selective Cell Control by Surface Structuring for Orthopedic Applications

E. Fadeeva *, S. Schlie, J. Koch and B. N. Chichkov

Laser Zentrum Hannover e.V., Hollerithallee 8, 30419 Hannover, Germany

Abstract

In this work we present an *in vitro* study of the influence of two types of femtosecond laser generated topographies on a titanium surface — grooves with different periodicities and "lotus-like" structures — on the behavior of human fibroblast and MG-63 osteoblast cells. We show that anisotropy in wetting of groove structures correlates well with contact guidance of cells. The "lotus-like" structured titanium surfaces show superhydrophobic properties and influence differentially osteoblast and fibroblast adhesion and growth. The proliferation of fibroblast cells is inhibited, whereas the proliferation of osteoblast cells is promoted. This technique for cell specific control offers promising perspectives in fabrication of new functionalized implants.

Keywords

Femtosecond laser fabrication, titanium, surface topography, contact angle, fibroblast, osteoblast

1. Introduction

In this study we report on *in vitro* investigations of the potential of femtosecond laser structuring technique for surface functionalization of orthopedic titanium implants.

The long term success of orthopedic implants requires a direct contact between living bone and implant surface without an intervening fibrous scar tissue. This unique bone-implant interface is termed "osseointegration" [1].

Titanium is a widely used material for orthopedic applications because of its good biocompatibility, mechanical performance and long-term durability. In the past, considerable efforts were made to provide better osseointegration of titanium implants using numerous surface modification techniques. Additive methods such as titanium plasma spraying [2], as well as subtractive methods such as acid etching [2–4], grit-blasting [5, 6] and laser manufacturing [7–10] have been used. The main

* To whom correspondence should be addressed. Tel.: +49 511 2788-378; Fax: + 49 511 2788-100; e-mail: e.fadeeva@lzh.de

Surface and Interfacial Aspects of Cell Adhesion
© Koninklijke Brill NV, Leiden, 2010

objective was to provide increased implant surface area and hence better potential for biomechanical interlocking. Unfortunately, previous research data do not completely explain the role of surface roughness in osseointegration and do not provide the ultimate topography solution for orthopedic applications.

In this study, we have examined the potential of femtosecond laser generated structures for better osseointegration, with a focus on cell type control. In essence, we investigated whether structured titanium surfaces could differentially influence osteoblast and fibroblast cells adhesion and growth. The goal of this approach is the development of a surface topography which is "attractive" for bone tissue cells — osteoblasts — and "unattractive" for fibrous scar tissue cells — fibroblasts.

Structuring of titanium surfaces was performed by femtosecond laser ablation. The evident advantage of laser technology over other structuring methods (e.g., lithography, chemical etching, plasma spraying, etc.) is the ability to produce user-defined topographies in a one-step process. Femtosecond laser processing has additional advantages over long-pulse laser processing, namely higher quality of micromachined structures, negligible material damage and a reduced heat-affected zone [11]. It has been shown that structuring with femtosecond lasers, in addition to user-defined designs, allows a large variety of self-organized surface topographies like spikes [12–14], laser-induced periodic surface structures (LIPSS) [15, 16], micro- and nanoroughness [17] that enables tuning of surface properties over a wide region [13–18]. Another advantage of femtosecond laser material processing is the ability of this technology to structure complex implant forms for *in vivo* applications.

In the present work we have used femtosecond laser ablation to produce two types of structures: groove structures with different periodicities and self-organized hierarchical micro- and nanostructures. Previously, we had demonstrated that spike structured topography on silicon could be used effectively for selective cell control [19–21]. Also we had found that some groove structures were effective for suppression of fibroblast growth [22]. In this study, we have performed investigations on two structure types on titanium surface to determine their effectiveness for cell-specific control for orthopedic applications.

2. Experimental

2.1. Materials and Structure Fabrication Methods

Commercially available pure titanium samples with dimensions 3 mm × 3 mm × 1 mm were used in this study. The samples were mechanically polished, further cleaned with acetone followed by methanol, and finally rinsed in distilled water in an ultrasonic bath.

Structuring of titanium surfaces was performed with a commercially available amplified Ti:Sapphire femtosecond laser system (Femtopower Compact Pro, Femtolasers Produktions GmbH, Austria). The system delivers sub-30-fs pulses at

800 nm central wavelength with a pulse energy of up to 1 mJ, and a repetition rate of 1 kHz.

An x–y motorized translation stage (Physik Instrumente GmbH, Germany) was used for sample positioning and translation. A computer controlled LCD element was used for laser pulse energy setting.

For the fabrication of groove structures a mask projection technique was applied. 250, 500, 750, 1000, 1250, and 1500 µm wide slit masks were imaged onto the sample surfaces with 50-times demagnification using a 50× microscope objective (Leica HCX PL APO L 50×/0.55 UVI), providing structures with 5, 10, 15, 20, 25, and 30 µm groove widths and periodicities of 10, 20, 30, 40, 50 and 60 µm, respectively. The depth of the structures is $\geqslant 2$ µm. Using the mask projection technique assures high quality of groove structures and a high processing speed. In comparison to structuring with a focused Gaussian beam, we are able to reduce processing time for the same groove pattern by a factor of 60. Uniform structuring of the large area (3×3 mm^2) with the required high resolution is possible using an autofocus system (INH200, Vistec, Germany). Confocal microscopy images of these groove structures are shown in Fig. 1.

For fabrication of hierarchical superimposed nano- and microstructures, titanium surfaces were uniformly irradiated with 500 femtosecond laser pulses at fluences of 20–70 J/cm^2. The samples were processed under ambient air conditions. Following femtosecond laser fabrication, topographies were analyzed with a scanning electron microscope (SEM). SEM images of these structures are shown in Fig. 2.

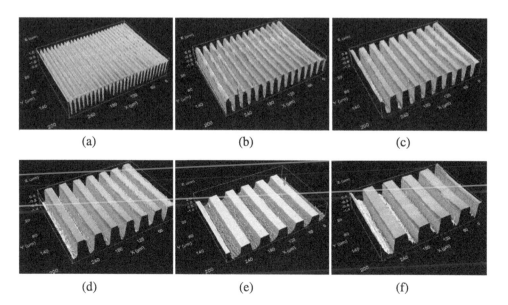

(a)   (b)   (c)

(d)   (e)   (f)

Figure 1. Confocal microscopy images of femtosecond laser fabricated groove surface structures: 5 µm (a), 10 µm (b), 15 µm (c), 20 µm (d), 25 µm (e), and 30 µm (f) groove width fabricated on titanium samples.

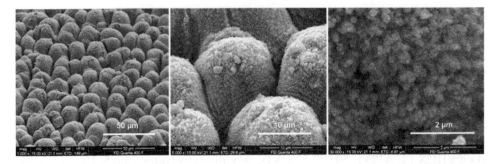

Figure 2. SEM micrographs (at different magnifications) of femtosecond laser-induced hierarchical nano- and micro-superimposed structures fabricated on titanium surfaces by laser treatment at a laser fluence of 50 J/cm^2.

Chemical characterization of unstructured and structured titanium samples was performed using energy dispersive spectrometry (EDS) with a Noran Voyager 3050 system (Noran Instruments, Inc., Middleton, WI, USA). It was found that after the femtosecond laser machining, oxygen content increased by 14.0 weight % on structured areas. In addition, trace amounts of silicon (Si) — 0.09 weight % — were found on the surface of unstructured as well structured titanium samples. The origin of silicon can be explained by the fact that a Si-containing powder was used for polishing.

2.2. Contact Angle Measurements

To characterize how structuring had changed the surface properties of the material, we carried out wettability tests on all structured sample surfaces. Measurements were done using a video-based optical contact angle measuring system (OCA 40 Micro, DataPhysics Instruments GmbH, Germany).

Since the groove structured surfaces are characterized by anisotropic wetting properties, the water contact angle measurements were performed in two directions — perpendicular and parallel to the grooves. Furthermore, top-view photographs of elongated droplets on groove structures were used to visualize the wetting anisotropy.

To measure the advancing (Θ_A) and receding (Θ_R) water contact angles on structured surfaces, the tilting plate method was applied. A water droplet was placed on the structured surface. Then the sample was tilted slowly while a CCD camera (with a maximal frame rate of 25 Hz at full resolution) recorded changes in the droplet's shape. Θ_A and Θ_R angles were measured using the last image before the droplet base started to move.

All contact angles were measured automatically using the software of the contact angle measuring system. Measurements were performed at normal atmospheric conditions using 10 µl water droplets.

2.3. Cell Culture

Before starting the cell experiments, titanium samples were sterilized under UV light for about 30 min. In order to investigate material effects on cell-specific responses, human fibroblasts and MG-63 osteoblasts were used. The human fibroblasts were received from the Medical High School of Vienna, Austria. The MG-63 osteoblasts were received from the Medical High School of Hannover, Germany. The cells were cultivated on the samples in 96-well plates filled with 200 μl of Dulbecco's Modified Eagle Medium (DMEM; Sigma, Taufkirchen, Germany) supplemented with 10% fetal calf serum and antibiotics (pH 7.4; osmolarity 300 ± 5 mosmol/l). Glass slides served as references. The cell plates were placed in a cell culture incubator (Heraeus, Hanau, Germany) in which a 95% air: 5% CO_2 atmosphere and 80% humidity were maintained.

2.4. Proliferation Assay

After various times of cultivation on laser structured titanium, the adherent human fibroblasts and MG-63 osteoblasts were trypsinized. To determine the cell density, the cell suspension was collected and centrifuged at $800g$ for 10 min. Then the cells were resolved in cell culture medium and counted using a Fuchs Rosenthal cell counter device. For a better comparison between the experiments, the cell densities were normalized as percent of the seeding density at time zero (9.7×10^4 cells/ml for fibroblasts and 1.13×10^5 cells/ml for osteoblasts). The results are given as average $\pm$ standard error of the mean.

2.5. Investigation of Cell Morphology and Orientation

Cell morphology and orientation were analyzed by fluorescence after nucleus and actin filaments staining using DAPI and phalloidin-Alexa 488, respectively (Molecular Probes, Invitrogen, Karlsruhe, Germany). After 24 h cultivation time, human fibroblasts and MG-63 osteoblasts grown on the samples were fixed by a 10 min incubation in phosphate buffered saline (PBS) containing 4% formaldehyde. The cells were permeabilised by incubation in PBS containing 0.3% Triton X-100 for 10 min. The chromatin in the nucleus was stained by incubation in PBS containing 1 μM DAPI for 10 min. After washing with PBS, actin filaments were stained with 0.6 U phalloidin-Alexa 488 dissolved in PBS for 1 h in the dark. For further analysis the cells were conserved in phosphate buffered saline.

The cells were observed with a fluorescence microscope (Nikon TE 2000-E, Nikon, Düsseldorf, Germany) at excitation wavelengths of 348 nm for DAPI and 488 nm for phalloidin-Alexa. Images were acquired using a CCD camera and software "E Z-C1 3.5" (Nikon, Düsseldorf, Germany).

Quantitative evaluation of the results was performed with ImageJ software. First, length and width of each single nucleus (L_n, W_n) and each single cell (L_c, W_c) were measured. As the scales are automatically given in pixels, each length and width was converted into μm. The results are given as mean $\pm$ standard error of the mean from 100 independent measurements.

3. Results and Discussion

The quality of orthopedic implant integration can be significantly improved by selective cell control on implant surfaces. For orthopedic applications, it is important that osteoblast adhesion and proliferation should be promoted whereas fibroblast adhesion and proliferation should be inhibited. One effective way to realize this is structuring of the implant surfaces.

In our previous work [19–22] it was demonstrated that fibroblast adhesion and growth were suppressed on structured surfaces in comparison to unstructured flat surfaces of the same material. Fibroblast morphology was changed as well, with cells presenting a more round shape. Since there were no detected alterations in the chemistry of structured surfaces in comparison to original unstructured surfaces using energy dispersive spectrometry (EDS) analysis, it was concluded that the altered cell behavior was related to surface topography. The next point was to find topography characteristics, which can be responsible for the observed altered cell behavior.

In the past, different topography types (e.g., grooves, dots, spikes, mesh and random roughness) have been extensively studied in view of their effects on the cells [23]. So it was attempted to relate different topography characteristics — depth, groove width and various roughness parameters — with resulting cell reactions. The obtained results are quite complex and do not provide a generalized view.

In our studies we characterized surface topography by wettability alteration which was produced by femtosecond laser microstructuring. For example, using this technique in [21] we were able to control the water contact angle on platinum in the range of 81°–158°. A clear correlation between the alteration in water contact angle and fibroblast proliferation was observed: greater decreases in wettability of structured surfaces in comparison to original flat surfaces resulted in lower fibroblast proliferation. Thus, the correlation between cell response (e.g., proliferation, morphology) and topography-induced contact angle changes reflects the influence of given topographies on cells.

General wetting laws for textured surfaces are well established and allow us to design topographies which promote or inhibit [24] wetting and cell adhesion. In cell-biomaterial surface interaction, cell adhesion is the first and very important step, which influences all subsequent processes such as proliferation, organization of the cytoskeleton, differentiation, etc. [25–27]. One notable example can be found in [30] where it was demonstrated that endothelial cells could be switched from growth to apoptosis simply by modifications of the surface available for cell adhesion. Therefore, if the surface available for cell adhesion is modified by structuring, the general cell response can be controlled. Note that in order to realize this approach one needs topographical surface features smaller than the cell size.

It is well known that proteins mediate cell adhesion to a biomaterial surface. Taking into account that the surface chemistry is not altered by structuring and surface topography features (up to 100 nm), which are generally much larger than the proteins sizes (only a few tens of nm), one can assume that protein adsorption is not affected by the surface topography. Indirect evidence for unchanged protein adsorption is the ability of some cell types (e.g., neuroblastoma cells) to attach and proliferate on the same topography that is "unattractive" for other cells (e.g., fibroblasts [20]). To clarify the effect of surface topography on protein adsorption, concentration, and conformation further investigations are required, which are beyond the scope of this paper.

In the present work, the influence of two types of structures (groove structure with different periodicities and self-organized superimposed nano- and micro-hierarchical structure) on osteoblast and fibroblast cells behavior was studied. Both types of structures were fabricated on titanium surface and characterized with water contact angle measurements. The water contact angle on an unstructured flat titanium surface was 80°. This relatively high contact angle can be attributed to a trace silicon (Si) contamination of titanium surfaces after polishing (see Section 2.1).

The first structure type — grooves — introduces anisotropy in wetting. The water contact angle data are given in Table 1. Characteristic pictures used for contact angle measurements are shown in Fig. 3. We found that apparent contact angles measured in plane perpendicular to the grooves were larger than the intrinsic value on the unstructured titanium surface. This is because the droplet is sticking to the edges of the grooves [28], which leads to stretching of the droplet in the other direction. The droplet elongation data estimated from the top-view images can also be found in Table 1. To summarize, structuring with 5 µm grooves introduces the strongest

Table 1.
Effect of groove structures on measured contact angle values

Structure type	Contact angle (°) perpendicular to the grooves	Contact angle (°) parallel to the grooves	Elongation = long droplet axis/short droplet axis
5 µm grooves	132	65	1.89
10 µm grooves	109	67	1.6
15 µm grooves	101	72	1.37
20 µm grooves	92	76	1.25
25 µm grooves	87	78	1.15
30 µm grooves	85	80	1.08

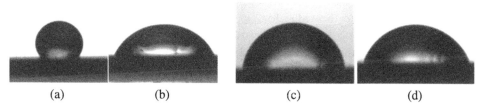

(a)	(b)	(c)	(d)

Figure 3. Water contact angles measured on 5 µm grooves perpendicular (a) and parallel (b) to the grooves. The same measurements on 30 µm grooves (c) and (d).

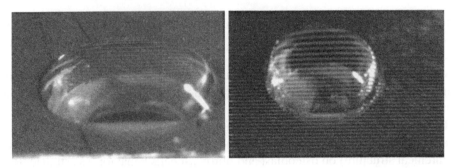

Figure 4. Pronounced water droplet elongation on 5 μm grooves (left) and weak elongation on 30 μm grooves (right).

anisotropic wetting, followed by 10 and 15 μm grooves with less pronounced wetting anisotropy. For structures with 20, 25 and 30 μm grooves, wetting anisotropy is found to be rather weak. Top-view photographs shown in Fig. 4 demonstrate this difference. The contact angles measured in plane parallel to the grooves are comparable to the intrinsic contact angle of the unstructured titanium.

The response of fibroblast cells on the groove structures is illustrated in Fig. 5(a)–(f). It can be seen that the groove structures of defined periodicity induce the so-called "contact guidance" [23] of cell response. Interestingly, the fibroblast contact guidance correlates well with wetting anisotropy. The groove structures characterized by pronounced wetting anisotropy introduce clear contact guidance, and structures with weak anisotropy introduce no contact guidance. The morphology of osteoblasts (Fig. 5(g)–(l)) on groove structures is similar to fibroblast morphology with the exception of osteoblast response to 20 μm wide grooves. While fibroblasts no longer "sense" the groove structures of this dimension, osteoblasts still reveal clear contact guidance. One possible reason for this distinction is the better ability of osteoblasts to conform to this surface structure which is analogous to a liquid droplet with lower surface tension remaining pinned on the edges of 20 μm groove structure. Another possible explanation is the difference in cell sizes.

For further insight, the cell sizes of fibroblasts and osteoblasts were quantified by measuring the length (L_c) and width (W_c) of the total cell body on control glass surface. It is found that fibroblasts are smaller in cell width (W_c) with an average width of 20.49 ± 0.96 μm, whereas osteoblasts present a width of 26.71 ± 1.31 μm. Concerning the cell length given as L_c, the value for fibroblasts is higher with 124.19 ± 6.36 μm than the value for osteoblasts with 111.12 ± 4.12 μm. These results suggest that the cell width is a critical parameter in the cell reaction on the groove structures. Further analysis is needed to determine correlations between the cell sizes and topography sizes, since the cell shape is very dynamic and can change under certain conditions [29].

As previously mentioned, the fibroblast cells are narrower than osteoblasts. Since the larger spacing between the grooves causes less pronounced cell orientation, the narrower fibroblasts do not respond to 20 μm grooves. These results show that

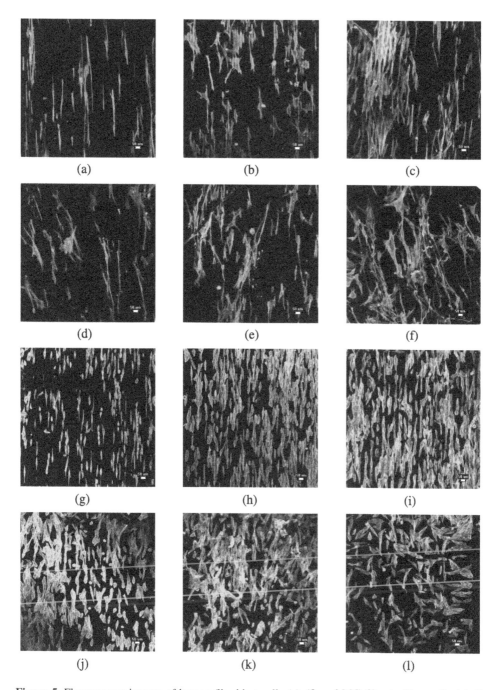

Figure 5. Fluorescence images of human fibroblast cells (a)–(f) and MG-63 osteoblast cells (g)–(l) cultivated on groove structures on titanium in dependence of groove width after 24 h cultivation time. Groove width: 5 μm (a), (g); 10 μm (b), (h); 15 μm (c), (i); 20 μm (d), (j); 25 μm (e), (k); 30 μm (f), (l).

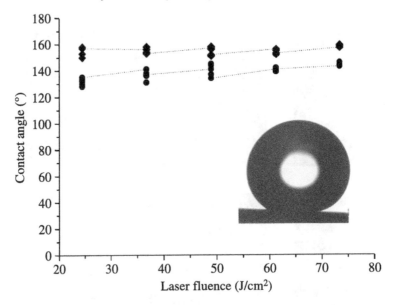

Figure 6. Advancing (rhombuses) and receding (circles) water contact angles measured on hierarchical micro- and nanostructures *versus* laser fluence used for fabrication. The insert shows the water drop on a tilted "lotus-like" structured titanium surface.

different sized groove structures can be employed to control cell orientation, depending on the cell type.

Another type of textured surfaces examined in this work were with self-organized superimposed nano- and micro-hierarchical structures (Fig. 2). These structures were fabricated by ablation with 500 overlapping laser pulses at a laser fluence of 25 J/cm² or higher. Increasing laser fluence leads to a minor increase in the size and density of microstructures, whereas the nanostructures remain nearly unchanged. The water contact angle data (see Fig. 6) show that this type of structure transforms the initially hydrophilic (water contact angle 80°) unstructured titanium surface into a water-repellent surface. These structured surfaces not only have high advancing water contact angles (up to 160°) but also a low contact angle hysteresis (⩽15°).

Here, fabrication of titanium water-repellent surfaces using a simple one-step processing technique is demonstrated, for the first time to the best of our knowledge. The cause of the superhydrophobic properties of the structured titanium surface is the hierarchical multiscale surface topography resembling natural superhydrophobic surfaces (e.g., lotus leaf) [31]. Similar surface topographies, as has been shown theoretically, are able to transform any substrate surface with a non-zero contact angle into a non-wetting surface [32]. The low contact angle hysteresis was found to be important for non-adhesive water-repellent properties of the surface [24]. The laser fabricated self-organized superimposed micro- and nano-hierarchical structure also meets the low hysteresis condition. The non-adhesiveness of water droplets to

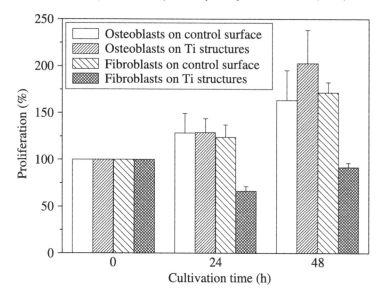

Figure 7. Osteoblast and fibroblast proliferation on unstructured control surface and on "lotus-like" structured titanium surface.

a structured titanium surface leading to water repellency is a result of a greatly reduced solid surface fraction being in contact with the droplet.

In our previous work, we had already tested fibroblast response to surface structures which modified the wettability properties from originally hydrophilic to hydrophobic [19–21]. In the present work, a biomaterial surface with extremely altered (superhydrophobic) surface properties was investigated for the first time.

The fibroblasts response to the "lotus-like" structured titanium surface is represented in Fig. 7. The diagram shows reduced fibroblast proliferation on the structured surface in comparison to the unstructured surface. The reason for the observed proliferation reduction can be attributed to the reduced titanium surface available for fibroblasts adhesion. It was previously observed [33] that fibroblasts had a tendency to stretch over complex surface structures than conforming to them. This was attributed to inflexibility of the cell cytoskeleton. We presume that fibroblasts are not able to conform themselves to very rough "lotus-like" structured titanium surface and, therefore, have a greatly diminished solid surface available for adhesion.

This suggestion is also supported by the morphological analysis of the cells. In contrast to the control, where the cells represent an elongated shape, their shape is different on the "lotus-like" structures (Fig. 8). In comparison to our previous observations of poor fibroblast spreading on spike structured silicon [20], on the "lotus-like" structured titanium surface fibroblast cells are not able to spread at all and have a round shape. This change in the cell shape is caused by poor adhesion to the surface. A clear correlation between poor adhesion and reduced proliferation has already been shown in [25].

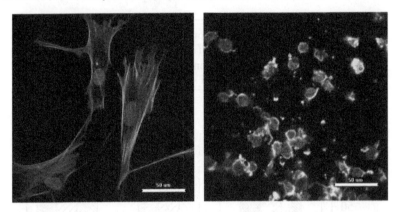

Figure 8. Fibroblast morphology on control glass (left) and "lotus-like" titanium surface (right).

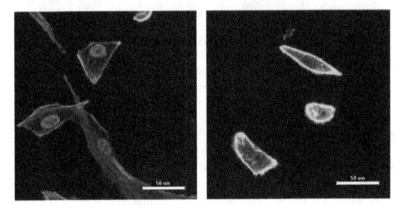

Figure 9. Osteoblast morphology on control glass (left) and "lotus-like" titanium surface (right).

The osteoblast response to the "lotus-like" titanium surface topography is completely different. Compared to the control glass surface, osteoblast proliferation on the structured Ti surfaces is increased (Fig. 7). Pictures of cell morphology (see Fig. 9) demonstrate that osteoblasts are able to spread on the "lotus like" surface structure. This indicates that in contrast to fibroblasts, osteoblast adhesion on the structured surfaces is promoted. It is possible that due to a more flexible skeleton, the osteoblast cells are able to conform to the structured surfaces and, therefore, have increased surface area available for adhesion.

4. Conclusion

The possibility of cell selective (fibroblast *versus* osteoblast) control of proliferation by structuring titanium surfaces using femtosecond laser ablation has been demonstrated. For the characterization of structured surfaces, wettability tests have been employed. It has been observed that topographies causing the greatest changes in surface wettability stimulate the most intense cell response. This knowledge can be used in biomaterial science for characterization of structures and their optimization. The superhydrophobic "lotus-like" titanium surface structures, inhibiting fibroblast and promoting osteoblast proliferation, are very promising for orthopedic applications.

Acknowledgements

This work was supported by the German Research Foundation, DFG, under the projects SFB599 "Sustaintable Bioresorbing and Permanent Implants of Metallic and Ceramic Materials" and Transregio Project TR37.

References

1. P. I. Branemark, B. O. Hansson, R. Adel, U. Breine, J. Lindström, O. Hallen and A. Ohman, *Scand. J. Plast. Reconstr. Surg. Suppl.* **16**, 1 (1977).
2. A. Gaggl, G. Schulters, W. D. Müller and H. Kärcher, *Biomaterials* **21**, 1067 (2000).
3. P. R. Klokkevold, R. D. Nishimura, M. Adachi and A. Caputo, *Clin. Oral. Implant. Res.* **8**, 442 (1997).
4. P. S. Vanzillotta, M. S. Sader, I. N. Bastos and G. de Almeida Soares, *Dental Mater.* **22**, 275 (2006).
5. A. Wennerberg, T. Albrektsson, C. Johansson and B. Andersson, *Biomaterials* **17**, 15 (1996).
6. R. V. Bathomarco, G. Solorzano, C. N. Elias and R. Prioli, *Appl. Surface Sci.* **233**, 29 (2004).
7. A. Joob-Fancsaly, T. Divinyi, A. Fazekas, Cs. Daroczi, A. Karacs and G. Peto, *Smart Mater. Struct.* **11**, 819 (2002).
8. G. Peto, A. Karacs, Z. Paszti, L. Guczi, T. Divinyi and A. Joob, *Appl. Surface Sci.* **186**, 7 (2002).
9. M. Bereznai, I. Pelsöczi, Z. Toth, K. Turzo, M. Radnai, Z. Bor and A. Fazekas, *Biomaterials* **24**, 4197 (2003).
10. M. Trtica, B. Gakovic, D. Batani, T. Desai, P. Panjan and B. Radak, *Appl. Surface Sci.* **253**, 2551 (2006).
11. B. N. Chichkov, C. Momma, S. Nolte, F. von Alvensleben and A. Tunnerman, *Appl. Phys. A* **63**, 109 (1996).
12. T. H. Her, R. J. Finlay, C. Wu, S. Deliwala and E. Mazur, *Appl. Phys. Lett.* **73**, 1673 (1998).
13. V. Zobra, L. Persano, D. Pisignano, A. Athanassiou, E. Stratakis, R. Cingolani, P. Tzanetakis and C. Fotakis, *Nanotechnology* **17**, 3234 (2006).
14. B. K. Nayak, M. C. Gupta and K. W. Kolasinski, *Appl. Phys. A* **90**, 399 (2008).
15. A. Y. Vorobyev, V. S. Makin and C. Guo, *J. Appl. Phys.* **101**, 034903 (2007).
16. A. Borowiec and H. K. Haugen, *Appl. Phys. Lett.* **82**, 4462 (2003).
17. A. Y. Vorobyev and C. Guo, *Appl. Surface Sci.* **253**, 7272 (2007).
18. A. Y. Vorobyev and C. Guo, *Phys. Rev. B* **72**, 195422 (2005).
19. E. Fadeeva, S. Schlie, J. Koch, A. Ngezahayo and B. N. Chichkov, *Phys. Status Solidi. A* **206**, 1349 (2009).
20. S. Schlie, E. Fadeeva, J. Koch, A. Ngezahayo and B. N. Chichkov, *J. Biomed. Appl.* doi:10.1177/0885328209345553.
21. E. Fadeeva, S. Schlie, J. Koch, B. N. Chichkov, A. Y. Vorobyev and Chunlei Guo, in: *Contact Angle, Wettability and Adhesion*, K. L. Mittal (Ed.), Vol. 6, pp. 163–171. VSP/Brill, Leiden (2009).

22. U. Reich, P. P. Mueller, E. Fadeeva, B. N. Chichkov, T. Stoever, T. Fabian, T. Lenarz and G. Reuter, *J. Biomed. Mater. Res.* **87B**, 146 (2008).
23. A. Curtis and C. Wilkinson, *Biomaterials* **18**, 1573 (1997).
24. D. Quere, A. Lafuma and J. Bico, *Nanotechnology* **14**, 1109 (2003).
25. E. Clark and R. O. Hynes, *J. Biol. Chem.* **271**, 14814 (1996).
26. F. G. Giancotti and E. Rousalahti, *Science* **285**, 1028 (1999).
27. B. M. Gumbiner, *Cell* **84**, 345 (1996).
28. Y. Chen, B. He, J. Lee and N. A. Patankar, *J. Colloid Interface Sci.* **281**, 458 (2005).
29. J. C. Adams, *Cell. Mol. Life Sci.* **58**, 371 (2001).
30. C. S. Chen, M. Mrksich, S. Huang, G. M. Whitesides and D. E. Ingber, *Science* **276**, 1425 (1997).
31. T. Sun, L. Feng, X. Gao and L. Jiang, *Acc. Chem. Res.* **38**, 644 (2005).
32. S. Herminghaus, *Europhys. Lett.* **52**, 165 (2000).
33. D. M. Brunette, in: *Titanium in Medicine*, D. M. Brunette, P. Tengvall, M. Textor and P. Thomsen (Eds), pp. 485–512. Springer, Berlin (2001).
34. R. N. Wenzel, *Ind. Eng. Chem.* **28**, 988 (1936).

Inorganic PVD and CVD Coatings in Medicine — A Review of Protein and Cell Adhesion on Coated Surfaces

Jürgen M. Lackner * and **Wolfgang Waldhauser**

Joanneum Research Forschungsges.m.b.H., Laser Center Leoben, Leobner Strasse 94,
A-8712 Niklasdorf, Austria

Abstract

The functionalization of biomaterials for implants becomes increasingly important for designing bioinert and bioactive surfaces to reduce the impact of implantation to human body (inflammation, encapsulation) and extend the lifetime of implants. Even pharmacological effects can be triggered by nanomaterials like thin films and nanoparticles in medical treatment. However, the systematic knowledge of the interactions between cells and artificial, inorganic materials is poor yet. Finding the decisive influences for high hemo-compatibility or osseointegration is very difficult. Surface chemistry including wetting behaviour, surface charge, homogeneity and functional groups as well as surface topography are some of the fundamental surface parameters defining the cell–surface interaction. Focusing on physical and chemical vapour deposited thin films and coatings, this review will provide for a better understanding of biocompatible coating materials like titanium- and carbon-based compounds and calcium phosphates.

Keywords

Physical vapour deposition, chemical vapour deposition, protein adhesion, cell adhesion, biocompatibility

1. Introduction

Medical devices that come in contact with biological media are fabricated using many different types of materials including metals, polymers, ceramics and natural materials [1]. Currently available biomaterials, however, do not provide optimal performance [2]. In clinical application, inflammation reactions due to inappropriate implant surface might add on top of the inflammatory process, already initiated by the surgical procedure. Then, e.g., non-mechanically stable connective tissue rather than bone is formed around a bone implant (artificial hip joint, dental implant, etc.) or blood clotting may occur around a stent. Less biologically perturbing surfaces may lead to a bone healing process and blood flow, which is similar to that in the absence of an implant. Designing such surfaces by various techniques is one

---

* To whom correspondence should be addressed. Tel.: +43 3842 81260 2305; Fax: +43 3842 81260 2310; e-mail: Juergen.Lackner@joanneum.at

Surface and Interfacial Aspects of Cell Adhesion
© Koninklijke Brill NV, Leiden, 2010

of the main goals of biomaterials research [3, 4]. However, the precise mechanisms involved in this cell–surface and protein–surface interactions are still unknown.

Generally, surfaces of synthetic biomaterials (e.g., polymers, metals and ceramics) are not bioactive themselves. Rather, surface bioactivity is provided by the proteins that adsorb to the biomaterial surface following exposure of the surface to biological fluids. The first molecules are water molecules reaching the surface in a biological milieu containing cells. When cells arrive at the surface, they 'see' a protein-covered surface whose bioactive protein layer has properties that were initially determined by the pre-formed water shells. To control cellular response, it is important to first understand how surface chemistry, surface topography and mechanical forces influence the formation of the adsorbed protein layer and the bioactive sites presented by this layer. Besides water, proteins and cells, biological model systems for biointerfaces (or surfaces) must contain amino acids, nucleic acids and lipids, peptides and DNA (segments), and tissue to describe the sequences occurring in the human body [5]. The recognition of these molecular-scale features is programmed into the molecules through the combination of their 3-D topographic architecture, the superimposed chemical architecture, and the dynamic properties [5]. Although the fundamental interactions occur on the molecular scale, there is an interesting and unique synergistic connection between the nanometer and the micrometer length scales [6–8], when cells are present.

Recent advances in topographical surface modification techniques such as electron beam lithography [9], photolithography [10], colloidal lithography [11, 12], laser ablation [13] and polymer demixing [14] have allowed the fabrication of surfaces with nano-sized features. In addition, the modification of the surface chemistry by coating with biomaterials or biomolecules of 'higher biocompatibility' helps to separate between bulk properties (mainly mechanical stability and load support) and surface properties (topography, surface chemistry, mechanical behaviour, hydrophilicity, surface charge, etc.). The variety of available coatings (thin films) and coating techniques is huge: Physical 'non-covalent' surface modification techniques include physisorption of molecules from solutions, the Langmuir–Blodgett–Kuhn technique for thin organic coatings and the gas phase deposition of organic and metal-like materials (e.g., physical vapour deposition, PVD). Chemical 'covalent' modifications are based on grafting and photo-reactions, self-organizing monolayers and plasma modifications (e.g., chemical vapour deposition, CVD, plasma polymerization).

PVD and CVD techniques are preferably used to produce thin and ultra-thin coatings ($\sim$10 μm down to nanometer thickness) of inorganic elements (metals, semiconductors) and compounds (ceramics) as well as carbon and some organic compounds. PVD coating requires vaporization of a solid material — the target. CVD coating starts from mixtures of reactive gases (precursors) which are decomposed thermally or plasma activated. In contrast to CVD techniques, the PVD processes are generally characterized by lower coating temperatures, larger numbers of possible coating and workpiece (= substrate) materials, the higher purity of the deposited coatings and a better environmental compatibility of the manufacturing process [15, 16]. The classical evaporation processes (e.g., thermal, arc, electron-beam and ion-beam evaporation, pulsed laser deposition) use thermal energy to evaporate the target material. However, sputtering processes use highly-energetic particles, which hit the target surface and strike particles out of the target. To reduce the coating temperature in CVD techniques, plasma enhancing (PE-CVD) is applied during film deposition. Furthermore, the plasma polymerization technique allows the organic thin film deposition (polymers) from gaseous or liquid monomers. The vaporized atoms, ions and clusters can undergo chemical reactions with atoms from the process gas atmosphere before they are deposited on the substrate surface.

Choosing the best technique for achieving the required properties demands high experience in materials science (physics, chemistry of thin films), medicine and biology. This interdisciplinary approach in biomaterial science needs explanation of medical biology for technical scientists (chemists, physicists, material scientists) and materials science and technology for biomedical staff. To establish ties between these scientific fields is the goal of this review article, targeted towards inorganic coatings deposited by PVD and CVD.

2. Artificial Surfaces in Contact with Biological Media — The Fundamental Processes

The response of cells coming in contact with biomaterials induces several, time-dependent steps (Fig. 1) [17]: Immediate water layer formation, protein adsorption, cellular adhesion and migration, proliferation and differentiation of cells. Most

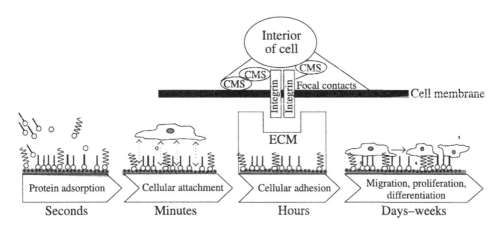

Figure 1. Schematics of the time-dependent protein and cell adhesion to biomaterial surfaces (adapted from [17]).

important for the advantageous biocompatible response is the concentration, composition and conformation of adsorbed proteins [18–21].

2.1. Water Adsorption

The first event on a biomaterial in contact with tissue or blood is the adsorption of inorganic ions and water molecules [22]. In liquid form, water molecules span hydrogen bridge-bonding between the H_2O molecules in less than 10×10^{-12} s, leading to extended three-dimensional networks of self-associated molecules. On contact with water the surface structure of hydrophobic materials (water contact angles exceeding 65° [23]) is not changed: it is the structure of water itself that is modified. An intermediate layer is formed by a dynamic process with continuous rearrangement of ordered microdomains. This layer is a few molecules thick and has less dense, but higher ordered structure than pure liquid water. On the other hand, the surface of hydrophilic materials is greatly modified by water and the hydrogen bonded water molecule network collapses. The density of water molecules increases.

2.2. Protein Adsorption

After adsorption of a water film, proteins are rapidly transported to the interface, selectively adsorbed and undergo conformation rearrangements. The adsorption is based on the following binding mechanisms of proteins to surfaces:

- Electrostatic bond formation between charged groups on the protein and oppositely charged surface sites (on highly polar surfaces, e.g., glass). Most proteins and surfaces carry charge, which depends on the pH and ionic strength of the surrounding solution. Attractive forces arise if the protein and surface have opposite charges [24]. If the charges are balanced over the surface of the protein molecule (isoelectric point), intramolecular electrostatic interactions favour a compact structure and the amount of adsorbed proteins is at its maximum. If there is an excess of charge, either positive or negative, intramolecular forces will lead to repulsion and expansion of the molecule, as well as protein unfolding and a less stable protein [24].

- Hydrogen bonding (on relative polar substrates). Hydrogen bonding is based on the reduction of the conformational entropy of the bonds in polypeptide chains, when the rotational mobility of the highly ordered, hydrogen-stabilized structure is decreased by intramolecular hydrogen bonds [25].

- Hydrophobic interactions (on non-polar substrates, e.g., polymers). Hydrophobic effects occur by spontaneous dehydration and subsequent aggregation of non-polar components in aqueous environment. They are based on the tendency of water to exclude non-polar groups which result in a large increase in entropy of the released water molecules. If either the protein or the contacting surface is hydrophobic, the entropic gain due to dehydration acts as a driving force for protein adsorption. Conversely, if both the protein and surface are hydrophilic, hydration of the surface is energetically favourable and removal of the water from the surface *via* adsorption of the protein is not favoured [26]. The heterogeneous and dynamic nature of proteins makes it difficult to define a protein as either hydrophobic or hydrophilic. Proteins generally contain both polar and non-polar regions [27]. The degree of hydrophobicity of a protein can also affect the extent of denaturation that occurs upon adsorption. If a protein adsorbs on a hydrophobic surface, it will rearrange its structure to expose the hydrophobic regions to the surface [28]. For a hydrophobic protein this may involve little or no denaturation, but for a hydrophilic protein, the internal polar regions can be exposed resulting in denaturation of the protein.

The relative strengths of these contributions depend on protein type (composition and functional groups) and on local environment (pH — isoelectric point of protein, concentration of ions, strength of solution, temperature) [5, 29]. The size and shape of proteins play an important role in adsorption as the number of binding sites will increase with increasing protein size.

In the multiprotein system of the human body, the competitive adsorption behaviour of proteins at surfaces is of great importance. The factors influencing competitive (as well as single protein) adsorption may be, in order of importance:

- the chemical surface properties (in terms of charge, hydrophobicity, available chemical functional groups and chemical surface homogeneity),

- the protein properties (such as electrical charge, hydrophobicity, available chemical functional groups, stability of the protein structure, interactions between proteins in the adsorbed layer, relative concentration in the bulk phase and molecular size) and

- the surface topography.

Although combining all these factors in a quantitative model is impossible, in general it has been found that preferential or selective adsorption occurs so that certain proteins may be enriched in the surface and *vice versa*. Surface chemistry, time of adsorption and protein type are major factors in determining the composition of the adsorbed layer. The Vroman effect [30–32] describes this competitive adsorption of protein mixtures, where smaller proteins by virtue of their higher diffusion coefficients adsorb first and then may be replaced by larger ones which have a greater affinity for a particular surface. Here certain proteins (integrins) act as cell receptors and have chemotactic or adhesive properties.

2.3. Cell Adhesion and Adsorption

Due to rapid protein adsorption and the relative long time it takes for cells to reach a surface, cell–surface interactions are usually influenced by the adsorbed protein layer [33]. Cell–surface contact involves initial attachment and adhesion followed by cell spreading and migration [34, 35], which occur prior to growth and integration of foreign materials in the body [32].

Cell–surface interactions are mediated through integrins that bind to specific amino acid sequences in extracellular matrix (ECM) molecules, such as RGD, rather than directly with the surface of the biomaterial [36–39]. These span the cell membrane and link the ECM to the cell's cytoskeleton. They are approximately 10 nm wide and 10–100 times more prevalent on the cell's surface than other receptors types [40]. In their inactive state, integrins are freely diffusive within the cell membrane until they encounter an available binding domain in the ECM. Upon ligand binding, integrins undergo a conformational change that leads to the possibility of binding of two groups of cytoplasmic proteins — those that biomechanically connect the integrins to the cytoskeleton and those that biochemically initiate or regulate intracellular signalling pathways. Through physical clustering of multiple integrins, more cytoplasmic proteins are transported to the adhesion site to increase its size, adhesion strength, and biochemical signalling activity [41, 42]. These larger, clustered structures of integrins and cytoplasmic proteins are commonly called focal contacts (FCs). FCs are scattered across the cell surface and are typically 0.25–0.50 µm wide and 2–10 µm long, though they arise from much smaller clusters. They separate the cell approximately 10–15 nm from the biomaterial surface. The structure of FCs is dynamic and heterogeneous, and these structures are motile and can assemble/disassemble, disperse and recycle according to the cell's need [43, 44]. Their formation, development, and disassembly are not only key activities for *cellular attachment* and adhesion but also appear to be central modulators in cell spreading and migration.

Compared to the cellular attachment phase, the following *cell adhesion* phase lasts longer and involves various proteins and molecules (Fig. 1). Among serum proteins, fibronectin and vitronectin are recognized as playing a major role in cell adhesion and movement *via* cell adhesion motifs [34, 35, 45–49], while albumin is non-adhesive for cells [50]. Other non-RGD-containing cell binding domains exist such as tyrosine–isoleucine–glycine–serine–arginine and isoleucine–lysine–valine–alanine–valine in laminin, arginine–glutamate–aspartate–arginine–valine in fibronectin and various heparin binding domains [51].

Surface characteristics regulate the number, size and dynamics of FCs through influencing the conformation of adsorbed proteins and adhering cells [52–55]. Anisotropic topographies (e.g., topographical grooves, chemically patterned stripes or curved surfaces) exert morphological as well as physicochemical features on cells at the same time, indicative of the complex environmental influence on cells. It is evident for designing biomaterial surfaces that cells use the nanotopography of the substrate for orientation and migration [56–60]. Surface structures with dimensions down to 1 µm are the guiding structures for cells and further influence the integrin expression, expression of other genes, and cell differentiation [61–65]. However, little is known about the exact mechanisms by which surface topography modulates cell adhesion and spreading. Additionally, topographical modifications often cause chemical modifications. Hence, it is difficult to independently control the chemistry and topography of surfaces.

2.4. Cell Migration, Proliferation and Differentiation

Intercellular communication soon after the cell adhesion is of high significance due to different mechanisms of initial cellular adhesion to a surface and long-term adhesion and differentiation. This is shown by a lack of statistical correlation between short-term adhesion (strength of cell attachment and early adhesion) and long-term adhesion (strength of cell–matrix interface) forces [66–68]. Interaction between the ECM and associated changes in the orientation of the cytoskeleton are crucial for cell metabolism and morphology due to actin–myosin tension structures [69]. Thus, the cellular migration rate is dependent on the cell type and its differentiation stage and the surface properties. FCs may strengthen the linkage between cell and ECM and also impair the ability to dynamically remodelling of the ECM and influence the migration rate. Cells with a low motility are characterized by a strong formation of FCs while motile cells form less adhesive structures.

3. Examples of Interactions of Cells and Artificial Surfaces

3.1. Blood Cells

Blood is composed of plasma, red blood cells (erythrocytes), white blood cells (leukocytes) and platelets (thrombocytes). The adhesion and adsorption of blood proteins and cells to foreign materials are the early events in two complex biological processes: complement activation and coagulation cascade pathways. The complementary system plays an important role in the body's defence mechanism against infections and 'non-self' elements while haemostasis [70] is the sum of mechanisms that allows blood to maintain an intact circulatory system following trauma by a local solidification (clot) and to be in a liquid or non-coagulated state in normal conditions.

The interaction of the biomaterial with blood starts with the adsorption of plasma proteins on its surface. The first proteins that are adsorbed on the biomaterial's surface are albumin, fibrinogen and fibronectin, which are then replaced by factors XII and high-molecular-weight kininogen [71]. All these proteins determine the cellular reaction of a biomaterial. It is the ratio of adsorbed albumin and fibrinogen that determines the hemocompatibility.

Platelets play the most important role in blood–material interactions. The adhesion is linked to protein adsorption (fibrinogen, γ-glubulin, von Willebrand factor, collagen, thrombospondin) [72]. Adsorption of albumin retards the adhesion and activation of platelets, while adsorption of fibrinogen promotes platelet adhesion and activation. Following platelet adhesion, the platelet release reaction takes place in the adhering platelets, releasing serotonin, thromboglobulin, platelet Factor IV, thromboxane B2, that lead to platelet aggregation on the surface. Hemodynamic conditions are of major importance in the adhesion of platelets to the vessel wall and in determining localization, growth and fragmentation of thrombi. Platelet adhesion to a surface is governed by two independent mechanisms: (1) the transport of platelets to the surface, which depends on the flow conditions and (2) the reaction of platelets with the surface, which depends on the nature of the surface and the adsorbed proteins. Platelet response following the contact of blood with an artificial surface is also influenced by diffusion and shear forces. Several authors report the close interaction of platelets and the electron transfer from the biomaterial surface [73, 74]: By reducing the density of states (the number of states at an energy level that are available to be occupied) in the corresponding energy region of fibrinogen, clotting of platelets can be reduced by inhibiting its polymerization to fibrin.

Red cells, the most numerous cells in blood, do not adhere to endothelium, but they can bind weakly to some non-endothelial materials without spreading. They contribute significantly to blood–surface interactions [71]. Rheological effects of red cells have also been described: collision of red cells with other blood cells and plasma proteins reduces their adsorption on the vessel walls.

Leucocytes, particularly neutrophils and monocytes, have a strong tendency to adhere to surfaces; they react with platelets, complement, coagulation and fibrinolytic systems, and they are major mediators of inflammatory response [75]. As a result of leucocyte adhesion, several reactions are initiated. These include platelet–platelet and platelet–leucocyte interactions, the detachment of adhered thrombi by the action of leucocyte proteases, the detachment of adhered platelets and adsorbed proteins by leucocytes and the release of leucocyte products which may give rise to both local and systemic reactions.

3.2. Bone Cells

The primary tissue of bone (osseous tissue) is formed mostly of calcium phosphate in the chemical arrangement termed calcium hydroxyapatite. Osteoblasts are the bone-forming cells, located on the surface of osteoid seams and making a protein mixture known as osteoid (primarily Type I collagen), which mineralizes to become bone. Osteocytes originate from osteoblasts which have migrated into and become trapped and surrounded by bone matrix which they themselves produce. Their functions include the formation of bone, matrix maintenance and calcium homeostasis. They have also been shown to act as mechano-sensory receptors — regulating the bone's response to stress and mechanical load. Osteoclasts are the cells responsible for bone resorption (remodeling of bone to reduce its volume).

Osteoblasts and osteoclasts are mainly responsible for the binding of biomaterials to bone, a two-step process of osteoconduction and osseointegration, which is pertinent to fracture and/or defect healing [76, 77]. The wetting of implant surfaces with blood during surgery allows the transformation of fibrinogen to fibrin and cell migration to the implant. Within seconds or minutes after tissue contact, a conditioning film is adhering, consisting of the glycoproteins (opsonins) albumin, fibrinogen, immunoglobulin G and complement components [78–80]. Thus, hydrophilic surfaces with high numbers of polar components seem to improve osteoblast cell attachment and matrix synthesis and also the osteogenic potency compared to hydrophobic surfaces [81–84]. The first cells attaching to the implanted material are monocytes. Their adhesion induces the release of cytokines that will favour later the adhesion and differentiation of osteoblast or mesenchymal stem cells on the surface [85, 86]. Expression of bone morphogenetic proteins triggers the differentiation of the mesenchymal stem cells. Osteoconduction comprises the dispersion of osteoblasts and the bone growth at the implant surface. Osteogenesis proceeds towards the implant by cell migration, contracting the initially formed fibrin tissue. Fully differentiated cells start to produce osteoid and mineralize, which should advantageously occur after the migrating cells have reached the implant surface to guarantee full implant surrounding by bone. Osseointegration is reached, after 60% of the implant surface is in direct contact with bone [78]. Finally, initially formed woven bone is remodelled to lamellar bone to withstand biomechanical strains.

4. Principles and Techniques of Vacuum and Plasma Coating

In vacuum, gas pressure is less than the ambient atmospheric pressure. Plasma is a gaseous environment where there are enough ions and electrons to provide appreciable electrical conductivity. Vacuum deposition is the deposition of a film or coating in a vacuum (or low-pressure plasma) environment. Vacuum in deposition processing increases the 'mean free path' for collisions of atoms and high-energy ions and helps reduce gaseous contamination to an acceptable level. When establishing plasma in a vacuum, the gas pressure plays an important role in the enthalpy, the density of charged and uncharged particles and the energy distribution of particles in the plasma. Plasma in high vacuum ($<10^{-2}$ to 10^{-3} mbar) provides a source of ions and electrons that may be accelerated to high energies in an electric field. These high-energy ions can be used to sputter a surface as a source of deposition material and/or bombard a growing film to modify the film properties. Plasma may also be used to 'activate' reactive gases and vapours in reactive deposition processes and to fragment the chemical vapour precursors.

In Physical Vapour Deposition (PVD) processes, material (element, alloy or compound) is atomistically vaporized from a solid or liquid source and transported as a vapour through a vacuum or low-pressure gaseous or plasma environment. When it contacts a surface, it condenses. PVD processes can be used to deposit films of compound materials (reactive deposition) by the reaction of depositing material with the gas in the deposition environment. Typically, PVD processes are used to deposit films with thicknesses in the range of a few nanometers to thousands of nanometers; however, they can also be used to form multilayer coatings, thick deposits and free-standing structures.

In vacuum evaporation (and sublimation), material from a thermal vaporization source reaches the substrate without collision with gas molecules in the space between the source and substrate. The trajectory of the vaporized material is 'line-of-sight'. Typically, vacuum evaporation takes place in a gas pressure range of 10^{-5} to 10^{-9} mbar, depending on the level of contamination being tolerated in the deposited film. For appreciable deposition rates, the vaporization temperature must be chosen to reach a vapour pressure of 10^{-3} mbar or higher. Typical vaporization sources are resistively heated stranded wires, boats or crucibles (for vaporization temperatures below 1500°C) or high-energy electron beams that are focused and rastered over the surface of the source material.

In sputter deposition, particles are vaporized from a surface (sputter target) by physical sputtering. In this non-thermal vaporization process, surface atoms are physically ejected by momentum transfer from an energetic bombarding particle that is usually a gaseous ion accelerated from plasma or an ion gun. Sputter deposition can be performed in a vacuum or low-pressure gas ($<10^{-2}$ mbar) where the sputtered particles do not suffer gas-phase collisions in the space between the target and the substrate. It can also be done in a higher gas pressure ($\sim 10^{-1}$ mbar), where energetic particles that are sputtered or reflected from the sputtering target are 'thermalized' by gas-phase collisions before they reach the substrate. The most common sputtering sources are the planar magnetrons, where the plasma is magnetically confined close to the target surface and ions are accelerated from the plasma to the target surface. The high sputtering rates attainable in magnetron sputtering allow reactive deposition of compound films as long as the sputtering target is not allowed to react with the reactive gas to form a low sputtering rate compound.

In arc vapour deposition, the vapour is formed by interactions of a low-voltage, high-current electric arc and a cathode or anode target in low-pressure gas. The usual configuration is the cathodic arc where the evaporation is from an arc that is moving over a solid cathodic surface. In the anodic arc configuration, the arc is used to melt the source material that is contained in a crucible. The vaporized material is ionized as it passes through the arc plasma to form charged ions of the film material. In the arc vaporization process, molten globules (droplets) can be formed and deposited on the substrate. To avoid this problem, a plasma duct may be used to bend the charged particles out of the line-of-sight of the source, and the droplets will be deposited on the walls of the duct.

Pulsed laser deposition (PLD) is a technique based on the laser evaporation of a solid material by high-energy pulsed laser beams. Gases can be used in PLD to increase scattering of the ionized vapour subsequent to laser evaporation (ablation) or to deposit reactive compounds (e.g., oxides, nitrides) after ablating metal target materials. As described for arc deposition, droplets and particulates are a major problem in PLD and requires efforts to decrease their size and number. In industrially-scaled PLD coating, several laser beams are employed and the resulting vapour cones are overlapped to achieve high deposition rates on large areas [87].

Ion plating uses energetic particle bombardment on the depositing film to modify and control the composition and properties and to improve surface coverage and adhesion. The depositing material may be vaporized by evaporation, sputtering, arc erosion or other vaporization source. The energetic particles used for bombardment are usually ions of an inert or reactive gas or ions of the depositing material. Ion plating can be done in plasma environment, where ions for bombardment are extracted from the plasma (plasma-immersion ion implantation and deposition, PIIID), or formed in a separate ion gun (ion beam assisted deposition).

Chemical vapour deposition (CVD) deposits atoms or molecules by decomposition of a vapour precursor species that contains the material to be deposited. The reduction for decomposition is normally accomplished using hydrogen at an elevated temperature. Decomposition is accomplished by thermal activation. Using plasma (e.g., radiofrequency or microwave discharges) allows the reduction or decomposition to be done at a lower temperature than using temperature alone (plasma-enhanced CVD, PE-CVD). The deposited material may react with gaseous reactive species in the system to produce compounds (oxides, nitrides). CVD processing is generally accompanied by volatile reaction by-products and those, along with unused precursor vapours and other processing gases, must be removed from the deposition system.

5. Thin Film Growth Effects on Protein and Cell Interaction

The interaction of proteins and cells to artificial surfaces is generally dependent on the surface chemical properties (in terms of charge, hydrophobicity, available chemical functional groups and chemical surface homogeneity) and the surface topography. Both surface properties can actively be influenced by PVD and CVD techniques for depositing inorganic and carbon-based thin films, as shown in the following.

5.1. Surface Chemistry

After evaporation or sputtering of target atoms, the vaporized plume of atoms and possibly ionized species expands towards the substrate surface. The use of process and reactive gases, as mentioned above, leads to energy loss by scattering and chemical reactions before the deposition and condensation step. The processes during deposition are highly dependent on the total ionic and kinetic energy distributions of the arriving vaporized or ionized atoms [88] (Fig. 2) as described below:

- Thermally-deposited species ($<10\,eV$) have very low implication on the surface structure: Physisorption, chemisorption and simply scattering off the surface with some loss of their kinetic energy (direct inelastic scattering) are commonly found for thermal atoms and ions. Dissociation and chemisorption to the surface (activated dissociative chemisorption) are the mechanisms occurring for molecules.

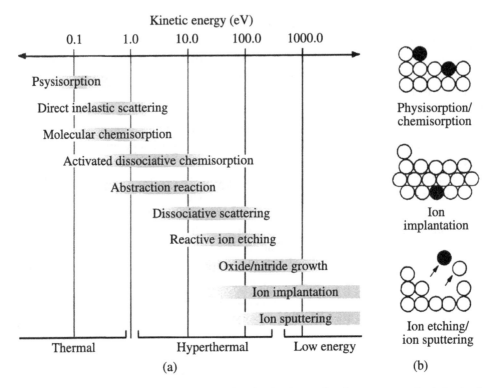

Figure 2. (a) Fundamental processes occurring in the interaction between vaporized (ionized) atoms/molecules and the solid surfaces. The shaded region in the background of the described processes designates the ion kinetic energy range over which the associated process is evident. (b) The schematics present simply the most important processes of this interaction (adapted from [88, 90, 91]).

- Hyperthermal species (>10 eV) can abstract atoms from surfaces (abstraction reactions) or dissociate and scatter if they are polyatomic (dissociative scattering). The ions also may react with the substrate atoms to create altogether new chemical species. These new chemical species may be volatile or readily sputtered, leading to surface etching (reactive ion etching). The ions may directly react with the surface atoms to form new species (e.g., oxide/nitride growth). Energy input in the surface can induce secondary collisions of substrate atoms which cause collision cascades and result in energy transfer far from the initial impact point. The effects of the impacting ion propagate further into the bulk region of the substrate and away from the impact point with increasing energy. Higher energy input (>100 eV) causes sputtering of surface atoms, ions or neutrals (ion sputtering), annealing of the surface, mixing of surface atoms, creation of unique surface topologies and defect formation.

- As the ion energy increases into the highest energy range known of processes (1000–10000 eV), the ion–surface interaction moves from predominantly nuclear–nuclear collisions between projectile and target to one that is primarily electronic in nature [89]. These low-energy ions can easily implant into the surface to form an alloy or doped material (ion implantation).

This overview allows to conclude that the energy input into the surface results in the introduction of lattice defects (interstitials, trapped process gas atoms, wrongly substituted atoms in the lattice, etc.) in the substrate surface and the growing film at low to medium deposition temperature (<70% of melting temperature). The missing atomic rearrangement in this temperature range finally results in chemically non-stoichiometric thin films, entailing the very advantageous possibility of controlling chemical film composition over a wide range. Changing deposition parameters only just at the end of deposition opens the door to different film bulk and surface properties.

While the predominant part of scientific work done on cell and protein interaction to PVD and CVD biocompatible films excludes the option of varying the deposition parameters from the investigation and, thus, impedes the comparison of the works due to unknown surface conditions, some (newer) works are dedicated to this topic. An overview of the literature for inorganic and carbon-based CVD and PVD films will be given in the following. For PVD and CVD coatings applied on FDA approved implants, the authors wish to refer to their review, recently published elsewhere [92].

First of all, the choice of applicable chemical elements for continuous contact to living cells is very limited. Hench and Ethridge [93] presented for metals in the oral cavity an overview of theoretical and practical biocompatibility (Fig. 3).

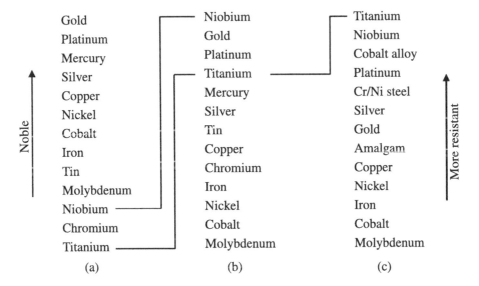

Figure 3. Resistance of materials in the oral cavity (according to [93]): (a) theoretical biocompatibility scale, (b) general passivation effect of the metals, (c) *in vivo* tested materials in the oral cavity.

(Self-)Passivation of the metal surfaces seem to be decisive, making titanium and niobium to be the most biocompatible *in vivo* tested materials. Expanding this list to other cellular surroundings in the human body and to the state-of-art and -science would lead in case of PVD and CVD manufacturable materials to a similar list including carbon and some nitride and oxide compounds of especially titanium and calcium phosphate.

5.1.1. Titanium-Based Coating Materials

Generally, titanium as metal is seen as a highly thrombogenic biomaterial [89] by the activation of the intrinsic system. However, it supports the growth of bone-forming cells and osteointegration, most probably due to protein adsorption and platelet activation with subsequent release of growth factors [94–96].

Nevertheless, the hemocompatibility improvements of titanium and the naturally grown (in air atmosphere) and artificially reared or deposited oxide layers formed on its surface were intensively studied by alloying as well as compound formation with oxygen, nitrogen, carbon and using doping elements: First of all, the thickness and properties of the titanium oxide films seem to have strong influences [97, 98], even though the influence of the crystallographic phase has not been understood. While the naturally grown oxide film on titanium exhibits low blood compatibility [94, 95], thicker oxide films generally prove to be very hemocompatible [99, 100], most probably due to the good separation of the metal from the biological surrounding. For example, oxygen ion implantation and titanium oxide layer formation was shown by Maitz and Shevchenko [101] to be advantageous for NiTi alloy (Nitinol) to reduce the nickel concentration at surfaces. Compared to virgin or helium ion implanted Nitinol, oxygen implanted NiTi results in lower adsorption of fibrinogen.

Tsyganov *et al.* [102] found rough trends for the hemocompatibility that crystal structures (rutile, anatase, brookite) and crystallite size of PIIID titanium oxide influence the activation of the plasmatic clotting system only to a minor degree. Implantation of P into titanium oxide was found to reduce the clot forming property, possibly due to hindering of the electron exchange between proteins and the modified surface. Huang *et al.* [103] found for PIIID Ti–O films better *in vivo* thromboresistant properties (low clotting and platelet adhesion) than for low-temperature isotropic carbon in long-term implantation. Chen *et al.* [104] studies revealed similar results for magnetron sputtered and subsequently thermal oxidized Ta-doped Ti–O coatings. Nitrogen addition to TiO_2 was found to reduce thrombocyte adhesion and fibrinogen adsorption [105] (optimal around $TiN_{0.4}O_{1.6}$). The behaviour was correlated with decreasing contact angle and inversely correlated with the increasing surface energy (especially the polar component).

For osseointegration, inflammatory reactions and cytotoxic effects in contact with osteoblasts, fibroblasts and epithelial cells were found to be absent for titanium oxide multi-component coatings, doped and alloyed with Ca, Si, Zr, C and N [106]. All films were able to support high rate of osteoblast proliferation and fast cell differentiation. Cell growth on Ca- and P-doped TiO_x films was higher than on control substrates. Fetal rat calvarial cells were able to secret bone matrix proteins and to deposit minerals in the collagenous matrix, demonstrating high level of bioactivity. Implantation studies using rat calvarian and hip defect models indicated early signs of bone formation on these coatings on titanium substrates. After 1 month, they found a close contact between the implant surface and cortical bone without any bone losses at the interface. After sodium PIIID in titanium, Maitz *et al.* [107] revealed for Na-enriched Ti surfaces bioactivation for osteoblasts, which was experimentally demonstrated by the precipitation of calcium phosphate and the rat bone marrow cell proliferation. In comparison, no bioactivity was found for untreated titanium.

Manso *et al.* [108] found that sputtered TiN coatings on Ti–Al–V alloy surfaces induce cell adhesion and proliferation, similar to titanium oxide. Chien *et al.* [109] showed the biocompatibility of TiN and Ti–Al–N for human gingival fibroblast cells. Proliferation and viability on these films were significantly higher than on control substrates. The *in vitro* expression of cytokines and the endothelial cell adhesion seeded on PE-CVD titanium carboxynitride coatings on polymers was investigated by Lehle *et al.* [110]. No toxic effects on human saphenous vein endothelial cells were found as well as an insignificant decrease in cytokine production and no essential variations in basal expression of adhesion molecules.

Potential heart valve duplex coatings, consisting of layers of Ti–O and Ti–N, were deposited onto biomedical Ti–alloy by Leng *et al.* [111]. The Ti–O layer was designed to improve the blood compatibility, while Ti–N was deposited to improve the mechanical properties. They found that the duplex coatings displayed better blood compatibility than low temperature isotropic carbon, higher microhardness and improved wear resistance than Ti–6Al–4V alloys. Major and coworkers [112–116] investigated titanium and titanium nitride (TiN) coatings grown by pulsed laser deposition on thermoplastic polyurethane foils for hemocompatibility applications. For films with ~50 nm thickness, they found higher probability of fibroblast cell adhesion to Ti than to TiN. The lack of the immunostimulator interleukin-1β activation proved the high biocompatibility of the highly elastic coatings.

High-energy chlorine implantation into TiN films leads to significant wear reduction of the polyethylene counterpart in artificial hip joint implants, if the lubricant was Hank's balanced salt solution [117]. After bovine serum albumin was added to the lubricant, a strong decrease of both friction and polymeric wear with a change from abrasive to adhesive wear mechanisms was observed for both implanted and non-implanted TiN coatings. The former case is explained by the additional formation of a titanium oxide layer on the TiN surface, while the latter is derived solely from albumin adsorption.

In summary, titanium-based coating materials exhibit high hemocompatibility and osteointegration, which increases with the thickness of the deposited layer due to the separation of, e.g., haemolytic substrate material from the biological surrounding. TiN adds higher mechanical and tribological stability to the implant's behaviour with only minor changes in biocompatibility. Alloying or doping these coating materials with Ca, P, Ta and other elements was shown to be advantageous, but requires more detailed research for understanding the mechanisms.

5.1.2. Carbon-Based Coating Materials

Carbon is a chameleon-like material due to its different molecular allotropes, based on its chemical binding structures in either two-dimensional hexagonal sp^2 or three-dimensional tetrahedral sp^3 hybridization. The three relatively well-known allotropes of carbon are amorphous carbon, graphite (full sp^2 hybridization) and diamond (full sp^3 hybridization). Fullerenes have a sp^2 graphite-like structure, but instead of purely hexagonal packing, they also contain pentagons (or even heptagons) of carbon atoms, which bend the sheet into spheres, ellipses or cylinders (carbon nanotubes).

Depositing carbon-based materials by PVD and CVD methods results at high temperatures ($\sim 1000°C$ by CVD) in the formation of diamond films or at lower temperatures in amorphous, 'diamond-like carbon' (DLC) films with a mixture of sp^2- and sp^3-hybridization and up to 45% hydrogen content. If hydrogen is present during deposition (e.g., in the precursor hydrocarbon gas or as water contamination in the deposition chamber), the hydrogen atoms are incorporated in the films and replace carbon–carbon bonds.

In the studies of Potocki *et al.* [118], diamond surfaces generally preserved cell integrity and promoted cell adhesion, viability and proliferation. In contrast to these bulk diamond materials, CVD diamond coatings are generally characterized by higher surface roughness due to the very symmetric structure of the small diamond crystals, of which they are formed. Their application as biocompatible materials was widely investigated with highly mixed and variable results of protein and cell adsorption and adhesion. For example, Tang *et al.* were among the first to examine cellular response to CVD diamond, and their study revealed that neutrophil adhesion to CVD diamond was equivalent to stainless steel [119]. Popov and coworkers found that osteoblast cells and human lung fibroblasts survived for extended time periods on CVD diamond [120, 121]. In contrast, there are several studies that concluded that CVD diamond does not support adhesion. Ariano *et al.* described very low survival rates (less than 10%) of neurons cultured on homoepitaxial CVD diamond, without pre-coating with a layer of laminin [122]. Furthermore, platelet aggregation was found to be minimal on CVD diamond [123, 124]. Finally, Jakubowski and Mitura *et al.* concluded that CVD diamond resists bacterial colonization better than titanium or steel [125, 126]. These highly variable results on cell interaction with CVD diamond may be due to phonotypic differences in the cell types studied, and/or differences in surface chemistry of CVD diamond (e.g., surface roughness, termination of elements on the surface, see below).

Besides diamond, DLC has emerged as a potential material in recent years due to its high hardness, low friction coefficient, high wear and corrosion resistance, chemical inertness, high electrical resistivity, infrared-transparency, high refractive index and excellent smoothness. Additionally, the biocompatibility of DLC was investigated in a large number of studies (see the reviews by Roy and Lee [127] and Hauert [128]). However, the dependence of its biocompatibility on its surface chemistry, atomic bond structure, H content, and dopant is not fully understood yet. Results of studies with proteins and cells will be shown exemplarily in the following.

Some studies revealed that the adhesion and spreading of cells on DLC surfaces was related to the bonding structure present on the surface and the ratio of sp^2/sp^3 [129]. It was observed that for magnetron sputtered a-C films (amorphous, non-hydrogenated amorphous carbon), the hemocompatibility improved with increasing sp^3 fraction [130]. However, on investigating the hemocompatibility of pulsed vacuum arc deposited ta-C (tetrahedral amorphous carbon) films, it was found that the hemocompatibility tended to improve with increasing sp^2/sp^3 ratio [131]. Similar behaviour was found for PIIID a-C:H (hydrogenated amorphous carbon) films [132]. Some investigations were performed to observe the dependency of hemocompatibility on the Raman D-band to G-band intensity ratio (I_D/I_G) of the DLC films. It was noticed that PIIID a-C:H coated TiNi alloy showed better blood compatibility at the minimum value of I_D/I_G [133]. Logothetidis *et al.* [130] compared the hemocompatibility of magnetron sputtered a-C, a-C:H and ta-C films. For the a-C:H films, the albumin/fibrinogen ratio was higher for films deposited at higher hydrogen atomic percent in plasma and positive substrate bias. Thus, these results imply that films with more polymeric component would exhibit better hemocompatibility. However, hydrogen-free films were found to result in increased osteoblast cell viability compared to hydrogenated carbon (a-C:H) films [134]. At the moment, there is no general theory describing the influence of the chemical bonding structure on the biocompatibility. Maybe, other properties related to the binding structure are of much higher importance.

DLC coatings generally grow very dense on substrate surface due to their amorphous structure and the missing of nano-sized pinholes as found in nano- and microcrystalline coatings between the cone-shaped crystals in columnar growth mode. Thus, separation from the biological surrounding can easily be achieved. For example, arc-discharge ion plated, 1 μm thick a-C:H on Ni–Ti orthodontic archwires reduced the amount of Ni release in harsh immersion tests by 80% and showed no influences on fibroblast cell growth [135]. On growing mouse peritoneal macrophages and mouse fibroblast on tissue-culture plates treated with a-C:H, no toxicological effects on the cells were observed [136]. a-C:H films grown by RF glow discharge promote cell attachment and normal cell growth rates [137]. *In vitro* studies of biocompatibility of RF PE-CVD a-C:H coatings showed no cytotoxic effects for mouse macrophages, human fibroblasts and osteoblast-like cells [138]. Linder *et al.* [139] found for RF PE-CVD grown a-C:H films high biocompatibility to human monocytes and macrophages *in vitro*.

Testing DLC coatings *in vivo* resulted in generally advantageous results. Tests with 1-year implantation of DLC-coated parts in rabbit's skeletal muscles did not reveal any delamination of the film [140]. Orthopaedic pins and screws coated with RF PE-CVD a-C:H showed incidences of connective tissue-capsule formation from fibrocytes and collagen fibers, both in subcutaneous tissue, bones and muscles one year after implantation into guinea pigs [141]. However, no phagocytic reaction and corrosion products in the walls of the capsule were noticed. After implantation of RF PE-CVD a-C:H coated CoCr cylinders in transcortical sites of sheep and intramuscular location of rats for three months, it was observed that the DLC-coated implants did not show any significant toxicological effect and were well tolerated in both bodies [142]. A comparative study of Zr, Ti, Al and DLC-coated implants in rats showed for Zr and DLC-coated Zr better tissue response than for the other metals [143]. DLC-coated steel was implanted inside fractured bone in the human body for 7 months and no corrosion, metal release or inflammation were observed [144].

Gutensohn *et al.* [145] analyzed the intensity of the platelet activation antigens CD 62p and CD63. They showed that the DLC coating of a stainless steel coronary stent resulted in a decrease in the concentration of both antigens, indicating low platelet activation on DLC and low tendency of thrombus formation. PIIID ta-C (tetrahedral amorphous carbon) coatings with $>80\%$ sp^3 hybridization showed in investigations of Tong *et al.* [146] kinetic clotting times longer than for Ti–6Al–4V and TiC. Furthermore, the haemolysis rate of ta-C is less than that of low temperature isotropic carbon.

Because of their amorphous structure, DLC films can be easily doped and alloyed with different elements. Several attempts were made to improve the hemocompatibility of DLC by incorporating suitable elements into it. Kwok *et al.* [147] investigated P-doped PIIID a-C:H films and obtained water contact angles of $\sim16.9°$ as well as minimum interaction with plasma proteins. This gives rise to preferential adsorption of albumin, which is a prerequisite for better hemocompatibility. Silicon doping generally lowers the electrical resistivity, work function, surface energy and I_D/I_G ratio in a-C:H films. As a consequence, improved hemocompatibility (reduced platelet attachment and enhanced human microvascular endothelial cell attachment) was reported by various authors [148–151]. On investigating the antithrombogenicity of fluorinated a-C:H films synthesized by RF PE-CVD, it was observed that the incorporation of F suppressed the platelet activation and adhesion to a significant extent [152]. Yang *et al.* [153] showed for nitrogen doped a-C:H:N films higher hydrophilicity, roughness, endothelial cell growth and anti-thrombotic properties than for a-C:H. a-C:H films with boron and nitrogen doping were investigated by Maitz *et al.* [154]. The CVD-deposited and RF-sputtered nitrogen-poor a-C:H:N films showed highest platelet adhesion, while nitrogen-rich and boron doped a-C:H had the highest ratios of adsorbed fibrinogen to albumin. Weak platelet adhesion was found for hydrogen-poor a-C:H, but the ratio of the two proteins was similar to the nitrogen-poor a-C:H:N film. A similar study on PIIID a-C:H:N films revealed 10% higher cell adhesion compared to the undoped, with a maximum at a N:C ratio of 0.2 [155]. This enhancement in cell adhesion was explained in terms of the presence of C–N and N–H bonds which seemed to promote the adsorption of proteins on the film surface. In summary, it seems to be highly advantageous for improving the hemocompatibility to dope or alloy DLC films with Si, F, P or N.

In contact to osteoblast cells, silicon doped PACVD a-C:H films were investigated by Bendavid *et al.* [156]. They found homogeneous and optimal tissue integration (fine cytoplasmic extensions) for pure and up to 22 at% Si doped coatings. Studies on RF PE-CVD deposited hydrogenated carbonitride coatings (a-C:H:N) with osteoblast and fibroblast cells revealed generally good biocompatibility of the coatings [157, 158]. Schroeder *et al.* [159] performed bone marrow cell culture tests on Ti-incorporated hydrogenated carbon films (prepared by a combination of RF PE-CVD and magnetron sputtering) and found increased bone cell proliferation and minimum activity of osteoclasts.

Addition of toxic elements like Cu and V to DLC films inhibited the growth and spreading of cells on their surfaces. Endrino *et al.* [134] found that a-C:H films with silver doping had a small initial negative effect on the osteoblast cell culture, but after 7 days the effect disappeared.

Limited research has been reported on metal–oxide containing DLC thin films. De Scheerder *et al.* [160] reported that the addition of SiO_x to DLC reduces inflammatory reactions on implanted stents. Dorner-Reisel *et al.* [161] showed that the addition of CaO to DLC decreased the water contact angle. Cell tests with mouse fibroblasts showed increased cell attachment and improved cell viability for the DLC–CaO films in comparison to DLC. Thorwarth *et al.* [162] studied the biocompatibility properties of DLC–TiO_x films on Ti–Al6–V4 alloy prepared by PIIID. *In vitro* biocompatibility tests showed promising results concerning proliferation and differentiation of human osteoblasts for DLC–TiO_x. Mixtures of DLC and TiO_x were synthesized by Amin *et al.* [163] using pulsed DC metal–organic plasma activated CVD (MOCVD. The biomimetic growth of amorphous apatite in simulated body fluid was found to be dependent on the TiO_x content of the films. UV light exposure prior to cell adhesion increased the growth rate of apatite formation significantly as a result of increased hydrophilicity of the surface.

As for the above described results relating the biocompatibility of DLC to intrinsic chemical properties, there is also lack of understanding of the influence of the wettability and surface energy on the hemocompatibility and osteointegration. Kwok *et al.* [147] indicated that higher surface energy of P-doped a-C:H films (hydrophilic behaviour) was associated with weak adsorption of proteins and preferential adsorption of albumin. Huang *et al.* [164] also pointed out that hydrophilic surfaces improved blood compatibility. However, while investigating the thrombogenicity of a coated catheter with hydrophilic properties, Leach *et al.* [165] observed that the hydrophilic coatings were equally thrombogenic as the uncoated ones. On the other hand, Jones *et al.* [166] reported that hydrophilic surfaces enhanced spreading of platelets and fibrinogen adsorption. They attributed better hemocompatibility of DLC surfaces to their low surface energy, hydrophobicity and smooth surface. Okpaluo *et al.* [167] noted that Si doping of a-C:H films lowered the surface energy resulting in improved blood compatibility. Saito *et al.* [168] also found that the antithrombogenicity increased with the increase in hydrophobicity of their fluorinated DLC films.

A further property — the interfacial tension between blood and biomaterials — has also been considered to understand the hemocompatibility. Cellular elements of blood maintain an interfacial tension of the order of 2–3 mN/m with blood plasma. Thus, it was suggested that the hemocompatibility can be improved if the interfacial tension between blood and biomaterials is in the same order of magnitude as the cellular elements of blood [169]. For carbon materials, e.g., the blood–material interfacial tension is between 12–20 mN/m for low temperature isotropic carbon, $\sim$12 mN/m for a-C:H, $\sim$8 mN/m for ta-C and $\sim$4 mN/m for P-doped a-C:H [147, 170]. Due to the lack of sufficient results on the interfacial tension, no clear tendency has been found yet.

In summary, DLC films exhibit high biocompatibility in different biological surroundings and can separate the bulk material of an implant from the biofluids around its surface. However, the large number of results obtained under very varying testing conditions and unknown surface properties (roughness, nanostructure, surface chemistry (termination of elements, etc.) do not allow yet to establish theories as to how the chemical binding, doping or mixing with nanoparticles may influence the biological behaviour.

5.1.3. Other Ceramic Based Compound Films

Although titanium- and carbon-based materials are the most important materials tested and partly used on implant surfaces, there are some other PVD and CVD coating materials of biocompatible importance.

Leng *et al.* [171] investigated the biomedical properties of tantalum nitride thin films. They showed that the blood compatibility was superior to other common biocompatible materials such as TiN, Ta and low temperature isotropic carbon. Amorphous silicon carbide films (α-SiC), deposited by PE-CVD were found to be highly advantageous for cell culturing [172], especially after treatment in NH$_4$F. Initially exposing the layer to KOH, the biocompatibility degraded and was similar to single crystal SiC. Poperenko *et al.* [173] sputtered non-stoichiometric chromium oxide films on stainless steel with varying oxygen contents and investigated the hemocompatibility. Thicker films were found to have lower platelet adhesion compared to thinner and the uncoated steel. However, the substrates started to corrode in human plasma.

Calcium phosphate phases such as hydroxyapatite (Ca$_{10}$(PO$_4$)$_6$(OH)$_2$, HA), carbonated hydroxyapatite and octacalcium phosphate Ca$_8$H$_2$(PO$_4$)$_6 \cdot$ 5H$_2$O, offer an interesting selection of biocompatible materials [174, 175]. They are well known for good osteoconductivity (the early stage of osteogenesis) as well as for direct binding to bone tissue *in vivo*. The alkaline phosphatase expression and parathyroid hormone response are generally higher in cultures grown in hydroxyapatite

(HA) than on titanium [176], and the *in vitro* formation of extracellular matrices was greater on Ca–P coatings than on titanium [177]. Implanted HA acts as a nucleation site and exhibits crystallographic properties in an epitaxial process of the newly developed structure. The calcium ions dissolve from the Ca–P surface, resulting in the deposition of a mineralized layer. This stimulates the bone cells to continue extracellular matrix (bonding zone) synthesis and calcification [178]. Several different deposition techniques have been utilized for the deposition of calcium phosphate and hydroxyapatite coatings including PVD and CVD techniques [179–182]. Thermal post-treatment between 400 and 1200°C of the deposited coatings has often been employed in order to increase the biomedically required crystallinity of the hydroxyapatite phase in Ca–P–O films [183–186], regardless of the deposition method. Yoshinari *et al.* [187] deposited 0.5 μm thin Ca–P films by magnetron sputtering of HA in combination with Ca-ion implantation by an electron beam gun. *In vivo* tests with dogs showed after 4 weeks implantation a higher percentage of bone contact in Ca–P films than on control substrates. Using ion beam assisted deposition (IBAD), it is possible to deposit hydroxyapatite (HA) coatings with a good adhesion and a low porosity and, thus, slow down the degradation *via* pores and cracks [188].

Veith *et al.* [189] investigated CVD grown HAlO compounds which are transparent, non-crystalline, glass-like material and in which oxygen as well as hydrogen are directly bound to aluminium [190]. The HAlO growth at 310°C led to micro-hill like topography and showed high human dermal fibroblast cell adhesion and proliferation comparable to single-crystalline silicon surfaces.

5.1.4. Controlling Surface Chemistry by Plasma Modification for Bond Termination

Several techniques for controlled attachment of proteins have been introduced, such as controlled drug release [191, 192] or immobilization of the bioactive substance to ensure short distance effects. Binding of cytokines is hard to achieve, because of employing complex chemistry in various sequential steps of protection and de-protection for covalent immobilization of proteins in their active conformation [193, 194]. This could also exert adverse effects *in vivo* [193]. No such effects arise when methods for stable attachment of bioactive substances to biocompatible substrates without prior chemical treatment are available. To date, physisorption is considered a phenomenon giving rise to unspecific adsorption [195]. So far such systematic studies on PVD and CVD coatings were performed for diamond and DLC.

Although diamond is inert, its surface can be covalently functionalized by, e.g., terminating bonds, resulting in highly controllable cellular adhesion by physisorption: Mesenchymal stem cells adhered in the work of Clem *et al.* [196] proliferated and showed differentiation on H-terminated nanocrystalline microwave CVD diamond, but not on F- or O-terminated. The termination of the different species was achieved by hydrogen, oxygen–helium and hydrogen–fluorine final plasma treatments with gradually reduced microwave power after CVD coating. The physicochemical characterization of the various surface treatments suggested that the hydrophobic/hydrophilic nature of the surface might be less critical for influencing mesenchymal stem cell behaviour than the immediate surface chemistry — H-terminated and F-terminated surfaces had similar water contact angles. Furthermore, they found that O treatment increases the hydroxyl to carbonate and carbonyl to aliphatic binding ratios. With F-treatment, the hydroxyl to carbonate ratio increases but the carbonyl to aliphatic ratio decreases. This behaviour is likely to indicate that fluorine is substituting for both the carbonate and carbonyl oxygen atoms on the surface, leaving hydroxyl species unaffected. The mechanism underlying the high degree of mesenchymal stem cell to H-terminated CVD diamond surfaces seems to be due to the intrinsic surface conductive properties of H-terminated surfaces [197–199]. O-terminated diamond is highly resistive while H-termination induces p-type surface conductivity even on undoped diamond. Anchoring is of high importance for mesenchymal stem cells — the inability to attach can limit cell survival.

Steinmüller-Nethl *et al.* [200] showed influences on the interaction of cytokines and growth factors (bone morphogenetic protein-2 (BMP-2)) to O- and H-terminated nanocrystalline CVD diamond. H-bonding to surface groups contributes significantly to the enthalpy of binding. In the case of O-terminated nanocrystalline CVD diamond, the activation energy necessary for re-orientation is considerably larger than in the case of H-terminated surfaces, where binding appeared to occur primarily through van der Waals interactions. Thus, strong BMP-2 immobilization was reached on O-terminated nanocrystalline CVD diamond only by physisorption without chemical cross-linking. O-terminated diamond provides favorable conditions for osteoblast, fibroblast and cervical carcinoma cell adhesion, cell spreading and cell viability [201, 202]. Ariano *et al.* [203] reported that rat hippocampal neurons as well as chick ciliary ganglion neurons plated on O-terminated diamond surface preserved their synaptic activity and Ca^{2+} current densities. Additionally, Michalikova *et al.* [202] found good cell spreading on larger terminated areas on microwave CVD nanocrystalline diamond and formation of elongated chains in narrower (line-like terminated) areas. Higher initial cell concentration enabled colonization of unfavourable H-terminated regions, which are surface conductive. However, no cellular patterns were formed in absence of fetal bovine serum. Hence, the cell selectivity is determined by the protein presence.

The plasma treatment of Si containing DLC films has a significant effect on their hemocompatibility as reported by Roy *et al.* [204]: CF_4, O_2, and H_2 plasma-treated RF PE-CVD grown a-C:H:Si films improved the albumin adsorption with time, while H_2 plasma-treated samples decreased the rate of fibrinogen adsorption. A longer clotting time was observed in case of O_2 plasma-treated Si–DLC films with extraordinary high albumin to fibrinogen ratio, which also minimized platelet adhesion and activation compared to other samples.

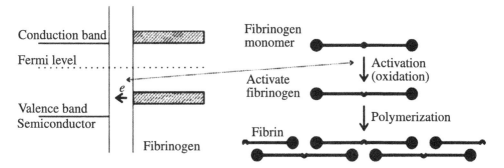

Figure 4. Schematic representation of improved clotting behaviour (minimization of clotting): A semi-conductor surface will inhibit electron transfer and, thus, activation and polymerization of fibrinogen to fibrin (according to [207]).

The above described differences in the protein and cell adhesion behaviours for O- and H-terminated surfaces could be due to the effect of surface charge transfer. The adsorption of proteins on biomaterials prior to cell adsorption is associated with the charge transfer to the biomaterials (see Fig. 4), and the adsorbed proteins undergo various conformational changes during electron exchange process. It has been found that fibrinogen has an electronic band gap of 1.8 eV. When fibrinogen comes in contact with a material whose Fermi energy level lies below the Fermi energy level of fibrinogen, charge transfer takes place from the valence band of fibrinogen to the material resulting in the adsorption of fibrinogen to form fibrin and finally thrombus. On the other hand, if fibrinogen comes in contact with the materials with higher Fermi energy level than that of fibrinogen, no charge transfer would occur from fibrinogen to the materials, which would suppress the oxidation of fibrinogen and thrombus formation [205]. It has, thus, been suggested that materials with lower electrical resistivity or smaller work function are more hemocompatible [139, 206].

5.2. Surface Topography

Surface nanotopography appears to affect cell interactions at surfaces and alter cell behaviour when compared to micro-topography [208–210]. Different physical relationships exist between cells and nano- *versus* cells and micro-scale surface features. The roughness and topography of PVD and CVD coatings are both dependent on the substrate surface itself as well as on the film growth and topography evolution. While the substrate surface is in most cases micro-structured, the film growth can lead either to nano- or microstructures, determined by the processes taking place during deposition. Two steps are important for the topography formation on the surface in dependence of the required coating thickness: (1) film nucleation for several nanometer thick films and (2) thick film growth for films > 10 nm thickness.

(1) Film nucleation. In general, the substrate will have a different chemical nature than the coating material. Thus, depositing atoms from the vapour phase cannot

immediately condense at the substrate, and are just adsorbed on the substrate, whereas condensation takes place by the formation of small clusters through the combination of several adsorbed atoms [211]. To become supercritical for nucleation, these clusters grow by direct deposition on the clusters and diffusion of adsorbed atoms from the surrounding. The lateral growth of cluster islands on the surface is limited when the clusters touch and start to coalesce to form new islands occupying an area smaller than the sum of the original two. Finally, the whole surface is covered with deposited atoms in large islands, leaving channels of exposed substrate. Secondary nucleation can fill these channels. The thickness of this initially deposited cluster film is one to a few atom layers, depending on the binding between deposited atoms and substrate surface atoms [212, 213]. If the diffusion is constrained (low temperature, high deposition rates), atoms can be buried at their impinging positions by subsequently deposited atoms, resulting in low-order (amorphous-like) layers [214]. All these nucleation and initial film growth mechanisms result in roughness of one to several atoms in the nanometer range and is of high importance only for very thin surface modifications.

(2) Film growth. The growth of thicker coatings (>10 nm) involves several basic processes such as shadowing, desorption, surface diffusion, bulk diffusion and recrystallization. All these basic processes can be quantified in terms of the characteristic roughness of the coating surface, the activation energies for surface and bulk diffusion, and the sublimation energy. For many materials, these energies are related and proportional to their melting point (T_m). Thus, various basic processes can be expected to dominate over different ranges of the substrate temperature (T_s) expressed in terms of its homologous temperature (T_s/T_m), leading to Structure Zone Models (SZMs). For modern energetic PVD and CVD coating techniques, the model of Messier *et al.* [215] is of highest importance, considering both the deposition temperature and the energy of impinging atoms. The kinetic/ionic energy of the deposited atoms/ions can trigger surface diffusion and, thus, increasingly replaces the Zone-1 growth mode features by the Zone-T at higher energies (Fig. 5(a)). These zones strongly influence the surface roughness and topography [216] (Fig. 5(b)) as discussed below.

- During film growth in the low T_s Zone-1 regime a non-dense structure with a fine fibre texture develops. Ad-atom mobilities are low and columns preserve the random orientation of the nuclei. The columns are generally not single grains, but are composed of smaller more equiaxed grains or can be completely amorphous. Surface roughness develops in fractal geometry [216]. Extensive porosity results from the wide angular distribution of the deposition flux and atomic shadowing and limited surface diffusion.

- Zone-T growth is influenced by grain coarsening during coalescence of small islands with large surface to volume ratios, while grain boundaries become

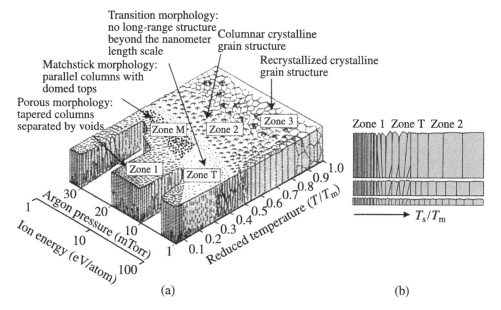

Figure 5. (a) Structure Zone Model (SZM) after Messier *et al.* for ion-assisted PVD (adapted from [215]) and (b) SZM schematically representing microstructural evolution of pure elemental films (adapted from [217]).

immobile in continuous films. Selection of the preferential growth orientation during coarsening is incomplete, thus, crystallites are nearly random or only weakly textured with a wide distribution of grain sizes. The orientation and size of individual crystallites will determine their behaviour during subsequent growth processes characterized by the competition among neighbouring grains. The consequence of this competitive growth is a continuous change in morphology, texture and surface topography as a function of film thickness. In thin films (up to ~100 nm), the microstructure consists of randomly orientated small grains, out of which V-shaped columns with the favoured orientations slowly emerge in thicker films and overgrow kinetically disadvantageous columns. Faceted column tops result in the above noted increasing surface roughness at increasing film thickness. Open column boundaries result from atomic shadowing.

- At still higher homologous temperatures in Zone-2 regime, bulk diffusion becomes significant. Zone-2 structures consist of columnar grains separated by distinct, dense, intercrystalline boundaries. Large grains with low surface and interface energy grow at the expense of smaller or unfavourably oriented grains. Normal grain growth is impeded if the grains have a strong texture, resulting in grain diameter of two to three times the film thickness [216]. The flat surface is decorated by grain boundary grooves.

- Zone-3 consists of equiaxed, recrystallized grains with a flat and bright surface. Fully activated bulk diffusion results in film growth close to the thermal equilibrium.

Systematic studies for obtaining surface roughness values of metal films deposited at different temperatures and particle energies have not been performed yet. (However, an overview is in preparation for publishing by our research group.) Thus, only approximate results for RMS surface roughness can be given for the different SZM zones: Amorphous coatings and Zone-1 coatings are quite smooth with up to 2–20 nm roughness (depending on the film thickness), when deposited on flat substrate surfaces (e.g., polished silicon wafers). Zone-T coatings have around 2 to 10 times higher RMS roughness, while Zone-2 coatings have values in-between these two ranges. Zone-3 coatings are extremely smooth, like Zone-1 coatings.

These roughness values are extremely small in comparison to most of the human and animal cells (eukaryotic cells), which are 1–100 μm in size [218]. Interwoven protein fibres in the extracellular matrix such as collagen or elastin have fibre diameters between 10 and 300 nm [219], being better comparable to the roughness of PVD and CVD coatings. Also integrin receptors with 10 nm diameter are of the same dimension as the surface roughness. Although focal contacts with 0.25–0.50 μm seem to be too large for the controlled interaction with the surface, they are formed of a large number of single integrin–surface contacts and, thus, could be highly advantageous for controlling cellular adhesion to artificial surfaces.

Nanotopography-specific effects on cellular behaviour have been demonstrated using a wide range of different cell types, showing effects on the chemical reactivity of implant surfaces and on the ionic and biomolecular interactions with the surface [220] (a review about this topic is also given in the current Issue of this journal by Anselme et al.). Proposed changes include enhanced wettability, altered protein adsorption and cell adhesion, likely involving both integrin and non-integrin receptors. In the case of PVD and CVD coatings, some studies were published including nanoscale topographic feature–cell adhesion dependencies. For titanium nitride films, the influence of the crystallinity and topography at constant surface chemistry on the growth of primary hippocampal neuronal cells was investigated by Cyster et al. [221]. Based on the findings of Ikegaya et al. [222], showing that microspikes produced by neurones could extend across appreciable distances to locate on surfaces of suitable attachment, a preferential neuronal network morphology was shown for the films with lower roughness values ($R_a \sim 25$ nm) and decreased size of topographical features. For the ion-plating coated films, a roughness of ~ 50 nm impedes the neuronal network growth. However, the presence of interstitial nitrogen rather than the difference in nanotopography influences the beginning of cell spreading. In contrast, Tsyganov et al. [102] found no significant influence of the roughness of PIIID and cathodic arc evaporated titanium and titanium oxide/oxynitride layers below 50 nm on platelet adhesion and activation of the clotting cascade. Nanoporous titanium surfaces (pore size ~ 3 nm), achieved by high-dose helium ion implantation to Nitinol (NiTi) surfaces, were shown to possess low hemocompatibility by high adsorption of fibrinogen. This result of Maitz and Shevchenko [223] is in contrast to the general higher endothelial cell growth on nanostructured polymers and ceramics [224–226]. The reason for Maitz's results seems to be the too low size of pores, one order of magnitude less than the length of a fibrinogen molecule and roughly the size of the globular domains of the molecule [227]. This almost excludes an extensive entanglement of the fibrinogen in the nanoporous surface. However, for domains, which can be captured in a pore, conformation changes may be sterically hindered, and the domains will be less accessible for antibodies. Thus, the nanoporous surfaces showed very low amount of denatured fibrinogen.

Cell behaviour in relation to grooves on surfaces has been extensively studied [228, 229] — however, not with applying thin films on the grooved surfaces. Nevertheless, all vapour deposition techniques depict the substrate surface defects on the film surface and, thus, could be applied in future for a biomodification of rough surfaces. *In vitro* studies on bone-derived cells grown on titanium surfaces of varying roughness have reported changes in osteoblast morphology, proliferation, differentiation, and production of soluble mediators [230–232], supporting the beneficial results found in clinical application. Most of the results revealed that on rough samples, the cells were parallel to the grooves caused by mechanical finishing; whereas on smooth surfaces, the cells grew with random orientation. In addition, low cell proliferation and spreading was observed on titanium surfaces of very high roughness [233]. Den Barber *et al.* [234] evaluated the effect of both parallel surface microgrooves and surface free energy on cell growth. Their findings indicated that physico-chemical parameters such as wettability and surface free energy influenced cell growth but played no measureable role in the shape and orientation of cells on microtextured surfaces.

Besides smooth films, PVD and CVD techniques allow the growth of three-dimensionally shaped films too. Such structures can be achieved by, e.g., applying glancing incidence of the straight directed vapour expansion, by seeding growth nuclei. Carbon nanotubes (CNTs) are, besides ZnO nanotubes, the most prominent three-dimensionally grown films, consisting of cylindrical graphene sheets of nanometer diameter and are commercially manufactured by using CVD methods with carbon containing gas mixtures. While fullerenes — spherical molecules consisting of mostly 60 carbon atoms — were reported to be cytotoxic [235], studies involving the biocompatibility of carbon nanotubes and carbon nanofibres (CNFs) held in some structure have been performed for the contact to bone by examining the adhesion and function of bone-forming osteoblasts [236–238]. These studies showed that CNTs and CNFs did not induce cytotoxic response and that the nano-phase CNFs demonstrate potential as orthopaedic material. Proper functionalization of nanomaterials — e.g., by strong acid treatment for more hydrophilic surfaces — and preventing dispersed short nanotubes did not cause any noticeable harmful effects to fibroblasts and increased their adhesion [239]. Furthermore, the high surface area of the functionalized nanomaterials was found to provide an ideal environment for adhesion and spreading of fibroblasts. The interactions of CNTs with other cells have been explored with neuronal cells [240], fibrinoblasts [241], antibodies and immune system [242]. Adhesion properties of osteoblasts, chondrocytes, fibroblasts and smooth muscle cells on nanocomposites of polymers and CNT have been widely explored [237, 243]. For example, fibroblasts spread on CNT surfaces flat on the surface and in all directions, which can be used to control their growth by aligning CNTs. Comparative studies between nanocomposites and materials normally used in orthopaedic implants (Ti–6Al–4V and CoCrMo) have shown a better adhesion of the osteoblasts on the nanocomposites [237, 243].

Cellular materials with CVD coatings can be realized as pure tantalum coatings on a low-density, reticulated vitreous carbon skeleton, derived from pyrolized polyurethane with specified cellular density and pore geometry [244]. Their biomedical application allows equivalent radiographic and histological appearance to highly biocompatible carbon fibre cages [245]. Strength and stiffness increase with decreasing porosity, which is typical of porous solid materials such as cancellous bone too [246].

Besides continuous coatings, also powders play an important role in medicine. On the one hand, powders could be formed by wear of implants or delamination of poorly-adhering films and, thus, should be bio-inert. In contrast, bio-activity of powders is advantageous in the treatment of diseases. For the latter case, nano- to micro-sized diamond powder — in contrast to bio-inert bulk diamond — was found to be bio-active, antioxidant and anticancerogenic in living organisms [247]. The diamond powder was synthesized by various techniques including RF and microwave PACVD. The high bioactivity influences gene expression and triggers oxidative stress, cellular stress and genotoxic stress in human. The effects possibly depend on the extension of the diamond surfaces. Nevertheless, other authors revealed bio-inertness of diamond powder particles [248–250].

Bioreactions induced by worn DLC particles were investigated *in vitro* in bone marrow cell cultures with particles delaminated from 500 nm thick DLC films. The cells were able to internalize most of the particles within a few days and the appearance of the cells after 7 days was no different from the control cultures with no particles. Furthermore, the addition of particles did not have any effect on the lysosomal activity of the cells nor on the proliferation or differentiation, indicating that no toxic or inflammatory reaction of the body to delaminated DLC particles can be expected [251].

In summary, the surface roughness modification has not been widely investigated for PVD and CVD coating materials yet. However, the available results stimulate more detailed research on free-standing three-dimensional structures like nanotubes to small nano-hills of coating materials and nanoporous coatings.

6. Conclusions and Trends for Future Research

Inorganic and carbon-based thin films deposited by PVD and CVD techniques possess a wide spectrum of different biomedically important properties from highly bio-inert to bioactive and from cytotoxic to highly biocompatible. Although biomedical R&D on such materials as titanium nitride, titanium oxide, diamond-like carbon, etc. started almost in the 1980's, the systematic knowledge about influences on the, e.g., protein and cell adsorption and adhesion behaviour is still lacking. As shown in this review, the influence of the outermost atomic layers on the surface is of tremendous importance, when proteins and cells come in contact with the artificial biomaterial surface. However, the confusion in the partly highly positive, partly negative effects of similar coating materials on the interaction to the biological surrounding is based on the missing information on the topmost surface condition. First steps were taken in the last years by targeted plasma treatments for bond termination, e.g., on diamond coatings.

From the authors' point of view, R&D on biocompatible coatings and coated medical products should be targeted towards understanding the influence of the surface condition on the biological effects — like protein and cell adhesion. Such finishing plasma treatments provide an optimized surface for subsequently applied surface modification techniques to finally develop fully biomimetic implants, combining several physical and chemical biological functionalization methods including nanostructuring, protein immobilization, drug delivery, etc. Achieving this goal of understanding and designing the biomaterial–tissue interface requires highly inter- and transdisciplinary work in medicine, physics, chemistry, biophysics, biochemistry and materials science.

Acknowledgements

Financial support for this work by the Austrian Federal Ministry of Traffic, Innovation and Technology, the Austrian Industrial Research Promotion Fund (FFG) in the frame of the Austrian Nanoinitiative and the MNT–ERA–NET program, the Government of Styria, and the European Union in the frame of the EFRE is highly acknowledged.

References

1. B. D. Ratner, A. S. Hoffman and J. E. Lemons, *Biomaterials Science: An Introduction to Materials in Medicine*, 2nd edn. Elsevier, Amsterdam (2004).
2. N. Weber, D. Bolikal, S. L. Bourke and J. Kohn, *J. Biomed. Mater. Res.* **68A**, 496 (2004).
3. K. Anselme, *Biomaterials* **21**, 667 (2000).
4. E. Tziampazis, J. Kohn and P. V. Moghe, *Biomaterials* **21**, 511 (2000).
5. B. Kasemo, *Surface Sci.* **500**, 656 (2002).
6. B. Kasemo and J. Gold, *Adv. Dent. Res.* **13**, 8 (1999).
7. B. Kasemo and J. Lausmaa, *CRC Critical Reviews in Biocompatibility* **2**, 335 (1986).

8. B. Kasemo and J. Lausmaa, *Osseointegration in Oral Rehabilitation*. Quintessence, London (1993).

9. A. Curtis, B. Casey, J. O. Gallagher, D. Pasqui, M. A. Wood and C. Wilkinson, *Biophys. Chem.* **94**, 275 (2001).

10. J. Meyle, K. Gultig, H. Wolburg and A. F. von Recum, *J. Biomed. Mater. Res.* **27**, 1553 (1993).

11. M. J. Dalby, M. O. Riehle, D. S. Sutherland, H. Agheli and A. S. G. Curtis, *J. Biomed. Mater. Res.* **69A**, 314 (2004).

12. F. A. Denins, P. Hanarp, D. Sutherland and Y. F. Dufrene, *Nano Letters* **2**, 1419 (2002).

13. B. Schleicher and S. K. Friedlander, *J. Appl. Phys.* **78**, 2424 (1995).

14. M. J. Dalby, M. O. Riehle, H. J. H. Johnstone, S. Affrossman and A. S. G. Curtis, *J. Biomed. Mater. Res.* **67A**, 1025 (2003).

15. F. H. Frey and G. Kienel, *Dünnschichttechnologie*. VDI-Verlag, Düsseldorf (1987).

16. P. Huber, D. Manova, S. Mandl and B. Rauschenbach, *Surf. Coatings Technol.* **156**, 176 (2002).

17. M. Jäger, C. Zilkens, K. Zanger and R. Krauspe, *J. Biomed. Biotechnol.*, Article ID 69036 (2007).

18. N. Nath, J. Hyun, H. Ma and A. Chilkoti, *Surf. Sci.* **570**, 98 (2004).

19. B.-Z. Katz, E. Zamir, A. Bershadsky, Z. Kam, K. M. Yamada and B. Geiger, *Mol. Biol. Cell* **11**, 1047 (2000).

20. P. A. Underwood and J. G. Steele, *J. Immunol. Meth.* **142**, 83 (1991).

21. D. D. Deligianni, N. Katsala, S. Ladas, D. Sotiropoulou, J. Amedee and Y. F. Missirlis, *Biomaterials* **22**, 1241 (2001).

22. E. A. Vogler, *Adv. Colloid Interface Sci.* **74**, 69 (1998).

23. E. A. Vogler, *J. Biomater. Sci., Polym. Ed.* **10**, 1015 (1999).

24. W. Norde, *Adv. Colloid Interface Sci.* **25**, 267 (1986).

25. P. R. van Tasse, P. Viot and G. Tarjus, *J. Chem. Phys.* **106**, 761 (1997).

26. W. Norde, in: *Biopolymers at Interfaces*, M. Malmsten (Ed.), pp. 21–45. Marcel Dekker, New York, NY (2003).

27. J. D. Andrade, V. Hlady and A. P. Wei, *Pure Appl. Chem.* **64**, 1777 (1992).

28. J. D. Andrade, in: *Surface and Interfacial Aspects of Biomedical Polymers*, J. D. Andrade (Ed.). Plenum Press, New York, NY (1985).

29. L. Vroman, *Seminars in Thrombosis and Hemostasis* **13**, 79 (1987).

30. L. Vroman and A. L. Adams, *Surf. Sci.* **16**, 438 (1969).

31. S.-Y. Jung, S.-M. Li, F. Albertorio, G. Kim, M. C. Gurau, R. D. Yang, M. A. Holden and P. S. Cremer, *J. Am. Chem. Soc.* **125**, 12782 (2003).

32. L. Vroman and A. L. Adams, *J. Biomed. Mater. Res.* **3**, 43 (1969).

33. B. Kasemo and J. Gold, *Adv. Dent. Res.* **13**, 8 (1999).

34. P. A. Underwood, J. G. Steele and B. A. Dalton, *J. Cell Sci.* **104**, 793 (1993).

35. N. A. Hotchin and A. Hall, *J. Cell Biol.* **131**, 1857 (1995).

36. T. A. Horbett, in: *Biopolymers at Interfaces*, M. Malmsten (Ed.), pp. 393–415. Marcel Dekker, New York, NY (2003).

37. C. C. Kumar, *Oncogene* **17**, 1365 (1998).

38. J. M. Whitelock and J. Melrose, in: *Wiley Encyclopaedia of Biomedical Engineering*. Wiley, New Jersey (2006).

39. J. G. Steele, G. Johnson and P. A. Underwood, *J. Biomed. Mater. Res.* **26**, 681 (1992).

40. B. Alberts, A. Johnson, J. Lewis, M. Raff, K. Roberts and P. Walter, *Molecular Biology of the Cell*. Garland Science, New York, NY (2002).

41. K. Burridge and M. Chrzanowska-Wodnicka, *Annu. Rev. Cell Dev. Biol.* **12**, 463 (1996).

42. B. Geiger, A. Bershadsky, R. Pankov and K. M. Yamada, *Nature Rev. Mol. Cell Biol.* **2**, 793 (2001).

43. L. B. Smilenov, A. Mikhailov, R. J. Pelham, E. E. Marcantonio and G. G. Gundersen, *Science* **286**, 1172 (1999).

44. S. K. Sastry and K. Burridge, *Expl. Cell. Res.* **261**, 25 (2000).

45. F. Grinnell and M. K. Feld, *J. Biol. Chem.* **257**, 4888 (1982).

46. B. G. Keselowsky, D. M. Collard and A. J. Garcia, *J. Biomed. Mater. Res.* **66A**, 247 (2003).

47. L. Hao and J. Lawrence, *J. Biomed. Mater. Res.* **69A**, 748 (2004).

48. J. Sottile, D. C. Hocking and K. J. Langenbach, *J. Cell Sci.* **113**, 4287 (2000).

49. A. J. Garcia and D. Boettiger, *Biomaterials* **20**, 2427 (1999).

50. D. A. Puleo and R. Bizios, *Bone Mineral* **18**, 215 (1992).

51. A. Rezania and K. E. Healy, *Biotechnol. Progr.* **15**, 19 (1999).

52. E. Zamir, B.-Z. Katz, S. Aota, K. M. Yamada, B. Geiger and Z. Kam, *J. Cell Sci.* **112**, 1665 (1999).

53. L. S. Goldstein and A. V. Philp, *Annu. Rev. Cell Dev. Biol.* **15**, 141 (1999).

54. A. Diener, B. Nebe, F. Luthen, P. Becker, U. Beck, H. G. Neumann and J. Rychly, *Biomaterials* **26**, 383 (2005).

55. T. A. Haas and E. F. Plow, *Curr. Opin. Cell Biol.* **6**, 656 (1994).

56. A. S. G. Curtis and C. Wilkinson, *Trends in Biotechnol.* **19**, 91 (2001).

57. M. J. Dalby, L. di Silvio, G. W. Davies and W. Bonfield, *J. Mater. Sci.: Mater. Med.* **11**, 805 (2000).

58. J.-H. Wang, C.-H. Yao, W.-Y. Chuang and T.-H. Young, *J. Biomed. Mater. Res.* **51**, 761 (2000).

59. R. G. Flemming, C. J. Murphy, G. A. Abrams, S. L. Goodman and P. F. Nealey, *Biomaterials* **20**, 573 (1999).

60. R. Barbucci, D. Pasqui, A. Wirsen, S. Affrossman, A. Curtis and C. Tetta, *J. Mater. Sci.: Mater. Med.* **14**, 721 (2003).

61. K. Matsuzaka, X. F. Walboomers, J. E. de Ruijte and J. A. Jansen, *Biomaterials* **20**, 1293 (1999).

62. X. F. Walboomers, H. J. Croes, L. A. Ginsel and J. A. Jansen, *J. Biomed. Mater. Res.* **47**, 204 (1999).

63. P. J. Brugge and J. A. Jansen, *Biomaterials* **23**, 3269 (2002).

64. G. B. Schneider, R. Zaharias, D. Seabold, J. Keller and C. Stanford, *J. Biomed. Mater. Res.* **69A**, 462 (2004).

65. M. Wieland, B. Chehroudi, M. Textor and D. M. Brunette, *J. Biomed. Mater. Res.* **60**, 434 (2002).

66. M. Bigerelle and K. Anselme, *J. Biomed. Mater. Res.* **72A**, 36 (2005).

67. M. Bigerelle and K. Anselme, *J. Biomed. Mater. Res.* **75A**, 530 (2005).

68. M. Bigerelle and K. Anselme, *Acta Biomater.* **1**, 499 (2005).

69. E. Martines, K. McGhee, C. Wilkinson and A. S. G. Curtis, *IEEE Trans. Nanobioscience* **3**, 90 (2004).

70. T. Halkier, *Mechanisms in Blood Coagulation, Fibrinolysis and the Complement System.* Cambridge University Press, Cambridge (1991).

71. K. C. Dee, D. A. Puleo and R. Bizios, *An Introduction to Tissue–Biomaterial Interactions.* Wiley, New York, NY (2003).

72. G. H. Rao and T. Chandy, *Bull. Mater. Sci.* **22**, 633 (1999).

73. P. N. Sawyer and J. W. Pate, *Surgery* **34**, 491 (1953).

74. P. Baurschmidt and M. Schaldach, *J. Bioeng.* **1**, 261 (1977).

75. M. B. Gorbet and M. V. Sefton, *Biomaterials* **25**, 5681 (2004).

76. J. Anderson, in: *Biomaterials Science*, 2nd edn, B. Ratner, A. Hoffmann, F. J. Schoen and J. E. Lemons (Eds). Elsevier, Amsterdam (2004).

77. J. E. Davies, *Int. J. Prosthodont.* **11**, 391 (1998).

78. T. Albrektsson and C. Johansson, *Eur. Spine J.* **10**, 96 (2001).

79. U. Meyer, T. Meyer and D. B. Jones, *J. Mater. Sci.: Mater. Med.* **9**, 301 (1998).

80. A. M. Moursi, R. K. Globus and C. H. Damsky, *J. Cell Sci.* **110**, 2187 (1997).

81. A. Rezania and K. E. Healy, *J. Biomed. Mater. Res.* **52**, 595 (2000).

82. M. Jäger, F. Urselmann and F. Witte, *J. Biomed. Mater. Res.* **86A**, 61 (2007).

83. S. A. Redey, M. Nardin and D. Bernache-Assolant, *J. Biomed. Mater. Res.* **50**, 353 (2000).

84. G. Zhao, Z. Schwartz and M. Wieland, *J. Biomed. Mater. Res.* **74A**, 49 (2005).

85. R. K. Schenk and D. Buser, *Periodontol.* **22**, 2200 (1998).

86. M. Centrella, T. L. Mc Carthy and E. Canalis, *Connective Tissue Res.* **20**, 267 (1989).

87. J. M. Lackner, Industrially-scaled hybrid pulsed laser deposition at room temperature, *Habilitation Thesis*, Orekop, Krakow (Poland) (2005).

88. L. Hanley and S. B. Sinnott, *Surface Sci.* **500**, 500 (2002).

89. J. F. Ziegler, J. P. Biersack and U. Littmark, *The Stopping and Range of Ions in Solids*. Pergamon Press, New York, NY (1985).

90. D. C. Jacobs, *J. Phys. Condens. Matter* **7**, 1023 (1995).

91. B. Rother and J. Vetter, *Plasmabeschichtungsverfahren und Hartstoffschichten*. Dt. Verlag für Grundstoffindustrie (1992).

92. J. M. Lackner and W. Waldhauser, *Jahrbuch der Oberflächentechnik* (2009), accepted.

93. L. L. Hench and E. C. Ethridge, *Biomaterials — An Interfacial Approach, Biophysics and Bioengineering Series*, Vol. 4, p. 18. Acedemic Press (1982).

94. J. Hong, J. Andersson, K. N. Ekdahl, G. Elgue, N. Axen, R. Larsson and B. Nilsson, *Thromb. Haemost.* **82**, 58 (1999).

95. C. H. Gemmell and J. Y. Park, in: *Bone Engineering*, J. E. Davies (Ed.), pp. 108–117. Em Squared, Toronto (1999).

96. J. Hong, A. Azens, K. N. Ekdahl, C. G. Granqvist and B. Nilsson, *Biomaterials* **26**, 1397 (2005).

97. B. Walivaara, B. O. Aronsson, M. Rodahl, J. Lausmaa and P. Tengvall, *Biomaterials* **15**, 827 (1994).

98. F. Zhang, Z. Zheng, Y. Chen, X. Liu, A. Chen and Z. Jiang, *J. Biomed. Mater. Res.* **42**, 128 (1998).

99. H. Nygren, P. Tengvall and I. Lundstrom, *J. Biomed. Mater. Res.* **34**, 487 (1994).

100. N. Huang, P. Yang, X. Chen, Y. Leng, X. Zeng, G. Jun, Z. Zheng, F. Zhang, Y. Chen, X. Liu and T. Xi, *Biomaterials* **19**, 771 (1998).

101. M. F. Maitz and N. Shevchenko, *J. Biomed. Mater. Res.* **76A**, 356 (2006).

102. I. Tsyganov, M. F. Maitz and E. Wieser, *Appl. Surf. Sci.* **235**, 156 (2004).

103. N. Huang, P. Yang, Y. X. Leng, J. Y. Chen, H. Sun, J. Wang, G. J. Wang, P. D. Ding, T. F. Xi and Y. Leng, *Biomaterials* **24**, 2117 (2003).

104. J. Y. Chen, Y. X. Leng, X. B. Tian, L. P. Wang, N. Huang, P. K. Chu and P. Yang, *Biomaterials* **23**, 2545 (2002).

105. I. A. Tsyganov, M. F. Maitz, E. Richter, H. Reuther, A. I. Mashina and F. Rustichelli, *Nucl. Instr. Meth. Phys. Res. B* **257**, 122 (2007).

106. D. V. Shtansky, N. A. Gloushankova, I. A. Bashkova, M. I. Petrzhik, A. N. Sheveiko, F. V. Kiryukhantsev-Korneev, I. V. Reshetov, A. S. Grigoryan and E. A. Levashov, *Surf. Coatings Technol.* **201**, 4111 (2006).

107. M. F. Maitz, R. W. Y. Poon, X. Y. Liu, M.-T. Pham and P. K. Chu, *Biomaterials* **26**, 5465 (2005).

108. M. Manso, S. Ogueta, J. Perez-Rigueiro, J. P. Garcia and J. M. Martinez-Duart, *Biomol. Eng.* **19**, 239 (2002).

109. C.-C. Chien, K. T. Liu, J. G. Duh, K. W. Chang and K. H. Chung, *Dental Mater.* **24**, 986 (2008).

110. K. Lehle, J. Buttstaedt and D. E. Birnbaum, *J. Biomed. Mater. Res.* **65A**, 393 (2003).

111. Y. X. Leng, P. Yang, J. Y. Chen, H. Sun, J. Wang, G. J. Wang, N. Huang, X. B. Tian and P. K. Chu, *Surf. Coatings Technol.* **138**, 296 (2001).

112. R. Major, J. Bonarski, J. Morgiel, B. Major, E. Czarnowska, R. Kustosz, J. M. Lackner and W. Waldhauser, *Surf. Coatings Technol.* **200**, 6340 (2006).

113. J. M. Lackner, W. Waldhauser, R. Major, B. Major, E. Czarnowska and F. Bruckert, *Proc. SPIE* **7005**, 70050Q (2008).

114. J. M. Lackner, W. Waldhauser, R. Major, B. Major, R. Kustosz and E. Czarnowska, in: *Proc. Annual Technical Conf./Society of Vacuum Coaters* (2006).

115. J. M. Lackner, W. Waldhauser, R. Berghauser, M. Kahn, F. Bruckert, R. Major and B. Major, in: *Proc. Annual Technical Conf./Society of Vacuum Coaters* (2007).

116. J. M. Lackner, W. Waldhauser, R. Kustosz, B. Major, R. Major and L. Major, *Jahrbuch für Oberflächentechnik* **62**, 390 (2006).

117. M. P. Gispert, A. P. Serro, R. Colaco, A. M. Botelho do Rego, E. Alves, R. C. da Silva, P. Brogueira, E. Pires and B. Saramago, *Wear* **262**, 1337 (2007).

118. S. Potocki, A. Kromka, J. Potmesil, Z. Remes, V. Vorlicek, M. Vanecek and M. Michalka, *Diamond Relat. Mater.* **16**, 744 (2007).

119. L. Tang, C. Tsai, W. W. Gerberich, L. Kruckeberg and D. R. Kania, *Biomaterials* **16**, 483 (1995).

120. C. Popov, W. Kulisch, M. Jelinek, A. Bock and J. Strnad, *Thin Solid Films* **494**, 92 (2006).

121. C. Popov, W. Kulisch, J. P. Reithmaier, T. Dostalova, M. Jelinek and N. Anspach, *Diamond Relat. Mater.* **16**, 735 (2007).

122. P. Arianom, P. Baldelli, E. Carbone, A. Gilardino, A. L. Giudice and D. Lovisolo, *Diamond Relat. Mater.* **14**, 669 (2005).

123. R. J. Narayan, W. Wei, C. Jin, M. Andara, A. Agarwal and R. A. Gerhardt, *Diamond Relat. Mater.* **15**, 1935 (2006).

124. W. Okroj, M. Kaminska, L. Klimek, W. Szymanski and B. Walkowiak, *Diamond Relat. Mater.* **15**, 1535 (2006).

125. W. Jakubowski, G. Bartosz, P. Niedzielski, W. Szymanski and B. Walkowiak, *Diamond Relat. Mater.* **13**, 1761 (2004).

126. K. Mitura, P. Niedzielski, G. Bartosz, J. Moll, B. Walkowiak and Z. Pawlowska, *Surf. Coatings Technol.* **201**, 2117 (2006).

127. R. K. Roy and K.-R. Lee, *J. Biomat. Mater. Res.* **83B**, 72 (2007).

128. R. Hauert, *Diamond Rel. Mater.* **12**, 583 (2003).

129. F. Z. Cui and D. J. Lee, *Surf. Coatings Technol.* **131**, 481 (2000).

130. S. Logothetidis, M. Gioti, S. Lousinian and S. Fotiadou, *Thin Solid Films* **482**, 126 (2005).

131. Y. X. Leng, J. Y. Chen, P. Yang, H. Sun, G. J. Wan and N. Huang, *Surface Sci.* **531**, 177 (2003).

132. J. Y. Chen, L. P. Wang, K. Y. Fu, N. Huang, Y. Leng, Y. X. Leng, P. Yang, J. Wang, G. J. Wan and H. Sun, *Surf. Coatings Technol.* **156**, 289 (2002).

133. Y. Chen and Y. F. Zheng, *Surf. Coatings Technol.* **200**, 4543 (2006).

134. J. L. Endrino, R. Escobar Galindo, H. S. Zhang, M. Allen, R. Gago, A. Espinosa and A. Anders, *Surf. Coatings Technol.* **202**, 3675 (2008).

135. Y. Ohgoe, S. Kobayashi, K. Ozeki, H. Aoki, H. Nakamori, K. K. Hirakuri and O. Miyashita, *Thin Solid Films* **497**, 218 (2006).
136. A. C. Evans, J. Franks and P. J. Revell, *Surf. Coatings Technol.* **47**, 662 (1991).
137. J. R. McColl, D. M. Grant, S. M. Green, J. V. Wood, T. L. Parker, K. Parker, A. A. Goruppa and N. St. J. Braithwaite, *Diamond Relat. Mater.* **3**, 83 (1993).
138. R. Butter, M. Allen, L. Chandra, A. H. Lettington and N. Rushton, *Diamond Relat. Mater.* **4**, 857 (1995).
139. S. Linder, W. Pinkowski and M. Aepfelbacher, *Biomaterials* **23**, 767 (2002).
140. M. Mohanty, T. V. Anilkumar, P. V. Mohanan, C. V. Muraleedharan, G. S. Bhuvaneshwar, F. Derangere, Y. Sampeur and R. Suryanarayanan, *Biomol. Eng.* **19**, 125 (2002).
141. E. Mitura, S. Mitura, P. Niedzielski, Z. Has, R. Wolowiec, A. Jakubowski, J. Szmidt, A. Sokolowska, P. Louda, J. Marciniak and B. Koczy, *Diamond Relat. Mater.* **3**, 898 (1994).
142. M. Allen, B. Myer and N. Rushton, *J. Biomed. Mater. Res.* **58B**, 318 (2001).
143. M. B. Gulielmotti, S. Renou and R. L. Cabrini, *Int. J. Oral Maxillofac. Implants* **14**, 565 (1999).
144. K. Zlynski, P. Witkowski, A. Kaluzny, Z. H. Has, P. Niedzielski and S. Mitura, *J. Chem. Vap. Deposition* **4**, 232 (1996).
145. K. Gutensohn, C. Beythien, J. Bau, T. Fenner, P. Grewe, R. Koester, K. Padmanaban and P. Kuehnl, *Thromb. Res.* **99**, 577 (2000).
146. H. H. Tong, O. R. Monteiro, R. A. MacGill, I. G. Brown, G. F. Yin and J. M. Luo, *Current Appl. Phys.* **1**, 197 (2001).
147. S. C. H. Kwok, J. Wang and P. K. Chu, *Diamond Relat. Mater.* **14**, 78 (2005).
148. P. D. Maguire, J. A. McLaughlin, T. I. T. Okpalugo, P. Lemoine, P. Papkonstantinou, E. T. McAdams, M. Needham, A. A. Ogwu, M. Ball and G. A. Abbas, *Diamond Relat. Mater.* **14**, 1277 (2005).
149. T. I. T. Okpalugo, A. A. Ogwu, P. D. Maguire and J. A. D. McLaughlin, *Biomaterials* **25**, 239 (2004).
150. X. He, W. Li and H. Li, *Vacuum* **45**, 977 (1994).
151. N. Huang, P. Yang, Y. X. Leng, J. Wang, H. Sun, J. Y. Chen and G. J. Wan, *Surf. Coatings Technol.* **186**, 218 (2004).
152. T. Saito, T. Hasebe, S. Yohena, Y. Matsuoka, A. Kamijo, K. Takhashi and T. Suzuki, *Diamond Relat. Mater.* **14**, 1116 (2005).
153. P. Yang, N. Huang, Y. X. Leng, Z. Q. Yao, H. F. Zhou, M. Maitz, Y. Leng and P. K. Chu, *Nucl. Instr. Meth. Phys. Res. B* **242**, 22 (2006).
154. M. F. Maitz, R. Gago, B. Abendroth, M. Camero, I. Caretti and U. Kreissig, *J. Biomed. Mater. Res.* **77B**, 179 (2006).
155. T. Yokota, T. Terai, T. Kobayashi and M. Iwaki, *Nucl. Instrum. Methods. Phys. Res. B* **242**, 48 (2006).
156. A. Benavid, P. J. Martin, C. Comte, E. W. Preston, A. J. Haq, F. S. Magdon Ismail and R. K. Singh, *Diamond Relat. Mater.* **16**, 1616 (2007).
157. C. Du, X. W. Su, F. Z. Cui and X. D. Zhu, *Biomaterials* **19**, 651 (1998).
158. R. Hauert, U. Müller, G. Francz, F. Birchler, A. Schroeder, J. Mayer and E. Wintermantel, *Thin Solid Films* **308/309**, 191 (1997).
159. A. Schroeder, G. Francz, A. Bruinink, R. Hauert, J. Mayer and E. Wintermantel, *Biomaterials* **21**, 449 (2000).
160. I. De Scheerder, M. Szilard, H. Yanming, X. B. Ping, E. Verbeken, D. Neerinck, E. Demeyere, W. Coppens and F. Van de Werf, *J. Invasive Cardiol.* **12**, 389 (2000).

161. A. Dorner-Reisel, C. Schürer, C. Nischan, O. Seidel and E. Müller, *Thin Solid Films* **420–421**, 263 (2002).

162. G. Thorwarth, B. Saldamli, F. Schwarz, P. Jürgens, C. Leiggener, R. Sader, M. Haeberlen, W. Assmann and B. Stritzker, *Plasma Process. Polym.* **4**, 364 (2007).

163. M. S. Amin, L. K. Randeniya, A. Bendavid, P. J. Martin and E. W. Preston, *Diamond Relat. Mater.*, in press.

164. N. Huang, P. Yang, Y. X. Leng, J. Wang, J. Y. Chen, H. Sun, G. J. Wan, A. S. Zhao and P. D. Ding, in: *Proc. 6th Asia Symposium on Biomaterial*, Chengdu, China (2004).

165. K. R. Leach, Y. Kurisu, J. E. Carlson, I. Repa, D. H. Epstein, M. Urness, R. Sahatjian, D. W. Hunter and W. R. Casteneda-Zuniga, *Radiology* **175**, 675 (1990).

166. M. I. Jones, I. R. McColl, D. M. Grant, K. G. Parker and T. L. Parker, *J. Biomed. Mater. Res.* **52**, 413 (2000).

167. T. I. T. Okpaluo, A. A. Ogwu, P. D. Maguire and J. A. D. McLauglin, *Biomaterials* **25**, 239 (2004).

168. T. Saito, T. Hasebe, S. Yohena, Y. Matsuoka, A. Kaija, K. Takahashi and T. Suzuki, *Diamond Relat. Mater.* **14**, 1116 (2005).

169. E. Ruckenstein and S. V. Gourisanker, *J. Colloid Interface Sci.* **101**, 436 (1984).

170. L. J. Yu, X. Wang, X. H. Wang and X. H. Liu, *Surf. Coatings Technol.* **128/129**, 484 (2000).

171. Y. X. Leng, P. Yang, J. Y. Chen, H. Sun, J. Wang, G. J. Wang, N. Huang, X. B. Tian and P. K. Chu, *Surf. Coatings Technol.* **138**, 296 (2001).

172. C. Iliescu, B. Chen, D. P. Peonar and Y. Y. Lee, *Sensors Actuators B* **129**, 404 (2008).

173. L. V. Poperenko, M. F. Maitz and M. V. Vinnichenko, *Funct. Mater.* **10**, 447 (2003).

174. W. Suchaneck and M. Yoshimura, *J. Mater. Res.* **13**, 94 (1998).

175. G. H. Nancollas, *Pure Appl. Chem.* **64**, 1673 (1992).

176. R. Massas, S. Pitarui and M. M. Weinreb, *J. Dent. Res.* **72**, 1005 (1993).

177. J. Y. Martin, Z. Schwartz, T. W. Hummert, D. M. Schraub, J. Slmpson, J. Lanktord Jr., D. L. Cochran and B. D. Boyan, *J. Biomed. Mater. Res.* **29**, 389 (1995).

178. J. E. G. Hulshoff, K. van Dijk, J. P. C. M. van der Waerden, J. G. C. Wolke, L. A. Ginse and J. A. Jansen, *J. Biomed. Mater. Res.* **29**, 967 (1995).

179. V. Nelea, H. Pelletier, M. Iliescu, J. Werckmann, V. Craciun, I. N. Mihailescu, C. Ristoscu and C. Ghica, *J. Mater. Sci.: Mater. Med.* **13**, 1167 (2002).

180. V. Nelea, H. Pelletier, P. Mille and D. Muller, *Thin Solid Films* **453–454**, 208 (2004).

181. J. M. Fernández-Pradas, M. V. García-Cuenca, L. Clèries, G. Sardin and J. L. Morenza, *Appl. Surf. Sci.* **195**, 31 (2002).

182. E. van der Wal, J. C. G. Wolke, J. A. Jansen and A. M. Vredenberg, *Appl. Surf. Sci.* **246**, 183 (2005).

183. M. Hamdi and A. M. Ektessabi, *J. Vac. Sci. Technol. A* **19**, 1566 (2001).

184. A. K. Lynn and D. L. DuQuesnay, *Biomaterials* **23**, 1937 (2002).

185. K. A. Gross, V. Gross and C. C. Berndt, *J. Am. Ceram. Soc.* **81**, 106 (1998).

186. R. J. Dekker, J. D. de Bruijn, M. Stigter, F. Barrere, P. Layrolle and C. A. van Blitterswijk, *Biomaterials* **26**, 5231 (2005).

187. M. Yoshinari, Y. Oda, T. Inoue, K. Matsuzaka and M. Shimono, *Biomaterials* **23**, 2879 (2002).

188. Z. S. Luo, F. Z. Cui, Q. L. Feng, H. D. Li, X. D. Zhu and M. Spector, *Surf. Coatings Technol.* **131**, 192 (2000).

189. M. Veith, C. Petersen, O. C. Aktas, W. Metzger, M. Oberringer, T. Pohlemann, M. Müller and S. Gerbes, *Mater. Letters* **62**, 3842 (2008).

190. M. Veith, K. Andres, S. Faber, J. Blin, M. Zimmer and Y. Wolf, *Eur. J. Inorg. Chem.*, 4387 (2003).

191. Y. Liu, K. de Groot and E. B. Hunziker, *Bone* **36**, 745 (2005).

192. A. Hartl, E. Schmich, J. A. Garrido, J. Hernando, S. C. Catharino and S. Walter, *Nature Mater.* **3**, 736 (2004).

193. A. Nanci, J. D. Wuest, L. Peru, P. Brunet, V. Sharma and S. Zalzal, *J. Biomed. Mater. Res.* **40**, 324 (1998).

194. K. M. Mclean, S. L. McArthur, R. C. Chatelier, P. Kingshott and H. J. Griesser, *Colloids Surf. B* **17**, 23 (2000).

195. S. Matsumoto, Y. Sato, M. Tsutsumi and N. Setaka, *J. Mater. Sci.* **17**, 3106 (1982).

196. W. C. Clem, S. Chowdhury, S. A. Catledge, J. J. Weimer, F. M. Shaikh, K. M. Hennessy, V. V. Konovalov, M. R. Hill, A. Waterfeld, S. L. Bellis and Y. K. Vohra, *Biomaterials* **29**, 3461 (2008).

197. V. Chakrapani, J. C. Angus, A. B. Anderson, S. D. Wolter, B. R. Stoner and G. U. Sumanasekera, *Science* **318**, 1424 (2007).

198. H. Kawarada, *Surf. Sci. Reports* **26**, 205 (1996).

199. F. Maier, M. Riedel, B. Mantel, J. Ristein and L. Ley, *Phys. Rev. Letters* **85**, 3472 (2000).

200. D. Steinmüller-Nethl, F. R. Kloss, M. Naja-Ul-Haq, M. Rainer, K. Larsson, C. Linsmeier, G. Köhler, C. Fehrer, G. Lepperdinger, X. Liu, N. Memmel, E. Bertel, C. W. Huck, R. Gassner and G. Bonn, *Biomaterials* **27**, 4547 (2006).

201. M. Kalbacova, M. Kalbac, L. Dunsch, A. Kromka, M. Vaněček, B. Rezek, U. Hempel and S. Kmoch, *Phys. Stat. Sol. B* **244**, 4356 (2007).

202. L. Michalikova, B. Rezek, A. Kromka and M. Kalbcova, *Vacuum* (2009), in print.

203. P. Ariano, P. Baldelli, E. Carbone, A. Gilardino, A. Lo Giudice, D. Lovisolo, C. Manfredotti, M. Novara, H. Sternschulte and E. Vittone, *Diamond Relat. Mater.* **14**, 669 (2005).

204. R. K. Roy, H. W. Choi, J. W. Yi, M.-W. Moon, K.-R. Lee, D. K. Han, J. H. Shin, A. Kamijo and T. Hasebe, *Acta Biomater.* **5**, 249 (2009).

205. L. J. Yu, X. Wang, X. H. Wang and X. H. Liu, *Surf. Coatings Technol.* **128–129**, 484 (2000).

206. S. D. Bruck, *Polymer* **16**, 25 (1975).

207. S. Mändl and B. Rauschenbach, *Surf. Coatings Technol.* **156**, 276 (2002).

208. K. J. Klabunde, J. Strak, O. Koper, C. Mohs, D. Park and S. Decker, *J. Phys. Chem.* **100**, 12141 (1996).

209. S. J. Wu, L. C. DeJong and M. N. Rahaman, *J. Am. Ceram. Soc.* **79**, 2207 (1996).

210. M. I. Baraton, X. Chen and K. E. Gonsalves, *Nanostruct. Mater.* **8**, 435 (1997).

211. C. A. Neugebauer, in: *Handbook of Thin Film*, L. I. Maissel (Ed.). McGraw-Hill, New York, NY (1983).

212. W. Ensinger, *Nucl. Instr. Meth. Phys. Res. B* **127/128**, 796 (1997).

213. H. A. Jehn, in: *Advanced Techniques for Surface Engineering*, W. Gissler and H. A. Jehn (Eds), p. 5. Kluwer, Dordrecht (1992).

214. H. Jehn, in: *Hartstoffschichten zur Verschleißminderung*, H. Fischmeister and H. Jehn (Eds). Deutsche Gesellschaft für Metallkunde e. V., Oberursel (1987).

215. R. Messier, A. P. Giri and R. A. Roy, *J. Vac. Sci. Technol. A* **2**, 500 (1984).

216. C. V. Thompson, *Annu. Rev. Mater. Sci.* **20**, 245 (1990).

217. I. Petrov, P. B. Barna, L. Hultman and J. E. Greene, *J. Vac. Sci. Technol. A* **21**, S117 (2003).

218. G. Bao and S. Suresh, *Nature Mater.* **2**, 715 (2003).

219. B. Alberts, A. Johnson, J. Lewis, M. Raff, K. Roberts and P. Walter, *Molecular Biology of the Cell*. Garland Science, New York, NY (2002).
220. G. Mendonca, D. B. Mendonca, F. J. Aragao and L. F. Cooper, *Biomaterials* **29**, 3822 (2008).
221. L. A. Cyster, K. G. Parker, T. L. Parker and D. M. Grant, *Biomaterials* **25**, 97 (2004).
222. Y. Ikegaya, Y. Itsukaichi-Nishida, M. Ishihara, D. Tanaka and N. Matsuki, *Neuroscience* **97**, 215 (2000).
223. M. F. Maitz and N. Shevchenko, *J. Biomed. Mater. Res.* **76A**, 356 (2006).
224. D. C. Miller, A. Thapa, K. M. Haberstroh and T. J. Webster, *Biomaterials* **25**, 53 (2004).
225. L. Polonchuk, J. Elbel, L. Eckert, J. Blum, E. Wintermantel and H. M. Eppenberger, *Biomaterials* **21**, 539 (2000).
226. T. J. Webster, R. W. Siegel and R. Bizios, *Biomaterials* **20**, 1221 (1999).
227. P. Cacciafesta, A. D. L. Humphris, K. D. Jandt and M. J. Miles, *Langmuir* **16**, 8167 (2000).
228. M. Ahmad, D. Gawronski, J. Blum, J. Goldberg and G. Gronowicz, *J. Biomed. Mater. Res.* **46**, 121 (1999).
229. L. Ponsonnet, V. Comte, A. Othmane, C. Lagneau and M. Charbonnier, *Mater. Sci. Eng. C* **21**, 157 (2002).
230. Z. Schwartz, C. H. Lohmann, J. Oefinger, L. F. Bonewald, D. D. Dean and B. D. Boyan, *Adv. Dent. Res.* **13**, 38 (1999).
231. K. Kieswetter, Z. Schwartz, T. W. Hummert, D. L. Cochran, J. Simpson and D. D. Dean, *J. Biomed. Mater. Res.* **32**, 55 (1999).
232. J. Y. Martin, Z. Schwartz, T. W. Hummert, D. M. Schraub, J. Simpson and J. Lankford Jr., *J. Biomed. Mater. Res.* **29**, 389 (1995).
233. K. Anselme, P. Linez, M. Bigerelle, D. Le Maguer, A. Le Maguer and P. Hardouin, *Biomaterials* **21**, 1567 (2000).
234. E. T. den Barber, J. E. de Ruijter, H. T. J. Smits, L. A. Ginsel, A. F. von Recum and J. A. Jansen, *J. Biomed. Mater. Res.* **29**, 511 (1995).
235. C. M. Sayers, J. D. Fortner, W. Guo, D. Lyon, A. M. Boyd, K. D. Ausman, Y. J. Tao, B. Sitharaman, L. J. Wilson, J. B. Hughes, J. L. West and V. L. Colvin, *Nano Letters* **4**, 1881 (2004).
236. E. Flahaut, M. C. Durrieu and M. Remy-Zolghadri, *J. Mater. Sci.* **41**, 2411 (2006).
237. T. J. Webster, M. C. Waid and J. L. Mckenzie, *Nanotechnology* **15**, 48 (2005).
238. K. E. Elias, R. L. Price and T. J. Webster, *Biomaterials* **23**, 3279 (2002).
239. Y. H. Yun, Z. Dong, Z. Tan, M. J. Schulz and V. Shanov, *Mater. Sci. Eng. C* **29**, 719 (2009).
240. T. Gabay, E. Jakobs and E. Bem-Jacob, *Physica A* **21**, 611 (2005).
241. M. A. Correa-Duarte, N. Wagner and J. Rojas-Chapana, *Nano Lett.* **4**, 2233 (2004).
242. N. N. Naguib, Y. M. Mueller and P. M. Bojuczuc, *Nanotechnology* **16**, 567 (2005).
243. R. A. Price, M. C. Waid, M. Karen and K. M. Haberstroh, *Biomaterials* **24**, 1877 (2003).
244. http.//wwww.plasmachem.com
245. V. M. Elinson, V. V. Sleptsov, A. N. Laymin, V. V. Potraysay, L. N. Kostuychenko and A. D. Moussina, *Diamond Relat. Mater.* **8**, 2103 (1998).
246. J. A. Mclaughlin, B. Meenan, P. Maguire and N. Jamieson, *Diamond Relat. Mater.* **5**, 486 (1996).
247. K. Bakowicz-Mitura, G. Bartosz and S. Mitura, *Surf. Coatings Technol.* **201**, 6131 (2007).
248. A. H. Lettington, *Carbon* **36**, 555 (1998).
249. R. Butter, M. Allen, L. Chandra, A. H. Lettington and N. Rushton, *Diamond Relat. Mater.* **4**, 857 (1995).
250. D. P. Dowling, P. Kola, K. Donelly, T. C. Kelly, K. Brumitt, L. Lloyd, R. Eloy, M. Therin and N. Weil, *Diamond Relat. Mater.* **6**, 390 (1997).
251. A. Schroeder, *PhD Thesis*, ETH Zürich (1999).

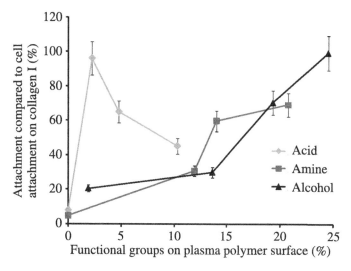

Figure 3. Attachment of primary keratinocytes to plasma-copolymerized surfaces at 24 h. Results are shown as a % of attached cells in comparison to cell attachment on collagen I, a favoured attachment surface. Monomers used to generate the acid, amine and alcohol functionalities are acrylic acid, allylamine and allyl alcohol, respectively. Data compiled from two studies [36, 49]. Bars represent standard error of the mean.

plasma power and the length of the polymerization time can all be utilised to control this density. Keratinocyte attachment to these surfaces was measured at 24 h, with collagen I providing a 'gold standard' control and TCPS a negative control. The cells were cultured using a keratinocyte medium which favoured keratinocytes over melanocytes [46]. Whilst TCPS provided an erratic control (for reasons already described) the data in Fig. 3 show that the plasma-copolymerized surfaces afforded a culture surface, without the requirement of collagen coating.

The attachment results from the series of acrylic acid plasma polymers were particularly exciting (Fig. 3) as the measured cell number at 24 h was comparable to that seen on collagen I. Optical microscopy of the cells on the 2.3% acid surface and collagen I revealed cell features consistent with good attachment, spreading and colony formation. On allylamine and allyl alcohol surfaces, cell attachment increased with increasing functionality, but for allylamine surface, cell attachment only reached 70% in comparison to the collagen I surface. On the plasma-polymerized allyl alcohol surfaces, cell attachment increased with surface alcohol. However, because of the nature of the plasma deposition process, in high alcohol surfaces, a small high binding energy peak was seen in the XPS C_{1s} core line, which may have arisen from a small number of carboxyl groups in the plasma deposit. In light of this first study it is possible that these carboxyls are responsible for the cell attachment observed on the high alcohol surfaces; a point which was never followed up. On the low acid surface, on extending the culture period, cells proliferated and spread to form confluent sheets. Similar behaviour was observed with collagen I, and after 7 days in culture the cells on the acid plasma

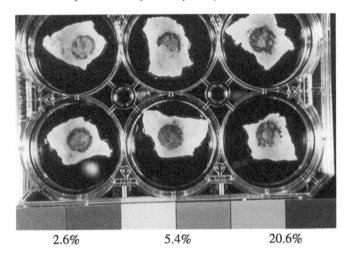

2.6%   5.4%   20.6%

Figure 4. Transfer of healthy keratinocytes from carrier substrates coated with acid plasma polymer films of different acid amounts (2.6, 5.4 and 20.6%). Transferred cells are stained (MTT-ESTA staining) purple to show metabolically active cells.[2]

polymer were at least as confluent as cells on the collagen control surface [63]. The cell morphology on the plasma polymer surfaces was as expected for normal healthy keratinocytes (i.e., comparable to keratinocytes grown on 'gold standard' collagen I).

Having demonstrated that plasma polymer surfaces could support keratinocyte cell attachment this work was extended to explore to what extent these surfaces could support the transfer of cells to wound beds [21]. Preclinical trials employed a model of an *in vivo* wound bed, an acellular, de-epidermalized human dermis (DED) that retained basement membrane antigens [34]. Simply put, a DED comprises skin in which the epidermis has been removed, which provides a convenient model for assessing the transfer of cells from a carrier surface to human dermis. Several different acrylic acid plasma copolymers were used in these studies which had initial surface acid values of 2.6%, 5.4% and 20.6%, as assessed by XPS (these % values were slightly reduced following ethylene oxide sterilization). Medical grade substrates were coated with these plasma polymers, and with collagen I as a control. Keratinocytes were cultured on these substrates for 24 h, before being placed cell side down onto the DED over four days for cell transfer to occur, after which the substrates were removed. The results showed successful cell transfer from the acid plasma-polymerized surfaces onto the DED (Fig. 4), with transfer from the 20.6% acid surface approximately 50% of that seen from a collagen-coated surface.

---

[2] Reproduced with permission of the authors and publishers from page 426 of D. B. Haddow, S. MacNeil and R. D. Short, 'A cell therapy for chronic wounds based upon a plasma polymer delivery surface', *Plasma Processes and Polymers* **3**, 419 (2006).

Close examination of the transferred cells showed the formation of an epidermal layer on the DED surface, with these cells closely attached to the basement membrane of the DED. Whilst at 4 days the epithelial layer was very thin, when maintained in culture, a well-organised epithelial layer of 5–10 cells thick developed.

3.2. Delivery of Keratinocytes

The above pilot study provided great optimism that a bandage could be fabricated for the treatment of burns, scalds and chronic wounds, where cells could be cultured and then delivered from the same bandage surface. The approach owes a lot to the original Howard Green CEA method, with the advantages of reducing the number of steps, e.g., the need to remove cells from the cultureware using dipase II and the placing of the cells onto a backing dressing to facilitate cell sheet handling. It also allowed the cells to be used sub-confluent, thereby reducing the culture time required after obtaining the tissue biopsy from the patient. With appropriate ethical approval, the work moved into a proof-of-concept trial involving six diabetic patients with long standing non-healing foot ulcers. Six diabetic patients presented with nine neuropathic ulcers (resistant to conventional therapy) were treated with weekly applications of autologous keratinocytes delivered from the acid-coated bandages. A more detailed description of these ulcers and the treatment regime are given elsewhere [64]. In general, these ulcers responded to autologous cell delivery by a gradual decrease in depth and size and the development of granulation tissue on the ulcer base followed by complete closure with epithelialization. Six of the nine treated ulcers responded to the treatment with complete healing after 6–17 weekly applications of cells (Fig. 5). One patient showed a significant decrease in ulcer size after 19 applications of cells but the ulcer subsequently deteriorated. One patient failed to respond to treatment despite a total of 24 applications and treatment was discontinued in one patient after only three applications of cells due to the development of Methicillin-Resistant *Staphylococcus aureus* infection (MRSA). There were no recurrences in the healed ulcers after 6 months of follow-up and no side-effects were noticed during or after the treatment.

Subsequently, in Zhu *et al.* this bandage technology was extended to seven patients — two with acute major burns and five with chronic non-healing wounds [65]. These chronic wounds included a patient (no. 3 in the trial) who had extensive chronic wounds on both legs following partial skin grafts after 28% flame burns, and a chronic wound patient (no. 4) who presented a wound of 60 years duration, developed while a prisoner of war in World War II. All these wounds were treated with applications of autologous keratinocytes delivered from plasma-polymerized acrylic acid surfaces deposited upon a medical grade polymer disc. The culture period was two days after seeding with keratinocytes. For the two burns patients, transfer of cells was onto donor sites and the treatment was reported to facilitate healing of grafted burns wounds. In the five patients with intractable chronic wounds (who presented a total of 9 wounds) it was reported that repeated applica-

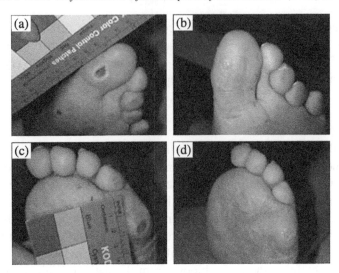

Figure 5. Neuropathic foot ulcer on patient 2 persistent for 2 years (a) and complete healing after eight applications of cells (b). Neuropathic foot ulcer on patient 3 persistent for 16 months (c) and complete healing after six applications of cells (d).[3]

tions of cells resulted in complete healing in 5 of the 9 ulcers. There was a major reduction in ulcer size for all other (4 of 9) ulcers, including patient 3 (98% healed). In the case of the war veteran it was reported that there was a partial healing of two ulcers and an improvement in quality of life. Reduction in ulcer size improved the wound conditions for two patients such that they were then considered suitable for conventional grafting and orthopaedic surgery, respectively. In the case of the latter patient, the versatility of the technology was demonstrated by virtue of its use 45 times to promote complete healing in one wound, and 22 times in another wound.

The technology was tested more rigorously with non-healing diabetic ulcers in a randomized, controlled, single-blind study [66]. Sixteen patients, presenting a total of 21 ulcers that were resistant to conventional therapy, were recruited through four diabetic centres in three cities. All 21 ulcers were treated, and it was reported that of these ten healed and eight improved. Two failed to respond, with one ulcer lost due to autoamputation. The analysis of the data focussed on 12 patients with 12 index ulcers who completed the treatment protocols — five were in the placebo group, and seven were in the active group. Of the 12, five ulcers healed completely and the remaining seven reduced by more than 50% in size. It was reported that complete healing took a median of ten active applications. The healing (in all 18 ulcers) did

---

[3] Reproduced with permission of the authors and publishers from page 428 of D. B. Haddow, S. MacNeil and R. D. Short, 'A cell therapy for chronic wounds based upon a plasma polymer delivery surface', *Plasma Processes and Polymers* **3**, 419 (2006).

not appear attributable to patient recruitment, nor to the cell-free carrier dressing, but to the delivery of the cultured cells.

3.3. Delivery of Melanocytes

Although at an early stage of development the group of MacNeil had reported the basis of a delivery vehicle, based on a plasma-polymerized layer of acrylic acid, for returning pigmentary function to autologous skin which is relevant for burns, as well as for patients with vitiligo and piebaldism [67, 68]. They describe a simple methodology for delivering cultured keratinocytes and melanocytes to the patient that is low risk. Although the work is still at preclinical stage, they show that using two types of media for cell expansion, Greens and a serum-free alternative M2 (melanocyte medium), there is reproducible transfer of physiologically relevant numbers of melanocytes capable of pigmentation from the co-culture of melanocytes and keratinocytes [69].

3.4. Plasma-Polymerized Surfaces to Culture Cells from the Cornea

The cornea is continually resurfaced by cells that migrate from the stem cell-rich limbus. If the limbal epithelial stem cells are depleted by injury or disease, neighbouring conjunctival epithelium covers the cornea causing vascularization, corneal opacity, severe patient discomfort and ultimately blindness [70, 71]. Limbal integrity is required for corneal transplantations to succeed as the loss of the limbus leads to the inability to re-epithelialize the transplanted cornea. In cases with loss of limbal population, small pieces of limbal tissue can be expanded on a variety of surfaces for transplantation onto the damaged eye. The limbal tissue can be taken from a donor or from the undamaged eye, expanded *in vitro* and transplanted [72]. As previously stated, for transplantation the most successful surface used for limbal epithelial cell expansion is HAM. Limbal cells have been shown to maintain some 'stemness' when cultured on HAM, although there is a steady decrease in stem cell properties as the cells grow from the limbal explant [73]. The cultured limbal epithelium is often transplanted onto the damaged eye along with the HAM, with the epithelial side facing away from the eye where it remains and becomes a substitute basal membrane for the epithelial cells. The use of HAM carries the risk of disease transmission so an alternative is to culture autologous cells onto recombinant human collagen constructs that mimic the cornea [50, 74]. If limbal integrity is lost in both eyes epithelial cells obtained from the oral mucosa can be used to resurface the cornea, where similar adult stem cells reside, thus eliminating the need for donor tissue with its inherent complications. Epithelial cells from the oral mucosa have been cultured upon thermoresponsive polyNiPAAm into a cell sheet that could be detached from the polymer by a decrease in temperature. The cell sheet was then transplanted onto the damaged eye leading to an improvement in visual acuity [75]. It was also noted that the epithelial cells from the oral mucosa changed their keratin protein expression to the one more similar to corneal epithelium, indicating that direct contact with the underlying stroma is important in controlling the cell behaviour [76]. This cannot be achieved whilst using HAM as a carrier or by using any carrier system, therefore making a carrier-free culture system extremely favourable.

Limbal cells have been cultured onto TCPS that has been coated with plasma-polymerized acrylic acid in the absence of serum, though still requiring the use of feeder cells [77]. However, this research did show that the acrylic acid surfaces performed better than TCPS surfaces in maintaining cell metabolic activity under serum-free conditions, indicating that the surface was vital in promoting protein adsorption and cell attachment.

The use of plasma polymer coatings for the expansion and delivery of epithelial cells to the cornea is currently the subject of research conducted by MacNeil and co-workers [78]. In a recent report the MacNeil group investigated a range of plasma polymer coatings and their ability to support *in vitro* expansion of human and rabbit cells both with and without serum. Successful cell transfer from acrylic acid plasma polymer coated contact lenses to a cell-denuded rabbit cornea model was also demonstrated. This research was then expanded to coating contact lenses with an acrylic acid based plasma polymer and the culture of a human corneal cell line on the lenses [79]. Studies were done to elucidate the ability of the culture cells to transfer to a cornea *in vitro*. They found that a surface with approximately 15% –COOH groups was able to promote cell attachment and transfer however, it was stated that an unstable plasma polymer surface was required to promote cell transfer. This unstable surface was generated by increasing the flow rate of acrylic acid whilst having a low power plasma. This resulted in a surface with less cross-linking and hence generated an unstable surface.

An innovative application of plasma polymers in cell delivery has recently been reported by Di Girolamo and colleagues [80]. Here the authors have taken a commercial extended wear contact lens (Ciba Vision Focus® Night & Day™) as substrate for expansion *ex vivo* of either limbal or conjunctival biopsies from three patients. Once the biopsies had reached confluence ($\sim$10 days) the contact lens with cell sheet was placed onto the patient's cornea, having first performed a superficial keratectomy to remove the abnormal epithelium. In each of the three cases reported the patient developed normal transparent corneal epithelium. On follow-up no recurrence of either conjunctivalization or corneal vascularization was found. This application is of extreme interest since the original reasoning behind the plasma polymer coating of the contact lens was to improve surface wettability and reduce the degree of lipid fouling due to the material's inherent hydrophobicity [81]. The Focus® Night & Day™ extended wear contact lens is an example of a new generation of high permeability silicone hydrogel contact lens, using monomers containing silicon and fluorine components and combining it with conventional hydrogel monomers. This generates a lens that can transmit very high levels of oxygen to the extent that the lenses can be worn for extended periods of time without removal [82]. The material enables high oxygen permeability and fluid transport but the biological–biomaterial interactions are poor due to the material's inherent hydrophobicity. To overcome these issues the lens surface is coated with a proprietary

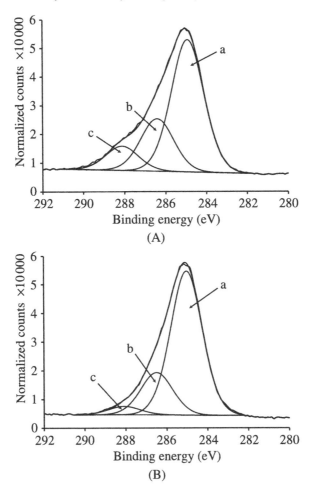

Figure 6. High resolution XPS spectra of C_{1s} peak of (A) Ciba Vision Focus® Night & Day™ contact lens and (B) allylamine plasma polymer where (a) represents $\underline{C}$–H, (b) represents $\underline{C}$–OH/R, $\underline{C}$–NR$_2$, $\underline{C}$=N and (c) represents $\underline{C}$=O, N–$\underline{C}$–O. A comparison indicates the two surfaces, although not identical, are similar in chemistry.

plasma polymer which, from the patent literature, is suspected to be similar to the one deposited from an amine-containing monomer (Fig. 6) [83]. Once again the mechanism(s) by which plasma polymer surfaces are able to support both *in vitro* cell expansion and *in vivo* cell delivery have yet to be fully elucidated.

4. Further Development of Technology

4.1. Improved Culture Surfaces

The use of plasma technology to deposit the cell attachment layer opens the scope to further engineer these surfaces to enhance both initial cell attachment and improve the quality of cell culture. The provision of surface functionalities affords the opportunity to immobilize biomolecules and a range of chemistries that have been described by Griesser and co-workers [1]. Perhaps the simplest is the use of surface amine groups using carbodiimide chemistry. However, carboxylic acid groups can also be used to immobilise biomolecules through accessible primary amines (in the biomolecule). Carbodiimide chemistry has been used to immobilise a peptide involved in cell attachment; the amino acid sequence Arg-Asp-Ser (RGD) [84, 85], as well as key biomolecules, such as extra cellular matrix molecules [86–88] and enzymes [89]. Recently, it has been reported how surface carboxylic acid groups may be used to immobilize the 9E10 antibody covalently. This antibody captures (any) myc-tagged biomolecule and in [90] the 9E10 antibody is used to immobilize the myc-tagged intercellular signalling molecule delta-like-1 (Dll1). The importance of this molecule is that it regulates (inhibits) cell differentiation and, by identifying a critical surface density of Dll1, it is possible to fabricate a culture surface for stem cells that permits cell expansion, without differentiation.

However, biomolecules do not necessarily need to be surface tethered. The plasma-polymerized layer may be used to 'present' a passively adsorbed biomolecule in an optimal conformation, for subsequent cell attachment. For example, the passive adsorption of vitronectin, collagen and immunoglobulin-G from single protein solutions [91] and heparin [92] onto plasma-copolymerized surfaces is described. The advantages of this approach are that only a very small amount of the (potentially) very expensive biomolecule is used; many of these biomolecules may be obtained from non-animal sources (e.g., recombinant proteins). Although more speculative, we could readily envisage the immobilization of small biomolecules within the depositing plasma polymer [93, 94]. This approach uses atmospheric plasma-deposition and the simultaneous atomization of the biomolecule into the plasma gas. As Heyse *et al.* show [95] from the intact incorporation of a green fluorescent protein (GFP) into plasma-polymerized coatings, it is possible by this approach to encapsulate 'intact' molecules within a growing plasma deposit.

A further advantage of plasma in the provision of enhanced culture surfaces is the ability to readily pattern the deposited layers, so different chemistries may be deposited on a single surface [96] whereby each surface region may be tailored to a specific cell type. A common problem, as described by MacNeil, is that in the co-cultures of two cell types (e.g., keratinocytes with melanocytes) one cell may readily overgrow the other [67]. Plasma patterning offers a potential route to overcome this problem, as demonstrated in [97] and later [98] which show single cell type island cultures that could readily be extended to co-cultures. Co-culture affords us the opportunity to return to the patient more than a single cell type and in certain cases one cell type is required to support the other.

Following the development of a clinically proven plasma polymer based delivery vehicle for keratinocytes MacNeil and colleagues have sought to refine the culture protocol further [99]. Seeking a xenobiotic-free protocol, this group examined keratinocyte co-culture with murine and human dermal fibroblasts and a fetal lung fibroblast cell line MRC-5 on plasma-polymerized acid-functionalized surfaces. Irradiated human and murine fibroblasts both performed equally well in supporting the initial keratinocyte expansion both with complete and serum-free media. Interestingly, keratinocyte differentiation was significantly less when grown serum-free with fibroblasts than when grown with serum. Delivery of keratinocytes from co-cultures to the human dermal wound bed model was as efficient as previously demonstrated in a study of monoculture keratinocyte transfer from serum and serum-free cultures [21].

Further research by MacNeil and colleagues examined the co-culture of keratinocytes and melanocytes cultured in complete and serum-free media [100]. A range of plasma polymer surfaces, produced from acrylic acid, allylamine or co-polymerization of both, supported the attachment and proliferation of both cell types. No preference to plasma polymer was found however, co-culture proliferation was superior to individual cultures. These studies are being further extended [101] and also pursued by other research teams [102].

4.2. Improved Delivery Surfaces

The mode of cell delivery from plasma-polymerized surfaces remains a matter of conjecture. It is likely that the highly acid-functionalized surface currently employed to deliver keratinocytes in the commercially available product Myskin™ is not particularly stable in the harsh environment of a wound bed, and the dissolution of this coating provides a gentle release mechanism of the cells. However, in circumstances where an intact monolayer of cells is required, a more sophisticated mechanism of delivery is required. In this instance something akin to a thermoresponsive polyNiPAAm surface which, under the action of temperature can release a cell monolayer, might function better [103]. The delivery of cell sheets utilizing cultureware coated with polyNiPAAm has been demonstrated for keratinocyte [104] and corneal epithelium [105] in addition to urothelium [106] and periodontal ligaments [107]. We have shown [108] that plasma can be used to deposit a functional plasma polymer (pp) NiPAAm layer that responds in the appropriate temperature range. The advantage of this approach is that it might facilitate the delivery of cell sheet layers, allowing (by successful deliveries) a thicker 'tissue-like' construct to be fabricated either *in vitro* or *in vivo*. *In vitro* demonstrated applications include the engineering of pulsatile cardiac tissue [109] and the potential for smooth muscle and more complex structures exists.

5. Conclusions

It can be seen that the use of plasma-polymerized surfaces for cell attachment and delivery is a maturing technology. From the fundamental application of these surfaces for single cell types such as in the case of keratinocytes the field has progressed. The delivery of different cell types, their use in the support of co-cultures, the development of xenobiotic and serum-free culture protocols and the immobilization of active biomolecules continue toward clinical application. Whilst the potential for plasma-polymerized, thermoresponsive surfaces and atmospheric plasma polymers and the simultaneous incorporation of active biomolecules remains to be fully explored, the future for plasma polymer surfaces is encouraging and a place for this technology in cell therapy is assured.

Acknowledgements

Prof. Rob Short and Dr. David Steele gratefully acknowledge Prof. Shelia MacNeil for her introduction to the field of cell therapy and Ms. Hollie Wickstein for her assistance in preparation of this manuscript.

References

1. K. S. Siow, L. Britcher, S. Kumar and H. J. Griesser, *Plasma Process. Polymers* **3**, 392 (2006).
2. B. D. Ratner, *J. Biomater. Sci.-Polym. Ed.* **4**, 3 (1993).
3. H. J. Griesser, R. C. Chatelier, T. R. Gengenbach, G. Johnson and J. G. Steele, *J. Biomater. Sci.-Polym. Ed.* **5**, 531 (1994).
4. C. M. Chan, T. M. Ko and H. Hiraoka, *Surface Sci. Reports* **24**, 3 (1996).
5. R. Daw, I. M. Brook, A. J. Devlin, R. D. Short, E. Cooper and G. J. Leggett, *J. Mater. Chem.* **8**, 2583 (1998).
6. L. P. Tang, Y. L. Wu and R. B. Timmons, *J. Biomed. Mater. Res.* **42**, 156 (1998).
7. M. C. Shen, L. Martinson, M. S. Wagner, D. G. Castner, B. D. Ratner and T. A. Horbett, *J. Biomater. Sci.-Polym. Ed.* **13**, 367 (2002).
8. J. J. A. Barry, M. Silva, K. M. Shakesheff, S. M. Howdle and M. R. Alexander, *Adv. Funct. Mater.* **15**, 1134 (2005).
9. R. Förch, A. N. Chifen, A. Bousquet, H. L. Khor, M. Jungblut, L. Q. Chu, Z. Zhang, I. Osey-Mensah, E. K. Sinner and W. Knoll, *Chem. Vapor Deposition* **13**, 280 (2007).
10. S. Kamath, D. Bhattacharyya, C. Padukudru, R. B. Timmons and L. P. Tang, *J. Biomed. Mater. Res.* **86A**, 617 (2008).
11. M. Zelzer, R. Majani, J. W. Bradley, F. Rose, M. C. Davies and M. R. Alexander, *Biomaterials* **29**, 172 (2008).
12. R. Vasita, K. Shanmugam and D. S. Katti, *Current Topics Medicinal Chem.* **8**, 341 (2008).
13. D. F. Williams, *J. Biomed. Eng.* **11**, 185 (1989).
14. D. F. Williams, *Biomaterials* **29**, 2941 (2008).
15. D. F. Williams, *J. Mat. Sci.* **22**, 3421 (1987).
16. D. B. Haddow, S. MacNeil and R. D. Short, *Plasma Process. and Polymers* **3**, 419 (2006).
17. B. D. Ratner, A. S. Hoffman, F. J. Schoen and J. E. Lemons, *Biomaterials Science: An Introduction to Materials in Medicine*, 2nd edn. Elsevier, Amsterdam (2004).
18. D. Steele and R. Short, in: *New Industrial Plasma Technology from the 3rd Varenna School, Italy*, H. Ikegami, N. Sato, A. Matsuda, M. Kuzuya, A. Mizuno, K. Uchino and Y. Kawai (Eds). Wiley-VCH (2009).
19. R. Lanza, R. Langer and J. Vacanti, *Principles of Tissue Engineering*, 3rd edn. Elsevier Academic Press, New York (2007).
20. A. Atala, S. B. Bauer, S. Soker, J. J. Yoo and A. B. Retik, *Lancet* **367**, 1241 (2006).

21. D. B. Haddow, D. A. Steele, R. D. Short, R. A. Dawson and S. MacNeil, *J. Biomed. Mater. Res.* **64A**, 80 (2003).

22. P. Deshpande, M. Notara, N. Bullett, J. T. Daniels, D. Haddow and S. MacNeil, *Tissue Eng. Part A* **15**, 1 (2009).

23. T. Nakamura, K.-I. Endo and L. J. Cooper, *Invest. Ophthalmol. Visual Sci.* **44**, 106 (2003).

24. H. Green, O. Kehinde and J. Thomas, *Proc. Natl Acad. Sci. USA* **76**, 5665 (1979).

25. N. O'Connor, J. Mulliken and S. Banks-Schlegel, *Lancet* **317**, 75 (1981).

26. R. Freshney, *Culture of Animal Cells: A Manual of Basic Technique.* Hoboken, New Jersey (2005).

27. N. Mohandas, R. M. Hochmuth and E. E. Spaeth, *J. Biomed. Mater. Res.* **8**, 119 (1974).

28. S. K. Chang, O. S. Hum, M. A. Moscarello, A. W. Neumann, W. Zingg, M. J. Leutheusser and B. Ruegsegger, *Med. Progr. Technol.* **5**, 57 (1977).

29. H. Yasuda, B. S. Yamanashi and D. P. Devito, *J. Biomed. Mater. Res.* **12**, 701 (1978).

30. A. W. Neumann, M. A. Moscarello, W. Zingg, O. S. Hum and S. K. Chang, *J. Polym. Sci.: Polym. Symp.* **66**, 391 (1979).

31. A. Bruil, L. M. Brenneisen, J. G. A. Terlingen, T. Beugeling, W. G. Vanaken and J. Feijen, *J. Colloid Interface Sci.* **165**, 72 (1994).

32. B. D. Ratner, T. Horbett, A. S. Hoffman and S. D. Hauschka, *J. Biomed. Mater. Res.* **9**, 407 (1975).

33. P. K. Chu, J. Y. Chen, L. P. Wang and N. Huang, *Mater. Sci. Eng. R-Reports* **36**, 143 (2002).

34. R. Dawson, N. Goberdhan, E. Freedlander and S. MacNeil, *Burns* **22**, 93 (1996).

35. H. Rennekampff, V. Kiessig and J. Hansbrough, *J. Surgical Res.* **62**, 288 (1996).

36. K. Smetana, J. Vacik and B. Dvorankova, in: *Proc. 36th Microsymposium on Macromolecules — High-Swelling Gels*, Huthig & Wepf Verlag, Prague, Czech Republic (1995).

37. M. D. Harriger, G. D. Warden, D. G. Greenhalgh, R. J. Kagan and S. T. Boyce, *Transplantation.* **59**, 702 (1995).

38. B. Dvorankova, K. Smetana, R. Konigova, H. Singerova, J. Vack, M. Jelnkova, Z. Kapounkova and M. Zahradnk, *Biomaterials* **19**, 141 (1998).

39. J. G. Steele, B. A. Dalton, G. Johnson and P. A. Underwood, *J. Biomed. Mater. Res.* **27**, 927 (1993).

40. K. Alitalo, E. Kuismanen, R. Myllyla, U. Kriistala, S. Asko-Seljavaara and A. Vaheri, *J. Cell Biol.* **94**, 497 (1982).

41. H.-S. Kim, X. Song, C. D. Paiva, C. Zhuo, S. C. Pflugfelder and L. De-Quan, *Exptl. Eye Res.* **79**, 41 (2004).

42. J. Zieske, V. Mason, M. Wasson, S. F. Meunier, C. J. Nolte, N. Fukai, B. Olsen and N. L. Parenteau, *Exptl. Cell Res.* **214**, 621 (1994).

43. T. Nakamura, K.-I. Endo, L. Cooper, N. J. Fullwood, N. Tanifuji, M. Tsuzuki, N. Koizumi, T. Inatomi, Y. Sano and S. Kinoshita, *Invest. Ophthalmol. Vision Sci.* **44**, 106 (2003).

44. B. Conley, J. Young, A. Trounson and R. Mollard, *Int. J. Biochem. Cell Biol.* **36**, 555 (2004).

45. S. Oh, H. Kim, Y. Park, H. W. Seol, Y. Y. Kim, M. S. Cho, S. Y. Ku, Y. M. Choi, D.-W. Kim and S. Y. Moon, *Stem Cells* **23**, 605 (2005).

46. R. M. France, R. D. Short, R. A. Dawson and S. MacNeil, *J. Mater. Chem.* **8**, 37 (1998).

47. R. M. France and R. D. Short, *J. Chem. Soc.-Faraday Trans.* **93**, 3173 (1997).

48. T. Sun, M. Higham, C. Layton, J. Haycock, R. Short and S. MacNeil, *Wound Repair Regeneration* **12**, 626 (2004).

49. N. Koizumi, L. Cooper, N. Fullwood, T. Nakamura, K. Inoki, M. Tsuzuki and S. Kinoshita, *Invest. Ophthalmol. Vision Sci.* **43**, 2114 (2002).

50. M. Griffith, R. Osborne, R. Munger, X. Xiong, C. Doillon, N. Laycock, M. Hakim, Y. Song and M. Watsky, *Science* **286**, 2169 (1999).

51. R. J. F. Tsai and S. C. G. Tseng, *Cornea* **13**, 389 (1994).

52. T. Ohki, M. Yamato, D. Murakami, R. Takagi, J. Yang, H. Namiki, T. Okano and K. Takasaki, *Gut* **55**, 1704 (2006).

53. H. Yasuda, *Plasma Polymerization*. Academic Press, London (1985).

54. A. P. Kettle, A. J. Beck, L. O'Toole, F. R. Jones and R. D. Short, *Composites Sci. Technol.* **57**, 1023 (1997).

55. N. Lopattananon, A. P. Kettle, D. Tripathi, A. J. Beck, E. Duval, R. M. France, R. D. Short and F. R. Jones, *Composites Part A* **30**, 49 (1999).

56. A. J. Beck, F. R. Jones and R. D. Short, *Polymer* **37**, 5537 (1996).

57. R. J. Ward, *PhD Thesis*, Department of Chemistry, University of Durham (1985).

58. L. O'Toole, A. J. Beck, A. P. Ameen, F. R. Jones and R. D. Short, *J. Chem. Soc.-Faraday Trans.* **91**, 3907 (1995).

59. K. Vasilev, A. Michelmore, H. J. Griesser and R. Short, *Chem. Comm.* 3600 (2009).

60. R. Daw, S. Candan, A. J. Beck, A. J. Devlin, I. M. Brook, S. MacNeil, R. A. Dawson and R. D. Short, *Biomaterials* **19**, 1717 (1998).

61. J. D. Whittle, R. D. Short, C. W. I. Douglas and J. Davies, *Chem. Mater.* **12**, 2664 (2000).

62. R. M. France, R. D. Short, E. Duval, F. R. Jones, R. A. Dawson and S. MacNeil, *Chem. Mater.* **10**, 1176 (1998).

63. D. B. Haddow, R. M. France, R. D. Short, S. MacNeil, R. A. Dawson, G. J. Leggett and E. Cooper, *J. Biomed. Mater. Res.* **47**, 379 (1999).

64. M. Moustafa, C. Simpson, M. Glover, R. A. Dawson, S. Tesfaye, F. M. Creagh, D. Haddow, R. Short, S. Heller and S. MacNeil, *Diabetic Medicine* **21**, 786 (2004).

65. N. Zhu, R. M. Warner, C. Simpson, M. Glover, C. Hernon, J. Kelly, S. Fraser, M. Brotherston, D. Ralston and S. MacNeil, *European J. Plastic Surgery* **28**, 319 (2005).

66. M. Moustafa, A. J. Bullock, T. M. Creagh, S. Heller, W. Jeffcoate, F. Game, C. Amery, S. Tesfaye, Z. Ince, D. B. Haddow and S. MacNeil, *Regenerative Medicine* **2**, 887 (2007).

67. J. Phillips, D. J. Gawkrodger, C. M. Caddy, S. Hedley, R. A. Dawson, L. Smith-Thomas, E. Freedlander and S. MacNeil, *Pigment Cell Res.* **14**, 116 (2001).

68. P. C. Eves, N. A. Bullett, D. Haddow, A. J. Beck, C. Layton, L. Way, A. G. Shard, D. J. Gawkrodger and S. MacNeil, *J. Invest. Dermatol.* **128**, 1554 (2008).

69. P. C. Eves, A. J. Beck, A. G. Shard and S. MacNeil, *Biomaterials* **26**, 7068 (2005).

70. V. S. Sangwan, *Bioscience Reports* **21**, 385 (2001).

71. J. T. Daniels, J. K. G. Dart, S. J. Tuft and P. T. Khaw, *Wound Repair Regeneration* **9**, 483 (2001).

72. R.-F. Tsai, L. M. Li and J. K. Chen, *New England J. Medicine* **343**, 86 (2000).

73. S. Kolli, M. Lako, F. Figueiredo, H. Mudhar and S. Ahmad, *Regenerative Medicine* **3**, 329 (2008).

74. W. Liu, K. Merrett, M. Griffith, P. Fagerholm, S. Dravida, B. Heyne, J. C. Scaiano, M. Watsky, N. Shinozaki, N. Lagali, R. Munger and F. Li, *Biomaterials* **29**, 1147 (2008).

75. K. Nishida, M. Yamato, Y. Hayashida, K. Watanabe, N. Maeda, H. Watanabe, K. Yamamoto, S. Nagai, A. Kikuchi, Y. Tano and T. Okano, *Transplantation* **77**, 379 (2004).

76. Y. Hayashida, K. Nishida, M. Yamato, K. Watanabe, N. Maeda, H. Watanabe, A. Kikuchi, T. Okano and Y. Tano, *Invest. Ophthalmol. Visual Sci.* **46**, 1632 (2005).

77. M. Notara, N. A. Bullett, P. Deshpande, D. B. Haddow, S. MacNeil and J. T. Daniels, *J. Mater. Sci.-Mater. Medicine* **18**, 329 (2007).

78. P. Deshpande, N. Bullett, M. Notara, J. T. Daniels, D. B. Haddow and S. MacNeil, *Proc. Conference of the International Society of Tissue-Engineering and Regenerative Medicine*, London, 1773 (2007).

79. P. Deshpande, N. Bullett, M. Notara, J. T. Daniels, D. B. Haddow and S. MacNeil, *Tissue Eng.: Part A* **15**, 2889 (2009).

80. N. D. Girolamo, M. Bosch, K. Zamora, M. Coroneo, D. Wakefield and S. Watson, *Transplantation* **87**, 1571 (2009).

81. F. P. Carney, W. L. Nash and K. B. Sentell, *Invest. Ophthalmol. Visual Sci.* **49**, 120 (2008).

82. P. C. Nicolson and J. Vogt, *Biomaterials* **22**, 3273 (2001).

83. R. C. Chatelier, L. Dai, H. J. Griesser, S. Li, P. Zientek, D. Lohmann and P. Chabrecek, *CIBA-Geigy AG and Commonwealth Scientific and Industrial Research Organization*, EP patent no. 1993/002420 and WO patent no. 1994/006485 (2004).

84. Y. Ito, M. Kajihara and Y. Imanishi, *J. Biomed. Mater. Res.* **25**, 1325 (1991).

85. T. I. Valdes, W. Ciridon, B. D. Ratner and J. D. Bryers, *Biomaterials* **29**, 1356 (2008).

86. Y. Kinoshita, T. Kuzuhara, M. Kirigakubo, M. Kobayashi, K. Shimura and Y. Ikada, *Biomaterials* **14**, 209 (1993).

87. Y. Kinoshita, T. Kuzuhara, M. Kirigakubo, M. Kobayashi, K. Shimura and Y. Ikada, *Biomaterials* **14**, 546 (1993).

88. A. Kishida, Y. Ueno, N. Fukudome, E. Yashima, I. Maruyama and M. Akashi, *Biomaterials* **15**, 848 (1994).

89. U. Konig, M. Nitschke, A. Menning, C. Sperling, F. Simon, C. Arnhold, C. Werner and H. J. Jacobasch, *Surface Coatings Technol.* **116**, 1011 (1999).

90. R. A. Walker, V. T. Cunliffe, J. D. Whittle, D. A. Steele and R. D. Short, *Langmuir* **25**, 4243 (2009).

91. J. D. Whittle, N. A. Bullett, R. D. Short, C. W. I. Douglas, A. P. Hollander and J. Davies, *J. Mater. Chem.* **12**, 2726 (2002).

92. D. E. Robinson, A. Marson, R. D. Short, D. J. Buttle, A. J. Day, K. L. Parry, M. Wiles, P. Highfield, A. Mistry and J. D. Whittle, *Adv. Mater.* **20**, 1166 (2008).

93. P. Heyse, R. Dams, S. Paulussen, K. Houthofd, K. Janssen, P. A. Jacobs and B. F. Sels, *Plasma Process. Polymers* **4**, 145 (2007).

94. M. G. Ortore, R. Sinibaldi, P. Heyse, S. Paulussen, S. Bernstorff, B. Sels, P. Mariani, F. Rustichelli and F. Spinozzi, *Appl. Surface Sci.* **254**, 5557 (2008).

95. P. Heyse, M. B. J. Roeffaers, S. Paulussen, J. Hofkens, P. A. Jacobs and B. F. Sels, *Plasma Process. Polymers* **5**, 186 (2008).

96. N. A. Bullett, R. D. Short, T. O'Leary, A. J. Beck, C. W. I. Douglas, M. Cambray-Deakin, I. W. Fletcher, A. Roberts and C. Blomfield, *Surface Interface Anal.* **31**, 1074 (2001).

97. E. Szili, H. Thissen, J. P. Hayes and N. Voelcker, *Biosensors & Bioelectronics* **19**, 1395 (2004).

98. A. L. Hook, H. Thissen, J. P. Hayes and N. H. Voelcker, *Biosensors & Bioelectronics* **21**, 2137 (2006).

99. A. J. Bullock, M. C. Higham and S. MacNeil, *Tissue Eng.* **12**, 245 (2006).

100. P. Eves, A. J. Beck, A. G. Shard and S. MacNeil, *British J. Dermatology* **152**, 856 (2005).

101. S. Rimmer, C. Johnson, B. Zhao, J. Collier, L. Gilmore, S. Sabnis, P. Wyman, C. Sammon, N. J. Fullwood and S. MacNeil, *Biomaterials* **28**, 5319 (2007).

102. C. J. Kirkpatrick, S. Fuchs, M. I. Hermanns, K. Peters and R. E. Unger, *Biomaterials* **28**, 5193 (2007).

103. T. Matsuda, *J. Biomater. Sci.-Polym. Ed.* **15**, 947 (2004).

104. M. Yamato, M. Utsumi, A. Kushida, C. Konno, A. Kikuchi and T. Okano, *Tissue Eng.* **7**, 473 (2001).
105. K. Nishida, M. Yamato, Y. Hayashida, K. Watanabe, N. Maeda, H. Watanabe, K. Yamamoto, S. Nagai, A. Kikuchi, Y. Tano and T. Okano, *Transplantation* **77**, 379 (2004).
106. Y. Shiroyanagi, M. Yamato, Y. Yamazaki, H. Toma and T. Okano, *Tissue Eng.* **9**, 1005 (2003).
107. M. Hasegawa, M. Yamato, A. Kikuchi, T. Okano and I. Ishikawa, *Tissue Eng.* **11**, 469 (2005).
108. N. A. Bullett, R. A. Talib, R. D. Short, S. L. McArthur and A. G. Shard, *Surface Interface Anal.* **38**, 1109 (2006).
109. T. Shimizu, M. Yamato, Y. Isoi, T. Akutsu, T. Setomaru, K. Abe, A. Kikuchi, M. Umezu and T. Okano, *Circulation Res.* **90**, E40 (2002).

Adhesion of Cells and Tissues to Bioabsorbable Polymeric Materials: Scaffolds, Surgical Tissue Adhesives and Anti-adhesive Materials

Shuko Suzuki * and **Yoshito Ikada**

Faculty of Medicine, Nara Medical University, Shijo-cho 840, Kashihara-shi, Nara 634-8521, Japan

Abstract

Interaction between the body and implanted materials is a very complex phenomenon, and understanding the concept of cell and tissue adhesion, which is the first event that occurs when a biomaterial is implanted, is of great importance. Currently, bioabsorbable materials are preferably used as biomaterials. They are only present in the body temporary, and promote guided cell and tissue growths for the required time. Hence, the choice of the material, depending on the application, becomes crucial. This article offers an overview of cell and tissue adhesion to three types of clinically-important bioabsorbable materials from a materials perspective: scaffolds, surgical tissue adhesives, and anti-adhesive materials.

Keywords

Bioabsorbable polymers, cell and tissue adhesion, scaffolds, surgical tissue adhesives, anti-adhesive materials

1. Introduction

A biomaterial is defined as a material used to replace part of a defective living system in close contact with the living system. Biomaterials are, therefore, constantly interacting with tissue or biological fluids. The biocompatibility required for biomaterials is associated with the bulk and surface characteristics of materials. Bulk properties involve chemical composition, mechanical properties, and design of the implants, whereas surface properties involve the material/tissue interface. The optimal surface property definitely depends on the specific application. There are many materials that meet their bulk biocompatibility requirements; however, not many of them possess ideal surface properties. Cell and tissue adhesion is one of the first events that occurs at the biomaterial surface, which leads to cell proliferation and differentiation. Hence, this determines the host responses towards the implant material.

---

* To whom correspondence should be addressed. E-mail: suzukis@naramed-u.ac.jp

Surface and Interfacial Aspects of Cell Adhesion
© Koninklijke Brill NV, Leiden, 2010

Currently bioabsorbable polymeric materials are preferably used due to the many advantages over non-absorbable materials, such as no necessity for secondary operation to remove the materials and long-term biocompatibility. Three types of clinically-important bioabsorbable polymers, natural and snthetic, are scaffolds, surgical tissue adhesives and anti-adhesive materials. In tissue engineering, scaffolds serve as a mechanical support for cell growth and differentiation, to which cell adhesion is essential. Surgical tissue adhesives are used as sealers, hemostatic agents, and adhesives, and are required to actively adhere to tissues to which these are applied. Anti-adhesive materials, on the other hand, are designed to prevent cell and tissue adhesion and work as barrier material in order to prevent post-operative adhesion. This article focuses on cell and tissue adhesion to bioabsorbable materials from the materials perspective. For more specific information on cell and tissue interaction with material surfaces, readers are referred to other articles in this Special Issue (e.g., "Proteins, water, and the initial attachment of mammalian cells to biomedical surfaces" P. Parthi, A. Golas and E. A. Vogler, "How substrate properties control cell adhesion. A physical–chemical approach" A. Carre and V. Lacarriere, "Cell/material Interfaces: influence of surface chemistry and surface topography on cell adhesion" K. Anselme *et al.*).

2. Basic Concepts of Cell and Tissue Adhesion to Biomaterials

When a biomaterial is implanted in the body, a cascade of biological events occurs. The host tissue response is induced by bulk and surface properties of the material in contact with the tissue. If the body recognizes the surface as a foreign body, fibrous encapsulation occurs which isolates the biomaterial from the nearest biological environment, occasionally leading to rejection of the implant. In the case of bioabsorbable materials, long-term biocompatibility is not necessary as they degrade and finally disappear from the body. In this section, normal wound healing processes and tissue reactions to biomaterials are briefly discussed.

2.1. Wound Healing Process

The natural wound healing process accompanies many changes in both the type and number of cells present at the site of soft tissue reparation [1]. The three processes that follow the injury are hemostasis, inflammation, and wound healing. The first stage (hemostasis) happens in a few minutes when blood platelets fill the injury site forming a temporary shield from the environment.

This is followed by the inflammation stage. Two important types of leukocyte cells (white blood cells) participate in this step. They are neutrophils and macrophages that migrate to the wound. Neutrophils kill the invading bacteria, while macrophages remove cellular and foreign debris from the wound site. New blood vessels are formed at the injury site. Fibroblasts then migrate and synthesize extracellular substances, such as collagen and glycosaminoglycans, which fill the wound and form a scar tissue.

The difference in the healing mechanism between hard and soft tissues is the unique ability of the hard tissue to regenerate rather than scarring [2]. Decalcification, resorption, and remodelling of necrotic bone occur to regenerate a new hard tissue [1]. Fibroblasts and osteogenic cells migrate and proliferate toward the injured site. Osteoblasts are evolved from the osteogenic cells near the bone and start to calcify the callus into the trabecular bone.

When a biomaterial is implanted, a similar wound healing process may occur. Inflammation, wound healing, and foreign-body response, as well as fibrous encapsulation, are generally considered as the typical host biological responses to the implant [3]. Inflammation is an essential biological occurrence at the implantation site before healing can begin. The normal reaction of biological systems to an introduced material is to biodegrade it or to attempt its digestion. Upon failing this, the alternative is encapsulation of the introduced material within a fibrous collagen capsule [3]. This foreign-body reaction involves foreign-body giant cells in addition to macrophages and fibroblasts that are already present at the wound site.

Once a fibrous capsule is formed, the biomaterial is isolated from the body. Vascularisation cannot occur inside the capsule; therefore, no cell growth is possible. If bacterial infection occurs inside this fibrous capsule, the infection may be prolonged as macrophages are unable to access the interior of the capsule and do the job for which they are designed. Excessive fibrous growth can result in tissue damage and ultimately in tissue death.

2.2. Reaction of Cell and Tissue to Biomaterials

When a material is implanted in the body, a series of events occur at the implant surface [4]. Firstly, water molecules come in contact with the implant material surface and form a water layer producing hydrated ions. The behaviour of water near the surface has been recognised to play an important role for the subsequent events [4, 5]. The various biomolecules, in particular proteins, come and interact with the hydrated surface. Initially, a range of proteins are adsorbed non-specifically on the surface. Then displacement by other proteins may occur. When cells arrive at the material surface *in vivo*, they encounter the adsorbed protein layer. If the proteins undergo denaturation, the cells will not attach or may even activate a macrophage response if they recognise the material as a foreign substance. This can lead to fibrous encapsulation. Therefore, cell and tissue interaction with the material surface is mediated by the adsorbed proteins [6, 7]. The types, amounts, and conformations of adsorbed proteins on the material surface determine the cell behaviour. Protein adsorption, on the other hand, is directly affected by the material surface properties. Figure 1 represents the cell adhesion dependence on the surface hydrophobicity/hydrophilicity balance observed using different types of materials [8]. The cell adhesion is enhanced if the material surface is intermediately hydrophobic or hydrophilic (water contact angle $\sim 70°$) [8]. Moderately hydrophobic surfaces are known to induce strong irreversible adsorption of proteins. This denatures their conformation and hence causes a loss in bioactivity. Very hydrophilic surfaces, on

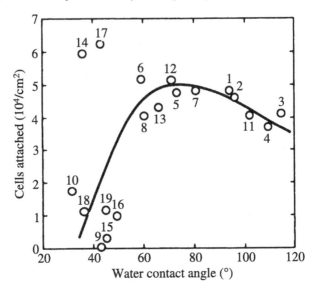

Figure 1. Effect of the water contact angle of surfaces on L cell adhesion at 37°C for 1 h. Film materials: 1, polyethylene; 2, polypropylene; 3, poly(tetrafluoroethylene); 4, tetrafluoroethylene-hexafluoropropylene copolymer; 5, poly(ethylene terephthalate); 6, PMMA; 7, nylon 6; 8, vinyl alcohol-ethylene copolymer; 9, poly(vinyl alcohol); 10, cellulose; 11, silicone; 12, polystyrene; 13, commercial plastic sheet for cell culture; 14, glass; 15, polyacrylamide-grafted polyethylene; 16, poly(acrylic acid)-grafted polyethylene; 17, fibrinogen-grafted polyethylene; 18, gelatin-grafted polyethylene; 19, albumin-grafted polyethylene (reproduced from [8]).

the other hand, inhibit protein adsorption [8]. As one example, grafting of water-soluble poly(ethylene oxide) (PEO) has been shown to decrease the interfacial free energy and the steric repulsion force between the PEO chains and proteins and it has been used for vascular graft applications [9].

Functional groups are often introduced to increase hydrophilicity, but electrostatic interaction can also occur between the opposite charges of the protein and the material surface [10, 11]. This is a very important interaction since amino acid residues of a protein are charged at any given pH. Kato *et al.* investigated protein adsorption onto surfaces with ionic and non-ionic groups and showed that protein adsorption onto the charged surfaces was governed by the electrostatic interaction [11]. The effect of functional groups on protein adsorption and endothelial cell growth was investigated using self-assembled monolayers (SAMs) of alkanethiolates on gold [12]. It was found that cell growth increased in the following order $-CH_2OH < -COOCH_3 < -CH_3 \ll -COOH$. However, tissue culture polystyrene (TCPS) showed higher cell growth than SAMs. TCPS contains oxygen species (12%) in various polar groups (i.e., as hydroxyl, carbonyl, carboxylic acid, and esters), as well as nitrogen (0.56%) [13]. The results of such cell growth studies seem to indicate that multiple functionalities provide a pronounced synergistic effect [12]. However, it needs to be mentioned that cell differentiation is another important factor to consider in the case of cell and tissue reaction to biomaterials.

3. Scaffolds

A scaffold plays a pivotal role in tissue engineering. It functions as an artificial extracellular matrix (ECM) where cells can attach, grow, and synthesize their own matrix. A few basic requirements that a scaffold needs to fulfill are: (1) high porosity and proper pore size, (2) high surface area, (3) bioabsorbable, (4) mechanically strong, (5) non-toxic to cells, and (6) support cellular activities. Both natural and synthetic polymers have been widely applied as scaffolds due to their easy control over bioabsorbability and processability. Natural polymers, such as collagen, gelatin, and fibrin, degrade enzymatically, whereas commonly used synthetic polymers, such as poly(α-hydroxyacid)s, hydrogels such as PEO, and polyurethanes (PUs), hydrolytically degrade or dissolve into the body fluid. Scaffolds should support cellular activities such as adhesion, growth, migration, and differentiation function. In contrast to natural polymers, which contain cell adhesive domains, synthetic polymers often require surface modification to improve interaction with cells.

3.1. Natural Polymers

The sponge form of collagen, which is the main component of ECM, has been extensively investigated as scaffold due to its excellent biocompatibility, bioabsorbability, and porous structure [14]. The degradation rate of collagen can be altered by enzymatic pre-treatment or cross-linking using various cross-linking techniques. Various tissues, especially soft tissues, have been regenerated by collagen scaffolds. Commercially available products include DuraGen (Integra, US) for repair of dura mater and Pelnac (Gunze, Japan) as graft for dermis defects. Despite many clinical successes, the main concerns which limit wide-spread usage of collagen are the potential pathogen transmission and immune reactions. Gelatin, which is denatured collagen, has less possibility of these concerns due to its processing methods (i.e., denaturation by alkaline or acid treatment). Because gelatin is water soluble and randomly coiled, high concentrations can be easily handled. Fibrin, which is the major component of blood clot, has been used as a gel for cell encapsulation. It has been reported that cells entrapped in the fibrin gel produced more collagen and elastin than those entrapped in collagen gel [15].

Naturally occurring polysaccharides and their derivatives, such as hyaluronate, alginate, chondroitin sulfate, chitosan and chitin, have been investigated for scaffold fabrication. Since these materials do not contain cell-adhesive domains in the molecules, they are used when minimal cell interaction is preferred, except for chitosan which contains large amounts of NH_2 groups.

3.2. Synthetic Polymers

Most extensively used synthetic polymers for scaffolds fabrication are poly(α-hydroxyacid)s, especially poly(lactic acid) (PLA), poly(glycolic acid) (PGA), and their copolymers (PLGA). A variety of techniques have been employed for processing three-dimensional porous scaffolds and have been reviewed [16, 17]. Although they have advantages over natural polymers such as easy control of bioabsorption rate and tunable mechanical properties, their surfaces are hydrophobic with contact angles higher than 70° and prevent cell and tissue adhesion. Without any treatments, these synthetic scaffolds do not absorb culture medium with the majority of their pores remaining empty. It is important to obtain a uniform distribution for initial cell seeding density throughout the scaffold volume for tissue regeneration [15]. Surface modification techniques have been often employed which will be discussed in the following section.

3.3. Modification of Scaffolds

Commonly utilized treatments for scaffolds to increase cell adhesion include prewetting with ethanol, hydrolysis with NaOH, oxidation with perchloric acid solution, oxygen or ammonia plasma discharge treatments, physical or chemical coating with hydrophilic polymers or cell adhesive proteins, and blending [15]. Blending of different polymers before scaffold fabrication can introduce new functional groups, but this will alter the bulk properties of the material. Plasma discharge treatment works better on two-dimensional surfaces, but uniform treatment is difficult for three-dimensional scaffolds which have surfaces inside the pores as well.

Alkaline or acid hydrolysis of these absorbable polymers increases hydrophilicity and hence higher cell adhesion can be achieved [18, 19]. Nam *et al.* investigated hepatocytes adhesion on PLGA (85:15) surfaces which were treated with NaOH [20]. They showed that the extent of adhesion was dependent on the NaOH treatment time. Smith *et al.* reported that etching of PLGA (50:50) scaffold with NaOH created nano-scale surface features [21]. This nano-topography reduced fibroblast adhesion but increased osteoblast adhesion. Aminolysis techniques have been used for modifying ester-containing polymers such as poly(3-hydroxybutyrate-co-3-hydroxyvalerate), poly(ε-caprolactone) (PCL), PLA and PUs [22–25]. In such treatments, the polymers are treated with solutions containing diamino compounds (e.g., diaminohexane). Despite their effectiveness, chemical methods have several disadvantages such as degradation of the bulk properties and decreases in molecular weight. It is also difficult to monitor and control the modification depth profile of the treated materials. In the case of PLGA, chemical treatments have been optimised for minimal degradation [23].

When cell adhesion proteins are coated on a biomaterial, proteins may be randomly folded upon adsorption on the surface, and as a result the receptor binding domains may not always be available. In addition, they may detach or be exchanged in biological environment. Covalent immobilization of proteins results in a substantial loss of activity due to the loss of active sites by bonding or reorientation of the biomolecules after bonding [26], and sterilization becomes more difficult [27]. On the other hand, covalent immobilization of short peptide sequences has been extensively investigated, as they are relatively more stable during modification process than large proteins [15]. Their surface density and orientation favorable for ligand-receptor interaction can be easily controlled. These short peptide sequences can be synthesized on a large scale in the laboratory more economically than whole

Table 1.
Cell adhesive peptides and ECM proteins which contain these peptides [28]

Peptide sequence	ECM protein
RGD	Fibronectin, vitronectin, laminin, collagen
YIGSR	Laminin
IKVAV	Laminin
REDV	Fibronectin
DGEA	Collagen
KQAGDV	Fibrinogen
VAPG	Elastin

proteins. Incorporation of these peptides in hydrogels, which have minimal cell adhesion capacity, will provide cellular interactive materials [28, 29].

Several peptides of ECM protein have been identified as specific and non-specific cell adhesive domains (Table 1). Among these, RGD (R: arginine, G: glycine, D: aspartic acid) sequence is the most widely used in tissue engineering, as it promotes cell adhesion and migration.

The RGD sequence has been immobilized onto collagen gels, fibrin gels and silk, and showed enhanced cell adhesion [30–32]. Alvarez-Barreto and Sikavitsas investigated the seeding of rat mesencymal stem cells on RGD-modified PLLA foams using oscillatory flow perfusion [33]. RGD dose dependence on the scaffold cellularity was observed. In bone tissue engineering, the RGD peptide could induce differentiation and growth of osteoblasts [34–36]. RGD-immobilized poly(α-hydroxyacid) scaffolds were also investigated for cartilage regeneration [37, 38].

Other examples of peptide immobilization onto biomaterials include YIGSR functionalized fluorinated ethylene propylene (FEP) [39], agarose hydrogel [40], IKVAV containing hyaluronic acid hydrogel [41] and KQAGDV and VAPG immobilized hydrogel [42]. Selective cell adhesion can also be achieved. For example, immobilized REDV interacts with endothelial cells, but not with platelets, fibroblast, or smooth-muscle cells [43] and KRSR promotes osteoblast adhesion but inhibits fibroblast adhesion [44].

For hard tissue applications, scaffolds composed of polymer and inorganic materials, such as micro- and nano-sized hydroxyapatite particles, have often been used. These composites have advantages of not only the improved mechanical properties but also enhanced protein adsorption that leads to osteoblasts adhesion and growth. Negatively charged groups are capable of chelating calcium ions that leads to calcium phosphate growth and bone bonding ability [45–48]. Incorporation of these groups on the scaffold surface is a popular strategy to improve the bone bonding ability. The surfaces containing carboxylic acid groups often require pre-treatment with $Ca(OH)_2$ or $CaCl_2$. An apatite layer or particles can be formed on the materials surface by immersing in the simulated body fluid (SBF) if functional groups are present on the surface, or by deposition from the highly concentrated SBF, and have been shown to improve the bone bonding ability [49, 50].

4. Surgical Tissue Adhesive Materials

Surgical adhesives have been widely used as sealing agents (to seal small holes in diseased soft tissues such as lung), hemostatic agents (to stop bleeding), and adhesives (to bond two separate tissues). They have many advantages over traditional techniques (i.e., sutures and staples) such as reducing operation times. The ideal tissue adhesive should fulfill the following requirements: (1) be liquid and rapidly curable when applied even in the presence of water, (2) be as pliable as the soft tissues when cured, and (3) be absorbable in the body. Three main types of such materials are available: naturally derived (such as fibrin glue), semi-synthetic (such as naturally derived polymer — aldehyde based), and synthetic (such as cyanoacrylates), and are briefly discussed in this section.

4.1. Naturally Derived Adhesives

4.1.1. Fibrin Glue
Fibrin glue is composed of human derived fibrinogen, thrombin, Factor XIII, and bovine aprotinin. The mechanism of fibrin clot formation from the solutions of fibrinogen (molecular weight of 360 kDa) and thrombin involves enzymatic and polymerization reactions as shown in Fig. 2.

Firstly, thrombin catalyzes partial hydrolysis of fibrinogen chains to form fibrin monomers. These monomers then react with one another in the presence of calcium ions to form polymers that eventually become three-dimensional gel network. Thrombin also activates Factor XIII into Factor XIIIa in the presence of calcium ions. Finally, Factor XIIIa covalently cross-links α and γ chains of different fibrin monomers in the matrix. The fibrin clot degrades by physiological fibrinolysis, which is initiated by a complex cascade of body's enzymes, in two weeks [51].

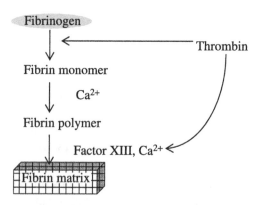

Figure 2. Mechanism of fibrin clot formation.

Fibrin glue has been commercially available in Europe and Japan for over 20 years, and in the USA since 1998, and is the most widely used tissue adhesive in a variety of surgical procedures. This has many advantages such as excellent biocompatibility, biodegradability, injectability, and *in situ*-producible scaffolds for tissue regeneration. Successful clinical applications of fibrin glue have been demonstrated in many fields, including cardiovascular surgery, neurosurgery, gastrointestinal surgery, vascular surgery, and plastic surgery [52–54].

There are, however, several disadvantages such as risk of virus infection due to human and animal blood derived products, low strength, and rapid loss in their strength [55]. In commercial products, careful donor selection, extensive plasma screening, and effective viral reduction steps during preparation have significantly reduced the possibility of viral transmission. No cases of serious viral transmission due to commercial fibrin glues have been reported [52]. However, the risk of virus infection is not completely free due to acquired immune deficiency and hepatitis.

Mussel-derived adhesive protein is another natural product which is known to be the most powerful natural adhesive that has flexibility and elasticity [56, 57]. In 1981, Waite and Tanzer discovered 3,4-dihydroxy-L-phenylalanine (DOPA) as a key component for wet-resistant adhesion in mussel adhesive proteins [58]. Non-toxic and non-immunogenic effects of mussel proteins have been reported [59–61]. More interestingly, these maintain their adhesion in wet environment, and adhere virtually to any type of synthetic and natural surfaces [62]. However, uneconomical extraction, which requires 10 000 mussels to obtain 1 g of one type of adhesive protein, and unsuccessful large scale production have limited its practical applications [56].

4.2. Semi-Synthetic Adhesives

4.2.1. Albumin-Based Glue
In BioGlue (BioGlue, CryoLife, Inc., Kennesaw, Georgia, USA), stoichiometrically equivalent doses of bovine serum albumin (45% solution) and glutaraldehyde (10% solution) are mixed in a custom cartilage delivery system [63]. It was approved by US Food and Drug Administration (FDA) in 1999 for use during surgical repair of acute thoracic aortic dissection, and in 2001 approved for general use as a hemostatic adjunct in cardiac and vascular surgery. It sets to a gel within 20–30 s and establishes a firm adhesive bond due to the reactive glutaraldehyde which covalently cross-links the albumin molecules to each other, as well as to the tissue proteins at the application site. In a recent study, Zehr reported detailed application techniques for the use of BioGlue in a variety of cardiovascular surgical cases and found that BioGlue was safe and effective when used properly [64].

However, the toxicity of unreacted and leached glutaraldehyde and dense postoperative gel structure have been a concern and have limited its application. Fürst and Banerjee have investigated the release of glutaraldehyde from BioGlue and its toxicity both *in vitro* and *in vivo* [65]. It was found that 100–200 µg/ml of glutaraldehyde was released from 1 ml of gelled BioGlue into 5 ml of saline, and the supernatant showed cytotoxicity to cultured human embryo fibroblasts (MRC5) and mouse myoblasts (C1C12). The *in vivo* animal study using rabbits showed that the application of BioGlue caused serious adverse effects to lung and liver tissues, whereas aortic tissues showed low to medium-grade inflammation. The reason for high resistance of aortic tissues was suggested that adventitial layer was less damaged during the clinical application on aortic dissections in contrast to the raw tissue surfaces of lung and liver, because the aortic tissue is composed of large amounts of ECM and low cell contents. Other *in vivo* studies also showed toxic effects of BioGlue to lung and nerve tissues, and low inflammation to the aortas. A high infection rate (13%) was observed in case of pediatric craniotomy and laminectomy when BioGlue was used [66]. It was suggested that extensive inflammatory response caused by BioGlue leads to expansion and weakening of the surrounding tissue, followed by secondary infection by bacterial growth.

4.2.2. Gelatin-Based Glue

Gelatin-resorcinol-formaldehyde (GRF glue, Cardial, Techinopole, Sainte-Etienne, France) is composed of two types of solutions: GR solution (gelatin and resorcinol) and F-activator (formaldehyde and glutaraldehyde). As in case of BioGlue, formaldehyde and glutaraldehyde react not only with gelatin and resorcinol, but also with amino groups of tissues, creating strong bonding between the resulting gel and the tissue. Since the introduction of GRF in 1979 for acute aortic dissection [67], it has been widely used except in the United States.

Toxicity of formaldehyde and glutaraldehyde, as well as resorcinol, is the major concern of this product. Conflicting data on the effect of GRF glue in aortic surgery can be seen in literature. Long-term stability of the dissected aorta has been reported, whereas other studies show late adverse events such as late re-operation and distal embolization [68–70]. It has been suggested that these differences arise from an excess use of F-activator. Kunihara *et al.* [71] used GRF glue with the minimal amount of F-activator for 21 cases out of 49 operated cases with Stanford type A acute aortic dissection within 48 h with poly(tetrafluoroethylene) felt strip for reinforcement at the outer suture line. When applied properly, GRF was clinically safe with regard to mid-term death or re-operation, but combination use of outer felt reinforcement was recommended.

Different cross-linking types of gelatin-based glues have been developed. One of them is a rapidly curable bioabsorbable glue consisting of gelatin and poly(L-glutamic acid) (PLGA) [72, 73]. Mixing of gelatin and PLGA aqueous solutions with water-soluble carbodiimide (WSC) forms a hydrogel within a few seconds through amide formation between amino groups of gelatin and carboxyl groups of PLGA, as shown in Fig. 3.

This gel gives higher bonding strength to various soft tissues than that of clinically used fibrin glues [72, 73]. Degradation of the gel did not cause any significant inflammation in the animal model. However, the gelatin aqueous solution, before adding WSC, spontaneously sets to a physical gel at temperature around 25°C and

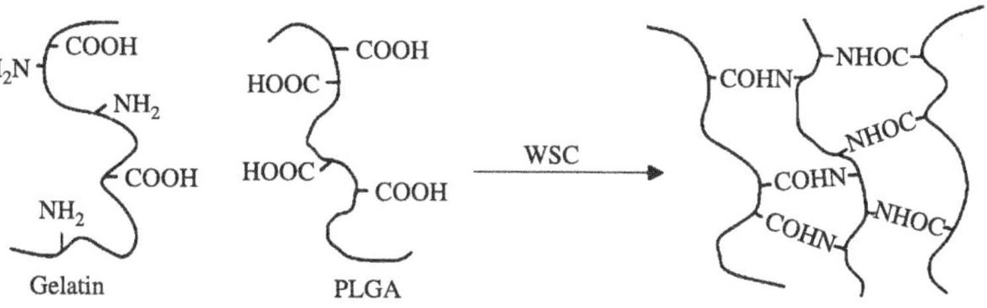

Figure 3. Schematic representation of gelatin-PLGA-WSC cross-linking system.

below. This was overcome by the addition of urea without altering the WSC cross-linking characteristics and tissue adhesion property [74].

Hemostatic capability of this gel was examined using an injured model with constant blood oozing on dog spleen [75]. In comparison with fibrin glue, the gelatin-PLGA gel gave a significantly higher success rate of hemostasis due to the strong adhesion of the glue and rapid gelation of the solution mixture. The reason for the strong adhesion was thought to be the direct reaction of WSC-activated PLGA with amino groups of the tissue. When the effectiveness of the gelatin-PLGA gel as lung air leak sealant was investigated, the gel was also found to adhere strongly to the lung surface and showed superior sealant effect compared to that of fibrin glue [76].

To eliminate the use of WSC, a glue composed of N-hydroxysuccinimide (NHS)-activated PLGA and gelatin was developed [77]. The NHS-activated PLGA was able to be stored for an extended time in a dry cool condition, without losing the cross-linking ability. It spontaneously formed a cross-linked gel when mixed with gelatin, and the bonding strength of this glue with natural tissues was higher than that of fibrin glue.

In another study, a gel composed of amine-modified gelatin and aldehyde-modified polysaccharides was synthesized. The amino content of gelatin was increased by reaction of gelatin with ethylenediamine in the presence of WSC [78]. Dextran and hydroxyethyl starch were oxidized by sodium periodate to convert 1,2-hydroxyl groups into dialdehyde groups. When the solutions of gelatin and modified dextran or hydroxyethyl starch were mixed, gel was formed by the formation of Schiff base between the amino groups of gelatin and the aldehydes of polysaccharides. As shown in Fig. 4, higher bonding strengths (max. 225 gf/cm^2) could be achieved as well as faster gelation time (2 s) by the increasing amine and aldehyde groups.

Sung *et al.* investigated several cross-linking agents (epoxy compound, WSC, and genipin) for gelatin as tissue adhesive and their properties were compared with those of GRF (gelatin-resorcinol-formaldehyde) and GRG (gelatin-resorcinol-glutaraldehyde) [79]. In the cytotoxicity tests using 3T3 fibroblasts, it was found that GRF, GRG, and gelatin cross-linked with an epoxy compound (GRE) were highly toxic, whereas gelatins cross-linked with WSC (and alginate, GAC) and genipin (GG) had no toxic effects on 3T3 fibroblast growth even when high concentration of cross-linkers (10 000 ppm) was used. GRE showed no bonding strength,

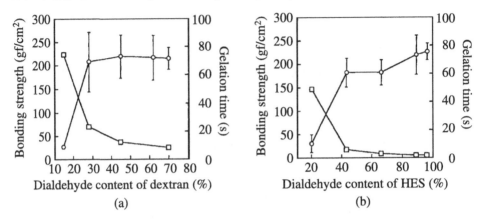

Figure 4. Bonding strength and gelation time of gelatin-polysaccharide gel as a function of dialdehyde contents (a) dextran and (b) hydroxyethyl starch (HES) (reproduced from [78]).

whereas GAC and GG showed 76.2 ± 11.2 and 69.5 ± 5.6 gf/cm^2, respectively. In comparison, GRF and GRG showed higher bonding strengths (137.9 ± 25.7 and 152 ± 23.6 gf/cm^2, respectively) and yielded stiffer gels. Therefore, they concluded that GAC and GG were preferable when minimal cytotoxicity and stiffness of adhesives were required.

4.2.3. Chondroitin Sulphate-Based Glue

To bind a regenerated cartilage tissue with an existing one, Wang and colleagues have recently developed a glue that is based on functionalized chondroitin sulphate with two organic groups: methacrylate and aldehyde [80]. The aldehyde groups react with amine groups of collagen in the native tissue and form covalent bonds. In addition, methacrylate groups solidify the implanted material by polymerization reaction. The glue showed strong mechanical bond to the native cartilage and did not damage the cells in the implants and the native cartilage.

4.3. Synthetic Adhesives

Cyanoacrylates are monomers containing esters of cyanoacrylic acid. These monomers polymerize when contacted with water or a weak base [81]. These create strong and flexible films as sealants to bond apposed wound edges. They have been mainly used for the external surfaces due to the toxicity of monomers and the degradation product which is formaldehyde. Butyl cyanoacrylate has been used only for small low-tension lacerations and incisions due to its poor tensile strength and brittleness, whereas octyl cyanoacrylate gives stronger and more flexible adhesives [82].

Poly(ethylene glycol) (PEG)-based hydrogels are flexible and adhere firmly even to wet tissues. AdvaSeal (Ethicon Inc., Johnson & Johnson Medical KK, Somerville, NJ, USA) is one of such products, which is derived from photopolymerization of two aqueous solutions consisting of PEG and oligotriethylene carbonate with acrylate ester end caps, in the presence of triethanolamine and eosin Y as a photoinitiator, using visible light from a xenon arc lamp. It has been clinically used for sealing of pulmonary air leakage in Europe [83, 84]. Tanaka *et al.* investigated its use for sealing the false channel in acute aortic dissection using mongrel dogs and found that AdvaSeal was effective for hemostasis and closure of the false lumen [85].

Another PEG-based sealant product is DuraSeal (Confluent Surgical Inc., Waltham, MA, USA). Two solutions, a PEG-ester and trilysine amine, which is a low molecular weight product of L-lysine, are mixed together and sprayed on the treatment site, which then cure to form an elastic hydrogel. DuraSeal is absorbed in four to eight weeks. The sealant was approved for use as an adjunct to sutured dural repair during cranial surgery by the US FDA and the European Commission (EC) in June 2003 and September 2005, respectively. It was also approved by the EC in September 2005 for use as an adjunct in the sealing of pleural air leaks during pulmonary resection.

5. Anti-adhesive Materials

Adhesion of internal organs to each other by the formation of fibrous tissues has often been observed during pelvic, abdominal and gynecological surgeries such as cesarean sections, colectomies, and hernia repairs [86]. In the healing process, swelling occurs that brings organs in close proximity in comparison to the normal condition. Adhesion is generally caused by fibrin deposition followed by fibrous tissue formation that binds these physiologically separated organs together. It is thought that post-operative adhesion occurs in 93–100% of patients who undergo transperitoneal surgery [87]. Tissue adhesion can cause serious post-surgical complications, such as small-bowel obstruction, infertility, chronic abdominal pain, and necessity of second operations.

Anti-adhesive materials have been developed to prevent tissue adhesion by providing a physical barrier between an injured site and the adjacent tissues. Although several non-absorbable synthetic materials such as silicone and poly(tetrafluoroethylene) (PTFE) have shown to be effective, bioabsorbable materials are preferred because of no need for secondary surgery to remove the non-absorbable materials and no requirement to consider long-term biocompatibility such as encapsulation of the material which evoke tissue adhesion.

An optimal anti-adhesive material should possess the following properties: (i) cover the wound site until the site is no longer susceptible to adhesion, (ii) disappear quickly after the required time to avoid foreign-body reactions to the material that can cause another adhesion site, and (iii) should have adequate mechanical properties for easy handling. The extent of adhesion formation depends on patient, and the extent and type of surgery performed. According to Rodeheaver and coworkers, at least 36 h of anti-adhesive material application was required to effectively reduce adhesion in healing of injured abdominal surfaces of rat using a silicone film as adhesive barrier [88]. Matsuda *et al.* also found that two-day application of anti-adhesive material, poly(vinyl alcohol) film, reduced incidence rate of adhesion to 18% for rats [89].

Seprafilm (Genzyme Corp., Cambridge, MA, USA) is composed of sodium hyaluronate and carboxymethyl cellulose. This film adheres to an injured site by absorbing water from the surroundings and separates the tissues during the post-operative healing phase. Seprafilm degrades and disappears in 7 days. A major problem of this film is that the handling is quite difficult because of its low mechanical properties when wet and brittleness.

Another commercially available product is Interceed (Ethicon Inc., Somerille, NJ, USA) which is a knitted fabric of oxidized regenerated cellulose. It is reported that Interceed prevents tissue adhesion effectively when blood contamination was avoided during the application. However, as complete blood clearing is not always possible in clinical situation, surgeons are not really keen to use this product. Moreover, increased adhesion formation has been observed when Interceed was applied to the area where blood cumulation could not be prevented [90].

Gelatin can be cross-linked by thermal, UV, or chemical means, and its degradation can be controlled easily by modifying the cross-linking degree. UV cross-linked gelatin films have been investigated for use as an anti-adhesive material [89]. As shown in Fig. 5, the *in vivo* degradation of cross-linked gelatin film was controlled by UV irradiation time. When gelatin films with different degradation rates were evaluated for anti-adhesion efficacy using rat adhesion model, it was found that the film which degrades in 3 days showed the best anti-adhesive property,

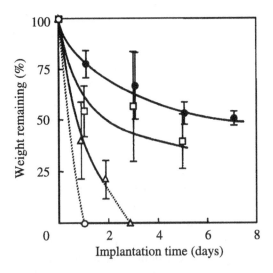

Figure 5. *In vivo* degradation of the cross-linked gelatin films with different UV irradiation times. Irradiation time: ○: 5 h; △: 10 h; □: 20 h and ●: 40 h. The films were implanted subcutaneously in rats, and remaining weights were evaluated after certain days. Error bar indicates the standard deviation (reproduced from [89]).

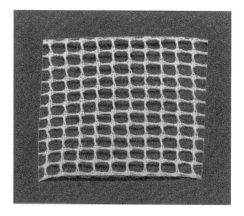

Figure 6. Photographic image of the gelatin sheet reinforced with PGA (reproduced from [91]).

whereas longer lasting films were less efficient due to the prolonged existence of the films causing the induced foreign-body reaction which resulted in tissue adhesion.

Gelatin sheet was also studied as a pericardial substitute [91, 92]. Adhesion between the heart and chest wall after open heart surgery can cause significant problems such as bleeding, injury to the heart or great vessels, and longer operation and perfusion times. The most commonly used pericardial substitute, PTFE, has problems such as severe inflammation and diffuse fibrosis. A thermally treated gelatin sheet, which degrades in 4 weeks, exhibited reduced formation of pleural and pericardial adhesion and inflammatory reaction compared to those with PTFE sheet in canine models [92]. Despite the effectiveness of the gelatin sheet as a pericardial substitute, there were two considerable drawbacks in its structural properties at the hydrated state: one is that the sheet was too malleable to handle, and the other was its low tensile strength which could not steadily hold sutures to the native pericardium.

To overcome these problems, the gelatin sheet was reinforced with bioabsorbable PGA mesh, as shown in Fig. 6 [91]. The modified sheet showed tenfold higher tensile strength at the suturing margin compared to the unmodified gelatin sheet, which was similar to that of PTFE. The effectiveness of this reinforced sheet in the prevention of adhesion was confirmed in a canine model.

A powder form of anti-adhesive material has been investigated. Solution or powder form has the advantage of easy application from the small incisions, such as ports in video-assisted operation, compared to those of films. Izumi *et al.* evaluated the anti-adhesive effect of cross-linked poly(γ-glutamic acid) (BioPGA, Meiji Seika Ltd, Tokyo, Japan) in a rat model [93]. Poly(γ-glutamic acid), a bioabsorbable polymer produced from strain of *bacillus subtilis*, can be cross-linked by γ-irradiation. When applied in the body, it forms a viscous hydrogel by absorbing the surrounding fluid. The anti-adhesive effect of cross-linked poly(γ-glutamic acid) was found to be superior to Seprafilm and Interceed.

6. Conclusion

Cell and tissue adhesion to biomaterials is the first event that occurs when materials are implanted, and this leads to the host responses. Since proteins adsorbed on the surface mediate cell adhesion and growth, the surface which provides the right types, amounts, and conformations of adsorbed proteins supports the desirable cell behaviour. In the case of scaffolds, natural polymers, such as ECM proteins, have advantages due to their processing of cell adhesive domains, whereas synthetic polymers often require surface modification to improve the interaction with cells. For surgical adhesives, biomaterials with strong bonding ability to tissues and less toxicity are still under investigation. For anti-adhesive materials, highly hydrophilic polymers or hydrogels with low protein adsorption have been employed to prevent cell and tissue adhesion. It is apparent that desirable cell and tissue responses to implants depend on the specific application, and efforts are still continuing to find suitable bioabsorbable polymers for many applications in tissue engineering.

References

1. L. L. Hench and E. C. Ethridge, *Biomaterials: An Interfacial Approach*. Academic Press, New York (1982).
2. V. Rosen and R. S. Thies, *The Cellular and Molecular Basis of Bone Formation and Repair*. R. G. Landes, Austin, TX (1995).
3. J. M. Anderson, in: *Biomaterials Science: An Introduction to Materials in Medicine*, B. D. Ratner, A. S. Hoffman, F. S. Schoen and J. E. Lemons (Eds). Academic Press, San Diego, CA (1996).
4. B. Kasemo and J. Gold, *Adv. Dental Res.* **13**, 8–20 (1999).
5. E. A. Vogler, *Adv. Colloid Interface Sci.* **74**, 69–117 (1998).
6. B. Kasemo, *Surface Sci.* **500**, 656–677 (2002).
7. C. J. Wilson, R. E. Clegg, D. I. Leavesley and M. J. Pearcy, *Tissue Eng.* **11**, 1–18 (2005).
8. Y. Tamada and Y. Ikada, in: *Polymers in Medicine II*, E. Cheillin, P. Giusti, C. Migliaresl and L. Nicolas (Eds), pp. 101–115. Plenum Press, New York (1986).
9. C. Mao, Y. Qiu, H. Sang, H. Mei, A. Zhu, J. Shen and S. Lin, *Adv. Colloid Interface Sci.* **110**, 5–17 (2004).
10. R. M. Shelton, A. C. Rasmussen and J. E. Davies, *Biomaterials* **9**, 24–29 (1988).
11. K. Kato, S. Sano and Y. Ikada, *Colloids Surfaces B* **4**, 221–230 (1995).
12. C. D. Tidwell, S. I. Ertel and B. D. Ratner, *Langmuir* **13**, 3404–3413 (1997).
13. B. A. Dalton, C. McFarland, T. R. Gengenbach, H. J. Griesser and J. G. Steele, *J. Biomater. Sci. Polym. Ed.* **9**, 781–799 (1998).
14. L. S. Nair and C. T. Laurencin, *Prog. Polym. Sci.* **32**, 762–798 (2007).
15. Y. Ikada, *Tissue Engineering: Fundamentals and Applications*. Elsevier, Oxford, UK (2006).
16. P. X. Ma, *Materials Today*, 30–40 (May 2004).
17. G. Chen, T. Ushida and T. Tateishi, *Macromol. Biosci.* **2**, 67–77 (2002).
18. W. L. Murphy and D. J. Mooney, *J. Am. Chem. Soc.* **124**, 1910–1917 (2002).
19. G. E. Park, M. A. Pattison, K. Park and T. J. Webster, *Biomaterials* **26**, 3075–3082 (2005).
20. Y. S. Nam, J. J. Yoon, J. G. Lee and T. G. Park, *J. Biomater. Sci. Polym. Ed.* **10**, 1145–1158 (1999).
21. L. L. Smith, P. J. Niziolek, K. M. Haberstroh, E. A. Nauman and T. J. Webster, *Int. J. Nanomed.* **2**, 383–388 (2007).

22. I. Keen, P. Broota, L. Rintoul, P. Fredericks, M. Trau and L. Grøndahl, *Biomacromolecules* **7**, 427–434 (2006).

23. T. I. Croll, A. J. O'Connor, G. W. Stevens and J. J. Cooper-White, *Biomacromolecules* **5**, 463–473 (2004).

24. Y. B. Zhu, C. Y. Gao, X. Y. Liu and J. C. Shen, *J. Biomed. Mater. Res.* **69A**, 436–443 (2004).

25. Y. B. Zhu, C. Y. Gao, X. Y. Liu and J. C. Shen, *Biomacromolecules* **3**, 1312–1319 (2002).

26. E. T. Kang, K. L. Tan, K. Kato, U. Uyama and Y. Ikada, *Macromolecules* **29**, 6872–6879 (1996).

27. A. S. Hoffman, in: *Biomaterials Science: An Introduction to Materials in Medicine*, B. D. Ratner, A. S. Hoffman, F. S. Schoen and J. E. Lemons (Eds). Academic Press, San Diego, CA (1996).

28. J. L. West, in: *Scaffolding in Tissue Engineering*, P. X. Ma and J. Elisseeff (Eds). CRC Press, Boca Raton, FL (2006).

29. M. W. Tibbitt and K. S. Anseth, *Biotechnol. Bioeng.* **103**, 655–663 (2009).

30. J. L. Myles, B. T. Burgess and R. B. Dickson, *J. Biomater. Sci. Polym. Ed.* **11**, 69–86 (2000).

31. J. C. Schense, J. Bloch, P. Aebischer and J. A. Hubbell, *Nature Biotechnol.* **18**, 415–419 (2000).

32. L. Meinel, V. Karageorgiou, S. Hofmann, R. Fajardo, B. Snyder, C. Li, R. Zichner, R. Langer, G. Vunjak-Novakovic and D. L. Kaplan, *J. Biomed. Mater. Res.* **71A**, 25–34 (2004).

33. J. F. Alvarez-Barreto and V. I. Sikavitsas, *Macromol. Biosci.* **7**, 579–588 (2007).

34. U. Hersel, C. Dahmen and H. Kessler, *Biomaterials* **24**, 4385–4415 (2003).

35. Y. Hu, S. R. Winn, I. Krajbich and J. O. Hollinger, *J. Biomed. Mater. Res.* **64A**, 583–590 (2003).

36. K. A. Corsi, E. M. Schwarz, D. J. Mooney and J. Huard, *J. Orthop. Res.* **25**, 1261–1268 (2007).

37. S. H. Hsu, S. H. Chang, H. J. Yen, S. W. Whu, C. L. Tsai and D. C. Chen, *Artif. Organs* **30**, 42–55 (2006).

38. H. J. Jung, K. Park, J.-J. Kim, J. H. Lee, K. O. Han and D. K. Han, *Artif. Organs* **32**, 981–589 (2007).

39. J. P. Ranieri, R. Bellamkonda, E. J. Bekos, T. G. Vargo, J. A. Gardella Jr. and P. Aebischer, *J. Biomed. Mater. Res.* **29**, 779–785 (1995).

40. R. Bellamkonda, J. P. Ranieri and P. Aebischer, *J. Neurosci. Res.* **41**, 501–509 (1995).

41. T. Y. Wei, W. M. Tian, X. Yu, F. Z. Cui, S. P. Hou, Q. Y. Xu and I.-S. Lee, *Biomed. Mater.* **2**, S142–S146 (2007).

42. B. K. Mann and J. L. West, *J. Biomed. Mater. Res. A* **60**, 86–93 (2002).

43. J. A. Hubbell, S. P. Massia, N. P. Desai and P. D. Drumheller, *Bio/Technology* **9**, 568–572 (1991).

44. K. C. Dee, T. T. Andersen and R. Bizios, *J. Biomed. Mater. Res. A* **40**, 371–377 (1998).

45. L. Grøndahl, F. Cardona, K. Chiem, E. Wentrup-Byrne and T. Bostrom, *J. Mater. Sci. Mater. Med.* **14**, 503–510 (2003).

46. S. Suzuki, L. Grøndahl, D. Leavesley and E. Wentrup-Byrne, *Biomaterials* **26**, 5303–5312 (2005).

47. O. N. Tretinnikov, K. Kato and Y. Ikada, *J. Biomed. Mater. Res.* **28**, 1365–1373 (1994).

48. S. Kamei, N. Tomita, S. Tamai, K. Kato and Y. Ikada, *J. Biomed. Mater. Res.* **37**, 384–393 (1997).

49. R. Zhang and P. X. Ma, *J. Biomed. Mater. Res.* **45**, 285–293 (1999).

50. R. Zhang and P. X. Ma, *Macromol. Biosci.* **4**, 100–111 (2004).

51. A. K. Gosain and V. B. Lyon, *Plastic Reconstructive Surgery* **110**, 1581–1584 (2002).

52. M. R. Jackson, *Am. J. Surgery* **182**, 1S–7S (2001).

53. T. Morikawa, *Am. J. Surgery* **182**, 29S–35S (2001).

54. R. Bitton and H. Bianco-Peled, in: *Handbook of Sealant Technology,* K. L. Mittal and A. Pizzi (Eds), pp. 513–531. CRC Press, Boca Raton, FL (2009).

55. I. C. Ennker, J. Ennker, H. D. Schoon, M. Rimpler and R. Hetzer, *Ann. Thorac. Surgery* **57**, 1622–1627 (1994).

56. H. J. Cha, D. S. Hwang and S. Lim, *Biotechnol. J.* **3**, 631–635 (2008).

57. H. Lee, B. P. Lee and P. B. Messersmith, *Nature* **448**, 338–341 (2007).
58. J. H. Waite and M. L. Tanzer, *Science* **212**, 1038–1040 (1981).
59. J. Dove and P. Sheridan, *J. Am. Dental Assoc.* **112**, 879 (1986).
60. D. A. Grande and M. I. Pitman, *Bulletin Hospital Joint Disease Orthopaedic Institute* **48**, 140–148 (1988).
61. C. Saez, J. Pardo, E. Gutierrez, M. Brito and L. O. Burzio, *Comparative Biochem. Physiol.* **98B**, 569–572 (1991).
62. H. Lee, S. M. Dellatore, W. M. Miller and P. B. Messersmith, *Science* **318**, 426–430 (2007).
63. H.-H. Chao and D. F. Torchiana, *J. Cardiovasc. Surgery* **18**, 500–503 (2003).
64. K. Zehr, *Ann. Thorac. Surgery* **84**, 1048–1052 (2007).
65. W. Fürst and A. Banerjee, *Ann. Thorac. Surgery* **79**, 1522–1529 (2005).
66. P. Klimo, A. Khalil, J. R. Slotkin, E. R. Smith, R. M. Scott and L. C. Goumnerova, *Operative Neurosurgery* **60**, 305–309 (2007).
67. D. Guilmet, J. Bachet, B. Goudot, C. Laurian, F. Gigou, O. Bical and M. Barbagelatta, *J. Thorac. Cardiovasc. Surgery* **77**, 516–521 (1979).
68. T. Kazui, N. Washiyama, A. H. Bashar, H. Terada, K. Suzuki, K. Yamashita and M. Takinami, *Ann. Thorac. Surgery* **72**, 509–514 (2001).
69. S. Fukunaga, M. Karck, W. Harringer, J. Cremer, C. Rhein and A. Haverich, *Eur. J. Cardiothorac Surgery* **15**, 564–570 (1999).
70. U. O. von Oppell, Z. Karani, A. Brooks and J. Brink, *J. Heart Valve Diseases* **11**, 149–257 (2002).
71. T. Kunihara, N. Shiiya, K. Matsuzaki, T. Murashita and Y. Matsui, *Ann. Thorac. Cardiovasc. Surgery* **14**, 88–95 (2008).
72. Y. Otani, Y. Tabata and Y. Ikada, *J. Biomed. Mater. Res.* **31**, 157–166 (1996).
73. Y. Otani, Y. Tabata and Y. Ikada, *J. Adhesion* **59**, 197–205 (1996).
74. Y. Otani, Y. Tabata and Y. Ikada, *Biomaterials* **19**, 2167–2173 (1998).
75. Y. Otani, Y. Tabata and Y. Ikada, *Biomaterials* **19**, 2091–2098 (1998).
76. Y. Otani, Y. Tabata and Y. Ikada, *Ann. Thorac. Surg.* **67**, 922–926 (1999).
77. H. Iwata, S. Matsuda, K., Mitsuhashi, E. Itoh and Y. Ikada, *Biomaterials* **19**, 1869–1876 (1998).
78. X. Mo, H. Iwata, S. Matsuda and Y. Ikada, *J. Biomater. Sci. Polym. Ed.* **11**, 341–351 (2000).
79. H.-W. Sung, D.-M. Huang, W.-H. Chang, R.-N. Huang and J.-C. Hsu, *J. Biomed. Mater. Res.* **46**, 520–530 (1999).
80. D.-A. Wang, S. Varghese, B. Sharma, I. Strehin, S. Fermanian, J. Gorham, D. H. Fairbrother, B. Cascio and J. H. Elisseeff, *Nature Mater.* **6**, 385–392 (2007).
81. A. Lauto, D. Mawad and L. J. R. Foster, *J. Chem. Technol. Biotechnol.* **83**, 464–472 (2008).
82. A. J. Singer and H. C. Thode, *Am. J. Surgery* **187**, 238–248 (2004).
83. W. R. Ranger, D. Halpin, A. S. Sawhney, M. Lyman and J. LoCicero, *Am. Surgeon* **63**, 788–795 (1997).
84. M. L. Lyman and A. S. Sawhney, in: *Proc. 5th World Biomaterials Congress*, p. 212, Toronto, Canada (1996).
85. K. Tanaka, S. Takamoto, T. Ohtsuka, Y. Kotsuka and M. Kawauchi, *Ann. Thorac. Surgery* **68**, 1308–1313 (1999).
86. J. W. Burnes, K. Skinner and J. Colt, *J. Surgery Res.* **59**, 644–652 (1995).
87. D. Menzies and H. Ellis, *Annals Royal College Surgeons England* **72**, 60–63 (1990).
88. E. S. Harris, R. F. Morgan and G. T. Rodeheaver, *Surgery* **117**, 663–669 (1995).
89. S. Matsuda, N. Se, H. Iwata and Y. Ikada, *Biomaterials* **23**, 2901–2908 (2002).
90. Interceed (TC7) Adhesions Barrier Study Group, *Fertil. Steril.* **51**, 933–938 (1989).

91. I. Yoshioka, Y. Saiki, K. Sakuma, A. Iguchi, T. Moriya, Y. Ikada and K. Tabayashi, *Ann. Thorac. Surgery* **84**, 864–870 (2007).
92. K. Sakuma, A. Iguchi, Y. Ikada and K. Tabayashi, *Ann. Thorac. Surgery* **80**, 1835–1840 (2005).
93. Y. Izumi, M. Yamamoto, M. Kawamura, T. Adachi and K. Kobayashi, *Surgery* **141**, 678–681 (2006).

Phagocyte Decisions at Interfaces

Virginie Monnet-Corti [a,b,c,d], Anne-Marie Benoliel [a,b,c], Anne Pierres [a,b,c] and
Pierre Bongrand [a,b,c,d,*]

[a] Laboratory Adhesion and Inflammation, INSERM UMR600, Parc Scientifique de Luminy,
163 Bd de Luminy, 13288 Marseille Cedex 09, France
[b] CNRS UMR 6212, France
[c] Université de la Mediterranée, France
[d] Assistance Publique - Hôpitaux de Marseille, France

Abstract

Phagocyte interaction with biomaterials plays an important role in inflammatory reactions. Numerous reports have shown that phagocyte behaviour at interfaces is strongly influenced by the nature and conformation of adsorbed biomolecules, which reflects, in an indirect way, the physico-chemical properties of underlying surfaces. Cell–surface interactions were thus thought to be governed by recognition events between membrane receptors and specific sites exposed by biomolecules. More recently, a growing number of reports have demonstrated that cells are also strongly influenced by physical or geometrical features such as surface rigidity or roughness and ligand topography. Here we review recent evidence supporting this concept and we describe recently disclosed cell dynamic properties that might be relevant to this phenomenon. Cell membrane movements at interfaces are expected to generate forces and deformations, and these phenomena are, in turn, expected to trigger or modulate signaling cascades through a number of processes including force-induced conformational change of membrane receptors, and alterations of in-plane movements of membrane molecules. It is suggested that a prerequisite to understanding cell behaviour at interfaces is to record all early molecular events triggered by membrane-to-surface approach and define cell decisions.

Keywords

Leukocyte, mononuclear phagocyte, undulation, mechanotransduction, rigidity, roughness, signaling

1. Introduction

A key requirement for the success of biomaterial implants is to achieve an adapted interaction with the immune system. Indeed, excessive immune cell activation will result in tissue damage and harmful consequences such as implant loosening. However, inhibition of immune cell function may favor infection and finally force implant removal. It is well accepted that so-called professional phagocytes,

---

* To whom correspondence should be addressed. Tel.: (+33) 491 82 88 52; Fax: (+33) 491 82 88 51; e-mail: pierre.bongrand@inserm.fr

Surface and Interfacial Aspects of Cell Adhesion
© Koninklijke Brill NV, Leiden, 2010

including granulocytes and mononuclear phagocytes, play a major role in inter-acting with foreign materials. Indeed, these cells may act as effectors of tissue or pathogen destruction, and influence the development of adaptive immune re-sponses either through antigen presentation to T-lymphocytes or through cross-talk with lymphocytes through soluble substances or contact interaction. Importantly, phagocyte interaction with foreign materials may lead to widely varying outcomes. Understanding the basic mechanisms of phagocyte decision-making at interfaces is, therefore, of considerable practical importance. In addition, a current challenge for cell biologists is to understand how cells integrate environmental signals to chose a particular fate concerning important processes such as survival, proliferation, differ-entiation, migration or secretion. The purpose of the present review is to describe recent advances strongly supporting the view that physical cues play an impor-tant role in phagocyte decisions at interfaces. We shall briefly review basic data on phagocyte function and biochemical control. Then we shall describe recently pub-lished results concerning cell response to physical messages, with an emphasis on professional phagocytes. Finally, we shall review potential mechanisms for signal generation at the cell–material interface in order to suggest possible explanations for phagocyte behavior at interfaces.

2. A Brief Summary of Phagocyte Properties

2.1. Professional Phagocytes

Phagocytosis is a basic function that is found in animals and vegetals, unicellu-lar and multicellular organisms. In higher organisms, phagocytosis is required to eliminate effete cells as well as foreign pathogens, and this function is essentially fulfilled by so-called professional phagocytes [1]. In mammals, a common myeloid precursor gives rise to blood monocytes and polymorphonuclear cells. The close relationship between these populations is exemplified by the known property of the HL60 promyelocytic line to differentiate into a pseudo-neutrophil or pseudo-monocyte when dimethylsulfoxide or phorbol myristate acetate, respectively, are added in culture medium. A crude simplification might be to describe neutrophils as powerful short-lived effectors that can very rapidly concentrate into inflamma-tion sites and destroy a number of phagocytizable particles. In contrast, monocytes are known to migrate towards peripheral tissues and differentiate into a variety of macrophage populations that will remain for months as fairly quiet resident cells and be ready to get activated and fulfill a number of functions. Several decades ago [2], cells composing the monocyte lineage were defined as mononuclear phago-cytes. Recently, there was a considerable interest in the capacity of myeloid cells to differentiate into dendritic cells the role of which is to capture foreign substances and present them to lymphocytes in order to initiate immune responses [3, 4].

2.2. Phagocyte Functions

2.2.1. Phagocytosis

Obviously, the capacity to ingest and destroy pathogens is an essential requirement for the survival of higher organisms. This important issue is the basis of a classification of pathogens into extracellular invaders, that evolved mechanisms for preventing ingestion, e.g., by damaging or repelling phacocytes, or by producing a tentatively anti-phagocytic hydrophilic coat. In contrast, intracellularly developing pathogens do not attempt to prevent phagocytosis, and they may even stimulate ingestion. However, they evolved a range of mechanisms for intracellular survival including inhibition of phagosome–lysosome fusion, adaptation to the acidic and highly aggressive phagosomal environment, or passage from endosomal vesicles to cytosolic compartment as performed by many viruses. This provides a rationale for a choice of phagocytes to follow different activation pathways.

2.2.2. Release of Bactericidal Mediators

While phagocytosis is a method of choice to destroy harmful phagocytizable particles without damaging surrounding tissues, this cannot be performed to eliminate too large particles such as parasites. Phagocytes encountering parasites may thus have to release bacteriolytic mediators and accept the risk of damaging surrounding tissues. Mediators include a number of enzymes, cationic polypeptides and reactive oxygen intermediates such as hydrogen peroxide, superoxide anion, singlet oxygen or hydroxyl radical [1].

2.2.3. Antigen Presentation

It has been known for decades that mononuclear phagocytes can act as accessory cells by capturing and processing foreign substances and presenting them to T-lymphocytes as oligopeptides bound to products of the major histocompatibility complex. Dendritic cells are most efficient in performing this function since they are thought to be the only cells capable of stimulating naïve T-lymphocytes. In addition to the expression of a particular set of membrane molecules, this function seemed related to a limitation of catabolic activity to keep antigenic fragments with a size grossly corresponding to decapeptides. This balance is nicely exemplified by experiments performed long ago by Biozzi and colleagues who showed that the macrophages from mice selected for high antibody responses displayed lower catabolic rate of ingested particles [5].

2.2.4. Tissue Homeostasis

In addition to aforementioned immunological functions, macrophages can strongly influence tissue homeostasis. Osteoclasts are members of the mononuclear phagocyte lineage with a key role in bone remodeling. In contrast, wound repair may be enhanced by macrophages through the release of suitable growth factors. Thus, there is a need for a tight regulation of macrophage function in order to ensure wound healing, starting from the elimination of damage tissues to the formation of new tissues [6].

In conclusion, macrophages have the capacity to exert an impressive range of functions and presumably they must chose between a number of pre-existing programs. Understanding and controlling this choice is clearly a major issue. Much experimental evidence has long supported the view that regulation was essentially due to biochemical signals.

3. Biochemical Regulation of Phagocyte Function

Resident phagocytes are usually fairly inefficient in performing the multiple tasks we have just mentioned. It has long been known that their functional capacity may be strongly enhanced by a number of processes called priming, stimulation, elicitation, or activation, resulting in the triggering of a particular program. Thus, while resident macrophages have a fairly low capacity to ingest and subsequently kill bacterial invaders, it has been recognized that bactericidal activity could be dramatically enhanced by specific mediators such as gamma interferon, formerly called macrophage activation factor. Increased ability to kill ingested pathogens was indeed considered as a suitable definition for macrophage activation. Further, while cytokines, such as gamma interferon, readily induced *bona fide* macrophage activation, incomplete stimulation with irritants could increase phagocytic activity without efficiently enhancing bactericidal capacity [7]. Numerous studies were performed with macrophages attracted and stimulated by the injection of substances such as thioglycollate into the peritoneal cavities of studied mice. These cells were usually called elicited macrophages [7]. More recently, numerous studies demonstrated the multiplicity of macrophage activation pathways and dependence on a particular kind of stimulus [6, 8]. Thus, conventional macrophage activation (called type I activation) resulting in increased capacity to destroy intracellular pathogens and release of inflammatory cytokines such as tumor necrosis factor α was triggered by gamma interferon and bacterial lipopolysaccharides [8]. Other mediators such as interleukin 4 or interleukin 13 increase phagocytic activity and expression of membrane receptors, such as mannose receptor, a process called alternate or type II macrophage activation [8, 9]. Finally, the resolution of inflammation and wound healing may require other activation states associated with specific cytokines with anti-inflammatory and growth-inducing properties such as interleukin 10 or transforming growth factor β [8, 10]. Similarly, it is known that the capacity of neutrophil granulocytes to release reactive oxygen intermediates is strongly enhanced by a preactivation process called priming, that can be induced by a range of mediators, such as tumor necrosis factor α [11]. Priming may thus be defined as induction of an increased capacity to respond to a number of stimuli. The representative examples we provided clearly illustrate the dominant view that phagocyte behaviour is strongly guided by a range of soluble biomolecules acting on specific membrane receptors.

While many regulatory messages involve soluble mediators, it is also well accepted that phagocyte function may be regulated by contact interaction with surface-bound specific molecules. Thus, while it is well established that monocyte adhesion makes them differentiate into macrophages, this may require the presence of specific matrix components on substrata [12]. Chemokines bound to the surface of activated endothelial cells can activate integrin receptors on leukocytes rolling on endothelial cells. Finally, while the role of specific ligands in phagocyte changes at interfaces was not always formally demonstrated [13, 14], it was often convincingly demonstrated that the influence of foreign surfaces on cell behaviour was mediated by the specific sites exposed by adsorbed biomolecules. Nonspecific physical–chemical properties of artificial surfaces may be translated into specific signals through conformational changes of adsorbed molecules [15]. This is the reason why inert biomaterials may trigger adverse foreign body reaction after being implanted. This outcome is often related to fibrinogen adsorption and subsequent exposure of specific sites recognized by phagocyte integrin receptors CD11bCD18 [16].

In conclusion, much evidence provided during decades has been consistent with the general hypothesis that phagocyte behavior might be essentially driven by specific interactions between membrane receptors and their ligands exposed by adsorbed biomolecules. More recently, a number of reports demonstrated the capacity of physical cues to influence cell behaviour at interfaces.

4. Cells Sense Physical and Topographical Properties of Surrounding Surfaces

There is now overwhelming evidence that cells sense the ridigity, rugosity and 2D-topography of underlying substrata and that they adapt their behaviour accordingly. We shall now give a few representative examples with an emphasis on phagocytes. As will become apparent, environment sensing is probably a common property of numerous cell populations, although preferences and decisions may differ between different cell types.

4.1. Rigidity

It has long been shown that adherent cells exert forces on substrata [17], which allows them to assay local rigidity. In many experiments, substratum rigidity was shown to influence short-term and long-term aspects of cell behaviour. Recently, murine fibroblasts were deposited on fibronectin-coated polyacrylamide gels displaying local discontinuities of reticulation and rigidity: cells were found to migrate towards harder areas [18]. This phenomenon was called durotaxis. Long term behaviour was also studied. It is well known that anchorage dependent cells cannot grow in a soft environment [19]. In addition, surface rigidity was recently shown to influence cell differentiation: human mensenchymal stem cells deposited on fibronectin-coated polymers of varying rigidity differentiated into neurons, myoblasts or osteoblasts when surface rigidity mimicked brain, muscle or bone, respectively [20]. Other authors studied the influence of substratum rigidity on cell spreading. A general finding was that cells spread more efficiently on harder surfaces [21]. This conclusion was also reached with macrophages [22], but neutrophils seemed less sensitive than other cell types to substratum rigidity [23]. A related finding is the observation that the ingestion of antibody-coated particles of varying rigidity by murine macrophages was more efficient when these particles were more rigid [24]. This result is consistent with the conventional view that macrophage spreading may be considered as frustrated phagocytosis. All these results are important since a number of experiments and models have supported the view that cell shape is an important determinant of cell behaviour [25–27]. See [28] for a review. In view of the current interest in cell response to substratum rigidity, two remarks are warranted. First, the definition of rigidity certainly deserves discussion. In most experiments, substrata are viewed as elastic media that may be conveniently described with a single parameter, i.e., the Young's or elastic modulus, expressed in units of Pa. However, since actual media are indeed viscoelastic and cells probe their environment dynamically, as discussed above, further experiments might show the need for a more precise definition of substratum properties relevant to interaction with living cells. Second, the experiments we described concerned cell interaction with 2D surfaces. As previously emphasized, these systems do not provide a satisfactory representation of cell behaviour in 3D environments [29].

4.2. Rugosity

Rugosity is also detected by adherent cells. Three decades ago, Rich and Harris [30] roughened a region of polystyrene (PS) dishes with a glass rod. Also, they treated polystyrene with sulphuric acid or they deposited palladium on PS to obtain more hydrophilic surfaces. Then they deposited fibroblasts, murine peritoneal macrophages or P388D1 murine phagocytes on these substrata. They concluded that macrophages displayed an anomalous preference for rough and hydrophobic surfaces, in contrast to fibroblasts. In other experiments, photolithographic techniques were used to obtain Perspex surfaces with linear steps of 5 μm depth. These steps seemed to interfere with protrusion formation by BHK cells, but not by neutrophils [31]. In later experiments, cells from the P388D1 murine macrophage line were deposited on silicon surfaces bearing nanogrooves of 30–282 nm depth [32]. This patterning appeared to increase cell spreading rate. In view of the aforementioned link between spreading and phagocytosis, it is interesting to note that nanocolumns of 100 nm diameter seemed to trigger endocytic attempts by deposited fibroblasts [33]. Actin polymerization and tyrosine phosphorylation were locally increased on these grooves, which demonstrated their capacity to influence cell organization. More recently, human neutrophils were deposited on polystyrene surfaces that had been locally roughened by mechanical scratching. Cell viability was lower in roughened areas. This was ascribed to an oxidative mechanism since cell damage was reduced by adding catalase [34]. Another experiment revealed the influence of 3D topography on phagocyte interaction with surfaces: polystyrene particles of varying shape were prepared and presented to phagocytes from the rat NR8383 or the murine J774 line. Using scanning electron microscopy, it was concluded that the macrophage choice to engulf a particle or to spread on its surface was determined by its size and the angle between its surface and the cell surface [35].

While it seems well demonstrated that surface rugosity influences cell behaviour, no general mechanism was shown to account for this finding. Interestingly, surface topography at the nanometer scale was found to influence interaction with proteins such as fibronectin, fibrinogen and albumin [36]. It is, therefore, warranted to study further the effect of surface rugosity on the efficiency of biomolecule adsorption and subsequent conformational changes.

4.3. Two-Dimensional Distribution of Receptor Ligands

The two-dimensional distribution of surface-exposed biomolecules is also important. This was very convincingly demonstrated by using a polymer-linking method to prepare surfaces coated with various surface densities of synthetic integrin ligands that were either homogeneously distributed, or organized as clusters of 5 to 9 of these molecules. Fibroblastic cells were then deposited on these surfaces. The clustering of integrin ligands was found to increase cell adhesion and migration [37]. The importance of ligand patterning was also demonstrated on macrophages: Two binding sites of β_1 integrins that are found on fibronectin, i.e., RGD and PHSRN amino-acid sequences (as represented with standard 1-letter code for aminoacids) were immobilized on poly(ethylene glycol)-based polymer surfaces and human monocyte-derived macrophages were deposited. The authors assayed the formation of multinucleated foreign-body giant cells formed by cell fusion. Interestingly, cell fusion occurred only if RGD and PHSRN peptides were suitably oriented on the same molecule [38].

In conclusion, it is now amply demonstrated that most cell populations, including phagocytes, can decode physical properties of the surfaces on which they are attached, and translate this information into a particular behaviour. It is, therefore, of utmost interest to analyze the mechanisms involved in this response to environmental properties. Two important questions must be addressed: (i) which signals are generated by cell–surface interaction? (ii) How are these signals processed by cells in order to make decisions? While no definite answer has been found to these questions, a growing amount of information is available. We shall select several representative examples. Other information can be found in a recent report [39].

5. Early Signaling at the Cell–Substratum Interface

Cell decisions are thought to be determined by intracellular signaling cascades and second messenger generation. Therefore, a reasonable way for unraveling the mechanisms of macrophage decision-making at interfaces consists of firstly recording all events generated during cell–surface encounter, and secondly elucidating the mechanisms used by cells to integrate all perceived messages. We shall now review significant data concerning the first point.

5.1. Early Events during Macrophage Encounter with Foreign Surfaces

While cells composing living organisms usually evolve in a three-dimensional environment, most reported experimental data concern cell interactions with surfaces. Although these data are not fully relevant to most actual situations, they are not entirely irrelevant: firstly, unicellular organisms often interact with surfaces during their normal life, and cells from higher organisms are known to retain many basic functions that appeared early during evolution. Secondly, even in higher organisms, monocytes must interact with blood vessel walls and alveolar macrophages must also reside at interfaces. Thus, in the present review, we shall essentially consider cell behaviour at interfaces. We shall first provide convenient orders of magnitude for a number of phenomena.

A typical phagocyte in suspension appears as a sphere of *ca.* 10 µm diameter surrounded with a plasma membrane studded with submicrometer-scale protrusions accounting for an excess surface area of 50–100% as compared to the area of a smooth sphere enclosing the same volume [40, 41]. After falling on a planar surface, the cell will flatten within seconds at a micrometer level [28, 42] (Fig. 1). During the following few minutes, a series of events will occur: the cell membrane will smooth at the submicrometer level, thus allowing molecular contact with the surface when they are separated by a gap lower than several tens of nanometers [40, 42, 43]. Simultaneously, lateral diffusion of membrane molecules will result in some kind of segregation. For example, the bulkiest molecules are expected to move out of regions of closest contact between membrane and surfaces [42, 44]. Cells may exhibit active spreading after minutes to hours [28]. This active process involves a

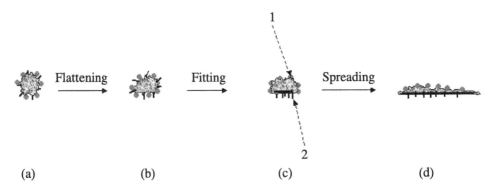

(a)   (b)   (c)   (d)

Figure 1. Sequential steps in cell interaction with a planar surface. A cell maintained in suspension (a) is usually fairly spherical with a wrinkled membrane. A few seconds after falling on a planar surface, it exhibits micrometer-scale flattening (b). During the following tens of seconds to minutes, the membrane displays significant smoothing at the submicrometer level, with increasing molecular fitting between the cell membrane and underlying surface. At the same times, bulky membrane components (1) diffuse out of the contact zone while receptors specific for substratum sites concentrate into the contact area (2) and a substantial thickening of the submembranar actin cytoskeleton is often apparent in adhesion area (c). In some cases, during the following minutes to hours cell exhibits active spreading (d).

growth of surface protrusions suchs as microvilli or lamellipodia that will bind to the surface and result in a dramatic increase of cell projected area.

The combination of in-plane diffusion of membrane molecules and transverse movements of the membrane will determine molecular contacts between membrane receptors and surface bound ligands. As a consequence of ligand-receptor association and exposure to forces generated by membrane movements, signaling cascades will be triggered. A typical example is a rise of cytosolic calcium concentration. This has been thoroughly investigated since (i) monitoring intracellular calcium was made possible with a number of specific fluorescent probes [45]. (ii) Dramatic (tenfold or more) calcium increases are very frequently observed after cell activation, and (iii) calcium is known to regulate a number of intracellular enzymes.

As exemplified in Fig. 2, macrophage contact with a surface often results in calcium rise, and this is certainly significant. Thus, calcium rise was shown to precede spreading by a few seconds [46] and calcium rise was shown to trigger spreading [47].

Note that calcium changes are only part of a cascade of signaling events usually related to the triggering of cell functions (see [48]).

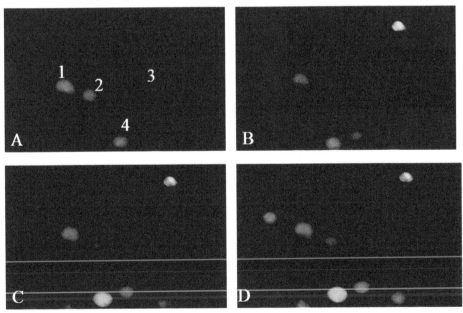

Figure 2. Cytosolic calcium rise in cells encountering a foreign surface. Human monocytic THP-1 cells were labeled with a combination of fluo-4 and fura-red calcium probes before being sedimented on fibronectin-coated surfaces at room temperature and observed with fluorescence microscopy. Since intracellular calcium increase results in decrease of the (red) fluorescence of fura-red and increase of the (green) fluorescence of fluo-4, a representative microscopical field is shown 164 s (A), 301 s (B), 592 s (C) and 800 s (D) after the onset of sedimentation. Cells (1) and (4) exhibit a marked calcium rise at 164 s, cell (2) exhibits a marked calcium rise at 301 s, while cell (3) exhibits a moderate calcium rise at 592 s. When experiments were done at 37°C, a calcium rise was observed quasi-instantaneously after cell–substratum contact.

The kinetics of these events may be strongly influenced by the presence of membrane receptor ligands on substrates. Thus, phagocytosis of a particle may be triggered within seconds after contact [49, 50]. Integrin receptors may be activated in less than a second after leukocyte contact with a surface exposing activatory chemokines [51]. However, when cells encounter surfaces bearing no specific phagocyte activator, calcium rise may occur after several tens of seconds [46, 52].

While it seems reasonable to assume that (i) signaling events are triggered by molecular contacts between cell and surface molecules, (ii) these contacts are dependent on membrane movements, and (iii) these movements are influenced by signaling, more details are required to reveal causal relationships between all these events. We shall sequentially describe membrane dynamics at interfaces and expected influence on signaling.

5.2. Membrane Dynamics at Interfaces

5.2.1. Cell Membranes Display Continual Deformation on a Timescale of a Few Seconds to Minutes
It is well known that cells display continual membrane deformation. They produce and retract cylindrical filopodia or sheet-like lamellipodia with a typical velocity of several tens of nanometers per second and a period of several tens of seconds [53]. The estimated force generated by these movements is on the order of a few tens of piconewton per filopodium or per micrometer of lamellipodium margin [39]. The characteristic period of protrusion–retraction cycles is on the order of several tens of seconds [54]. These movements are thought to play an important role in cell spreading on a timescale of several minutes. Also, bonds formed between membrane receptors and surface ligands are thought to be subjected to forces as a consequence of membrane movements. This is important since forces were shown to influence the maturation of cell–surface contacts and enhance the formation of specific structures such as focal adhesions [55] in a timescale of several minutes. However, while these phenomena may be highly relevant to the initiation of long-term cellular events such as cell differentiation, they are probably too slow to account for rapid decisions such as ingesting bacteria within a few seconds after contact [49], sticking to foreign surfaces in blood flow [51], or stopping and retracting a protrusion bumping against a foreign surface with a velocity of several tens of nanometer per second.

5.2.2. Membrane Undulations with a Frequency within the Hertz Range Were Evidenced on Many Cell Populations
It has been known for more than three decades that red blood cell membranes display continual transverse oscillations with a frequency of several hundreds of hertz and amplitude of several hundreds of nm [56]. While a theoretical model had been elaborated to account for thermal fluctuation of inert surfaces with sufficiently low tension and bending rigidity [57], some experiments suggested the involvement of metabolically active processes in red cell membrane oscillations [57, 59]. The occurrence of membrane oscillations in nucleated cell membranes was demon-

strated in other experiments. Using a combination of photometry and dark field microscopy, it was reported that nucleated cells such as fibroblasts, lymphocytes and monocytes displayed surface fluctuation of 20–30 nm amplitude and frequency on the order of 1 Hz [60]. More recently, atomic force microscopy revealed transverse fluctuations of 5 Hz frequency and 1–2 nm amplitude, corresponding to 10–60 pN forces, in murine fibroblasts [61]. Force fluctuations of 20–80 pN amplitude were also detected on human foreskin fibroblasts. Note that much higher oscillation amplitudes were reported on muscle cells [62], but the movements of muscle cells might not be expected to be representative of other cell types. Interference reflection microscopy (IRM) provides a powerful means of estimating the distance between a cell membrane and a planar surface with about 10 nm vertical resolution [62, 63]. This method was used to study the height fluctuations of membranes of phagocytes approaching glass surfaces. Murine J774 macrophage-like cells displayed vertical undulations of 5–10 nm amplitude. The authors concluded that these deformations were fairly consistent with a passive Helfrich model, although active processes could not be excluded [65]. Comparable results were found in a study made on the interaction between human THP-1 monocytic cells and fibronectin-coated surfaces (Fig. 3). An interesting point was that the amplitude of membrane undulations increased during the first minutes following cell–surface approach and a concomitant tightening of interaction and exclusion of some bulky membrane molecules from contact areas was observed [42]. This suggests that cells sensed the presence of a nearby surface and generated a response.

In conclusion, nonmuscle nucleated cells such as fibroblasts or phagocytes display membrane undulations with nanometer amplitude and 1 Hz frequency. Associated force fluctuations are on the order of tens of piconewtons. We shall now review

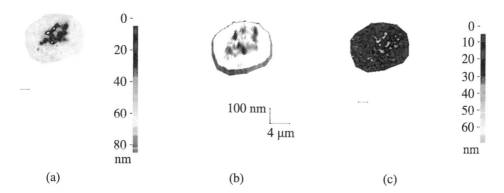

Figure 3. Cell membrane undulations in contact with foreign surfaces. Human monocytic THP-1 cells were made to sediment on fibronectin-coated surfaces at 37°C on a microscope stage with continuous observation with Interference reflection microscopy to image the distance between the cell membrane and the surface with a few nanometer accuracy. (a) — Color-coded display of the distance between the cell membrane and the surface. (b) — 3D image of the bottom cell surface. (c) — Color-coded display of the amplitude of vertical membrane undulation expressed as the standard deviation of the height of individual pixels in 8 sequential measurements performed with 1 Hz frequency.

experimental evidence demonstrating that these orders of magnitude can generate or modulate signals at the single molecule level. Additional examples can be found in a recent paper [39].

5.3. Relationship between Membrane Dynamics at Interfaces and Signaling

5.3.1. Cell Membrane Receptors Are Known to Initiate Signaling Cascades Through Two Major Nonexclusive Mechanisms

(i) *Conformational changes.* An important example is provided by receptors associated to trimeric G proteins. When a ligand binds to a receptor, the resulting conformational change releases a GTP-bound trimer, which dissociates and activates target enzymes (such as adenylate cyclases or phospholipases) through diffusing subunits. Leukocyte receptors for chemokines or chemotactic factors represent a typical example. Forces in the tens of piconewtons range generated by membrane undulations retracting from a surface after bond formation may change receptor conformation. Indeed, it was recently demonstrated [66] that a force of 2 pN might change the conformation of the integrin-associated molecule talin, thus exposing a binding site specific for vinculin, a cytosplasmic protein. Also, it has been shown that a lipid bilayer surface tension increase on the order of 0.01 N/m might open a bacterial ionic channel embedded in this bilayer [67]. Estimating at 10 nm diameter the channel structure, the disrupting force generated by the surface tension increase would be 100 pN.

Another related effect of forces would be to induce a displacement of a receptor or a receptor moiety perpendicularly to the membrane. Such a piston model was suggested to account for the function of a bacterial receptor for aspartic acid [68]. Using paramagnetic electron resonance, the authors provided evidence for a 0.1 nm displacement of a receptor component. In view of previous reports, such a small displacement was deemed sufficient to allow the receptor to form new interactions with intracellular components. More recently, it was suggested that the rigidity of the molecular complex formed by T-cell receptors and associated molecules was consistent with such a piston-like model [69, 70].

(ii) *Receptor clustering.* Many receptors are associated with a tyrosine kinase [71] that may be a domain of their intracytoplasmic region (many receptors for growth factors such as PDGF bear a so-called receptor tyrosine kinase) or a molecule interacting with their cytoplasmic region (thus, T-cell CD4 or CD8 co-receptors are associated with intracellular p56lck, belonging to the Src kinase family). It is thought that receptor multimerization will allow a kinase to encounter and phosphorylate its target tyrosine. This will result in recruitment of adapter proteins (with SH2 domains specific for the newly phosphorylated tyrosine) and other kinases, thus leading to the formation of a multimolecular signaling complex. This complex sequence of events is highly dependent on proximity relationships between potentially interacting molecules. Therefore, it is not surprising that signaling events can be influenced by a topographical redistribution of membrane molecules leading to a local concentration of selected molecular species in membrane domains. These mechanisms were intensively inves-tigated on the model of T-lymphocyte activation. The importance of lipid microdomains and their linking with cytosketal elements was recently re-viewed [72].

5.3.2. Cell Membrane Dynamics Provides a Link between Substratum Physical Properties and Signaling Events

There are a number of well documented mechanisms providing a link between membrane movements and signaling cascades at interfaces as initiated by afore-mentioned basic phenomena. Here are important examples.

(i) The contact between ligands and receptors is obviously dependent on membra-ne–substratum alignment. This is clearly exemplified by electron microscopic studies of macrophages adhering to smooth phagocytic targets [40]. The ap-parent distance between surfaces typically varies between a few nanometers and hundreds of nanometers, which is much larger than the length of many ligand–receptor couples: a macrophage medium affinity Fc receptor of about 8 nm length and IgG antibodies of about 16 nm length bound to a surface will be able to interact only if they are separated by less than about 24 nm. Thus, the frequency of ligand–receptor encounters is, at least parly, determined by membrane undulations. It is certainly warranted to explore more thoroughly the mechanisms involved in sub-micrometer fitting of cell membranes to newly encountered surfaces [28, 42, 43].

(ii) The forces generated by membrane motions may decrease [73] or increase [74, 75] the duration of ligand–receptor encounters. This may play a dominant role in cell response, as suggested in models of T-lymphocyte activation [70, 76]. In addition, as indicated above, these forces may initiate receptor conformational changes.

(iii) Membrane undulations are expected to strongly influence the redistribution of membrane molecules in contact areas as a result of two mechanisms. First, as described in a recent theoretical model [77], undulations may help molecules to hop between cytoskeletal barriers opposing diffusion. Second, transient approach between a membrane and a surface is expected to expel bulky mem-brane molecules such as leukocyte CD43 or CD45 membrane molecules of 40–45 nm length [44, 78, 79]. Since cell membranes are crowded with mole-cules [80], this may, in turn, concentrate smaller molecules into areas of close interaction. As emphasized above, this reorganization is expected to strongly influence signaling events. Indeed, some theoretical models of T-lymphocyte activation proposed that the exclusion of bulky phosphatases from regions of close membrane apposition might trigger signaling cascades by increasing the phosphorylation status of key proteins [81].

6. Conclusion: Open Issues

The purpose of this review was to provide a brief summary of available knowledge that was felt relevant to cell decision-making at interfaces. The major problem now is to integrate this information. This involves (i) finding causal relationships between cell reorganization at interfaces and signaling cascades, and (ii) relating these cascades to the triggering of functional events. As recently emphasized [39], we are still far from this goal. A major reason is that signaling cascades are not linear sequences of events resulting in a well-defined outcome. Firstly, a cell decision results from a substantial amount of messages. Indeed, a single kinase may trigger different phenomena depending on activation kinetics [82]. This means that cells display some kind of memory [83] allowing them to integrate a number of non-concomitant events. Cells are also responsive to the localization of stimuli, as exemplified by chemotaxis [84]: an eukaryotic cell can sense a chemoattractant gradient and start immediately in the appropriate direction, in contrast with bacteria that seem to change direction at random and adapt turning frequency to the time-dependence of surrounding chemoattractant concentration. Secondly, messages are not independent since signaling cascades act as complex and non-linear networks. Thirdly, it would be useful to know how long a cell decision is bound to last. Indeed, much recent evidence suggests that even cell differentiation is a quite reversible process [85].

Therefore, it is suggested that it would be most informative to focus on early cell events at interfaces and try to relate (i) cell structural changes, (ii) signaling events, (iii) fairly irreversible and biologically significant decisions such as mediator secretion or phagocytosis.

Acknowledgement

The project on which the present review is based was supported by a grant from the 'Fondation de l'Avenir'.

References

1. R. S. Flannagan, G. Cosio and G. Grinstein, *Nature Rev. Microbiol.* **7**, 355 (2009).
2. Z. A. Cohn and B. Benson, *J. Expl. Med.* **121**, 153 (1965).
3. R. M. Steinman and J. Banchereau, *Nature* **449**, 419 (2007).
4. C. Auffray, M. H. Sieweke and F. Geissmann, *Ann. Rev. Immunol.* **27**, 669 (2009).
5. D. Mouton, Y. Bouthillier, J. C. Mevel and G. Biozzi, *Ann. Immunol. (Paris)* **135D**, 173 (1984).
6. D. M. Mosser and J. P. Edwards, *Nature Rev. Immunol.* **8**, 958 (2008).
7. Z. A. Cohn, *J. Immunol.* **121**, 813 (1978).
8. D. M. Mosser, *J. Leukocyte Biol.* **73**, 209 (2003).
9. F. O. Martinez, L. Helming and S. Gordon, *Ann. Rev. Immunol.* **27**, 451 (2009).
10. S. Barrientos, O. Stojadinovic, M. S. Golinko, H. Brem and M. Tomic-Canic, *Wound Rep. Reg.* **16**, 585 (2008).
11. K. Onnheim, J. Bylund, F. Boulay, C. Dahlgren and H. Forsman, *Immunology* **125**, 591 (2008).
12. G. Kaplan and G. Gaudernack, *J. Expl. Med.* **156**, 1101 (1982).

13. F. Laurent, A. M. Benoliel, C. Capo and P. Bongrand, *J. Leukocyte Biol.* **49**, 217 (1991).

14. A.-F. Petit-Bertron, C. Fitting, J.-M. Cavaillon and M. Adib-Conquy, *J. Leukocyte Biol.* **73**, 145 (2003).

15. J. Vitte, A. M. Benoliel, A. Pierres and P. Bongrand, *Eur. Cells Mater.* **7**, 52 (2004).

16. W. J. Hu, J. W. Eaton and L. Tang, *Blood* **98**, 1231 (2001).

17. A. K. Harris, P. Wild and D. Stopak, *Science* **208**, 177 (1980).

18. C. M. Lo, W. H. Bang, M. Dembo and Y. L. Wang, *Biophys. J.* **79**, 144 (2000).

19. J.-J. Yang, J.-S. Kang and R. S. Krauss, *Mol. Cell Biol.* **18**, 2586 (1998).

20. A. J. Engler, S. Sen, H. L. Sweeney and D. E. Discher, *Cell* **126**, 677 (2006).

21. R. J. Pelham and Y. L. Wang, *Proc. Natl. Acad. Sci. USA* **94**, 13661 (1997).

22. S. Féréol, R. Fodil, B. Labat, S. Galiacy, V. M. Laurent, B. Louis, D. Isabey and E. Planus, *Cell Motility Cytoskel.* **63**, 321 (2006).

23. T. Yeung, P. C. Georges, L. A. Flanagan, B. Marg, M. Ortiz, M. Funaki, Z. Nahir, W. Ming, V. Weaver and P. A. Janmey, *Cell Motility Cytoskel.* **60**, 24 (2005).

24. K. A. Beningo and Y. L. Wang, *J. Cell Sci.* **115**, 849 (2002).

25. J. Folkman and A. Moscona, *Nature* **273**, 345 (1978).

26. J. Meyers, J. Craig and D. J. Odde, *Current Biol.* **16**, 1685 (2006).

27. S. R. Neves, P. Tsokas, A. Sarkar, E. A. Grace, P. Rangamani, S. M. Taubenfeld, C. M. Alberini, J. C. Schaff, R. D. Blitzer, I. I. Moraru and R. Iyengar, *Cell* **133**, 666 (2008).

28. A. Pierres, A. M. Benoliel and P. Bongrand, *Eur. Cells Mater.* **3**, 31 (2002).

29. E. Cretel, A. Pierres, A. M. Benoliel and P. Bongrand, *Cell. Mol. Bioeng.* **1**, 5 (2008).

30. A. Rich and A. K. Harris, *J. Cell Sci.* **50**, 1 (1981).

31. P. Clark, P. Connolly, A. S. G. Curtis, J. A. T. Dow and C. D. W. Wilkinson, *Development* **99**, 439 (1987).

32. B. Wocjiak-Stothard, A. Curtis, W. Monaghan, K. MacDonald and C. Wilkinson, *Expl. Cell Res.* **223**, 426 (1996).

33. M. J. Dalby, C. C. Berry, M. O. Riehle, D. S. Sutherland, H. Agheli and A. S. G. Curtis, *Expl. Cell Res.* **295**, 387 (2004).

34. S. Chang, Y. Popowich, R. S. Greco and B. Haimovich, *J. Vascular Surg.* **37**, 1082 (2003).

35. J. A. Champion and S. Mitragori, *Proc. Natl. Acad. Sci. USA* **103**, 49305 (2006).

36. M. Conti, G. Donati, G. Cianciolo, S. Stefoni and B. Samori, *J. Biomed. Mater. Res.* **61**, 370 (2002).

37. G. Maheswari, G. Brown, D. Lauffenburger, A. Wells and L. G. Griffith, *J. Cell Sci.* **113**, 1677 (2000).

38. W. J. Kao, D. Lee, J. C. Schense and J. A. Hubbell, *J. Biomed. Mater. Res.* **55**, 79 (2001).

39. A. Pierres, V. Monnet-Corti, A.-M. Benoliel and P. Bongrand, *Trends Cell Biol.* **19**, 428 (2009).

40. J. L. Mège, C. Capo, A. M. Benoliel and P. Bongrand, *Biophys. J.* **52**, 177 (1986).

41. S. Dewitt and M. Hallett, *J. Leukocyte Biol.* **81**, 1160 (2007).

42. A. Pierres A. M. Benoliel, D. Touchard and P. Bongrand, *Biophys. J.* **94**, 4114 (2008).

43. A. Pierres, P. Eymeric, E. Baloche, D. Touchard, A. M. Benoliel and P. Bongrand, *Biophys. J.* **84**, 2058 (2003).

44. M. Soler, C. Merant, C. Servant, M. Fraterno, C. Allasia, J. C. Lissitzky, P. Bongrand and C. Foa, *J. Leukocyte Biol.* **61**, 609 (1997).

45. G. Grynkiewicz, M. Poenie and R. Y. Tsien, *J. Biol. Chem.* **260**, 3440 (1985).

46. B. A. Kruskal, S. Shak and R. F. Maxfield, *Proc. Natl. Acad. Sci. USA* **83**, 2919 (1986).

47. E. J. Pettit and M. B. Hallett, *J. Cell Sci.* **111**, 2209 (1998).

48. J. Jongstra-Bilen, A. Puig Cano, M. Hasija, H. Xiao, C. I. E. Smith and M. I. Cybulsky, *J. Immunol.* **181**, 288 (2008).

49. E. Evans, *Cell Motility Cytoskel.* **14**, 544 (1989).

50. F. Niedergang and P. Chavrier, *Current Opinion Cell Biol.* **16**, 422 (2004).

51. R. Shamri, V. Grabovsky, J. M. Gauguet, S. Feigelson, E. Manevich, W. Kolanus, M. K. Robinson, D. E. Staunton, U. H. von Andrian and R. Alon, *Nature Immunol.* **6**, 497 (2005).

52. M. Horoyan, A. M. Benoliel, C. Capo and P. Bongrand, *Cell Biophys.* **17**, 243 (1990).

53. H. G. Döbereiner, B. Dubin-Thaler, G. Giannone, H. S. Xenias and M. P. Sheetz, *Phys. Rev. Lett.* **93**, 108105 (2004).

54. B. J. Dubin-Thaler, G. Giannone, H. G. Döbereiner and M. P. Sheetz, *Biophys. J.* **86**, 1794 (2004).

55. C. G. Galbraith, K. M. Yamada and M. P. Sheetz, *J. Cell Biol.* **159**, 695 (2002).

56. F. Brochard and J. F. Lennon, *J. Physique* **36**, 1035 (1975).

57. W. Helfrich and R. M. Servuss, *Nuovo Cimento* **3D**, 137 (1984).

58. S. Levin and R. Korenstein, *Biophys. J.* **60**, 733 (1991).

59. S. Tuvia, A. Almagor, A. Bitler, S. Levin, R. Korenstein and S. Yedgar, *Proc. Natl. Acad. Sci. USA* **94**, 5045 (1997).

60. A. Y. Krol, M. G. Grinfeldt, S. V. Levin and A. D. Smilgavichus, *Eur. Biophys. J.* **19**, 93 (1990).

61. B. Szabo, D. Selmeczi, Z. Koïnyei, E. Madarasz and N. Rozlosnik, *Phys. Rev. E* **65**, 041910 (2002).

62. J. Domke, W. J. Parak, M. George, H. E. Gaub and M. Radmacher, *Eur. Biophys. J.* **28**, 179 (1999).

63. A. S. G. Curtis, *J. Cell Biol.* **20**, 199 (1964).

64. C. S. Izzard and L. R. Lochner, *J. Cell Sci.* **21**, 129 (1976).

65. A. Zidovska and E. Sackmann, *Phys. Rev. Letters* **96**, 048103 (2006).

66. A. del Rio, R. Perez-Jimenez, R. Liu, P. Roca-Cusachs, J. M. Fernandez and M. P. Sheetz, *Science* **323**, 638 (2009).

67. S. I. Sukharev, W. J. Sigurdson, C. Kung and F. Sachs, *J. Gen. Physiol.* **113**, 525 (1999).

68. K. M. Ottemann, W. Xiao, Y.-K. Shin and D. E. Koshland, *Science* **285**, 1751 (1999).

69. Z.-Y. Sun, K. S. Kim, G. Wagner and E. L. Reinherz, *Cell* **105**, 913 (2001).

70. Z. Ma, P. Janmey and T. H. Finkel, *FASEB J.* **22**, 1002 (2008).

71. Y. Yang, S. Yuzawa and J. Schlessinger, *Proc. Natl. Acad. Sci. USA* **105**, 7681 (2008).

72. A. Viola and N. Gupta, *Nature Rev. Immunol.* **7**, 889 (2007).

73. G. I. Bell, *Science* **200**, 618 (1978).

74. W. E. Thomas, E. Trintchina, M. Forero, V. Vogel and E. Sokurenko, *Cell* **109**, 913 (2002).

75. B. Marshall, M. Long, J. W. Piper, T. Yago, R. P. McEver and C. Zhu, *Nature* **423**, 190 (2003).

76. T. W. McKeithan, *Proc. Natl. Acad. Sci USA* **92**, 5042 (1995).

77. F. L. H. Brown, *Biophys. J.* **84**, 842 (2003).

78. A. I. Sperling, J. R. Sedy, N. Manjunath, A. Kupfer, B. Ardman and J. K. Burkhardt, *J. Immunol.* **161**, 6459 (1998).

79. O. Leupin, R. Zaru, T. Laroche, S. Müller and S. Valitutti, *Current Biology* **10**, 277 (2000).

80. T. A. Ryan, J. Myeres, D. Holowka, B. Baird and W. W. Webb, *Science* **239**, 61 (1988).

81. N. G. Burroughs, Z. Lazic and P. A. van der Merwe, *Biophys. J.* **91**, 1619 (2006).

82. M. Ebisuya, K. Kondoh and E. Nihsida, *J. Cell Sci.* **118**, 2997 (2005).

83. J. W. Locasale, *Plos One* **2** (7), e627 (2007).

84. S. H. Zigmond, *J. Cell Biol.* **75**, 606 (1977).

85. C. V. Laiosa, M. Stadtfeld, H. Xie, L. de Andres-Aguayo and T. Graf, *Immunity* **25**, 731 (2006).

Soluble Amyloid-β Protein Aggregates Induce Nuclear Factor-κB Mediated Upregulation of Adhesion Molecule Expression to Stimulate Brain Endothelium for Monocyte Adhesion

Francisco J. Gonzalez-Velasquez, J. Will Reed, John W. Fuseler, Emily E. Matherly, Joseph A. Kotarek, Deborah D. Soto-Ortega and Melissa A. Moss [*]

Department of Chemical Engineering, University of South Carolina, Columbia,
South Carolina 29208, USA

Abstract

The adhesion of circulating monocytes to cerebrovascular endothelium has the potential to contribute to the pathogenesis of Alzheimer's disease (AD). In AD, the amyloid-β protein (Aβ) assembles to form fibrils that deposit within both the brain parenchyma and the cerebrovasculature. Soluble Aβ aggregates, intermediates on the pathway to mature fibril formation, are responsible for the stimulation of brain microvascular endothelial cells for increased adhesion of monocytes. However, a related role for soluble Aβ aggregates in the upregulation of endothelial cell surface adhesion molecules that support monocyte adhesion remains to be demonstrated. The current study establishes that upregulation of vascular cell adhesion molecule-1 (VCAM-1) and intercellular adhesion molecule-1 (ICAM-1) on the surface of human brain microvascular endothelial cells is selectively induced by soluble Aβ_{1-40} aggregates, while unaggregated monomer and mature fibril fail to elicit a substantial increase in adhesion molecule expression. Aβ induced changes in VCAM-1 and ICAM-1 expression are transient and are mediated by NF-κB activation. Both VCAM-1 and ICAM-1 participate in Aβ_{1-40} aggregate stimulated endothelial-monocyte adhesion. VCAM-1 plays an essential role in this cell adhesion event, while ICAM-1 presents a less pronounced contribution. These results delineate a selective role for soluble Aβ aggregates in the stimulation of brain endothelium *via* the NF-κB mediated upregulation of cell surface adhesion molecules VCAM-1 and ICAM-1 to augment monocyte adhesion, the first step for infiltration of monocytes across the blood–brain barrier where they can exacerbate AD pathogenesis.

Keywords

Alzheimer's disease, amyloid-β protein, brain microvascular endothelium, cerebral amyloid angiopathy, intercellular adhesion molecule-1, monocytes, nuclear factor-κB, vascular cell adhesion molecule-1

---

[*] To whom correspondence should be addressed. Department of Chemical Engineering, University of South Carolina, 2C02 Swearingen Engineering Center, Columbia, SC 29208, USA. Tel.: 803-777-5604; Fax: 803-777-0973; e-mail: mossme@cec.sc.edu

Surface and Interfacial Aspects of Cell Adhesion
© Koninklijke Brill NV, Leiden, 2010

Abbreviations

Aβ amyloid-β protein;

AD Alzheimer's disease;

CAA cerebral amyloid angiopathy;

BBB blood–brain barrier;

BSA bovine serum albumin;

DABCO 1,4-diazabicyclo[2.2.2]octane;

DAPI 4′,6′-diamidino-2-phenylindole dihydrochloride;

D-PBS Dulbecco's phosphate buffered saline;

DLS dynamic light scattering;

FBS fetal bovine serum;

HBMVEC human brain microvascular endothelial cell;

HIV human immunodeficiency virus;

ICAM-1 intercellular adhesion molecule-1;

IκB inhibitor of κB;

IL-1 interleukin-1;

IL-6 interleukin-6;

NF-κB nuclear factor-κB;

R_H hydrodynamic radius;

SEC size exclusion chromatography;

TNF-α tumor necrosis factor-α;

VCAM-1 vascular cell adhesion molecule-1.

1. Introduction

Cerebrovascular endothelial cells exhibit low levels of adhesion molecule expression and are joined by tight junctions [1]. These unique characteristics restrict the adhesion and subsequent transmigration of circulating immune cells to protect the brain from peripheral inflammatory responses. This blood–brain barrier (BBB), which is important to the normal functioning of the brain, is compromised in diseases that affect the central nervous system. The sequestration of circulating cells is observed in the vicinity of multiple sclerosis lesions [1–3], following ischemic stroke [1, 4–6], during cerebral malaria [7–9], and in conjunction with human immunodeficiency virus (HIV)-associated dementia [10, 11]. Similarly in Alzheimer's disease (AD), cerebrovascular deposits of the aggregated form of the amyloid-β protein (Aβ), collectively known as cerebral amyloid angiopathy (CAA), are associated with an increased adhesion of monocyte cells [12–15]. Circulating monocytes have the potential to transmigrate and subsequently transform into activated microglia [16, 17]. In fact, transgenic animal models, which exclusively develop cerebrovascular Aβ deposits, exhibit elevated levels of activated microglia within the brain [18] that correlate with the amount of CAA [15]. Monocyte-derived microglia have been observed to colocalize with parenchymal Aβ deposits where they can contribute to disease progression [17].

As in the periphery, the accumulation of circulating immune cells within the cerebrovasculature is facilitated by the upregulation of endothelial cell surface adhesion molecules [1]. In AD brain, upregulation of endothelial vascular cell adhesion molecule-1 (VCAM-1) [19] and intercellular adhesion molecule-1 (ICAM-1) [20–22] has been reported, and this upregulation is localized to areas of the cerebrovasculature that are in close proximity to Aβ deposits. Similar changes in cerebrovascular adhesion molecule expression reported for other central nervous system disorders have been linked to disease progression. Increases in vascular P-selectin, VCAM-1, and ICAM-1 are associated with demyelinating lesions of multiple sclerosis [2, 3]. These adhesion molecules define early stages of lesion formation [1], and the resulting early accumulation of monocytes correlates with subsequent demyelination [2]. Accordingly, the success of corticosteroid therapy in multiple sclerosis has been ascribed, in part, to its ability to regulate endothelial adhesion molecule expression [2]. Elevated expression of vascular adhesion molecules, including P-selectin, E-selectin, VCAM-1, and ICAM-1, observed following ischemic stroke facilitates the recruitment of leukocytes, which are proposed to contribute to post-ischemic injury [1, 4–6]. As such, anti-adhesion therapies reduce infarct volume in animal models [4–6, 23]. ICAM-1 is upregulated within the brain microvasculature of cerebral malaria patients and animal models of the disease [7–9]. It also colocalizes with the sequestration of infected erythrocytes [8, 9], whereas administration of anti-ICAM-1 releases sequestered erythrocytes [9]. In addition, genetic mutations within ICAM-1 correlate with susceptibility to cerebral malaria [9]. Increased levels of expression of VCAM-1 and E-selectin are found in the brains of patients with HIV-associated dementia [10] and facilitate the passage of infected monocytes across the BBB to contribute to brain infection and dementia [10, 11].

In AD, changes in the expression of cerebrovascular adhesion molecules may be a result of direct interaction between the endothelium and Aβ. Aβ has been shown to stimulate cultured brain microvascular endothelial cells for increased adhesion of monocytes [24–27] as well as subsequent monocyte transendothelial migration [24, 26–28]. In addition, treatment of brain endothelial cultures with Aβ upregulates expression of VCAM-1 and ICAM-1 [26]. Thus, understanding the interactions between Aβ and endothelial cells will provide insight into the contribution of vascular lesions to the pathogenesis of AD. Recently, we reported that Aβ stimulation of brain microvascular endothelial monolayers for monocyte adhesion is selectively elicited by soluble aggregates formed from Aβ_{1-40}, which appear as intermediates along the aggregation pathway between the monomeric protein and the fibrillar aggregates that deposit in the cerebrovasculature. In contrast, unaggregated monomer and mature fibrillar aggregates exhibited negligible activity [24, 29]. This result suggests that soluble Aβ aggregates, which are present both prior to cerebrovascular Aβ deposition and in equilibrium with deposited fibrillar aggregates, may stimulate the upregulation of inducible endothelial cell surface adhesion molecules that mediate the firm adhesion of monocytes, namely VCAM-1 and ICAM-1 [30, 31].

In the current study, we examine the role of NF-κB mediated upregulation of adhesion molecule expression in the Aβ stimulation of human brain microvascular endothelial cell (HBMVEC) monolayers for increased adhesion of monocytes. Immunocytochemistry is employed to demonstrate that endothelial cell surface adhesion molecules VCAM-1 and ICAM-1 are upregulated in a time-dependent and transient manner by soluble Aβ aggregates, while unaggregated monomer and mature fibril elicit less pronounced changes in adhesion molecule expression. These results parallel the selective stimulation of HBMVECs by soluble Aβ aggregates for monocyte adhesion [24]. Aβ upregulation of VCAM-1 and ICAM-1 expression is prevented by inhibition of NF-κB activation. Furthermore, antibodies that block functionality of VCAM-1 and ICAM-1 reduce Aβ stimulated increases in HBMVEC-monocyte adhesion. A complete abrogation of adhesion is observed in the presence of anti-VCAM-1 but not anti-ICAM-1. Together, these results provide evidence that soluble Aβ aggregates stimulate cerebrovascular endothelium for the adhesion of circulating monocytes *via* NF-κB mediated upregulation of VCAM-1 and ICAM-1, where VCAM-1 imparts an essential function.

2. Experimental

2.1. Materials

Aβ_{1-40} peptide was obtained from AnaSpec (San Jose, CA). Thioflavin T, RPMI 1640 medium, fetal bovine serum (FBS), sodium bicarbonate, penicillin, streptomycin, glutamine, glycine, pyruvate, β-mercaptoethanol, hydrocortisone, gelatin, Dulbecco's phosphate buffered saline (D-PBS), Calcein-AM, and 1,4-diazabicyclo [2.2.2]octane (DABCO) were obtained from Sigma (St. Louis, MO). Bovine serum albumin (BSA) was purchased from EMD Chemicals (Gibbstown, NJ). Carbobenzoxy-L-leucyl-L-leucyl-L-leucinal (Z-Leu-Leu-Leu-CHO or MG-132) was obtained from Calbiochem (San Diego, CA). CS-C Complete Medium, CS-C Serum-Free Medium, gentamycin, and CS-C Attachment Factor were purchased from Cell Systems (Kirkland, WA). Collagen was obtained from PureCol INAMED (Fremont, CA). Recombinant human tumor necrosis factor-α (TNF-α) was obtained from Promega (Madison, WI). Paraformaldehyde was purchased from Fisher (Fair Lawn, NJ). Mouse monoclonal anti-human VCAM-1 (CD106) and mouse monoclonal anti-human ICAM-1 (CD54) antibodies were obtained from R&D Systems (Minneapolis, MN). Rabbit anti-NF-κB-p65 (anti-Rel A) was obtained from Rockland (Gilbertsville, PA). Donkey normal serum, Cy3-conjugated donkey anti-rabbit antibody, and Cy3-conjugated donkey anti-mouse antibody were purchased from Jackson Immunoresearch (West Grove, PA). 4′,6′-diamidino-2-phenylindole dihydrochloride (DAPI) was purchased from Molecular Probes (Eugene, OR).

2.2. Preparation and Characterization of $A\beta_{1-40}$ Monomer and Aggregate Species

Lyophilized $A\beta_{1-40}$ peptide was stored desiccated at $-20°C$ until reconstitution as described previously [24]. Briefly, $A\beta_{1-40}$ peptide was reconstituted (2 mg/ml) in 50 mM NaOH and pre-existing aggregates were removed by size exclusion chromatography (SEC) on a Superdex 75 HR10/30 column (GE Healthcare, Piscataway, NJ). Concentrations of SEC-isolated $A\beta_{1-40}$ monomer were determined using UV absorbance with a calculated extinction coefficient of 1450 l/mol/cm [32]. Isolated $A\beta_{1-40}$ monomer was used fresh or stored at 4°C for up to 24 h.

$A\beta$ aggregates were prepared from isolated monomeric protein as described previously [24]. Briefly, monomeric $A\beta_{1-40}$ (100–200 µM) was agitated vigorously at 25°C in the presence of NaCl (3–250 mM) and 40 mM Tris-HCl (pH 8.0) for 4–20 h. Aggregation was monitored by thioflavin T fluorescence as described previously [24] by diluting an aliquot into thioflavin T (10 µM) and evaluating fluorescence using a Perkin-Elmer LS-45 luminescence spectrometer (Waltham, MA) (excitation = 450 nm, emission = 470–500 nm) with baseline (thioflavin T) subtraction. Unpurified $A\beta_{1-40}$ reaction mixtures containing monomer, soluble aggregates, and fibril were formed at low salt concentrations (3–15 mM) and shorter incubation times (4–12 h). Aggregate size was assessed as described previously [24] via hydrodynamic radius (R_H) measurements using a DynaPro MSX dynamic light scattering (DLS) instrument (Wyatt Technology, Santa Barbara, CA) and is reported as the intensity-weighted mean R_H derived from the predominant peak of the regularized histogram (Dynamics software, Wyatt Technology). Unpurified $A\beta_{1-40}$ reaction mixtures were used within 1 h of preparation. Soluble $A\beta_{1-40}$ aggregates within these reaction mixtures were separated from fibril via centrifugation and from unaggregated monomer via SEC on Superdex 75. Concentrations of SEC-isolated $A\beta_{1-40}$ soluble aggregates, expressed in monomer units, were determined using UV absorbance corrected for light scattering as described previously [32]. Isolated soluble $A\beta_{1-40}$ aggregates were characterized via R_H measurements and used for experimentation within 1 h of purification. $A\beta_{1-40}$ fibrils, the formation of which was optimized using a high salt concentration (250 mM) and longer in-

514   *F. J. Gonzalez-Velasquez* et al.

Surface and Interfacial Aspects of Cell Adhesion (2010) 509–530

cubation times (24–48 h), were isolated *via* centrifugation (15 min, $18\,000 \times g$), resuspended in 40 mM Tris-HCl (pH 8.0), and stored at 4°C for up to 1 week.

2.3. Cell Culture and Treatment

ACBRI 376 primary HBMVECs (Cell Systems, Kirkland, WA) and the human pre-monocytic cell line THP-1 (American Type Culture Collection, Rockville, MD), used in prior studies of $A\beta_{1\text{-}40}$ activation of brain microvascular endothelium, were maintained and prepared for experiments as described previously [24, 29]. HBMVECs, which were purchased at passage 2 and used between passages 4 and 8, were grown on surfaces coated with CS-C Attachment Factor. HBMVECs were sustained in gentamycin (50 µg/ml) supplemented CS-C Complete Medium and experiments were performed in gentamycin (50 µg/ml) supplemented CS-C Serum-Free Medium containing hydrocortisone (550 nM). THP-1 cells were sustained in supplemented RPMI 1640 medium containing penicillin (100 Units/ml), streptomycin (100 µg/ml), glutamine (1 mM), sodium bicarbonate (0.3%), pyruvate (1 mM), β-mercaptoethanol (0.05 mM), and FBS (10%). Both cell cultures were maintained at 37°C in a humid atmosphere of 5% CO_2 and 95% air.

HBMVEC monolayers were prepared for experiments as previously described [24]. Briefly, endothelial cells were seeded at a density of 5×10^5 cells/ml onto a cell culture surface coated with gelatin (2.0%) and collagen (100 µg/ml): 96-well flat-bottom tissue culture plates (Cellstar, Monroe, NC) for adhesion assays or 22 × 22 mm glass coverslips (Costar, Corning Inc., Acton, MA) for immunocytochemistry. HBMVEC monolayers were sustained until confluence (5–6 days). Confluent monolayers exhibited a transendothelial electrical resistance (400–600 Ω/cm^2) [29] similar to that reported for other BBB models [33], permeability coefficients (1.1×10^{-5}–3.6×10^{-5} cm/s) [29] within the range expected for an intact BBB model [33], and the localization of occludin and zonula occludin-1 at cell borders [29] confirming the presence of tight junction complexes characteristic of endothelial cells of the BBB [34–36]. In addition, the absence of immunocytochemical staining for muscle α-actin, a protein expressed by vascular smooth muscle cells, demonstrates the purity of the HBMVEC cultures [29].

Confluent monolayers were activated with $A\beta_{1\text{-}40}$ for 24 h prior to immunocytochemical evaluation of adhesion molecule expression or assessment of monocyte adhesion. The desired preparation of $A\beta_{1\text{-}40}$ (unaggregated monomer, isolated soluble aggregates, resuspended fibril, or an unpurified reaction mixture containing monomer, soluble aggregates, and fibril) was diluted into media to achieve a final concentration of 0.5 or 5 µM. Induction of HBMVEC toxicity within a treatment time of 24 h was previously shown to require $A\beta_{1\text{-}40}$ concentrations greater than 10 µM [24]; thus, the employed $A\beta_{1\text{-}40}$ treatment concentrations are sub-toxic. Parallel treatment with an equivalent dilution of buffer into media served as a negative control, while monolayers treated with TNF-α (10 Units/well) served as a positive control. For adhesion molecule blockage, HBMVEC monolayers that had been activated *via* 24 h Aβ_{1-40} treatment were subsequently treated for 1 h with mouse monoclonal anti-human VCAM-1 (30 µg/ml) or mouse monoclonal anti-human ICAM-1 (10 µg/ml) antibodies. For inhibition of NF-κB activation, HBMVEC monolayers were first treated with 3 µM MG-132 for 30 min followed by activation *via* 24 h Aβ_{1-40} treatment in the presence of MG-132. Treatment with 3 µM MG-132 in the absence of Aβ_{1-40} served as an additional negative control. Each treatment was performed with two or three repetitions.

As described previously [24], THP-1 cells were labeled using Calcein-AM. Briefly, cells (2 × 10^6 cells/ml) were incubated in the presence of Calcein-AM (10 µM) for 30 min. Labeled cells were pelletted, washed, and suspended in D-PBS at a concentration of 2 × 10^5 cells/ml.

2.4. Adhesion Assay

Adhesion assays were performed following 24 h Aβ_{1-40} treatment. As described previously [24], activated monolayers were washed and adhesion of Calcein-labeled THP-1 cells (2 × 10^4 cells/well) was allowed to proceed for 30 min (37°C, 5% CO$_2$). Nonadherent cells were removed by gently washing 3 times with D-PBS, and the number of adherent cells was assessed by Calcein fluorescence employing a BioTek Synergy 2 microplate reader (Winooski, VT) (excitation = 485 ± 10 nm; emission = 535 ± 12.5 nm) with baseline (D-PBS) subtraction. Results are reported as the percentage of adherent cells. Each treatment was performed in six replicates, and values are reported as the mean ± standard error.

2.5. Immunocytochemistry

Immunocytochemistry to visualize cell surface expression of VCAM-1 or ICAM-1 was performed following 24 h Aβ_{1-40} treatment. Immunocytochemistry to determine the time course for cell surface expression of VCAM-1 or ICAM-1 was performed following Aβ_{1-40} treatment times ranging from 1 h to 48 h. Immunocytochemistry to visualize cellular localization of NF-κB was performed following 30 min Aβ_{1-40} treatment. As described previously [24], activated monolayers were washed with D-PBS, fixed with paraformaldehyde (3.0%), permeabilized with Triton X-100 (0.1%) and glycine (0.01 M), and blocked with normal donkey serum (5%) in BSA (1%). Cells were incubated overnight at 4°C with the respective primary antibody, rabbit polyclonal anti-NF-κB-p65 (1:200), mouse monoclonal anti-human VCAM-1 (1:1000), or mouse monoclonal anti-human ICAM-1 (1:100). Monolayers were then blocked with normal donkey serum (5%) in BSA (1%) and bound primary antibodies were detected with Cy3-conjugated donkey anti-rabbit secondary antibody (1:120) or Cy3-conjugated donkey anti-mouse secondary antibody (1:200) overnight (4°C). Monolayers were additionally stained with DAPI (1:5000). Samples were mounted on slides with DABCO solution for visualization under a Zeiss LSM confocal microscope (Carl Zeiss, Thornwood, NY) using a plan-apochromat 63×/1.4 oil DIC immersion objective (Carl Zeiss). Images of ten different fields were acquired for each replicate of each HBMVEC treatment. Micrographs were visualized and converted to TIFF documents using LSM 5 Image Browser software (Carl Zeiss) and gamma adjusted using Adobe Photoshop 7.0 (Adobe Systems Incorporated, San Jose, CA).

2.6. Optical Image Analysis

ICAM-1 and VCAM-1 cell surface expression were quantified using optical image analysis. Color images were separated, and the red channel representing the fluorescently labeled adhesion molecules was isolated and converted to 8-bit monochrome images. The fluorescently-labeled adhesion molecules in the monochrome image field were thresholded as regions of interest, and the total area occupied by fluorescently-labeled adhesion molecules expressed on the cells within the field was calculated from the integrated morphometry subroutine of MetaMorph Image Analysis software (MDS Analytical Technologies, Downingtown, PA). Using the software's inclusive threshold subroutine, the lower threshold limit was set to the maximal value of adhesion molecule fluorescence observed in the activated cells and the maximal upper limit was set at 255. These limits as well as values of image brightness, contrast, and gamma function were kept constant for all images analyzed. Using the software's optical calipers, measurements were refined by setting boundary conditions for acceptance of an adhesion molecule fluorescent signal. This procedure minimized the contributions of any non-specific or background auto-fluorescence. Background fluorescence was negligible in untreated HBMVECs compared to experimentally treated cells.

2.7. Statistical Analysis

Statistical analysis was performed with a one-way ANOVA using Prism 5 software (GraphPad Software, La Jolla, CA). Dunnett's test for multiple comparisons was used to identify groups with means significantly different from both the negative control and the unpurified $A\beta_{1-40}$ reaction mixture treatment. $P < 0.05$ was considered significant.

3. Results and Discussion

3.1. $A\beta_{1-40}$ Upregulates Cell Surface Adhesion Molecule Expression in HBMVEC Monolayers

Changes in the expression of HBMVEC adhesion molecules VCAM-1 and ICAM-1, members of the immunoglobulin family of adhesion molecules that are reported to support the firm adhesion of monocyte cells [30, 31], were assessed by performing immunocytochemistry with HBMVECs grown to confluence under conditions that promote barrier properties. HBMVEC monolayers prepared in this manner exhibit a transendothelial electrical resistance and permeability coefficients similar to those observed for other BBB models and display the formation of endothelial tight junctions [24, 29]. Untreated quiescent HBMVEC monolayers expressed very low levels of both VCAM-1 (Fig. 1(a)) and ICAM-1 (Fig. 1(d)), demonstrating low cell surface expression as well as the absence of cytoplasmic pools of these proteins. These results are in agreement with previous studies that have reported low basal levels of VCAM-1 [37–40] and ICAM-1 [37, 39–43] expression in endothelial cells isolated from human brain tissue. In the current study, only a fraction of cells demonstrated staining for VCAM-1, while ICAM-1 staining exhibited a punctate pattern that was detected on most cells. Similarly, higher levels of basal ICAM-1 expression relative to VCAM-1 expression have been reported previously for other HBMVEC cultures [38–40].

Upon stimulation of HBMVEC monolayers with 5 μM of an unpurified $A\beta_{1-40}$ reaction mixture containing monomer, soluble aggregates, and fibril, a significant upregulation of both VCAM-1 (Fig. 1(c)) and ICAM-1 (Fig. 1(f)) was observed. An increase in the expression of VCAM-1 (Fig. 1(b)) and ICAM-1 (Fig. 1(e)) was also observed following treatment with the cytokine TNF-α, known to upregulate expression of these cell surface adhesion molecules in human brain endothelial cells [37, 40, 41, 43]. These results confirm the findings of Kalra and coworkers, in which an ELISA assay was employed to observe increases in the expression of VCAM-1 and ICAM-1 in HBMVECs following treatment with an uncharacterized preparation of $A\beta_{1-40}$ [26]. The current results extend these observations by using immunocytochemistry to visualize the $A\beta$ induced adhesion molecule expression on the cell surface. Upregulated expression of both VCAM-1 and ICAM-1 displayed an evenly distributed but punctate staining pattern resembling that observed for basal levels of ICAM-1 expression. A similar pattern of granular staining evenly distributed across the cell surface has also been described for the expression of both VCAM-1 [38] and ICAM-1 [41] following stimulation of HBMVECs with cytokines or lipopolysaccharide. The even distribution of this expression pattern is reflective of the role of these upregulated adhesion molecules in extracellular adhesion, rather than in maintenance of cell–cell contacts within the monolayer, while the punctate nature of the staining may indicate adhesion molecule clustering and/or localization to microvilli, both of which facilitate binding to monocyte cell surface ligands [44].

The upregulation of cell surface adhesion molecules has also been observed following the treatment of endothelial cells of brain origin with other disease-associated proteins. Oncostatin M, a member of the interleukin-6 (IL-6) cytokine family which is detected in multiple sclerosis lesions, selectively upregulates the surface expression of ICAM-1, but not VCAM-1, in primary cultures of human cerebral endothelial cells [39]. Angiotensin II, reported to play a role in ischemic stroke, induces elevated surface expression of ICAM-1 in primary cultures of rat brain endothelial cells [45]. Vasoactive endothelins [40] and C-reactive protein [46], also implicated in stroke disorders, stimulate upregulation of both VCAM-1 and ICAM-1 on HBMVECs and mouse brain endothelial cells, respectively. The HIV-

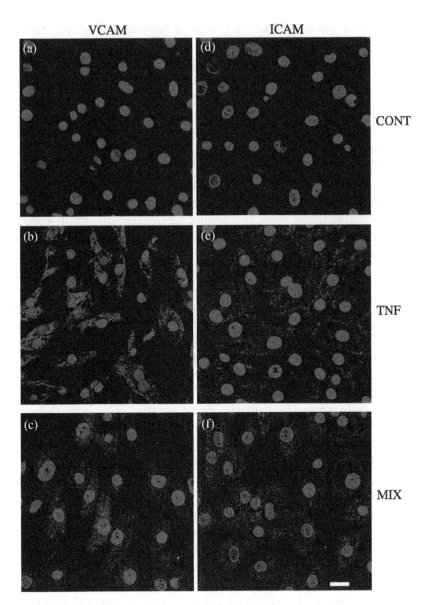

Figure 1. Effect of unpurified A$\beta_{1\text{-}40}$ reaction mixture on HBMVEC surface expression of VCAM-1 and ICAM-1. HBMVEC monolayers grown to confluence on glass coverslips were incubated for 24 h alone (control) (CONT, a, d), with 10 Units/ml TNF-α (TNF, b, e), or with 5 μM unpurified A$\beta_{1\text{-}40}$ reaction mixture ($R_H = 46$ nm) (MIX, c, f). Immunofluorescence staining was performed for VCAM-1 (VCAM, a, b, c) or ICAM-1 (ICAM, d, e, f) (red) in conjunction with nuclear (DAPI) (blue) staining as described in the *Experimental* section. Negligible background was observed in the absence of VCAM-1 or ICAM-1 primary antibodies (not shown). Images shown are representative of three independent experiments for which ten separate images were acquired from each of two replicates of individual HBMVEC treatments. Images are shown relative to a scale bar of 20 μm.

associated protein gp120, which is secreted during HIV infection, is capable of upregulating HBMVEC expression of both VCAM-1 and ICAM-1 [47]. Similarly, membrane associated and secreted proteins from *P. falciparum*-infected erythrocytes, which are present in cerebral malaria, increase expression of ICAM-1 on the surface of HBMVECs [48]. Thus, like other diseases of the central nervous system, inflammatory processes associated with AD may be due, in part, to elevated expression of endothelial adhesion molecules within the cerebrovasculature induced by a disease-related protein.

3.2. Soluble $A\beta_{1-40}$ Aggregates Selectively Stimulate Increases in HBMVEC Adhesion Molecule Expression

We previously reported that the activation of HBMVEC monolayers for increased monocyte adhesion could be induced only by soluble $A\beta_{1-40}$ aggregates and not by unaggregated monomer or mature fibril [24]. To determine whether soluble $A\beta_{1-40}$ aggregates also selectively upregulate the expression of endothelial adhesion molecules, surface expression of VCAM-1 and ICAM-1 was evaluated for HBMVEC monolayers incubated with unaggregated $A\beta_{1-40}$ monomer, mature $A\beta_{1-40}$ fibril, or isolated soluble $A\beta_{1-40}$ aggregates. Following treatment of HBMVEC monolayers with either 5 µM unaggregated $A\beta_{1-40}$ monomer, isolated *via* SEC, or 5 µM mature $A\beta_{1-40}$ fibril, isolated *via* centrifugation, cell surface expression of VCAM-1 (Fig. 2(b, c)) and ICAM-1 (Fig. 2(f, g)) was increased only marginally over basal levels of expression (Fig. 2(a, e)). In contrast, a 10-fold lower concentration of soluble $A\beta_{1-40}$ aggregates, depleted of mature fibrils *via* centrifugation and resolved from unaggregated monomer *via* SEC, induced the expression of both VCAM-1 (Fig. 2(d)) and ICAM-1 (Fig. 2(h)) at a level similar to that observed following treatment with the unpurified $A\beta_{1-40}$ reaction mixture. Measurements of the total area occupied by fluorescently labeled adhesion molecules were employed to quantify the cell surface expression of VCAM-1 and ICAM-1. Fluorescent area measurements confirmed the pronounced increase in both VCAM-1 and ICAM-1 expression induced by 0.5 µM soluble $A\beta_{1-40}$ aggregates as well as unchanged expression of these adhesion molecules following HBMVEC treatment with 5 µM $A\beta_{1-40}$ monomer (Fig. 3). Treatment of HBMVEC monolayers with 5 µM $A\beta_{1-40}$ fibril elicited a comparatively small, although significant, increase in the expression of both VCAM-1 and ICAM-1, indicating that these mature aggregates might exhibit a low level of activity. Alternatively, small amounts of soluble aggregates that have dissociated from mature fibrils may be responsible for this activity. In a previous study, we reported that only soluble $A\beta_{1-40}$ aggregates, but not $A\beta_{1-40}$ monomer or mature fibril, were capable of activating HBMVEC monolayers for increased adhesion of monocyte cells [24]. The $A\beta$ induced expression of HBMVEC adhesion molecules parallels the $A\beta$ stimulation of HBMVEC monolayers for monocyte adhesion, providing additional evidence for the physiological activity of soluble $A\beta$ aggregates in the activation of cerebrovascular endothelium.

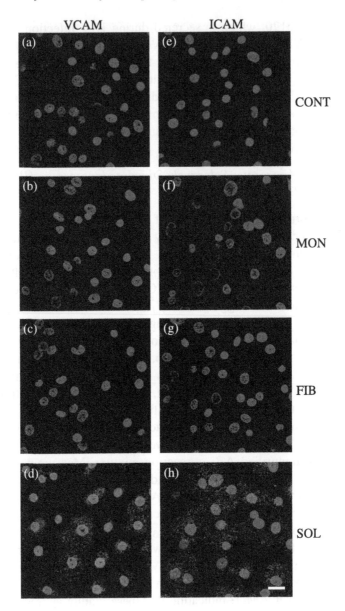

Figure 2. Effect of $A\beta_{1-40}$ monomer, fibril, and soluble aggregates on HBMVEC surface expression of VCAM-1 and ICAM-1. HBMVEC monolayers grown to confluence on glass coverslips were incubated for 24 h alone (control) (CONT, a, e), with 5 μM SEC-isolated $A\beta_{1-40}$ monomer (MON, b, f), with 5 μM $A\beta_{1-40}$ fibril isolated *via* centrifugation (FIB, c, g), or with 0.5 μM SEC-isolated soluble $A\beta_{1-40}$ aggregates ($R_H = 54$ nm) (SOL, d, h). Immunofluorescence staining was performed for VCAM-1 (VCAM, a–d) or ICAM-1 (ICAM, e–h) (red) in conjunction with nuclear (DAPI) (blue) staining as in Fig. 1. Images shown are representative of three independent experiments for which ten separate images were acquired from each of two replicates of individual HBMVEC treatments. Images are shown relative to a scale bar of 20 μm.

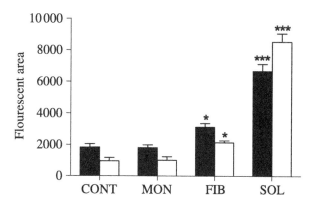

Figure 3. Comparison of VCAM-1 and ICAM-1 expression following $A\beta_{1-40}$ monomer, fibril, and soluble aggregate treatment of HBMVECs. HBMVEC monolayers were grown to confluence and incubated as in Fig. 2. Immunofluorescence staining was performed as in Fig. 2, and the quantity of cell surface adhesion molecules expressed was determined by optical image analysis as described in the *Experimental* section. Results are reported as the total area occupied by fluorescently labeled VCAM-1 (closed bars) or ICAM-1 (open bars) and reflect the average of measurements taken from six randomly selected images per treatment, acquired from duplicate samples. Results are representative of three independent experiments. $*P < 0.05$; $***P < 0.001$. Error bars indicate standard error.

Soluble $A\beta$ aggregates, including protofibrils, oligomers, and $A\beta$-derived diffusible ligands that are present both prior to $A\beta$ deposition and in equilibrium with deposited fibrillar aggregates, have been shown to inhibit long-term potentiation [49, 50], induce synaptic loss [49, 50], and impair learning and memory [51, 52]. Soluble $A\beta$ aggregate upregulation of adhesion molecules within the cerebrovasculature initiates the recruitment of circulating monocytes, which have the potential to transform into activated microglia. Recruitment of circulating monocytes to AD brain is further promoted by the presence of chemokines [53]. While local microglia can become activated in AD, increasing evidence suggests that a significant proportion of activated microglia originate from blood-derived monocytes [16, 17]. Furthermore, it is suggested that the accelerated rate of microglia turnover associated with AD results in an elevated proportion of monocyte-derived microglia within the brain [17, 54]. Activated microglia can release cytokines capable of inhibiting long-term potentiation [55] and inducing neurotoxicity [56]. Microglia may also contribute to amyloid plaque deposition by stimulating neuronal $A\beta$ production *via* cytokine release [55, 56] and facilitating $A\beta$ assembly *via* concentration of the protein within cytoplasmic channels [17]. In fact, monocyte-derived microglia have been identified in association with amyloid plaques [17, 54]. Thus, findings from the current study contribute to increasing evidence that supports a principal role for soluble $A\beta$ aggregates in AD pathogenesis [49, 50]. In addition, these results are in agreement with observations that chemokine-dependent microglial accumulation, which likely occurs through the recruitment of blood-derived monocytes, begins prior to the deposition of fibrillar amyloid plaques [53].

3.3. $A\beta_{1-40}$ Aggregate Stimulation of HBMVEC Adhesion Molecule Expression is Transient

In human brain endothelial cultures, a transient upregulation of cell surface adhesion molecules has been reported following cytokine stimulation [38, 41, 57]. To determine whether $A\beta_{1-40}$ aggregates also induce transient expression of cell surface adhesion molecule expression in HBMVECs, surface expression of VCAM-1 and ICAM-1 was evaluated at 1 h, 5 h, 10 h, 24 h and 48 h following treatment of HBMVEC monolayers with 5 μM of an unpurified $A\beta_{1-40}$ reaction mixture. $A\beta$ induced expression of VCAM-1 and ICAM-1 was observed to follow a similar time-course. At 1 h, no change in either VCAM-1 or ICAM-1 was observed (data not shown). A treatment time of 5 h increased expression over basal levels (Fig. 4(a, e)), with a more pronounced increase observed for VCAM-1 expression (Fig. 4(i)). Significant increases ($P < 0.001$) in expression became apparent following 10 h of treatment (Fig. 4(b, f, i)). $A\beta$ stimulated expression of VCAM-1 and ICAM-1 peaked near 24 h (Fig. 4(c, g, i)). At 48 h, levels of expression exhibited a significant decline (Fig. 4(d, h)) and returned to basal values (Fig. 4(i)), demonstrating the degradation of newly synthesized protein and the absence of VCAM-1 and ICAM-1 storage within cytoplasmic pools.

The observed time-course is similar to that reported by Wong and Dorovini-Zis for TNF-α and interleukin-1 (IL-1) stimulation of VCAM-1 [38] and ICAM-1 [41] in HBMVECs. However, a more rapid increase in the $A\beta_{1-40}$ induced expression of both VCAM-1 and ICAM-1 was reported by Kalra and co-workers for primary cultures of HBMVECs [26]. Using an ELISA to evaluate adhesion molecule expression, upregulation of VCAM-1 and ICAM-1 was observed as early as 30 min and a plateau in expression was observed between 4 h and 16 h. Because measurements at longer incubation times were not examined, the transient nature of VCAM-1 and ICAM-1 could not be ascertained. Variable time-courses for adhesion molecule upregulation have also been reported following treatment of primary brain endothelial cell cultures with cytokines. For example, upregulated expression of VCAM-1 following TNF-α stimulation has been reported to peak as early as 12 h [38], while other studies have demonstrated that peak expression does not occur until after 72 h of cytokine exposure [57]. These discrepancies may result from differences in the isolation of brain endothelial cell cultures among the studies. In fact, other researchers have noted a minimal contamination of freshly isolated cultures with both pericytes and glial cells [26, 58], which could contribute to endothelial activation *via* production of cytokines.

Other disease-associated proteins are known to transiently upregulate cell surface adhesion molecules in primary endothelial cell cultures derived from human brain tissue. HIV-associated gp120 protein [59], multiple sclerosis-associated cytokine oncostatin M [39], and membrane-associated and excreted proteins expressed by infected erythrocytes of cerebral malaria [48] all induce the transient upregulation of ICAM-1. Thus, a common mechanism may be involved in stimulation of

Printed and bound by CPI Group (UK) Ltd, Croydon, CR0 4YY

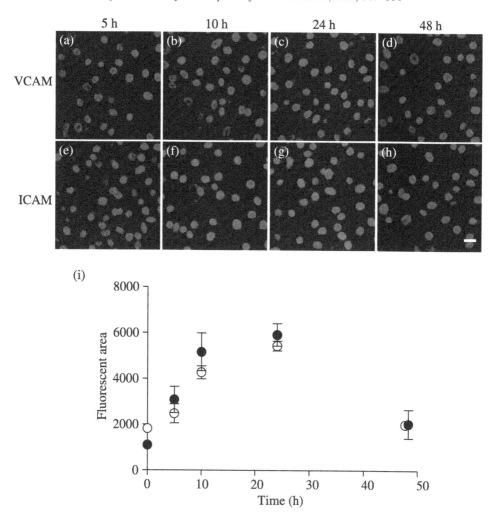

Figure 4. Time course of Aβ$_{1-40}$ aggregate-induced HBMVEC surface expression of VCAM-1 and ICAM-1. HBMVEC monolayers grown to confluence on glass coverslips were incubated with 5 μM unpurified Aβ$_{1-40}$ reaction mixture ($R_H = 56$ nm) for 5 h (a, e), 10 h (b, f), 24 h (c, g), or 48 h (d, h). Immunofluorescence staining was performed for VCAM-1 (VCAM, a–d) or ICAM-1 (ICAM, e–h) (red) in conjunction with nuclear (DAPI) (blue) staining as in Fig. 1, and the quantity of cell surface adhesion molecules expressed (i) was assessed as in Fig. 3 and reported as the total area occupied by fluorescently labeled VCAM-1 (●) or ICAM-1 (○) at each incubation time. Images shown are representative of ten separate images acquired from each of three replicates of individual HBMVEC treatments. Calculated fluorescent areas reflect the average of measurements taken from six randomly selected images per incubation time, acquired from triplicate samples. Results are representative of two independent experiments. Images are shown relative to a scale bar of 20 μm. Error bars indicate standard error.

cerebrovascular endothelium for cell surface adhesion molecule expression by both cytokines and disease-associated proteins.

3.4. NF-κB Mediates Aβ$_{1-40}$ Aggregate Upregulation of HBMVEC Adhesion Molecule Expression

The transient activation of cytokine induced endothelial cell surface adhesion molecule expression results from the involvement of NF-κB signaling pathways. NF-κB simultaneously stimulates the transcription and translation of both adhesion molecules and the inhibitor of κB (IκB). The latter binds to NF-κB to reverse its activation [60–64]. We have observed that NF-κB activation is required for Aβ$_{1-40}$ aggregate stimulation of HBMVEC monolayers for monocyte adhesion (data not shown). To determine if NF-κB activation is also required for Aβ$_{1-40}$ aggregate stimulation of HBMVEC adhesion molecule expression, the NF-κB inhibitor MG-132 was employed. MG-132 impedes the catalytic subunit of the proteasome [65] to prevent proteasomal degradation of IκB, which is required to unmask the NF-κB nuclear localization signal for activation, translocation to the nucleus, and DNA binding [60–64]. MG-132 effectively prevents Aβ$_{1-40}$ aggregate activation and nuclear translocation of HBMVEC NF-κB (Fig. 5(a, d)).

When HBMVEC monolayers were treated with MG-132 prior to and during stimulation with 5 μM unpurified Aβ$_{1-40}$ reaction mixture, the level of expression for both VCAM-1 (Fig. 5(e)) and ICAM-1 (Fig. 5(f)) was increased only marginally over basal levels of expression and was markedly less pronounced than the expression observed in the absence of MG-132 treatment (Fig. 5(b, c)). This result confirms a regulatory role for NF-κB activation in the Aβ induced upregulation of HBMVEC adhesion molecules VCAM-1 and ICAM-1. A direct role for NF-κB activation has also been demonstrated for the stimulation of ICAM-1 expression in HBMVECs following treatment with HIV-associated protein gp120 [59] or membrane-associated and excreted proteins expressed by infected erythrocytes of cerebral malaria [48]. In addition, NF-κB activation parallels the upregulation of ICAM-1 by ischemic stroke-associated protein angiotensin II in primary rat brain endothelial cells [45]. Thus, NF-κB signaling pathways are a common regulator of the stimulation of cerebrovascular endothelium for adhesion molecule expression by a number of disease-associated proteins.

3.5. Aβ Stimulation of HBMVEC Monolayers for Increased Adhesion of Monocytes is Mediated by Upregulated Expression of HBMVEC Adhesion Molecules

While the mechanism by which circulating monocytes adhere and transmigrate through the BBB is not fully elucidated, endothelial VCAM-1 and ICAM-1 are known to mediate firm adhesion of circulating monocytes within the periphery *via* recognition, respectively, of β_1 integrin $\alpha_4\beta_1$ (or VLA-4) or β_2 integrins $\alpha_L\beta_2$ (or LFA-1) and $\alpha_M\beta_2$ (or Mac-1) expressed on the monocyte cell surface [30, 31]. To delineate the role of VCAM-1 and ICAM-1 upregulation in the Aβ$_{1-40}$ aggregate stimulation of HBMVEC monolayers for monocyte adhesion, Aβ stimulated HBMVEC monolayers were incubated with anti-VCAM-1 or anti-ICAM-1

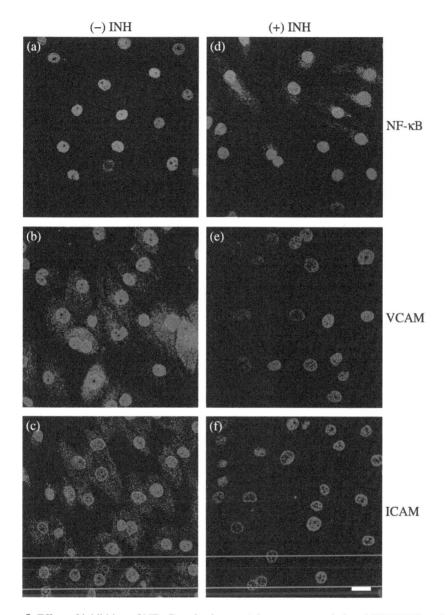

Figure 5. Effect of inhibition of NF-κB activation on Aβ_{1-40} aggregate-induced HBMVEC surface expression of VCAM-1 and ICAM-1. HBMVEC monolayers grown to confluence on glass coverslips were incubated with 5 μM unpurified Aβ_{1-40} reaction mixture ($R_H = 56$ nm) in the absence ((−) INH, a–c) or presence ((+) INH, d–f) of 3 μM MG-132, an inhibitor of NF-κB activation. Immunofluorescence staining was performed for NF-κB (NF-κB, a, d), VCAM-1 (VCAM, b, e), or ICAM-1 (ICAM, c, f) (red) in conjunction with nuclear (DAPI) (blue) staining as in Fig. 1. Images shown are representative of two independent experiments for which ten separate images were acquired from each of three replicates of individual HBMVEC treatments. Images are shown relative to a scale bar of 20 μm.

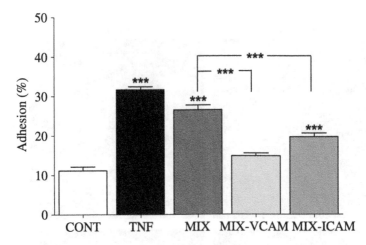

Figure 6. Functional role of VCAM-1 and ICAM-1 in $A\beta_{1-40}$ aggregate stimulation of HBMVEC monolayers for monocyte adhesion. HBMVEC monolayers grown to confluence in 96-well plates were activated for 24 h alone (CONT), with 10 Units/ml TNF-α (TNF), with 5 μM unpurified $A\beta_{1-40}$ reaction mixture ($R_H = 65$ nm) (MIX), with 5 μM unpurified $A\beta_{1-40}$ reaction mixture ($R_H = 65$ nm) followed by 1 h treatment with anti-VCAM-1 (MIX-VCAM), or with 5 μM unpurified $A\beta_{1-40}$ reaction mixture ($R_H = 65$ nm) followed by 1 h treatment with anti-ICAM-1 (MIX-ICAM). Activated monolayers were washed to remove residual protein, and adhesion of Calcein-labeled THP-1 monocytes (2×10^4 cells/well) was assessed *via* fluorescence as described in the *Experimental* section. *** $P < 0.001$. Results are representative of three independent experiments performed with six repetitions. Error bars represent standard error.

to specifically block adhesion molecule function prior to assessment of monocyte adhesion. When HBMVEC monolayers were treated with 5 μM unpurified $A\beta_{1-40}$ reaction mixture, a 2.4-fold increase was observed in the adhesion of THP-1 monocytes (Fig. 6), in agreement with previous observations of $A\beta$ stimulation of HBMVEC monolayers for monocyte adhesion [24, 26, 27]. When $A\beta$ stimulated HBMVEC monolayers were incubated with anti-VCAM-1, levels of monocyte adhesion were decreased to basal levels. Incubation of $A\beta$ stimulated HBMVEC monolayers with anti-ICAM-1 also decreased monocyte adhesion; however, a less pronounced 45% decrease in the augmented adhesion was observed. Thus, while both VCAM-1 and ICAM-1 play a role in the adhesion of monocytes to $A\beta$ activated cerebrovascular endothelium, VCAM-1 mediates a more significant portion of monocyte recruitment.

These results are in agreement with other studies that report a role for VCAM-1 and ICAM-1 in the adhesion of monocytes to brain endothelium. VCAM-1 has been shown to be involved in the adhesion of monocytes to monolayers of IL-1 activated rat cerebrovascular endothelial cells [66] and monolayers of endothelial cells isolated from a rat model of multiple sclerosis [66], as well as vessels within brain tissue sections prepared from mouse [67] and monkey [68] models of HIV-associated dementia. ICAM-1 is reported to be involved in the adhesion of monocytes to rat cerebrovascular cells activated with IL-1 [66, 69] or TNF-α [69]. Among studies that have compared the relative roles of these two adhesion molecules, VCAM-1 and ICAM-1 exhibited similar contributions in the adhesion of monocytes to IL-1 activated rat cerebrovascular endothelial cell monolayers [66], while VCAM-1 was observed to play a greater role than ICAM-1 in the adhesion of mononuclear cells, including monocytes and lymphocytes, to TNF-α stimulated human cerebral endothelial cells [57].

A role for VCAM-1 in monocyte adhesion to Aβ activated HBMVEC was also reported by Kalra and coworkers [26], and a role for ICAM-1, but not VCAM-1, was demonstrated for monocyte adhesion to Aβ activated primary rat brain microvascular endothelial cells [27]. In contrast, VCAM-1 did not contribute to monocyte transmigration across Aβ stimulated monolayers comprised of primary rat brain microvascular endothelial cells [27], and ICAM-1 did not contribute to monocyte transmigration across either rat [27] or human [26] brain microvascular endothelial monolayers stimulated by Aβ. However, a role for VCAM-1, but not for ICAM-1, has been demonstrated for the transendothelial migration of monocytes across monolayers of endothelial cells isolated from a rat model of multiple sclerosis [66]. Thus, individual adhesion molecules make unique contributions to different steps of monocyte infiltration into the brain, which may be both stimulus- and species-specific. Elucidating these contributions to monocyte recruitment in human AD brain will provide insight into the potential benefits of anti-inflammatory therapies.

4. Conclusion

The upregulated expression of endothelial cell surface adhesion molecules and the concurrent increase in the recruitment of circulating cells to the brain are associated with the pathogenesis of several diseases affecting the central nervous system, including AD. In AD, small soluble Aβ aggregates that appear along the aggregation pathway and are present prior to Aβ deposition as well as in equilibrium with deposited fibrillar aggregates have been implicated in both neuronal and vascular aspects of disease progression. Using an *in vitro* BBB model comprised of primary HBMVEC monolayers, the current study demonstrates that soluble Aβ_{1-40} aggregates, but not unaggregated monomer or mature fibril, induce the upregulation of cell surface adhesion molecules VCAM-1 and ICAM-1 in cultures of endothelial cells isolated from human brain microvasculature (Figs 2 and 3). This finding extends previous results that demonstrate a role of soluble Aβ_{1-40} aggregates in the activation of cerebrovascular endothelial monolayers, including the stimulation of monocyte adhesion and transmigration as well as increased permeability. Soluble Aβ_{1-40} aggregate stimulation of endothelial cell surface adhesion molecule expression is transient (Fig. 4), with expression reaching a maximum value near 24 h and declining thereafter, and is mediated by the activation of NF-κB (Fig. 5). These

Surface and Interfacial Aspects of Cell Adhesion (2010) 509–530

observations are characteristic of the upregulation of brain endothelial adhesion molecule expression by cytokines and other disease-associated proteins. Upregulated expression of both VCAM-1 and ICAM-1 participates in the adhesion of monocytes to $A\beta_{1-40}$ activated HBMVEC monolayers, with VCAM-1 playing a predominant role (Fig. 6). Together, these findings represent one step in elucidating the mechanism by which $A\beta$ aggregates facilitate the recruitment of circulating monocytes into AD brain. Future experimentation employing transgenic models of AD will be needed to further confirm the pathogenic role of monocytes recruited *via* $A\beta$ activation of brain endothelium. Discerning additional contributions of endothelial cell surface adhesion molecules within this process through both *in vitro* and *in vivo* experimentation will provide insight into the potential of anti-adhesion therapies, which have shown promise in other central nervous system disorders.

Acknowledgements

This work was made possible, in part, by a Beginning Grant-In-Aid (0565387U) from the American Heart Association, Mid-Atlantic Affiliate to MAM and by NIH Grant Number P20 RR-016461 from the National Center for Research Resources. Its contents are solely the responsibility of the authors and do not necessarily represent the official views of the NIH.

References

1. F. Gavins, G. Yilmaz and D. N. Granger, *Microcirculation* **14**, 667 (2007).
2. E. E. McCandless and R. S. Klein, *Expert Rev. Mol. Med.* **9**, 1 (2007).
3. K. A. Brown, *Int. Immunopharmacol.* **1**, 2043 (2001).
4. C. J. Frijns and L. J. Kappelle, *Stroke* **33**, 2115 (2002).
5. G. H. Danton and W. D. Dietrich, *J. Neuropathol. Exp. Neurol.* **62**, 127 (2003).
6. G. del Zoppo, I. Ginis, J. M. Hallenbeck, C. Iadecola, X. Wang and G. Z. Feuerstein, *Brain Pathol.* **10**, 95 (2000).
7. N. H. Hunt and G. E. Grau, *Trends Immunol.* **24**, 491 (2003).
8. H. Fujioka and M. Aikawa, *Microb. Pathog.* **20**, 63 (1996).
9. S. J. Chakravorty and A. Craig, *Eur. J. Cell Biol.* **84**, 15 (2005).
10. C. L. Maslin, K. Kedzierska, N. L. Webster, W. A. Muller and S. M. Crowe, *Curr. HIV Res.* **3**, 303 (2005).
11. J. Rappaport, J. Joseph, S. Croul, G. Alexander, L. Del Valle, S. Amini and K. Khalili, *J. Leukoc. Biol.* **65**, 458 (1999).
12. M. Yamada, Y. Itoh, M. Shintaku, J. Kawamura, O. Jensson, L. Thorsteinsson, N. Suematsu, M. Matsushita and E. Otomo, *Stroke* **27**, 1155 (1996).
13. M. L. Maat-Schieman, S. G. van Duinen, A. J. Rozemuller, J. Haan and R. A. Roos, *J. Neuropathol. Exp. Neurol.* **56**, 273 (1997).
14. T. Uchihara, H. Akiyama, H. Kondo and K. Ikeda, *Stroke* **28**, 1948 (1997).
15. J. Miao, M. P. Vitek, F. Xu, M. L. Previti, J. Davis and W. E. Van Nostrand, *J. Neurosci.* **25**, 6271 (2005).

16. C. Kaur, A. J. Hao, C. H. Wu and E. A. Ling, *Microsc. Res. Techniq.* **54**, 2 (2001).

17. R. G. Nagele, J. Wegiel, V. Venkataraman, H. Imaki, K. C. Wang and J. Wegiel, *Neurobiol. Aging* **25**, 663 (2004).

18. J. Miao, F. Xu, J. Davis, I. Otte-Holler, M. M. Verbeek and W. E. Van Nostrand, *Am. J. Pathol.* **167**, 505 (2005).

19. H. Akiyama, T. Kawamata, T. Yamada, I. Tooyama, T. Ishii and P. L. McGeer, *Acta Neuropathol.* **85**, 628 (1993).

20. M. M. Verbeek, I. Otte-Holler, P. Wesseling, D. J. Ruiter and R. M. de Waal, *Acta Neuropathol.* **91**, 608 (1996).

21. J. Apelt, J. Lessig and R. Schliebs, *Neurosci. Lett.* **329**, 111 (2002).

22. E. M. Frohman, T. C. Frohman, S. Gupta, A. de Fougerolles and S. van den Noort, *J. Neurol. Sci.* **106**, 105 (1991).

23. T. J. DeGraba, *Neurology* **51**, S62 (1998).

24. F. J. Gonzalez-Velasquez and M. A. Moss, *J. Neurochem.* **104**, 500 (2008).

25. J. A. Rhodin, T. N. Thomas, L. Clark, A. Garces and M. Bryant, *J. Alzheimers Dis.* **5**, 275 (2003).

26. R. Giri, Y. Shen, M. Stins, S. Du Yan, A. M. Schmidt, D. Stern, K. S. Kim, B. Zlokovic and V. K. Kalra, *Am. J. Physiol. Cell Physiol.* **279**, C1772 (2000).

27. C. Humpel, *Curr. Neurovasc. Res.* **5**, 185 (2008).

28. R. Giri, S. Selvaraj, C. A. Miller, F. Hofman, S. D. Yan, D. Stern, B. V. Zlokovic and V. K. Kalra, *Am. J. Physiol. Cell Physiol.* **283**, C895 (2002).

29. F. J. Gonzalez-Velasquez, J. A. Kotarek and M. A. Moss, *J. Neurochem.* **107**, 466 (2008).

30. H. Beekhuizen and R. van Furth, *J. Leukoc. Biol.* **54**, 363 (1993).

31. R. M. Faruqi and P. E. DiCorleto, *Br. Heart J.* **69**, S19 (1993).

32. M. R. Nichols, M. A. Moss, D. K. Reed, W.-L. Lin, R. Mukhopadhyay, J. H. Hoh and T. L. Rosenberry, *Biochemistry* **41**, 6115 (2002).

33. M. A. Deli, C. S. Abraham, Y. Kataoka and M. Niwa, *Cell. Mol. Neurobiol.* **25**, 59 (2005).

34. U. Kniesel and H. Wolburg, *Cell. Mol. Neurobiol.* **20**, 57 (2000).

35. P. Ballabh, A. Braun and M. Nedergaard, *Neurobiol. Dis.* **16**, 1 (2004).

36. C. Forster, *Histochem. Cell Biol.* **130**, 55 (2008).

37. M. Vastag, J. Skopal, Z. Voko, E. Csonka and Z. Nagy, *Microvasc. Res.* **57**, 52 (1999).

38. D. Wong and K. Dorovini-Zis, *Microvasc. Res.* **49**, 325 (1995).

39. K. Ruprecht, T. Kuhlmann, F. Seif, V. Hummel, N. Kruse, W. Bruck and P. Rieckmann, *J. Neuropathol. Exp. Neurol.* **60**, 1087 (2001).

40. R. M. McCarron, L. Wang, D. B. Stanimirovic and M. Spatz, *Endothelium* **2**, 339 (1995).

41. D. Wong and K. Dorovini-Zis, *J. Neuroimmunol.* **39**, 11 (1992).

42. E. F. Howard, Q. Chen, C. Cheng, J. E. Carroll and D. Hess, *Neurosci. Lett.* **248**, 199 (1998).

43. D. C. Hess, T. Bhutwala, J. C. Sheppard, W. Zhao and J. Smith, *Neurosci. Lett.* **168**, 201 (1994).

44. J. D. van Buul and P. L. Hordijk, *Transfus. Clin. Biol.* **15**, 3 (2008).

45. H. Q. Liu, X. B. Wei, R. Sun, Y. W. Cai, H. Y. Lou, J. W. Wang, A. F. Chen and X. M. Zhang, *Life Sci.* **78**, 1293 (2006).

46. J. Zhang, Y. C. Rui, P. Y. Yang, L. Lu and T. J. Li, *Life Sci.* **78**, 2983 (2006).

47. M. F. Stins, Y. Shen, S. H. Huang, F. Gilles, V. K. Kalra and K. S. Kim, *J. Neurovirol.* **7**, 125 (2001).

48. A. K. Tripathi, D. J. Sullivan and M. F. Stins, *Infect. Immun.* **74**, 3262 (2006).

49. M. D. Kirkitadze, G. Bitan and D. B. Teplow, *J. Neurosci. Res.* **69**, 567 (2002).

50. D. M. Walsh and D. J. Selkoe, *J. Neurochem.* **101**, 1172 (2007).

51. J. P. Cleary, D. M. Walsh, J. J. Hofmeister, G. M. Shankar, M. A. Kuskowski, D. J. Selkoe and K. H. Ashe, *Nat. Neurosci.* **8**, 79 (2005).

52. S. Lesne, M. T. Koh, L. Kotilinek, R. Kayed, C. G. Glabe, A. Yang, M. Gallagher and K. H. Ashe, *Nature* **440**, 352 (2006).

53. J. El Khoury, M. Toft, S. E. Hickman, T. K. Means, K. Terada, C. Geula and A. D. Luster, *Nat. Med.* **13**, 432 (2007).

54. J. Wegiel, H. Imaki, K. C. Wang, J. Wegiel, A. Wronska, M. Osuchowski and R. Rubenstein, *Acta Neuropathol.* **105**, 393 (2003).

55. M. Sastre, T. Klockgether and M. T. Heneka, *Int. J. Dev. Neurosci.* **24**, 167 (2006).

56. I. Blasko, W. Lederer, H. Oberbauer, T. Walch, G. Kemmler, H. Hinterhuber, J. Marksteiner and C. Humpel, *Dement. Geriatr. Cogn. Disord.* **21**, 9 (2006).

57. B. A. Kallmann, V. Hummel, T. Lindenlaub, K. Ruprecht, K. V. Toyka and P. Rieckmann, *Brain* **123** (Pt 4), 687 (2000).

58. M. F. Stins, F. Gilles and K. S. Kim, *J. Neuroimmunol.* **76**, 81 (1997).

59. M. F. Stins, D. Pearce, F. Di Cello, A. Erdreich-Epstein, C. A. Pardo and K. S. Kim, *Lab. Invest.* **83**, 1787 (2003).

60. M. S. Hayden and S. Ghosh, *Cell* **132**, 344 (2008).

61. T. Collins, M. A. Read, A. S. Neish, M. Z. Whitley, D. Thanos and T. Maniatis, *FASEB J.* **9**, 899 (1995).

62. R. De Martin, M. Hoeth, R. Hofer-Warbinek and J. A. Schmid, *Arterioscler. Thromb. Vasc. Biol.* **20**, E83 (2000).

63. I. M. Verma, J. K. Stevenson, E. M. Schwarz, D. Van Antwerp and S. Miyamoto, *Genes Dev.* **9**, 2723 (1995).

64. A. S. Baldwin Jr, *Annu. Rev. Immunol.* **14**, 649 (1996).

65. V. J. Palombella, O. J. Rando, A. L. Goldberg and T. Maniatis, *Cell* **78**, 773 (1994).

66. S. Floris, S. R. Ruuls, A. Wierinckx, S. M. van der Pol, E. Dopp, P. H. van der Meide, C. D. Dijkstra and H. E. De Vries, *J. Neuroimmunol.* **127**, 69 (2002).

67. H. S. Nottet, Y. Persidsky, V. G. Sasseville, A. N. Nukuna, P. Bock, Q. H. Zhai, L. R. Sharer, R. D. McComb, S. Swindells, C. Soderland and H. E. Gendelman, *J. Immunol.* **156**, 1284 (1996).

68. V. G. Sasseville, W. Newman, S. J. Brodie, P. Hesterberg, D. Pauley and D. J. Ringler, *Am. J. Pathol.* **144**, 27 (1994).

69. R. M. McCarron, L. Wang, A. L. Siren, M. Spatz and J. M. Hallenbeck, *Am. J. Physiol.* **267**, H2491 (1994).